Industrial Gases

Industrial Gases

N.A. Downie
Partner
Maran & Co.
Guildford, UK

BLACKIE ACADEMIC & PROFESSIONAL
An Imprint of Chapman & Hall
London · Weinheim · New York · Tokyo · Melbourne · Madras

Published by Blackie Academic and Professional,
an imprint of Chapman & Hall, 2–6 Boundary Row, London SE1 8HN, UK

Chapman & Hall, 2–6 Boundary Row, London SE1 8HN, UK

Chapman & Hall GmbH, Pappelallee 3, 69469 Weinheim, Germany

Chapman & Hall USA, 115 Fifth Avenue, New York, NY 10003, USA

Chapman & Hall Japan, ITP-Japan, Kyowa Building, 3F, 2–2–1 Hirakawacho, Chiyoda-ku, Tokyo 102, Japan

DA Book (Aust.) Pty Ltd, 648 Whitehorse Road, Mitcham 3132, Victoria, Australia

Chapman & Hall India, R. Seshadri, 32 Second Main Road, CIT East, Madras 600 035, India

First edition 1997

Typeset in 10/12 Times by Type Study, Scarborough
Printed in Great Britain at the University Press, Cambridge

ISBN 0 7514 0352 0

A catalogue record for this book is available from the British Library

Library of Congress Catalog Card Number: 96-86566

Printed on acid-free text paper, manufactured in accordance with ANSI/NISO Z39.48-1992 (Permanence of Paper).

Preface

The natural world is a wonderful place but much of the natural world would be inaccessible or invisible to us without the products of industry; for example air and road travel, underwater breathing apparatus and television. It is not simply the products of industry, however, that can enrich our lives; the activities and techniques of industry can, too.

The story of gases in industry is just one of a hundred interlinked subjects needed to describe the totality of modern industry. The subjects of this book, nitrogen, oxygen, carbon dioxide, argon and others, are inextricably woven into the fabric of modern manufacturing. Almost no branch of industrial processing is untouched by them. Gases begin their role close to primary extractive industries, for example in the processing of ore. The manufacture of materials from those as common as steel to such exotics as semiconductor-grade indium phosphide is now dependent on gases. The conversion of those materials into useful products and services involves a wide range of applications of gases. Gases now affect almost all fields of human activity: what we eat, what we drink, how we travel, what we read, our health and environment.

This book is about the industrial gases that have a key role in general industry. Methane or propane gas used for heating purposes could be easily replaced by other heat sources. Other gases, such as ethylene, are intermediates in chemical manufacture and exist only fleetingly in pipes deep within specialised chemical processing complexes. Although these are mentioned and described, the main thrust of enquiry will be directed towards those gases which are not simply burnt but which are visibly produced and used in industry at large.

Nitrogen, the biggest industrial gas worldwide, is today still mainly made by liquefying and then distilling air. This process is now very much more efficient than in former times. There remain, however, the intrinsic inefficiencies attached to cooling air to −195°C to liquefy it prior to separation. Despite many improvements in this process, and the considerable advantages of scaling, it continues to require many times the thermodynamic separation energy. Hence there are now important challenges being made to cryogenic separation. Pressure swing absorption units have been practical for many years and are now edging into the marketplace, although their thermodynamic efficiency is still not high. An even bigger challenger may be waiting in the wings: membrane separation. Membranes are probably the simplest possible way of separating gases: a filter passes the molecules required and sieves out the ones not required. Although a process as simple as this is possible, it is not yet truly practical and economic. These and other promising new processes for making gases will steadily grow more important as new applications for gases put new demands on production techniques.

New applications are constantly being found for gases and the whole balance of gas production has been shifted in recent years. Chapter 2 examines how nitrogen has now outstripped oxygen in volume terms in many countries. Although the steel industry and chemicals still take vast volumes, newer areas of application have opened up. The rapid freezing of foodstuffs with liquid nitrogen is a large and fast-growing sector, and the semiconductor industry has seen a meteoric rise in demand, becoming the most important consumer of gases in some countries. The future will also hold big changes, and Chapter 4 looks at what new applications may become important and how these will shape the industry and the technology used. A glossary of words that are used in a particular sense in the gas industry, or are unique to it, is included at the end of the book.

There is a lot of sophistication in the gases industry. Despite this, its basis is still firmly grounded in the kind of physics that students learn early in university or, in the UK, in the final years of a good school. Such students have, I believe, all the necessary scientific tools to take on a whole field of industrial science. They will find from this book that the part of industrial science that relates to industrial gases is easily comprehensible. A firm basic grounding in science is, however, not quite sufficient. There are many simple facts that are very useful to know for practical purposes but are difficult to fit into a systematic, scientific frame. It is impossible to give a comprehensive encyclopaedia of these in a short and wide-ranging book like this. Nevertheless, I have included a fair number of hints, tips and anecdotes that fall into this category, most from my own personal experience, which I hope will be useful.

The industrial gases industry is now one of the largest in the world with sales of $20bn in 1992. More than a million tonnes of industrial gases are used every day. Gases have put men on the moon and helped to uncover the secrets of the atomic nucleus. They are used in almost every conceivable sector of industry in every corner of the world. Despite this, gases are largely a hidden industry. Only rarely are gases directly used. As a result, relatively few people, even within the quarter of a million who work in the industry, know much about them. The secrets of how oxygen is made by distilling liquid air, for example, or how to make diamonds from a gas mixture deserve a wider audience. I know that I have enjoyed uncovering these secrets over the years and I hope that this book will enable the reader to appreciate them, too.

Acknowledgements

The author is indebted to a large number of people for their help in the preparation of this wide-ranging book – starting with my forbearing family and my partners in Maran & Co. Space does not permit me to list everyone, but I should at least list the organisations from which they came. (Please note, however, that responsibility for the opinions and facts given are entirely mine.) Space also forbade the use of all the illustrations and other information that people kindly provided; please accept my apologies if your contribution was not used.

Accles & Pollock
AFROX Company
AGA Company
Air Liquide
Air Products
AIRCO Company
ARC Machines
Atlas Copco
Balzers-Pfeiffer
Bergbau Forschung
Birwelco
BOC Group
British Meteorological Office
British Steel Corporation
British Technology Group
Broeder Mueller Company
Cambridge Fluid Systems
Cameron Balloons
CEA-LETI R&D institute, Grenoble, France
Ceodeux Company
CERN R&D institute, Geneva
CHAM
Chesterfield
Courtaulds
Cryoplant
DESY R&D Institute
DOW Chemical
DRAX Torches
DuPont
Edwards High Vacuum
Epichem
Eurometers
FLUOR Corp.
Gardner Cryogenics
GEC Semiconductors
Guildford Museum
Hale Hamilton Valves
Hawker Siddeley Switchgear
High Energy Physics Group, Imperial College, London
Hoechst
HYT Corp
IBM
ICI
IMEC R&D Institute, Leuven, Belgium
IMI Marston
Institute for Aviation Medicine, Farnborough

JEMI UK
JESSI Semiconductor R&D Programme, Munich
Linde
Liquid Air
Luxfer
Madder Scientific/Systech
Malaysia Oxygen Company
Matheson
Messer Griesheim
Millipore
Monsanto
Nippon Sanso
Ohmeda
Oxford Plasma Technology
PALL Filter
Pilkington
PRAXAIR
Puritan-Bennett
SAES
Siemens
Sulzer Brothers
Swagelok
Tetronics
Texas Instruments
TI Chesterfield
Tohoku University, Japan
Union Carbide Corp.
University of Edinburgh
University of St Andrews
University of Surrey
US Steel Corp.
VG Instruments
Wacker
Wormald International

Contents

1 An introduction to gases and the gases industry

1.1 The discovery of gases

A gas is the simplest form of matter. In the kinetic theory of matter, a gas is simply a group of freely moving small spheres whose only interaction with each other and other matter is via elastic collisions. It is in effect matter having neither constant shape nor volume, and substantially invisible and intangible.

An invisible and intangible form of matter seems at first to be an unlikely concept. However, it arose surprisingly early in history. The concept was already well established in classical Greek thought. That the Greeks recognised air as an actual substance distinct from vacuum is clear since they talk about objects such as clouds, water or the earth being formed by the condensation of the air (Wightman, 1950). Perhaps analogy between the behaviour of gases and the behaviour of (readily observable) liquids helped understanding, along with the observation of the visible behaviour of smoke- or mist-laden air.

The use of air-handling machinery, for example bellows, was also known to the Greeks. Perhaps as far back as the 2nd century BC the Greeks had developed musical organs driven by bellows. Being experts with mechanisms using water developed for their clepsydra or water clocks, it is thought that the Greeks also devised pipe organs with ingenious 'hydraulus' mechanisms employing water in a cistern to stabilise the air pressure delivered (Elvin, 1971).

The need for intense heat in the early metals, ceramic and glassware industries led to knowledge of the use of low-pressure air in furnaces, craft knowledge that was passed on down the centuries. Alchemists steadily advanced the state of practical gases technology and by medieval times several different gases were recognised in addition to air (although at this stage only air had any industrial importance). Illustrations in some of the very first printed books, showing bellows, valves and pipework, give an idea of the state of the technology after the Middle Ages (Agricola, 1556). Lacking a scientific approach, however, and without suitable instruments, gas-handling equipment or basic scientific knowledge, progress was slow.

Major advances were at last made in studying the properties of air in the 1600s. Robert Boyle, in 1662, published experiments with a mercury barometer which quantified the change of pressure of air with volume.

However, accurate experiments on the change of pressure of air with temperature had to wait for Gay-Lussac in 1802.

More rapid progress in understanding gases came with the rise of chemistry as a science. Joseph Black might be given credit for making advances in this direction in 1755. He published experiments which recognised at least two gases as distinct, 'ponderable' substances, rather than modified air: the oxygen in air and the most frequent combustion product, CO_2. However, more profound discoveries were not possible until improvements were made in gas handling by Priestley and others from around 1770 onwards.

All the early chemists made important discoveries in gases around this time (Ackroyd, 1935; Lavoisier, 1952). Priestley alone, for example, could claim the discovery, or at least the first correct identification, of the oxides of nitrogen, SO_2, HCl, SiF_4, ammonia and CO. In 1789 Lavoisier wrote his systematic *Elementary Treatise on Chemistry* which is recognisably a modern chemistry book and dedicates a large proportion of its text and illustrations to gases and gas handling, which was by then already well-developed. By the turn of the century oxygen, nitrogen, HCl, HBr, Cl_2 and many other gases were all fully described and routinely made in chemical laboratories.

Three of the best-known chemical pioneers – Priestley, Scheele and Lavoisier – could each claim to have been the discoverer of oxygen, around 1775. The use of the reversible chemical reaction of mercury with oxygen (Gibbs, 1967; Anon, 1980) pointed the way to the first commercial oxygen separation using barium. To Lavoisier must go the credit for providing a basic list of irreducible substances (chemical 'elements') and the naming of oxygen. Oxygen means 'acid producer' in Greek. Oxygen was the first gas, other than air, to be used on a large scale in industry, although 'inflammable air' (hydrogen) had been discovered in 1767 by Cavendish.

The advance of the atomic theory of chemistry, led by John Dalton, enabled the chemical formulae of many gases to be identified correctly by the 1800s. This in turn fed back to chemistry because Avogadro had predicted that the density of any gas would be precisely proportional to its molecular weight. The use of Avogadro's concept thus enabled the molecular formulae of gases and, eventually, the atomic weights of all elements to be precisely determined. However, although Avogadro first described his molecular theory of gases in 1811, it was half a century or so later before the theory was accepted.

Eventually a 'virtuous circle' became established in which one scientific discovery led to another or to another technical development which in turn led to another scientific discovery. By 1860 the physics and chemistry of the majority of industrial gases had been worked out. For example, the concept of a 'perfect' or 'ideal' gas – one in which the molecules are spherical and undergo only simple elastic collisions according to kinetic theory – was introduced in this era of history and the ideal gas equation formulated. This neatly encapsulated the formulae of Boyle, Gay-Lussac and Avogadro, and simplified the kind of calculations needed in industry.

In parallel with this scientific development were technical developments due largely to the requirements of the rising steampower industry: by the mid-nineteenth century gases could be handled not just at near-atmospheric pressure but at low pressure or high pressure up to 10 bar (e.g. Anon, 1989).

Helium and argon were discovered late in the nineteenth century. Terrestrial helium was discovered by Hildebrand in 1891 (Anon, 1995) through its optical emission spectrum, the helium being formed extremely slowly by alpha particle emission in minerals containing radioactive elements such as uranium. Helium had ingeniously been predicted more than 20 years earlier by detecting its emission spectrum in the sun. Lord Rayleigh observed as early as 1882 the slight differences in the density of gases prepared by different routes. Work by Sir William Ramsay established argon in 1894, preparation being by reacting combusted air (i.e. nitrogen + inerts) with magnesium. Ramsay went on to discover other inert gases.

Air distillation and the technology of cryogenics, which were being developed at the end of the nineteenth century, allowed the discovery of neon and the heavier 'rare' gases, krypton and xenon: an example of a virtuous circle operating. One science (low temperatures) led to a technology (air distillation) which led to another science (rare gases). Neon was found in the vapour above an early sample of liquid argon, whilst krypton was found by its optical emission spectrum. Xenon was found soon afterwards by distilling the crude krypton. By 1898 all the stable rare gases had been discovered. The considerable scientific feat that this represents should perhaps be emphasised; the discovery of part per million components in the air, when those components are undetectable to human senses and completely inert chemically, must rank as a major achievement.

By this point in history many other gases had not been discovered. However, the remaining undiscovered gases, with one or two important exceptions,[1] are not important in industry today. The stage was now set for industrial gases to blossom into a major industry on a firm scientific and technical footing.

1.2 Early use of gases in industry

1.2.1 Coal gas

Coal gas constituted the first major industrial use of gas: as early as 1600 coal was being heated in retorts and the gas given off burnt. In fact, the gas was a

[1] One class of gases that the Victorian chemists totally failed to discover was the chlorofluorcarbons, which had to wait until Midgely's work in the 1930s. Midgely, whilst working for General Electric, discovered that, despite being made from some of the most reactive elements known, the CFCs had an extraordinary level of inertness.

by-product of the requirement for coke for iron smelting; UK charcoal supplies being exhausted (Williams, 1981; Buchanan, 1992).

At first the gas was very impure and used only on a small scale. The basic components, hydrogen and carbon monoxide, were adulterated with ammonia, sulphur compounds and hydrocarbons even as heavy as benzene. However, great strides were swiftly taken in the technology and then in the distribution and use of coal or town gas. By 1800 a major industry had arisen: heating coal to make coal gas which was then used for lighting. The later invention of the 'gas mantle' improved the efficiency of gas lighting and extended the technical lifetime of the technique well into the twentieth century. The mantle was actually a thorium oxide-coated grid which when heated emitted a slightly greenish white light.

Many other early industrial processes also evolved gases as by-products but most of these were unpleasant or even dangerous rather than immediately useful. For example, the biggest Victorian era chemical industry, the alkali/soap industry, relied at first on the infamous LeBlanc process. This involved adding sulphuric acid to salt, leading to the emission of copious amounts of hydrogen chloride (heating with limestone and charcoal completed the conversion into crude sodium carbonate and produced another pollutant, a calcium sulphide sludge). After serious pollution incidents, the world's first anti-pollution body, the UK Alkali Inspectorate, was set up in 1863 (e.g. Anon, 1993).

1.2.2 Oxygen and acetylene

Limelight was one of the first applications for an industrial gas, in this case oxygen. An oxygen-intensified gas flame could be used to heat rods of calcium oxide (lime) to very high temperatures. At these temperatures, in excess of those achievable with air alone, the lime gave off a powerful white light far superior to that obtained using coal gas. 'Limelight', as the system was known, was much more satisfactory for illuminating large halls, such as theatres, than oil or gas lamps.

Limelight used oxygen derived from sources such as the barium oxide Brin process. This process was based on the production of barium peroxide by roasting barium oxide in air at 590°C, then raising the temperature to 870°C. At 870°C the peroxide formed decomposes back into oxide, releasing more or less pure oxygen which can then be cooled and compressed into steel gas cylinders. Although crude, the process was ingenious in that it required no continuous input of raw materials other than air and energy. Its active process was one of reversible absorption, making it superior to an irreversible process such as heating of nitrates.

At the turn of the century in Germany, Carl von Linde developed the liquid air distillation process for oxygen. This so revolutionised the production of oxygen that it was quickly adopted elsewhere, for example in the UK and the USA. His work was paralleled by that of Georges Claude in France. Both men founded companies which are today large and successful industrial gases firms: the Linde company of Germany and the USA, and the L'Air Liquide company of France. Now, at last, oxygen was available in large quantities at low prices.

Meanwhile another gas business had been growing up around acetylene. Gas lamps, particularly those with an incandescent mantle, are an effective source of light. However, the town gas of the day could only be provided via a pipe from the gasworks, and thus portable gas lighting and lighting outside of towns was impossible. The development of the calcium carbide/acetylene lamp was therefore an important innovation (Figure 1.1). Acetylene burns with a flame which is intrinsically incandescent, much more so than any other hydrocarbon, and so is eminently suitable for lighting. Portability was assured because acetylene can be derived simply by dripping water onto calcium carbide chips.

Oxy/acetylene arose out of the combination of these two Victorian gas industries, aided by the discovery that acetylene could be safely stored in large quantities in cylinders once it had been dissolved into an acetone absorbed on a porous inert packing. It caught on rapidly after the First World War and delivery networks of oxygen and acetylene cylinders were built up all over the world. On the back of this business most of today's major gas companies grew to be large and multinational corporations: companies such as BOC, L'Air Liquide, Union Carbide and Linde (then in both the US and Germany).

1.2.3 Oxygen for steel

In the intense rearmament era after the First World War, demand for steel was intense. Linde in Germany realised that oxygen could enhance the productivity of steel plants and devised processes for Bessemer converters. These processes required enormous quantities of oxygen, much larger volumes than had been produced hitherto, and resulted in rapid improvements in the efficiency and scale of liquid air plants. Plants in the range of a few tens of tonnes per day became quite common. Driven by this immense demand for its products, industrial gas businesses began to expand rapidly and to resemble today's huge industry.

1.3 Properties of gases

Gases are studied by physicists interested in kinetic theory, by chemists interested in their reactions, and by mechanical engineers interested in engines. Standard textbooks provide useful subsets of gases science and know-how aimed at satisfying the study needs of these groups. Many good

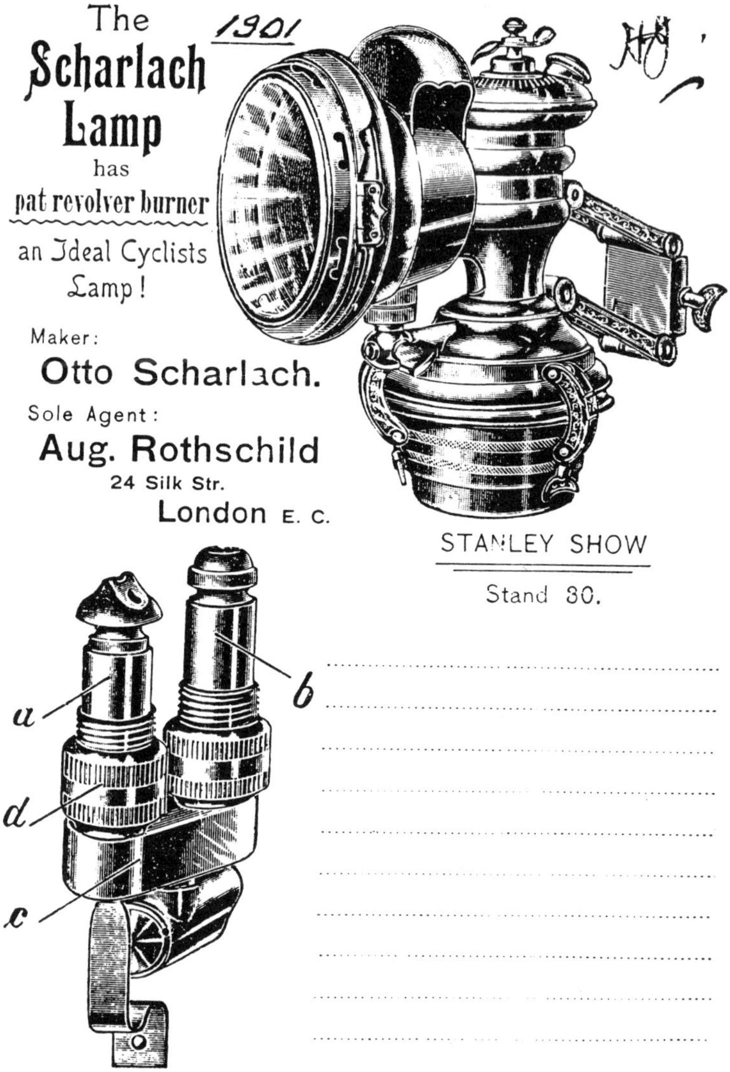

Figure 1.1 Acetylene bicyle lamp, *c*. 1900 (courtesy of Matthew Alexander, Guildford Museum). a, b: alternative burners; c: rotating body allowing instant burner change; d: union nut attaching burner.

standard works are available (see Further Reading) and the sections in this chapter are intended only as reminders of physical principles. The standard works should be consulted for details. Moore (1978) on physical chemistry, together with Flowers and Mendoza (1970) on kinetic theory, Zemanski (1968) on thermodynamics and Douglas *et al*. (1984) on flow are good introductions. A technical manual on chemical engineering, like Perry

(1984) is essential for accurate, detailed calculations, as are compilations of data such as are included in the *CRC Handbook of Chemistry and Physics*, *Tables of Physical and Chemical Constants* (Kaye and Laby, 1995) and specialist gas data publications.

1.3.1 Ideal gases and the energy stored in a compressed gas

The ideal gas equation describes the relationship between the equilibrium pressure (P) exerted by a known mass of ideal gas confined in volume V at temperature T

$$PV = nRT$$

where n is the number of gram-molecules of gas, R is 8.31 J K^{-1} (the gas constant) and T is the absolute temperature in degrees Kelvin.

The ideal gas equation approximately describes the behaviour of many gases at ordinary temperatures and pressures, and is still today enormously useful in industrial calculations.

Using the ideal gas equation, the energy (E) stored in a compressed gas can be calculated

$$E = \int P\mathrm{d}V = nRT \log \mathrm{e}(V_f/V_i) = P_f V_f \log \mathrm{e}(P_i / P_f)$$

where P_f and V_f are final pressure and volume, and P_i and V_i are initial pressure and volume. The energy stored in compressed gases is only occasionally apparent in everyday life. There are some gadgets that literally replace springs with gas. Examples of this are 'gas struts', actually gas confined within a piston and cylinder, which use pressures of 50 bar or so to support swing doors, e.g. the tailgate of a car.

The energy that can be stored in compressed gas can be released very suddenly with immense, potentially explosive, power, something which should always be borne in mind when designing gas systems and components. This is demonstrated routinely in the testing of gas cylinders (Figure 1.2). Occasionally a cylinder is tested to destruction by pumping it up to twice, or more, its rated pressure. The result is a very violent explosion which twists the very tough cylinder into contorted shapes and may blow pieces a considerable distance. The energy stored in a standard cylinder of 8 m^3 at, say, 400 bar pressure is of the order

$$E = 6PV = 5 \text{ MJ}$$

equivalent to about 500 g trinitrotoluene (TNT). By contrast, a hydraulic test of a similar cylinder will give rise only to a small split at a weak point, with no great release of acoustic or other energy. A compressed liquid, such as water, contains very little energy by comparison to compressed gas because of its low compressibility.

Energy storage by means of gas pressure has been employed industrially

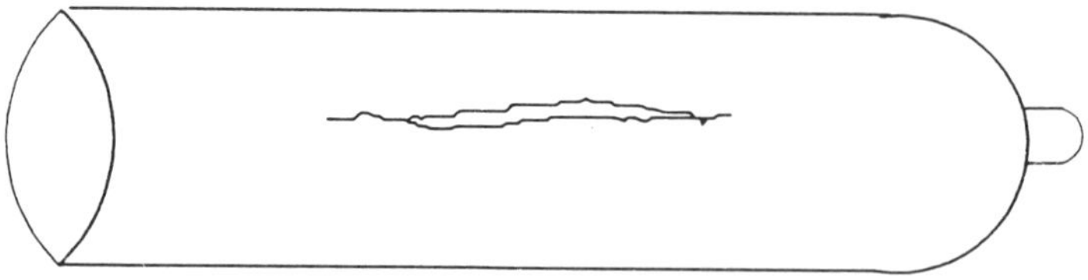

(a)

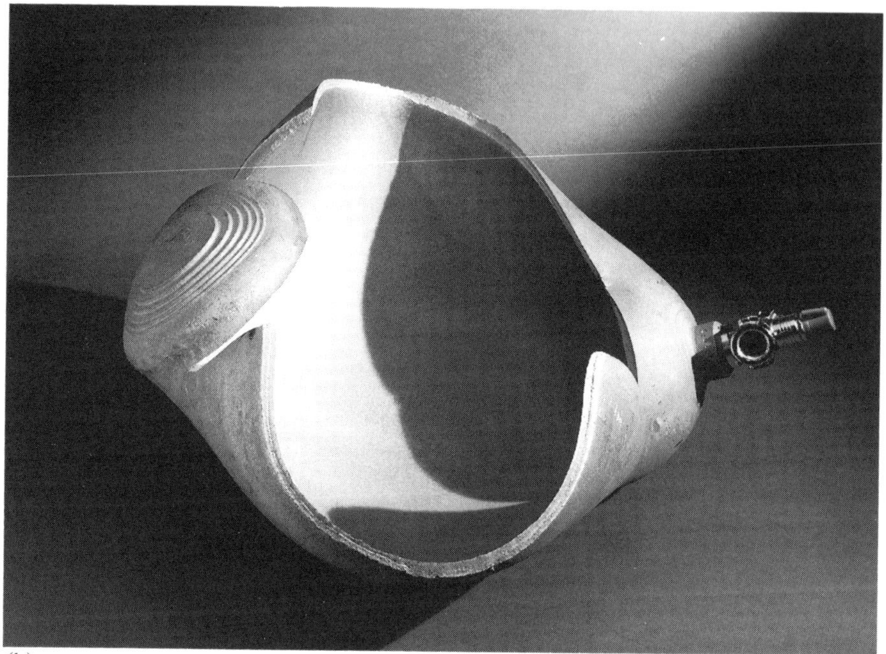

(b)

Figure 1.2 (a) Gas cylinder after hydraulic burst test. Note typical longitudinal split. (b) Gas cylinder after pneumatic test. Note gross but basically ductile failure. Few fragments have been emitted. Photograph from the National Physical Laboratory. © Crown copyright 1996. Reproduced by permission of the Controller of HMSO.

in a number of ways, for example in diesel engine starting or for gyroscope power in guided missiles. In the future other storage systems might be tried on a larger scale. Realistic schemes have been devised, for example for using off-peak electricity during the night, to pump high-pressure air from a turbo-compressor into underground caverns. During the day the air is released through a turbine generator. Perhaps surprisingly, such a scheme might be thoroughly competitive with more familiar schemes involving water pumping. There is considerable experience with fuel gas storage at up to 70 bar pressure in underground caverns excavated from salt deposits by dissolving the salt around a borehole (Melvin, 1988).

1.3.2 Real gases: deviants from the ideal

The ideal gas law is a good guide to the behaviour of gases unless the gases are at a density comparable to that of the liquid gas or the temperature is close to liquefaction. The empirical compressibility factor Z can be used in gas science to allow for deviations from the ideal gas law by introducing it as follows

$$PV = ZnRT$$

$Z = 1$ for an ideal gas. Tabulations of Z values exist for many gases over a wide range of pressure and temperature; the result of many years of repetitive measurements in laboratories around the world. For example, at around 100 bara pressure and room temperature, the approximate real gas Z values for air, hydrogen and oxygen are

air	0.984
hydrogen	1.062
oxygen	0.936

The closeness of these values to 1 shows that for the 'permanent' gases at pressures up to 100 bar the ideal gas equation is still a good approximation. Rather than relying on extensive tables of empirical data such as Z values, a number of gas equations based on physical models have been developed. In addition to giving a deeper understanding of gas behaviour, these are often convenient in calculation. First, being continuous functions, they can be used for algebraic manipulations. Second, even where a computer is being used for calculations, it is convenient to input a few physical model parameters rather than tediously long tables of Z values. One of the most useful model-based gas equations of state is that of van der Waals

$$(P - a/V^2)(V - b) = nRT$$

The ideal gas equation applies to gas molecules that are very small, almost point-like, elastic spheres which do not interact unless they collide. The van der Waals equation corrects two of the simplifying assumptions of the ideal gas equation by allowing the molecules to occupy a finite volume, represented by the constant b, and by allowing them to be attracted to each other slightly, the attractive force being represented by the constant a.

Gas mixtures. It is important to be able to predict the behaviour of mixtures of gases and one of the most powerful concepts in mixtures is Dalton's law of partial pressures. This law follows from the consideration of the ideal gas equation, for molecules of gases that behave like the elastic spheres of the ideal gas both when similar and different molecules collide.

Dalton's law is that the pressure due to a mixture of gases can be obtained by adding the 'partial pressure', P_x, of each gas, i.e.

$$P_1V = N_1RT$$

$$P_2V = N_2RT$$

$$P_3V = N_3RT$$

and so on, where N_x is the number of moles of each gas in the volume V. P_{tot}, the pressure exerted by the mixture, is simply the sum of the partial pressures.

Dalton's law is subject to similar but more complex corrections than the van der Waals equation to allow for different forces between different pairs of molecules. For example, a hydrogen molecule will have a smaller intermolecular attractive force for another hydrogen molecule than it has for a fluorine molecule: H_2 and F_2 explode on contact. For dilute solutions of one gas in another, however, Dalton's law is generally very accurate and it can be used as a guideline for the behaviour of mixtures of condensible vapours and liquefiable gases where the ideal gas equation itself is not applicable.

In practice, Dalton's law is a valuable principle and guideline but empirical measurements are essential at higher pressures and where easily liquefied gases are concerned. In particular, predictions about condensation from non-dilute mixtures are often erroneous (see, for example, the comment on the 50/50 nitrous oxide/oxygen medical gas mixture ENTONOX, section 3.9.1).

In addition to Dalton's law for pressure, there are other approximate rules for estimating other properties of gases which can in many cases avoid tedious experiments by interpolation between the properties of pure gases. Interpolation is least reliable for mixtures of gases where one or more of the gases is near to its boiling point or at high pressure, or where the gases are in roughly equal parts. However, for gases at high temperatures or low pressures, or for dilute mixtures, interpolation is acceptably accurate.

Interpolation of mixture properties is most accurately carried out for density (ρ), thermal conductivity (k), viscosity (μ) and thermal capacity (C_p) using the following approximate formulae

$$\rho = \rho_1 V_1 + \rho_2 V_2 + \ldots$$

$$K = \frac{K_1\rho_1^{0.33}V_1 + K_2\rho_2^{0.33}V_2 + \ldots}{\rho_1^{0.33}V_1 + \rho_2^{0.33}V_2 + \ldots}$$

$$\mu = \frac{\mu_1\rho_1^{0.5}V_1 + \mu\rho_2^{0.5}V_2 + \ldots}{\rho_1^{0.5}V_1 + \rho_2^{0.5}V_2 + \ldots}$$

$$C_p = \frac{C_{p1}\rho_1V_1 + C_{p2}\rho_2V_2 + \ldots}{\rho_1V_1 + \rho_2V_2 + \ldots}$$

where V_1, V_2, V_3, etc., are the volume fractions of gases 1, 2 and 3.

These interpolations each apply a slightly different curve of property calculated versus the molecular proportion of gas 1 in gas 2. The rationale for the (linear) density calculation follows from Dalton's law. The rationale for the thermal capacity calculation follows from noting that C_p is the thermal capacity per unit mass and therefore $C_p \rho V$ gives the thermal capacity per unit volume.

For thermal conductivity and viscosity the story is much more complicated and the expressions above are very rough. Many mixtures of gases, for example, show a viscosity or thermal conductivity above that of either component. In these cases it is insufficient to know the parameters of each gas on its own: parameters related to the interaction of gas 1 with gas 2 are needed. Accurate formulae are available but involve many of these gas 1–gas 2 interactions as additional parameters. For cases needing good values of these properties, it is necessary to consult detailed learned literature for these interaction parameters and/or to obtain experimental measurements.

1.3.3 The finite size of gas molecules

It is perhaps surprising that the finite size of gas molecules, and their finite number in a given macroscopic volume of gas, is important in various effects which are important industrially, e.g.

- the transition of flow from fluid to molecular-type flow in small orifices and in vacuum systems;
- the rate of interdiffusion of gas molecules;
- the Brownian motion of dust particles in gases, and hence the performance of filters of very small (> 0.1 μm) particles.

Avogadro's number, today defined as the number of atoms in 12 g of carbon (or 22.41 of an ideal gas at normal temperature and pressure), is 6.022×10^{23}.

1.3.4 Heat capacities

The heat capacities of all ideal gases should be identical: energy in an ideal gas is contained only in the kinetic energy of the molecules and this should increase with temperature in the same way for all molecules, irrespective of mass.

However, real imperfect gases differ quite markedly from each other in heat capacity, and some vary with temperature. It turns out that molecules which are significantly non-spherical can store energy not just in their translational motion but also in their rotational and vibrational motion. The plausibility of this can be enhanced if a gas is envisaged as consisting of N tiny billiard balls tied by a string to another N neighbours. If the string is very long (and does not get tangled up) then no effect is noticed, except that the gas has $2N$ particles and twice as much kinetic energy. However, as the string is shortened it begins to constrain the relative motion of the two tethered molecules and energy is reduced somewhat but energy will go into the stretching of the string. If the string is made so short that the balls are overlapping, then the kinetic energy reduces to that of N particles and the behaviour of a gas of N spherical particles is seen.

In an ideal gas, translational kinetic energy from thermal random motion will be roughly evenly distributed amongst the molecules and is proportional to temperature. The Boltzmann constant k times the absolute temperature (in K) gives the scale of this energy per molecule. The translational kinetic energy will be $1/2\,kT$ for each molecule or, for an Avogadro's number of molecules (1 'gramme mole' or 'mole'), $1/2\,NkT = 1/2\,RT$, where R is the gas constant, around 8.31 J K^{-1}.

In an ideal monatomic gas the gas will have a specific heat (at constant pressure, which is normally the most relevant specific heat), C_p, of $5/2\,R$. The specific heat at constant volume, C_v, a less useful quantity, is given by $3/2R$. This corresponds to $R/2$ per 'degree of freedom' i.e., in this case, the three orthogonal elements of translational kinetic energy at $R/2$ each plus R for the energy expended as the gas expands doing work on its containing walls in the constant pressure situation. For two atoms the gas will also store energy in rotations in two possible orthogonal axes and the expected C_p is then $7/2R$. On this basis, for polyatomic molecules the addition of another axis of rotation raises C_p to $8/2R$.

Energy stored by compression of the atomic bonds within the molecule can also give rise to another degree of freedom for each independent mode of vibration. (For $2N$ molecules with very loose strings, the result is obviously $10/2R$.) It turns out that quantum mechanics affects the specific heats of gases by allowing only quantised energy levels in both rotation and vibration of molecules. The rotation levels are closely spaced so that few differences are seen between 'billiard ball' behaviour and real atomic behaviour. However, vibration energy levels are less closely spaced, perhaps several tenths of an electron volt apart, although the larger the molecule, the more closely spaced they are. This means that ordinary thermal levels of energy, around 1/50 eV, can partially excite some of vibrational modes, depending on temperature, and this leads to higher C_p values than expected on the basis of the above analysis and the variation of heat capacity with temperature, especially for large molecules.

To summarise, molar heat capacities at constant pressure (C_p) of typical gases at room temperature approximate to

monatomic	ideally $5/2\ R$, e.g. He $= 2.50R$
diatomic	ideally $7/2\ R$, e.g. $H_2 = 3.47R$
polyatomic	ideally $8/2\ R$, e.g. $H_2S = 4.11R$

These approximations break down for most large polyatomic molecules. For example, ideally butane would have a molar heat capacity of $8/2R$, actually it is $11.73R$. This is because polyatomics such as butane are not rigid bodies but something between rigid bodies and a 'cloud' of individual particles.

The adiabatic expansion and compression of gases is an important process in gas technology and one that is governed by the ratio C_p/C_v, often designated γ or g. The specific heat ratio can be calculated from C_p by taking $C_v = C_p - R$.

1.3.5 Bernoulli and viscous gas flow

The designer of typical pipework systems does not need to know all the intricacies of aerodynamics, although for laminar flow aerodynamic principles are actually rather simple, but a basic knowledge of the relation of gas pressure and flow is very useful.

For many purposes gas flow can be considered to be laminar, i.e. the gas flows smoothly, without turbulent eddies, and is incompressible, i.e. the density of the gas varies only slightly because the absolute pressure varies only slightly. Under such conditions trails of smoke travel in straight lines which are approximately 'streamlines' (mathematically streamlines are lines parallel to the instantaneous velocity of the gas).

One of the simplest and most useful relations in gas flow is known as Bernoulli's equation. It applies for incompressible laminar flow and describes the pressure changes accompanying velocity changes along streamlines

$$P + 1/2\rho v^2 = K$$

where P is pressure, ρ is density, v is the stream velocity along the streamline and K is a constant. As a simple example of this equation's application, consider the effect of a smooth constriction in a pipe: to maintain the mass flow constant the flow will have to speed up through the constriction. It might naively be expected that the pressure will build up at the constriction. In fact the pressure will, according to the above equation, drop within the constriction and then recover back to its original pressure again. This can be understood if a single element of fluid is followed along its journey through the constriction: in order to accelerate to a higher stream velocity, there must be a downward pressure gradient before the constriction. Conversely, in order to decelerate afterwards, there must be an upward pressure gradient and hence there must be a lower pressure in the middle of the constriction. This simple principle is behind the working of the venturi flow meter.

Bernoulli's equation can also be used for pressure drop across an orifice plate, provided that the orifice is large enough that the pressure drop is small compared to line pressure (i.e. the flow is effectively incompressible). In the case of an orifice plate, of course, the pressure dropped is not recovered as the gas leaves the orifice but is lost in frictional effects as a result of the turbulence downstream.

1.3.6 Poiseuille's equation

At lower flow rates in pipes viscous flow predominates and the simple formula of Poiseuille gives pressure drops accurately. Considering the simple situation of a volume rate of flow Q of gas of viscosity u down a long circular pipe of constant radius R, with pressure difference ΔP along length L, the following relation applies

$$Q = \frac{\pi}{8}\,\frac{1}{u}\,R^4\,\frac{(P_1 - P_2)}{L}$$

Alternatively, in terms of the mean fluid velocity v and giving pressure drop as the dependent

$$\Delta P = 8uLv/R^2$$

These formulae can both be described as Poiseuille's equation. However, they should only be applied for the case of viscous flow. To know if flow is viscous, the Reynolds number (section 1.3.7) should always be calculated.

Motion of dust particles in gases. Another of the simpler formulae in laminar flow is for the calculation of drag on solid bodies in a gaseous medium. The Stokes formula is exact for spheres and is commonly used for the calculation of the motion of small particles of contaminating dust of any shape in gases

$$\text{drag} = 6\pi\mu R U_o$$

where μ is visocity, R is the radius of the sphere and U_o is the sphere speed relative to the gas. A particle will accelerate under gravity until it reaches a speed at which the drag is equal to its weight, the so-called 'terminal velocity'. This is given by

$$4/3\ \pi\rho' R^3 = 6\pi\mu R U_o$$

or

$$U_o = (2\rho' R^2)/(9\mu)$$

where ρ' is the effective density, i.e. the particle density less the fluid density, thus correcting for buoyancy.

These formulae will only work, of course, for particles which do not go so fast as to enter turbulent flow, which begins to have effects past Reynolds numbers of 0.2. For air, with particles of density 2.65 (e.g. sand), typical settling velocities are 2.7 m per day for 10 μm particles and proportionally slower for smaller particles. In air at 1 bar the Reynolds limit is approached by 200 μm particles, drifting at $12\,mm\,s^{-1}$. However, the Reynolds limit is more rapidly approached in denser high pressure gases: 50 μm particles would be on this limit at 100 bar.

1.3.7 Turbulent and laminar flow

Laminar flow aerodynamics, although simple in principle, are often complex in practice. Turbulent flow aerodynamics are complex in principle and practice. Fluid flow in general involves rotation of fluid elements in vortices. The simpler systems, such as Poiseuille flow, are 'irrotational', a simplification which fails to describe the conspicuous rotation seen in turbulent flow. The most useful thing that can be done is to learn about the simpler principles together with some rules of thumb and empirical equations. Armed with these many pipework problems can be understood and approximately, if not accurately, predicted.

Liquid flow, like gaseous flow, can also be easily calculated if it is laminar or viscous but is less tractable if it is turbulent. An easy trap to fall into is to assume that liquid flow, because liquids have much higher viscosities than gases, is invariably viscous. Athough liquids do commonly show laminar flow, cryogenic liquids are frequently exceptional. The low viscosity of cryogens, all less than 1/5 that of water, means that they often flow with turbulence under normal conditions.

1.3.8 Compressible flow

With the added dimension of compressibility, fluid dynamics is so complex that the best that can be done is to enumerate the basic equations for simple systems and then leave most practical calculations on the complex situations seen in practice to computer models. A few situations can satisfactorily be calculated analytically while in addition a few useful rules of thumb, albeit of limited applicability, exist.

One useful rule of thumb, for example, concerns the flow of gas through a narrow orifice. Consider an orifice restricting the flow of gas to the atmosphere (this situation pertains to the real situation for some gas cylinders which are fitted with an orifice to limit the maximum flow rate). As the flow tends to increase to near sonic speeds according to the Bernoulli equation, the flow through an orifice becomes 'choked' or 'critical', i.e. a decrease in the downstream pressure will have no effect on the flow because the influence of the downstream pressure can only propagate at the speed of sound and therefore cannot pass back through the orifice. In this simple situation the flow rate can be expressed as follows

$$\text{mass flow} \simeq A\rho_0 C$$

where C is the velocity of sound, $C = (\gamma P_0/\rho_0)^{0.5}$.

Hence

$$\text{mass flow} \simeq A(\gamma P_0 \rho_0)^{0.5} \simeq kAP_0(1/T_0)^{0.5}$$

Where P, ρ and γ are pressure, density and specific heat ratio (C_p/C_v) of the gas in the reservoir ($_0$), A is the orifice area and k is a constant.[2] That the formula should involve the ratio C_p/C_v (constant pressure/constant volume specific heats) should be no surprise as rapid expansion across an orifice is clearly adiabatic.

Compare the above with the non-compressible equivalent formula for an orifice

$$\text{mass flow} \simeq kAP_0(P_0 - P)^{0.5}$$

Here the downstream pressure P is important as the flow is subsonic and downstream events can affect upstream flow rates. k is a constant related to the ratio orifice area/'*vena contracta*' area, which is roughly constant. '*Vena contracta*' is a reference to the shrinking of the cross-sectional area of a stream issuing from an orifice.

Pipe flow with a high pressure drop is also sometimes calculable by formula. This may be needed, for example, when using very small capillary sample pipes or at low pressure. The relatively simple formula for this is

$$\text{mass flow} \simeq k(R^5(P_1^2 - P_2^2)/L)^{0.5}$$

where k is a constant depending on the gas and its temperature, and L is the length of the pipe. (Compare this with Poiseuille's equation

$$Q = \frac{\pi}{8}\,\frac{1}{u}\,R^4\,\frac{(P_1 - P_2)}{L}$$

If mass flow rate is desired, then density, of which an average is $(P_1 + P_2)$, must multiply the formula and Poiseuille then gives

$$\text{mass flow} \simeq k'/u\ R^4(P_1 - P_2)(P_1 + P_2)/L = k'/u\ R^4(P_1^2 - P_2^2)/L$$

where k' is another constant.)

Dimensionless groups and Reynolds number. Finally, before leaving simple numerical formulae and admitting that further problems require numerical solution with a computer, the possibilities of a dimensional analysis of a gas flow problem should be examined. Dimensional analysis, even where it does not give answers, can restrict the amount of experimentation which may be necessary and suggest ways in which a problem can be scaled down to suit laboratory experiment. Dimensional analysis may well

[2] A more exact formula replaces γ with $\gamma(2/(\gamma+1))^{(\gamma+1)/(\gamma-1)}$.

provide the required cross-check on the realism of computer-generated results.

Dimensionless groups have proved useful in much of physics but especially so in fluid dynamic problems. The most commonly used dimensionless group is the Reynolds number. This expresses the balance of viscous and dynamic force on elements of a gas stream. A high Reynolds number means that dynamic forces, the mass inertia forces of the flow, exceed the viscous forces within the gas. The result will generally be turbulent flow. A low Reynolds number means that the viscous forces far exceed inertia effects and the resultant flow pattern will be indistinguishable from that of treacle. The Reynolds number, Re, is defined as

$$\text{Re} = \text{inertia forces/viscous forces}$$

$$\text{Re} \simeq \rho V^2 L^3 / u V L^2$$

i.e.

$$\text{Re} = \rho V L / u$$

where ρ is the density, V is the mean velocity, L is the characteristic length and u is the viscosity.

For pipework $\text{Re} << 2000$ will lead to laminar flow whilst for $\text{Re} >> 2000$ turbulent flow will occur. However, the boundary between the flow regimes is not clear-cut and flow in the whole region around $\text{Re} = 2000$ may be difficult to predict. If a laboratory experiment is to be used to model full-sized equipment, and flow will occur near the turbulent/laminar flow boundary, then by designing an experiment where the Reynolds number will be similar to that at full-scale realistic results should be obtained.

Design of a linear-flow element demonstrates a simple application of the Reynolds number. Suppose the flow in a pipe is V per second. If the Reynolds number is 10 000, turbulent flow would be expected, giving a characteristic pressure drop depending on the square of the flow rate, Q^2. If that single piece of tube is replaced by 10 000 small diameter tubes adding up to the same cross-sectional area, then, at the same flow rate Q, the mean velocity will be the same but the pressure drop will vary linearly with the flow rate Q, because the Reynolds number will be only 100, with viscous flow predominating.

Linear-flow elements are designed to impose a linear-flow versus pressure curve on a gas flow by ensuring viscous flow. They do this by the use of a multiplicity of small gas passages. Reynolds number calculation reveals why this tactic works. Basically, by restricting the flow in this way, the effects of dynamic inertia are made small compared to the large viscous drag required to force the gas through the tiny passages.

Dimensionless groups in general make comparisons between the forces on an element of gas e.g. inertia, viscosity, gravity, pressure, etc. Other common dimensionless groups are:

- Nusselt number, hD/k: heat input/heat lost from conduction, relating to heat transfer;
- Prandtl number, Cu/k: kinematic viscosity/thermal diffusivity, i.e. the ratio of diffusion of momentum to diffusion of heat, relating to forced convection (contains only fluid properties and is substantially constant ($\simeq 0.74$) in many gases under normal conditions);
- Grashof number, $gD^3b\Delta T\rho^2/u^2$: buoyancy force/viscous force, relating to natural convection;

where h = coefficient of heat transfer, D = diameter, k = thermal conductivity, C = specific heat, u = viscosity, g = gravity, b = thermal expansion coefficient, ΔT = temperature difference and ρ = density.

Turbulent flow in pipework. With large Reynolds numbers pipework can display very predictable turbulent flow. This situation is described by another useful rule of thumb by Darcy. Darcy's (or Fanning's) rule is for well-developed turbulent flow in pipework and is an approximate rule analogous to Poiseuille's accurate equation

$$\Delta P = fLv^2\rho/r$$

where f is the empirical friction factor, v is the mean velocity, ρ is the medium density and L is the length of pipe. At low Reynolds numbers the factor f is equal to 8/Re, which will give Poiseuille's equation as before.

For a range of Reynolds numbers from 2000 to 100 000 and smooth pipes

$$f = 0.079/(\mathrm{Re}^{0.25})$$

For rougher pipes and higher Reynolds number, the friction factor f will depend on the roughness ratio $k/2r$, where k is the roughness size, but is independent of the Reynolds number. Values are typically

$k/2r$	f
0.04	0.016
0.01	0.01
0.04	0.07
0.001	0.005
0.0004	0.004
0.0001	0.003

If the value of f predicted by the roughness ratio is less than that of the smooth pipe formula, then the smooth pipe formula is used.

It is interesting to note that with the f factor constant, resistance to turbulent flow is independent of the fluid viscosity. Even a 'superfluid', a fluid of zero viscosity, thus shows resistance to flow at high flow rates. Cryogenic liquid gases, such as liquid nitrogen, as noted in more detail in section 3.7.2, have low viscosity whilst liquid helium can actually be considered as a mixture of superfluid and a normal fluid of low viscosity. In practice both of these frequently show large resistance to flow, contrary to what might be expected from their viscosity, because at fairly ordinary pumping speeds their flow is turbulent.

1.3.9 Computational fluid dynamics

Computer numerical models, such as the popular FLUENT and PHOENIX codes, are now generally available in industry and there are versions that can be run on desktop 'workstation' personal computers. These simulate gas behaviour by setting up finite elements of a gas and applying fundamental equations to the motion and state of these elements, with boundaries defined by the geometry of the parts contacting the gas. They are capable of simulating flow problems, including turbulence, compressibility and thermal effects, and will even allow calculations for reacting gases, such as in combustion. Most of these situations are impossible to calculate analytically. Computer models allow fairly accurate calculation of the performance of components of geometry which would not be possible using analytic methods, even for simple conditions such as laminar flow.

The simplest computer models for steady laminar flow simply provide solutions to Poisson's equation, with a few boundary conditions (e.g. $V = 0$ for $y < 0$ and $V = 100$ for $y > 10$).

$$\nabla^2 V = \mathrm{f}(x)$$

$$\nabla^2 V = \frac{\partial^2 V}{\partial x^2} + \frac{\partial^2 V}{\partial y^2} + \frac{\partial^2 V}{\partial z^2}$$

where lines of constant V are streamlines and $\mathrm{f}(x)$ is simply zero in the simplest case. (In the past, steady laminar flow streamlines have been plotted using electric fields across conductive paper, a kind of analog computer apparatus. This is possible because electric fields also obey Poisson's equation. The boundary conditions are applied by means of a battery and pieces of metal foil.)

Even the more complex computer models still have a relatively simple basis, such as the general partial differential equation

$\partial(\rho\phi)/\partial t$	$+ \nabla(\rho\phi\mathbf{V})$	$- \nabla(\Gamma_\phi \nabla\phi)$	$= S_\phi$
transient term	convection term	diffusion term	source term

where ρ is the density of the medium, t is time, $\mathbf{V}$ is the velocity vector, Γ_ϕ is an exchange coefficient, S_ϕ is a source term (e.g. pressure gradients $-\partial P/\partial x$, $-\partial P/\partial y$ and $-\partial P/\partial z$ if ϕ is a velocity component) and ϕ is the dependent variable to be computed (e.g. the velocity component V_x, $\phi = 1$ gives the equation $\partial\rho/\partial t + \nabla(\rho\mathbf{V}) = 0$, the continuity equation, which simply says that matter is neither created nor destroyed in the flow pattern).

More complex equations are needed for simulating multiple-phase flows.

The most sophisticated computer models can solve problems with other complications such as chemical reaction, thermal radiation, transonic or supersonic flow and porous media.

Like the simple Poisson equation these equations are solved by first deciding on boundary conditions, i.e. where the edge of the gas containment is, velocities, densities or pressures of gas sources, etc. Then gas equations of state must be entered into the model. In many cases a simple formulaic equation of state, such as the van der Waals model, may be applied. The differential equations must then be turned into algebraic equations which can be solved numerically with computer programs.

Such differential equations are applied to each finite volume element of a three-dimensional grid, chosen to be small compared to the scale of the phenomenon studied, and applied to successive time intervals, chosen to be short compared to the characteristic time of the phenomenon. The computer must simulate all the forces acting on the volume element (pressure on each face, turning moments) and calculate the reaction of the element (acceleration, change of temperature and volume). The influence of that element on its neighbours must then be calculated. Finally, the results must be displayed in a meaningful way to the operator to show flow behaviour. This is illustrated in Figure 1.3, which shows an example of a display from a CFD calculation of turbulence at a T-junctions in a gas pipeline.

The different computer programs, such as PHOENICS and FLUENT, differ in the detail of their programming and in how they approach the transformation from differential to algebraic equations. PHOENICS, for example, uses the 'finite-volume' method developed in Los Alamos, USA, and Imperial College, London. It straightforwardly applies a grid and integrates each differential equation for each volume element to give an algebraic equation which is then solved.

The workload on the computer is considerable even with small grids and few time slices, and quickly grows as the grid is divided more finely. Many computational tricks have to be used in order to use computational fluid dynamics (CFD) on desktop computers. Desktop computers are sufficient for many problems but it should be noted that quite large and expensive parallel-processor computers are needed for problems that are apparently only fractionally more difficult than those solved on the desktop computer.

There is a considerable amount of skill in applying CFD successfully and economically. For example, by choosing a nonlinear grid or two dimensions instead of three, computation time can be greatly reduced. In problems with rotational symmetry where turbulence is absent or unimportant, for example, a cylindrical grid, using only two dimensions, will be possible. In problems of steady-state flow, the first transient term can be omitted.

The use of computer modelling of flow is now almost mandatory for any

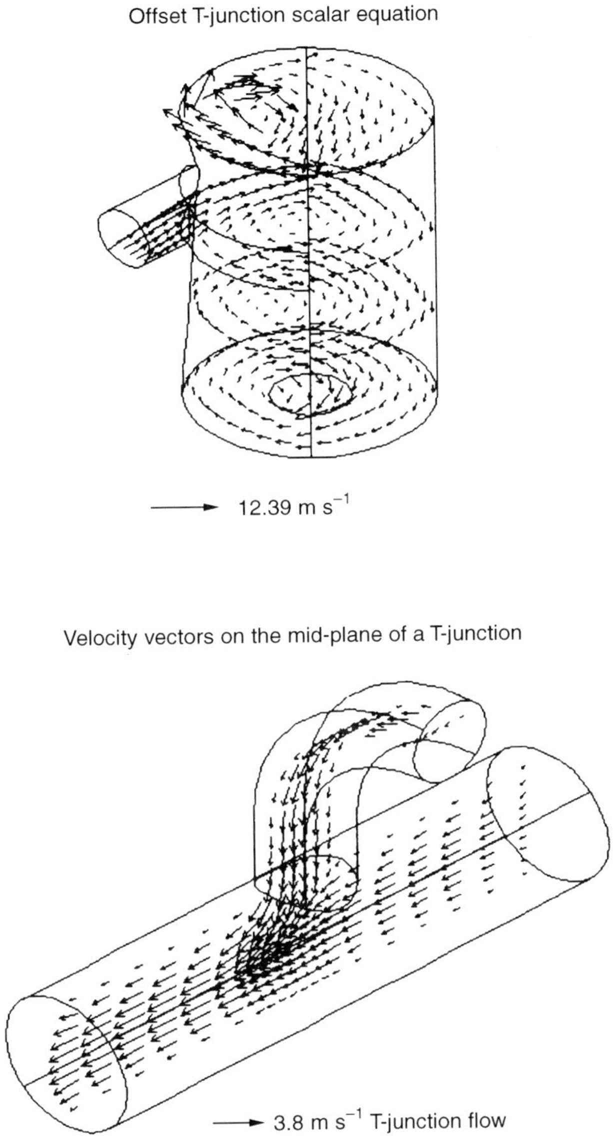

Figure 1.3 Example of CFD calculation of turbulence at a T-junction in a gas pipeline (courtesy of CHAM, London).

gas apparatus that is at all critical. However, when using computer-generated results a cross-check of some kind, from experiment, from simplified theory or perhaps even from a different computer model, may well be useful. Wherever possible, at least the order of magnitude of a reported quantity should be cross-checked.

1.3.10 Diffusion

According to kinetic theory the diffusion constant D of an ideal gas is given by

$$D = 1/3\ LV_{\text{m}}$$

where L is the mean free path (see section 1.3.11) and V_{m} is the mean molecular speed. Substituting standard formulae for L and V_{m}, the diffusion constant is therefore given by

$$D = (kT/6M)^{0.5}\frac{1}{nS}$$

where n is the molecular number density, per unit volume, k is Boltzman's constant, T is absolute temperature, M is molecular mass and S is the molecular cross-sectional area. The diffusion constant is thus inversely proportional to the pressure (the number density being proportional to pressure at ordinary pressures), i.e. $D = D_0P_0/P$.

Many time-varying diffusion problems may be solved by application of the diffusion equation, which in one dimension is

$$\partial C/\partial t = D\partial^2 C/\partial x^2$$

This is derived from the steady-state Fick's law, which defines the diffusion constant D as $J = -D\partial C/\partial x$, where J is the flux of molecules diffusing and $\partial C/\partial x$ is the concentration gradient.

The behaviour of mixtures in gas cylinders is a good example of how diffusion can achieve mixing over time. Suppose a cylinder of gas is filled with two gases more or less perfectly separated at time zero. The lighter gas is at the top, the heavier at the bottom. The mixture will slowly form at the interface and the progress of the mixed band through the cylinder will spread to a width proportional to the square root of the time passed. This curious time behaviour is a consequence of the mathematics of the diffusion equation.

The diffusion equation frequently has solutions of the form

$$C = C_0 \exp(-x^2/aDt)\ /\sqrt{(Dt)}$$

where a is a constant. These solutions show the characteristic diffusion equation property that the width over which a gas has diffused increases as the square root of time.

A typical 5 m^3 200 bar gas cylinder will be fully mixed for most purposes after a few months. However, inspection of the above equation will reveal a simple way to reduce the mixing time dramatically: simply lie the cylinder on its side. The gases will then take up the same positions with the lighter gas on top of the heavier gas. However, in this case the distance over which diffusion must take place is much shorter, a factor of around 10 less. With a characteristic distance 1/10, diffusion will take place in $(1/10)^2$ or 1/100 of the time, a day or so for a typical cylinder.

The calculation of diffusion in textbooks often gives an example of diffusion at atmospheric pressure, which may give an illusory idea of the rapidity of diffusion in all situations. Diffusion, because it is inversely proportional to pressure, is much slower at the higher pressures often used in industry. In a gas cylinder at 1 bara, for example, the textbook example will give 2 m diffusion distance in 1 h. However, with full cylinder at 200 bar, the diffusion distance for 1 h is only 14 cm and, furthermore, this distance only widens as the square root of time, so 200 h is needed for the equivalent diffusion state at 200 bar. Figure 1.4 shows an example of diffusional mixing in a gas cylinder.

It is sometimes asked whether a gas mixture, particularly a light/heavy mixture like sulphur hexafluoride in hydrogen, once formed will unmix: the answer is basically that it will not. The science of statistical mechanics, which explains why this is so, is well-described in standard texts, although popular books such as Ruelle (1991) may deepen understanding. Provided the constituents are not near their boiling point, and cannot therefore condense, a gas mixture will remain mixed. In fact, many mixtures with condensible components below their normal boiling point will not unmix either because the partial pressure of the condensible component is below 1 bar, the reference pressure for the boiling point.

There is a certain amount of gravitational separation in gas mixtures but in normal earth gravity this is entirely negligible

$$P_i = P_{i0} \exp(gh/kT)$$

where P_i is the partial pressure of the ith gas, g is the acceleration due to gravity, h is the height above sea level (at which $P_i = -P_{i0}$), k is Boltzmann's constant and T is the absolute temperature.

$P_1 = 100$ bar H_2, $P_2 = 100$ bar Ar

e.g.

$$\Rightarrow P_1/P_2 = 1 \text{ at the bottom of the cylinder, } P_1/P_2 = 1.001 \text{ at the top.}$$

In high-speed centrifuges this is no longer true and today the most efficient method yet discovered for the separation of uranium isotopes is gravitational separation of UF_6 using centrifuges.

1.3.11 Molecular flow: small orifices and vacuum systems

There are now many industrial processes employing gases at low pressure. Many are in the processing of microelectronic circuits, although metal/plastic coatings, glass coatings, engineering chemical vapour deposition (CVD) coatings, flatscreen displays and solar cells are also users of low pressures. Books on vacuum techniques are recommended (e.g. Dennis and Heppell, 1968) for a detailed account of the behaviour of gases at high-vacuum pressures.

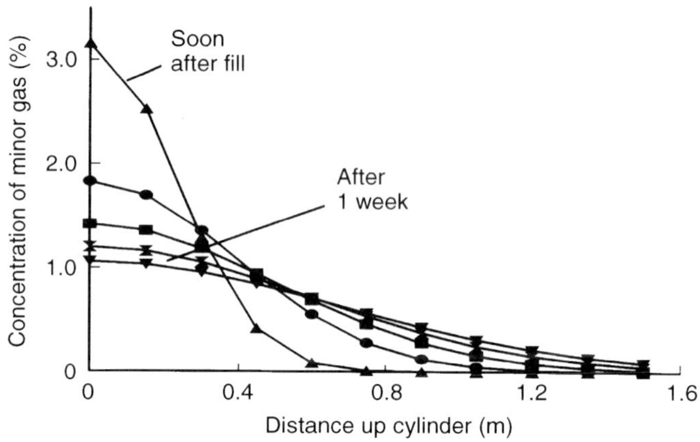

Figure 1.4 Diffusional mixing in a gas cylinder.

However, many processes operate in the 'no-man's-land' between pressures where true vacuum technique applies and pressures where gases behave more or less as they do at atmospheric pressure. An example of this is the evacuation of gas cylinders before filling them (for high purity grades). In this pressure region it is useful to know about vacuum as well as atmospheric pressure characteristics.

Gases in ordinary size pipes and vessels (an inch or so) switch from more or less familiar viscous or turbulent flow to the radically different regime of 'molecular flow' at around 0.1 millibar. Below this pressure gas molecules may travel from one side of the pipe to the other without colliding, giving radically different flow characteristics. The crucial parameter is the 'mean free path' i.e. the average distance gone before a molecule meets another molecule.

The mean free path in air at atmospheric pressure is just 0.05 μm. However, the mean free path is inversely proportional to pressure and at a typical industrial vacuum pressure, say 1/10 000 of an atmosphere, the mean free path is about 5 mm.

The mean free path is given approximately by

$$L \simeq 1/((\sqrt{2})n\sigma)$$

where n is the number of molecules per m^3 (proportional to pressure at lower pressures) and σ is the collision cross-section of the molecule, approximately $16\pi R^2$, where R is the molecular radius.

For a long pipe with steady viscous flow at normal pressures, the mass flow rate is given by the Poiseuille equation, as above

$$Q = kR^4 Pav(P_1 - P_2)/L$$

The molecular flow equation for low-pressure gases is

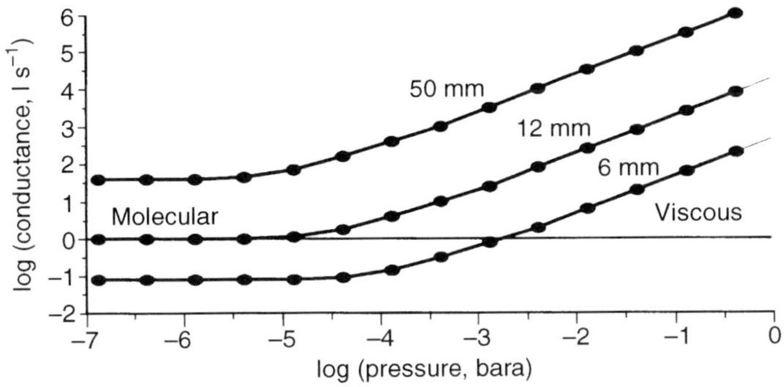

Figure 1.5 Flow through a pipe showing transition from molecular to fluidic flow.

$$Q = hR^3(P_1 - P_2)/L$$

The molecular flow regime involves a much lower resistance to flow than the viscous calculation gives. As the pressure sinks very low, the molecules can travel without collision along a long length of pipe, so this phenomenon is perhaps unsurprising. However, the lower power in pipe radius means that moving to a larger pipe does not offer as much advantage as in viscous flow. The absence of an average pressure term reflects the fact that the net flow, Q, is simply the difference between the molecules leaving the high-pressure end and those leaving the low-pressure end.

The evacuation of a cylinder to millitorr (10^{-6} bar) pressures involves a transition from molecular to fluidic flow (Figure 1.5). If an approximately constant volume flow rate from the vacuum pump is assumed, then, in the absence of resistance to flow from pipes, the cylinder will follow an exponential fall in pressure

$$P = P_0 \exp(-Vt/S)$$

where S is the pumping speed in l s^{-1}, V is the cylinder volume in litres and t is the time in s. However, the effective pumping speed, S, is affected by the resistance to flow, which is usually expressed in vacuum problems in terms of 'conductance'. The conductance of a pipe, C, in l s^{-1}, gives the flow with a differential pressure ΔP (traditionally given in torr) as follows

$$Q = C\Delta P$$

Q is thus the volume flow in Torr l s^{-1}. It is convenient to define C in this way because then the effective capacity S_{eff} of a pump connected via a pipe of conductance C can be simply calculated as

$$1/S_{\text{eff}} = 1/C + 1/S$$

If the pressure is higher than around 10^{-2} Torr, the conductance, C, is not constant, in fact it increases exponentially. This results in a steeper initial pressure drop when S_{eff} is high, followed by a slower exponential decrease reflecting the fact that S_{eff} is small, being dominated by the large term C. (In practice a curve of this sort is distorted below a certain pressure by the outgassing of water vapour absorbed onto the container and pipework walls. In our example, to achieve high vacuum the cylinder must be baked to 100°C or more.)

1.3.12 Heat transfer in gases

Heat transfer is a substantial branch of chemical engineering in its own right and standard texts such as Fraas and Ozisik (1965) and the usually substantial sections in books on chemical engineering, such as Perry (1984), are suggested for a deeper understanding. However, there are a number of simple situations which frequently arise in the use of gases where straightforward calculations are useful. Many of these involve the transfer of heat from solid surfaces to the gas and vice versa. Heat transfer in gases takes place via three mechanisms: radiation, convection and conduction. Before starting the detailed solution of any problem, the relative importance of the three mechanisms should be roughly estimated

- Radiation typically depends more on the nature of the solid surfaces than on the gas between them. It is usually unimportant for shiny polished surfaces, especially at low temperatures.
- Whether convective heat transfer can take place depends on whether effective circulation of the gas can occur, either forced with a fan or via natural convective flow. In a geometry with a hot surface above a cold one, natural convective heat transfer will be very low.
- Conductive heat transfer is the most readily calculated of the three mechanisms. It depends only on the geometry of the surfaces and their temperatures, and the nature of the gas. Only in high vacuum does conductive transfer reduce to zero, although in many problems it may not be as important as the other two mechanisms.

1.3.13 Thermal conductivity and viscosity of gases

For normal pressures the thermal conductivity of a gas, like the viscosity, is independent of pressure. This arises as follows. The thermal conductivity, K, of a substance is the constant K in Searl's equation for one-dimensional heat flow along a bar

$$\mathrm{d}Q/\mathrm{d}t = K\mathrm{d}T/\mathrm{d}x$$

where $\mathrm{d}Q/\mathrm{d}t$ is the heat transferred along the x direction. The rate at which heat (molecular kinetic energy) is carried along is clearly proportional to the heat capacity, C_v/N, of each molecule, the molecular velocity, V_m, and the number of molecules per unit volume, n. Less obvious perhaps is the effect of the mean free path to which the thermal conductivity is also proportional. The mean free path, L, is proportional to $1/(nS)$, where n is the number of molecules per unit volume and S is the molecular cross-section.

Hence

$$K \simeq V_m C_v \, nL/N = C_v \, nV_m \, /(nNS) \simeq C_v V_m/(NS)$$

In fact, the accurate formula for ideal gases is

$$K = C_v V_m \, /(3\sqrt{2}NS)$$

where K is the conductivity, C_v is the specific heat per mole, V_m is the mean molecular speed, N is Avogadro's number and S is the molecular collision cross-section, i.e. the area around a molecule's centre where a collision will occur. The only variable in the equation is the velocity, V_m, which is simply related to the absolute temperature by

$$V_m = (3/2 \; kT/m)^{0.5}$$

i.e.

$$K = \frac{1}{2\sqrt{3}} \, (kT/m)^{0.5} C_v \, /(NS)$$

which is manifestly independent of pressure. In fact, between 1 mbar and about 100 bar, a range of pressures from vacuum to the highest normal gas pressure, the conductivity of a gas, like the viscosity, is remarkably constant. This has important implications for thermal insulation: a Dewar vacuum insulated flask, for example, must employ a high enough vacuum that this constancy of the thermal conductivity breaks down. The thermal conductivity of a gas decreases with molecular weight and size (Table 1.1).

Table 1.1 Thermal conductivity of common gases

Gas	Thermal conductivity $W\,m^{-1}\,K^{-1}$ (0°C)
Air	2.41×10^{-2}
CO_2	1.45×10^{-2}
N_2	2.40×10^{-2}
O_2	2.45×10^{-2}
H_2	16.82×10^{-2}
He	14.22×10^{-2}
Ne	4.65×10^{-2}
Ar	1.63×10^{-2}
Kr	0.87×10^{-2}
Xe	0.52×10^{-2}
CH_4	3.02×10^{-2}
C_3H_6	1.56×10^{-2}
C_4H_8	1.34×10^{-2}
CF_2Cl_2	0.85×10^{-2}

Kinetic theory for the viscosity of gases gives the result

$$u = 1/(3\sqrt{2})mV_m/S$$

where u is the viscosity, m is the molecular mass, V_m is the mean molecular speed, and S is the collision cross-section (the area within which molecular vectors are collision paths). The only variable in the equation is the velocity, V_m, which is simply related to the absolute temperature by

$$V_m = \sqrt{(3/2\ kT/m)}$$

Hence

$$u = 1/(2\sqrt{3})\sqrt{(kTm)}/S$$

Kinetic theory thus predicts that the viscosity of an ideal gas does not vary with pressure. In fact, for real gases at ordinary pressures, the viscosity does indeed depend very little on pressure. Between 1 mbar and about 100 bar, a range of pressures from vacuum to the highest normal gas pressure, the viscosity of most gases is remarkably constant. Even though the gas at 100 bar has 100 000 times as many molecules per cm^3 as it has at 1 mbar, the viscosity, and hence the frictional resistance to motion, is the same. An additional factor that must be taken into account, however, is the greater likelihood of turbulence at higher pressures because the Reynolds number is proportional to density and hence pressure.

The formula above also predicts that gas viscosity should vary with $\sqrt{T}$. This is inaccurate and a correction to this formula (from Sutherland, see Kaye and Laby (1995)) gives

$$u \simeq BT^{3/2}/(T + C)$$

where B and C are constants.

Small heavy molecules are expected to have higher viscosities than large light ones and this is borne out in practice, with the noble gases all showing high viscosities and larger hydrocarbons low viscosities (Table 1.2).

The high viscosity of the monatomic gases has an impact on the use of gases as thermal insulators. In tiny cells, such as in plastic foam, the best thermal insulating gas will be that with the lowest thermal conductivity, normally a heavy polyatomic gas. However, in larger cells convection will take place, with the less viscous gases, such as the heavy polyatomics, convecting more and providing a poorer insulation than those of high viscosity.

1.3.14 Vapour pressure of liquids

Many gases are used in liquid form either under pressure or at low temperature. The liquid (or solid) form of a substance, if kept in a container with an ullage space, will eventually reach a dynamic equilibrium with the

Table 1.2 Viscosity of common gases

Gas	Viscosity ($\mu N\,s\,m^{-2}$) (0°C)
Air	17.3
CO_2	13.6
N_2	16.6
O_2	19.5
H_2	8.4
He	18.7
Ne	29.8
Ar	21
Kr	23.4
Xe	21.2
CH_4	10.3
C_3H_6	7.9
C_4H_8	7.0

gaseous form of that substance. This equilibrium pressure, the saturated vapour pressure, is a straightforward function of temperature only.

Normally when a liquid is heated gently molecules simply leave the liquid surface for the gas phase and evaporation is said to take place. If a liquid is heated more vigorously bubbles form in it when the saturated vapour pressure of the liquid exceeds the gas pressure and the liquid is said to boil. The converse process of condensation of the gas occurs during cooling and can also take place either in the body of the gas or at a surface.

In mixtures the temperature at which condensation takes place (the dew point) is cooler, often by several degrees, than the temperature at which bubbles form (the bubble point). For example, the bubble point of liquid air is 78.75 K, whilst the dew point is 81.75 K. Bubbles forming at the bubble point are rich in the lighter component whilst the liquid condensing at the dew point is rich in the heavier component of the mixture. (In the case of air condensing on a liquid nitrogen line, for example, the liquid forming on the outside of the line will be rich in oxygen, a phenomenon that can give rise to accidents if oxygen-rich liquid air drips on to organic material below.)

The boiling or condensing of a gas is a complex phenomena, with clean liquids and gases frequently persisting in states far from equilibrium. Pure liquids may be heated far above their boiling point, or superheated. When boiling finally occurs in a superheated liquid it is sudden and violent. Pure gases may be cooled far below their boiling point, when condensation may be also be very sudden and the vapour is said to be supersaturated.

Both supersaturated vapours and superheated liquids have been used for the detection and tracking of nuclear particles: the minute amount of ionisation caused by the passage of a high-energy charged nucleon is sufficient to act as a series of nuclei for condensation or boiling. If conditions are correct, nuclear particle tracks are revealed by rows of droplets in the

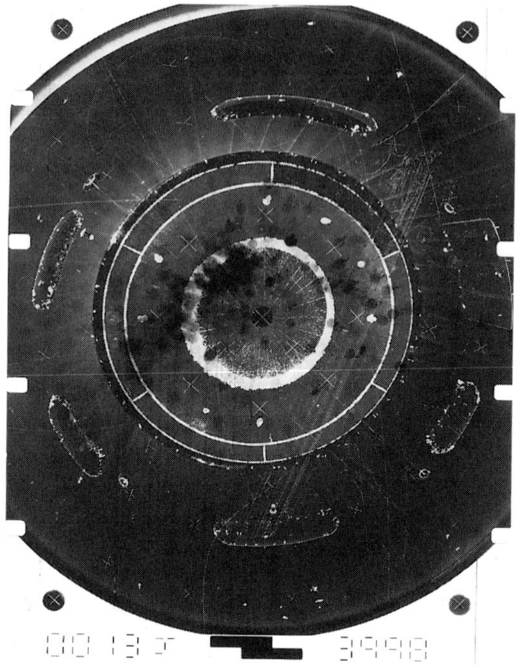

Figure 1.6 Liquid hydrogen bubble chamber (BEBC at Cern, Geneva) showing boiling around the expander piston (centre) and charged fundamental particle trails of microscopic bubbles, curved due to the large applied magnetic field. Sometimes neon was mixed with the LH to increase the cross-sections for the particle collisions (courtesy of High Energy Physics Group, Imperial College, London).

vapour in the case of the 'cloud chamber'. The early development (1912) by Wilson (see Enge, 1966) of the cloud chamber enabled many important nuclear physics discoveries. In the case of the 'bubble chamber' the tracks are revealed by rows of bubbles in the liquid, typically liquid hydrogen. The hydrogen bubble chamber was the most important particle physics analyser from the 1950s to the late 1970s (Enge, 1966; Perkins, 1972) (Figure 1.6). It enabled the foundations of the quark theory of matter to be laid. Supersaturated vapours and superheated liquids have important industrial applications, too (see section 3.7.2 for further discussion of boiling).

The saturated vapour pressure of many substances can be approximated to an exponential function of the form[3]

[3] This form can be understood in a simple way as follows. There is a spread of molecular speeds in a liquid, with only a few molecules having enough speed to escape the attractive forces of the liquid. The fraction with enough speed depends on temperature. The velocity distribution follows a normal distribution pattern to most statistical distributions and is given by an exponential of the form

$$f(v) = \exp(-v^2/KT)$$

Taking the fraction with energy 1/2 Mv^2 greater than the 'escape energy', E_a, gives the exponential pressure law above.

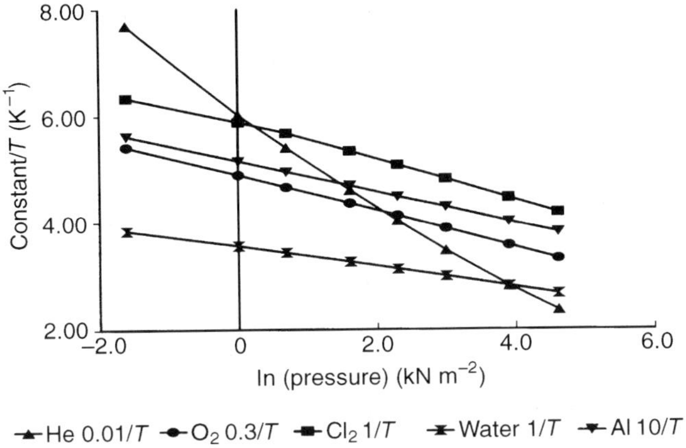

Figure 1.7 Vapour pressures of different substances.

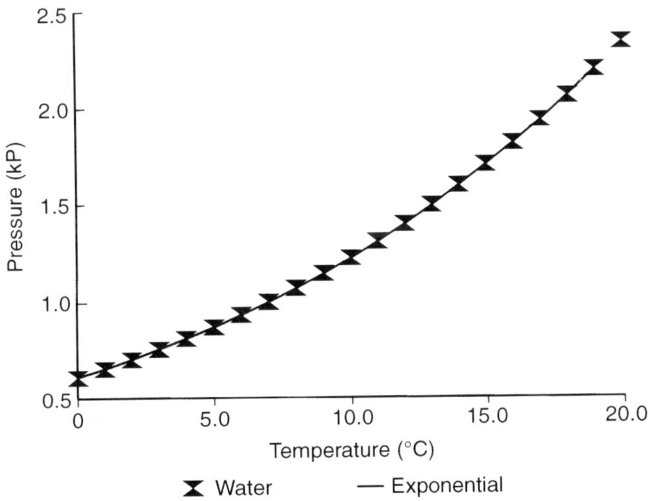

Figure 1.8 Vapour pressure of water.

$$SVP = C\exp[-E_a/(kT)]$$

or

$$\ln(SVP) = \ln C - E_a/KT$$

where C and E_a are constants peculiar to the substance, and k is Boltzmann's constant. In Figure 1.7 $\log_e$ (pressure) is plotted against a constant/T. This plot should be a straight line if this law is followed. For most substances, the exponential in $1/T$ is a remarkably good approximation. The temperatures vary in Figure 1.7 from the intense furnace heat of the 2620 K boiling point of aluminium to the ultracold behaviour of helium near absolute zero but deviations from the law are not great. Helium is a very exceptional substance, however, and is clearly following a slightly different law.

Figure 1.7 shows the vapour pressure of water from 0 to 20°C and also demonstrates the exactness of fit of the exponential function.

1.3.15 Sublimation and supercritical fluids

Most everyday substances are solid at low temperature, melt when heated and vaporise when heated more strongly. Referring to Figure 1.8, a P–T or 'phase diagram', most substances follow a line such as BB.

Amongst common industrial gases this standard behaviour is also followed, with a few exceptions. Solid carbon dioxide (dry ice) 'sublimes', going directly from solid to gas, as does acetylene. They have no liquid phase at atmospheric pressure, following a line such as AA on the phase diagram, below the triple point. At pressures higher than the triple point, liquid phases will form for these gases, however, so that, for example, carbon dioxide at 10 bar or more forms a liquid phase, a fact which is used to advantage when storing carbon dioxide in bulk (section 3.4.2).

Carbon dioxide is the most common gas which sublimes at atmospheric pressure. Solid acetylene is regarded as dangerous, being subject to detonation, so is not normally utilised as a solid but it sublimes at 189 K. A slightly more unusual gas, sulphur hexafluoride, sublimes at 209 K, while there are other more reactive and unusual gases which also sublime. It should be noted, however, that the triple points of gases such as nitrous oxide, neon and even argon are close to atmospheric pressure and modest pressure reductions can result in them solidifying, perhaps blocking up apparatus unexpectedly (Table 1.3).

At high temperatures substances follow a line such as CC in the phase diagram (Figure 1.9), passing from solid into a state which is dense but has no surface tension, i.e. a supercritical fluid. Above the critical temperature

Table 1.3 Triple points of common gases

Gas	Pressure (bara)	Temperature (K)
Nitrogen	0.125	63
Carbon dioxide	5.1	217
Nitrous oxide	0.87	182
Oxygen	0.00152	54
Butane	4×10^{-6}	135
Hydrogen	0.07	14
Acetylene	1.2	191
Argon	0.68	84

Helium will not solidify under its own vapour pressure so no triple point is possible. Solid helium only exists at above 25 bar pressure and below 0.8 K.

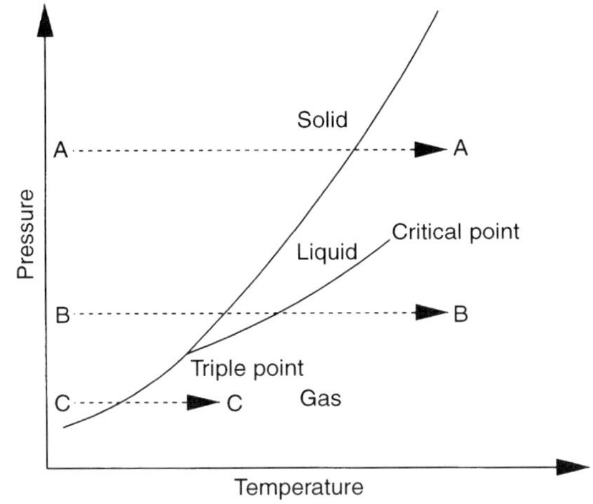

Figure 1.9 Schematic *P*/*T* diagram for argon.

no liquid can exist, but instead the substance forms a supercritical fluid or 'dense gas' (Table 1.4).

As liquids are heated and the average energies of their constitutent molecules grow larger, their surface tension, heat of vaporisation and viscosity grow smaller, and their compressibility increases. This is intuitively reasonable on an atomic model of a liquid, in which the higher thermal energies force an increase in molecular separation and a resultant decrease in intermolecular forces.

As the critical point is approached, the surface tension approaches zero, as does the heat of vaporisation, while the viscosity rapidly modulates downwards by three or more orders of magnitude from liquid to gaseous values. The compressibility also changes rapidly near the critical point, increasing to very large values just below the critical point but falling again to the larger values typical of gases well above T_c (e.g. Reid, 1977).

Supercritical fluids have some important properties. They are dense enough to hold appreciable amounts of other substances in solution and can be used as solvents in which surface tension is entirely absent. This enables them, for example, to be extracted from porous material (carrying dissolved substances away) without causing the collapse of pores. For this reason methane, methanol vapour, steam and carbon dioxide, *inter alia*, are important industrial solvents under supercritical conditions.

Carbon dioxide has a critical temperature conveniently near room temperature, 30°C. It was the first substance whose near-critical properties were investigated and can be used as the basis of a demonstration. A suitably small amount of liquid CO_2 partially filling a sealed strong (critical pressures are often in the 30–70 bara region, CO_2 is 74 bara) glass vessel (behind a safety screen) can simply be warmed through its critical point with a gentle

Table 1.4 Critical points of common gases

Gas	Pressure (bara)	Temperature (K)
Nitrogen	34	126
Carbon dioxide	74	304
Oxygen	51	110
Butane	38	425
Hydrogen	13	33
Helium	2.3	5

source of heat such as a hair-drier. Under the correct conditions the liquid meniscus disappears, while a misty appearance ('critical opalescence') can be seen: the consequence of rapid spatial variations in density, which should be expected as the compressibility of the fluid becomes very large because micro pressure changes are able to cause macro density changes.

Supercritical fluid extraction. The high compressibility of supercritical fluids means that the supercritical fluid density can be modulated hugely by relatively small changes in applied pressure. It is this that allows efficient continuous extraction processes, supercritical fluid extraction (SFE), to be designed (Kirk-Othmer (1982) and Irani and Funk (1977) both describe the process well). A small increase in pressure is used to raise density to obtain solvent action, while precipitation is carried out by a small reduction in pressure, when the low density achieved causes solubilities to fall dramatically.

CO_2 is one of the most popular substances for use in supercritical fluid extraction because of its combination of convenient properties – good solvent, supercritical at normal temperatures and reasonable (100 bar or so) pressures, low toxicity – as well as its low cost. It is used on an industrial scale for the preparation of decaffeinated coffee. The undesired stimulant compound, caffeine, is extracted from ground beans by SFE by steeping to dissolve the caffeine in high-density supercritical CO_2 followed by the release of the caffeine to waste and recovery of CO_2 by expansion to low density.

1.3.16 Dissolved gases in liquids

In many situations, the very small amount of gas that typically dissolves in a liquid is not important. However, there are situations where the dissolution of gases in liquids is a matter of great consequence in nature and of considerable importance in industry. The small amount of oxygen that will dissolve in water is vital for the life of aquatic creatures, while the similarly small amount of nitrogen that can dissolve in human blood is a serious difficulty for divers. Historically, the different solubilities of oxygen and nitrogen in water enabled crude oxygen concentrators to be constructed.

More recently, gases dissolved in hydrocarbons became very important as the oil industry matured and the very large volumes of gases held in most crude oils, including methane, ethane, ethylene, propane and butanes, are now extracted and put to useful purposes. In industry many separations or purifications of gases rely on the absorbtion of one gas from the mixture by means of a spray of liquid or the trickling of liquid over a packed tower.

Gases dissolve in liquids to a greater or lesser extent, depending on their chemical compatibility. For example, nitrogen, an inert, nonpolar molecule, dissolves only very slightly in water. By contrast carbon dioxide is a polar molecule which is attracted to water molecules and in fact reacts to a small extent with water to form carbonic acid. As a result, carbon dioxide is much more soluble than nitrogen in water: at 15°C about 1 volume of CO_2 dissolves in 1 volume of water, as opposed to 0.023 volumes of N_2.

The theory for the dissolution of gases in liquids is fairly simple. Many, particularly the more sparingly soluble gases, follow Henry's law up to 100 or 200 bar. Henry's law states that the pressure or partial pressure of a gas in equilibrium (P) with a liquid is proportional to the molar concentration (X) of that gas, i.e.

$$P = KX$$

Figure 1.10 shows the curves for common gases in water, demonstrating the validity of Henry's law.

The 'constant', k, in Henry's law varies with temperature and that variation can frequently be modelled by an exponential factor of the form

$$K_r = K_0 \exp(-H/RT)$$

where H is related approximately to the energy required to extract a mole of gas from the solvent.

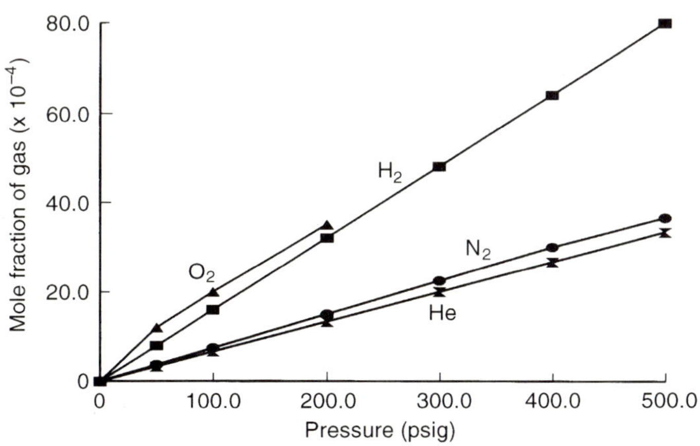

Figure 1.10 Solubility of common gases in water.

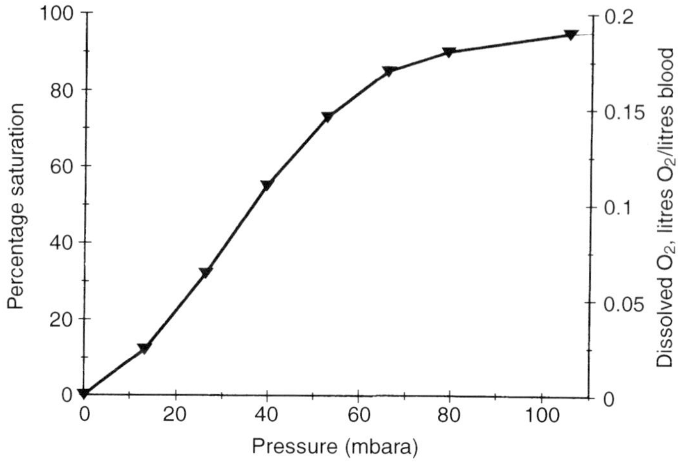

Figure 1.11 Oxy-haemoglobin saturation curve.

A more limited situation, but one which may be useful, is that of the vapour pressure of mixed liquids. These often adhere to the more stringent Raoult's law, which gives the constant of proportionality as the vapour pressure of the pure gas. This applies only in a few cases where the forces of attraction and repulsion between the two types of constituent molecule are closely similar, for example mixtures of atomic isotopes such as H_2O and D_2O (heavy water) or neighbouring members of an organic homologous series such as butanol and propanol.

For nitrogen in water and similar cases, the forces of attraction between the gas molecule and the liquid molecules are weak. Where the forces are larger and a weak chemical bond can be considered to have been formed, the straight line of Henry's law response is replaced by a curve with a plateau, clearly indicating the formation of a more or less stoichiometric chemical compound. A typical curve shows the amount absorbed steadily rising with gas pressure until the amount added corresponds to saturation of the absorber compound. After that point the amount absorbed increases only slowly with gas pressure. The solubility of oxygen in blood is an example of this (Figure 1.11).

A lot of experimental work has been done on the solubility of gases in liquids. Much data are of poor quality, however, observations having been made with apparatus of widely varying construction and effectiveness, and the data are often quoted in awkward units. A good account is given in Fogg and Gerrard (1991), which tabulates and graphs a lot of valuable data. Amongst this data the high solubility of carbon dioxide in oxygenated hydrocarbon solvents is striking, as is the solubility of oxygen in fluorocarbons.

Finally, it is worth noting that gases dissolve in solids as well as liquids. In the case of solids, however, it is much more difficult to reach the equilibrium solution because of the slow nature of the diffusion of gases through solids. The case of most interest is that of thin plastic films, where gases dissolve in the film and then diffuse through (see sections 2.1.7, 2.2.11 and 3.8.12).

1.4 Some specific gases

Countless organic compounds are possible and millions of inorganics. However, there are only a few hundred compounds which exert a vapour pressure of more than 1 bar at 20°C, and qualify for the name 'gas'. Furthermore, few of these compounds have importance in industry. A brief introduction to a few of them follows.

Nitrogen

Molecule:	formula	N–N
	weight	28.01
	% in air	78
Physical:	boils	77 K
	melts	63 K
	critical temperature	126 K
Liquid properties:	density	809 g l^{-1}
	viscosity	152 μN s m^{-2}
	tonnage cost	$15
	heat of evaporation	199 J g^{-1}
Reactions:	mainly inert	
	H_2 500 bar catalyst	
	iron nitrides at 1000°C	
	Mg burns	

Discovered by Daniel Rutherford in 1772, nitrogen is the principal gas in air (78%). Its first major industrial use lay in the manufacture of ammonia, used within the chemical industry as an intermediate for many compounds but especially nitrogenous fertilisers.

Oxygen

Molecule:	formula	O–O
	weight	32.0
	% in air	21

Physical:	boils	90 K
	melts	54 K
	critical temperature	155 K
Liquid properties:	density	$1141\ g\ l^{-1}$
	viscosity	$195\ \mu N\ s\ m^{-2}$
	tonnage cost	$15
	heat of evaporation	$213\ J\ g^{-1}$
Reactions:	burns/oxidises most elements and all organic compounds	
	C_2H_2 hottest flame	
	iron burns	
	alcohols forms aldehydes/acids	

Discovered in the mid 1700s, oxygen is the most important constituent gas of air (21.6%). Its first major industrial use was for limelight, intensifying gas flames until they could be used to heat calcium oxide (lime) to white heat. Liquid oxygen is the only common liquefied gas to be coloured; it is pale blue.

Argon

Molecule:	formula	Ar
	weight	40
	% in air	1
Physical:	boils	87 K
	melts	84 K
	critical temperature	151 K
Liquid properties:	density	$1403\ g\ l^{-1}$
	viscosity	$260\ \mu N\ s\ m^{-2}$
	tonnage cost	$500
Reactions:	inert	

Discovered at the end of the Victorian era, argon only became important with the rise of highly controlled smelting processes and with the growth in demand for high quality arc welding. Arc welding with powder-coated sticks, the powder forming a fluid slag or flux or evolving CO_2 or similar gases to shield the molten weld pool from oxidation, had been devised around in 1911 or 1912. However, the quality in these processes was not particularly good and patents were taken out in the 1930s for argon/hydrogen and similar inert gas mixtures. The full development of the potential for argon for welding only occurred after the Second World War, however, with the development of new processes using other gases added to argon and the arrival of inexpensive argon in large quantities.

Carbon dioxide

Molecule:	formula	N–N
	weight	28.13
	% in air	300 ppm
Physical:	sublimes	195 K
	critical temperature	303 K
Liquid properties	density	$1150\,g\,l^{-1}$
(at 10 bar):	viscosity	$240\,\mu N\,s\,m^{-2}$
	tonnage cost	$15
Reactions:	mainly inert	
	H_2 500 bar catalyst produces CO	
	Iron 1000°C mildly oxidising	

Discovered in the mid-1700s, carbon dioxide has a high heat capacity and sublimes directly from solid to gas at atmospheric pressure. It was therefore used early in solid form (dry ice) as a refrigerant.

Acetylene

Molecule:	formula	CH=CH
	weight	26
Physical:	sublimes	189 K
	critical temperature	308 K
Liquid properties:	density	$423\,g\,l^{-1}$
	viscosity	
	tonnage cost	$1500
	(liquid acetylene is a dangerous, explosive, liquid)	
Reactions:	highly exothermic when it reacts	
	oyxgen gives hottest flame	
	many active chemicals have addition reactions, e.g. with HCl to give vinyl chloride	

Discovered as the result of adding water to calcium carbide chips in the 1800s, the gas is still mostly prepared in that way today, although petroleum-derived acetylene is important, too.

Methane

Molecule:	formula	CH_4
	weight	16

Physical	boils	112 K
	melts	91 K
	critical temperature	191 K
Liquid properties:	density	423 $g\,l^{-1}$
	viscosity	119 $\mu N\,s\,m^{-2}$
	tonnage cost	$15
Reactions:	mainly inert except for combustion	

Methane was recognised as the flammable principle of 'marsh gas' early on and as a constituent of natural gas and petroleum in the 1850s but was treated largely as an undesirable by-product of petroleum until the 1950s. Many people are familiar only with one gas – the methane or propane that they burn for heating and cooking. Some of the non-heating applications of these gases are discussed in section 3.2.2.

Hydrogen

Molecule:	formula	H–H
	weight	2.05
Physical	boils	20 K
	melts	14 K
	critical temperature	33 K
Liquid properties:	density	71 $g\,l^{-1}$
	viscosity	13 $\mu N\,s\,m^{-2}$
	tonnage cost	$150
Reactions:	burns with invisible flame	
	saturation of unsaturated organics reactions	
	metal oxides reduced to metal	

At first named 'dephlogisticated' air by Priestley, hydrogen was well-known and fully investigated by the end of the 1700s.

Ethylene

Molecule:	formula	$CH_2{=}CH_2$
	weight	28
Physical	boils	169 K
	melts	104 K
	critical temperature	282 K
Liquid properties:	density	568 $g\,l^{-1}$
	viscosity	170 $\mu N\,s\,m^{-2}$
	tonnage cost	$50

Reactions: reactive hydrocarbon
H_2 saturation to ethane
HCl, etc., addition reactions with many chemicals
polymerisation
forms polyethylene ethylene/propylene at high pressure and with catalysts

1.4.1 Special gases

The term special gases, as it applies to the industrial gases industry, refers to any gas prepared and delivered in small quantities, up to a few tonnes or so, and of an unusual level of purity or in unusual mixtures. The most important classes of special gases are described here.

Semiconductor gases. These include hydride gases for deposition (like silane, SiH_4) and chlorine and fluorine-containing gases for etching and mixtures of these. They all have to be prepared to exceedingly high purity specifications.

Calibration mixtures. These comprise all those mixtures of gases used for setting up gas and other analysers. They vary in simplicity from simple pure gases, in individually checked cylinders, to mixtures of ten hydrocarbons used for calibrating gas chromatographs in the petrochemical industry. One of the newest areas of use for these gases is that of calibration for automotive exhaust gas analysers. Now that most countries have compulsory testing for engine exhaust, there are tens of thousands of analysers needing calibration for statutory purposes.

Ultrapure gases. These are needed for many purposes. Any scientific work needs to begin with ultrapure gases, whilst semiconductor and nuclear work often needs extraordinary purity.

Rare gases. These are sometimes known as noble gases and are those in Group VIII of the Periodic Table. Argon is supplied as a regular bulk or industrial gas but the others are supplied by special gas divisions or companies in small quantities. Helium and sometimes neon are available in liquid form, for use as cryogens. In addition to their other uses, all these gases are used in the lighting industry for filling glass bulbs.

Gaseous chemicals. These are basically those reactive chemicals whose vapour pressure is such as to require gas-handling techniques. They are mainly used as intermediates in specialised chemicals such as pharmaceuticals. Gas companies traditionally deliver small quantities of these compounds via their special gases divisions, with large (tens of tonnes plus) quantities being delivered by the chemical manufacturers direct. Chlorosilanes, phosgene, ammonia, ethylene oxide, propylene, butanes/butenes, chlorine, hydrogen sulphide, sulphur dioxide, methyl chloride and dimethylamine are some of the most common examples.

Refrigerants. These are gases which liquefy under modest pressures and are typically more or less inert and non-flammable (ammonia and ethylene are examples of gases used as refrigerants which do not fit this description, however). A range of available boiling points allows refrigerator machinery to be constructed with a wide range of cold-end temperatures. Although hitherto chlorofluorocarbons (CFCs) were the most important refrigerants, because of their bad environmental effects (stratospheric ozone depletion) they are now fast disappearing. In their stead, hydrofluorocarbons and HCFCs, such as HCFC-134a, are now growing in use.

1.5 The gases industry

1.5.1 Which gases matter?

It is difficult to find precise figures for the amounts of gases that are used in industry, so diverse are the uses to which they are put, but the following figures should give at least the magnitudes involved. The majority of the gases industry turnover, $10 billion or so (but larger if usage internal to companies were included), is concentrated in a few dozen key gases. Table 1.5 shows some of the most important industrial gases and their approximate US production (~30% of world production on average) in millions of tonnes (various sources). A number of gases are significant in the turnover they bring because although they are produced in rather small quantities measured in tonnage, they sell for very high prices, e.g. gases such as argon, silanes, acetylene, sulphur dioxide and nitrous oxide.

It may seem curious that some of the most important gases in tonnage terms are not those which might be ordinarily thought of as important in industry. In fact the list in Table 1.5 is distorted because enormous tonnages of gases such as CO or hydrogen are used but only a very small percentage of usage is recorded in statistics. If the full usage were included, then the list in Table 1.5 would look even more skewed towards gases apparently

Table 1.5 USA production of industrial gases (millions of tonnes)

Ammonia	19
Nitrogen	10
Propane	5
Oxygen	20
Carbon dioxide	3
Hydrogen	7
Ethylene	28
Propylene	14
1,2-dichloroethane	5
Butane	1.3
Helium	1
Argon	0.5
Cl_2	11

For comparison methane production = 200 megatons per annum.

unimportant in general industry. However, many of the largest tonnages are completely consumed within the pipework of a single chemical plant. For example, ethylene is produced in immense quantities by the cracking of petroleum. However, in many cases, the ethylene is both produced and used within the same plant for production, for example of vinyl chloride or alcohol.

The so-called 'merchant' gases, those produced and then sold to other companies or industries, are more interesting because of their interaction with industry at large. Small changes in process economics within the chemical industry can result in vast changes in the tonnages of non-merchant gases. For example, the use of ethylene increased about three-fold in the USA between 1977 and 1981 due to new petrochemical process changes.

Many of the uses of these internally consumed non-merchant gases are too specialised to be discussed in this volume. Similarly, heating applications of fuel gases such as methane and coal-gas, whose use could be supplanted by other heat sources, are not discussed. The remaining gases are truly industrial and are the subject of this book. They are used all over industry and are an irreplaceable component of many and varied processes.

Some of these gases seemed esoteric when they were discovered but now have important practical applications. An example of this is silane (SiH_4). This was made in small amounts from the 1850s onwards but was, as prepared then, highly unstable, normally spontaneously decomposing or igniting in seconds after only a few cubic centimetres had been prepared. Today, silane is no laboratory curiosity but is made, shipped and used in tonnage quantities as a result of enormous strides in the purity of silane and carefully designed production and utilisation plant.

1.6 Corporate structure of the industrial gases industry

The major players in the manufacture and supply of industrial gases today are mostly long established companies with global businesses dating back to 1900 or so. They are large multinational, or arguably global, businesses.

The nineteenth century saw the the peak of colonialism. Technically advanced nations in Europe and the USA administered or strongly influenced vast territories in Africa, Asia and the Americas. The gas companies of these nations expanded using the colonial areas or spheres of influence of their countries of origin in the early years of the twentieth century, before most colonial links began to be severed in the middle of the century.

The supply of compressed oxygen gas from air distillation, together with acetylene, was the root of their business expansion. Compressed oxygen was needed in modest quantities in the developing economies of all the former colonial areas. These requirements were a good match to the capabilities of the growing multinational gas companies. In most cases, the size of the gas demand was insufficient to be worth the developing country setting up its own firm, particularly when set against the considerable technical expertise needed.

The L'Air Liquide company of France and the BOC company of the UK are two of the largest players who fit this picture precisely. Air Products is a major player established more recently during the 1940s when US wartime requirements allowed rapid expansion. More regional, but still very large concerns, are gas companies such Nippon Sanso of Japan and Messer Griesheim of Germany. Messer Griesheim and the Union Carbide Linde Gases Division were the relatively small gas divisions of the chemical giants Hoechst and Union Carbide Company (UCC), respectively. Union Carbide have recently (1992) split off the gases division as PRAXAIR.[4] Linde of Germany is now a rather rare example of a basically national industrial gas company.

In Japan, gas for steel, until recently the biggest gas application, has often been manufactured by the steel companies themselves. This has meant that Japanese gas companies have been smaller than multinationals from the West. However, companies such as Nippon Sanso have grown much bigger in recent years and now have many ventures offshore. Nippon Sanso and the other Japanese companies still have some way to go, however, before they reach the global scale of operations achieved by the majors.

[4] The Union Carbide name is linked in the public mind with the Bhopal chemical plant accident of 1984 (see Jones, 1988). Particularly for a company supplying gases, the term 'gases' often having emotive 'poisonous' overtones, this is bad. In 1990 a proposed UCC plant in Scotland attracted so much bad publicity that the company withdrew the proposal despite the fact that central Scotland already has two companies operating gases plants and has massive petrochemical works of much greater safety import. The name change to PRAXAIR may well benefit UCC.

Many of the gas majors have very significant non-gases operations. BOC has substantial interests in vacuum pumps, distribution of refrigerated food and other products, and in hospital healthcare areas such as anaesthetics and medical equipment. Air Products and Chemicals, as its name suggests, is also a large speciality chemical supplier, e.g. in fluorinated chemicals. AGA has a large engineering wing, whilst Linde of Germany is a major manufacturer in Germany and the UK of fork-lift trucks.

There appear to be many more gas companies than there really are worldwide because of the habit of many of them of retaining the old name of companies long ago merged or taken over. For example, in the USA, AIRCO and Liquid Air turn out to be none other than BOC and L'Air Liquide respectively. Many comparatively small regional gas suppliers have been taken over by the handful of multinationals in recent years, again further reducing the number of alternative suppliers: in most markets, in effect, industrial gases are an oligopoly.

In several small newly industrialised countries, there are joint venture companies formed between the gas majors. For example, the Malaysia Oxygen Company is a joint venture between BOC and L'Air Liquide. In other countries, such as Turkey, there is a joint venture between a large conglomerate local company involved in other industrial activities with a multinational gas supplier: in effect, a marriage between local knowledge and influence, and multinational technology and know-how.

1.6.1 A typical industrial gas company

Although the industrial gases business is largely an oligopoly, the small number of companies involved do conform at least partially to a common pattern. A typical industrial gas company:

- is multinational;
- has a significant non-gas operation (see above);
- is not R&D (research and development) intensive, with 2% or less of turnover in R&D for its gases business: thus industrial gases lie half way between general engineering (~1%) and research-intensive sectors, like semiconductors and pharmaceuticals, which have >10% of turnover in R&D;
- has a significant distribution business: except for pipeline business with very low gas prices, all gases sold must be distributed to the user; the costly specialised cryogenic tank trucks and cylinder filling plants all add to the size and importance of this aspect of their operations;
- has a fairly high capital intensity: large and expensive production units, the specialised transport fleet and the need, because customers rarely supply their own, for a large inventory of cylinders (and cryogenic tanks) all dictate rather large capital requirements;

- has an average to high turnover per employee: although the turnover per employee is large in gas production, distribution of gas, especially the cylinder business, is labour intensive;
- has an average growth rate: although industrial gases is a mature industry in most respects, there are still significant growth areas in the application of gases; as a supplier to the shrinking heavy industries, gases sees some losses but in relatively new areas such as water treatment and food preserving there is strong growth; similarly, there are compensating gains and losses geographically, with fast-growing South-East Asia economies providing growth to compensate for losses elsewhere.

A typical industrial gas company is internally organised at least partly as a 'matrix'. The matrix has tonnage, bulk, compressed and special gas business units along the x-axis. These business units have largely separate staff for marketing, sales and finance, but also use the services of semi-autonomous central service departments. Departmental functions such as engineering, gas applications development, R&D, analytical laboratories, purchasing, stores, personnel and so forth, form the units along the the y-axis (see Table 1.6).

There is furthermore, within the operation of many of the companies, an additional degree of 'matrix' operation in that a particular project may draw on staff from various functions and from different business units.

The picture is further complicated by the multinational operation of the gas industry. World centres for financial services, legal services and R&D are operated by most of the gas companies. The world central R&D facility is where expensive research and analytical tools like electron microscopes or mass spectrometers are usually located. Basic topics in gas science

Table 1.6 Matrix organisation of a typical gas company

	Business units	Bulk	Tonnage	Compressed	Special
	Dedicated functions	Marketing Sales Finance Transport	Marketing Sales Finance	Marketing Sales Finance Transport	Marketing Sales Finance Transport
Shared functions				Cylinder management	
Engineering		√	√	√	√
Applications		√	√	√	√
Development		√	√	√	√
Research		√	√	√	√
Gas analysis		√	√	√	√
Purchasing		√	√	√	√
Safety		√	√	√	√
Personnel		√	√	√	√

technology as it applies to production and safety, and long-term work, are carried out centrally. The central facility is complemented by small local R&D laboratories. The small local laboratories concentrate on the applications of gas and the specific requirements of local users of gases.

1.6.2 Monopoly in the gas industry

There are frequently, and occasionally with some justification, cries of 'monopoly' against the gas companies. In some smaller countries there is effectively only one supplier, whilst in others there may be two or three who appear to collude on prices. Even within the massive economic area of the European Community monopoly investigations have been carried out on gas supply.

The basic problem is that gas is difficult to transport long distances at low cost and gas plants are subject to scaling laws giving big savings for big plants. In addition, in areas of intense industrial activity there are pipeline networks for common gases like nitrogen. The supply of industrial gases, except for special gases, is thus technically most efficiently operated as a monopoly. However, a monopoly is most undesirable from the point of view of fair and free trade, and governments have sought to try to reconcile the conflicting technical and commercial requirements. Perhaps the best compromise is an oligopoly combined with the ready availability of government controls if necessary.

In the UK especially, recent years have seen the denationalisation or 'privatisation' of many national utility organisations who have a 'natural monopoly' e.g. the network-based telephone, electricity, water and (methane) gas companies. In each case, along with the privatisation went a new but small government body whose job is to regulate the new near-monopoly private companies to ensure fair-trade practices, i.e. to inject the influences which competition would have supplied.

Whether these regulator bodies and the privatisation programme is successful will be a judgement for history. Their effectiveness is currently the subject of controversy with many of the new privatised utilities at least temporarily making large amounts of profit for their employees and shareholders, relative to much higher risk manufacturing companies. However, perhaps these could provide a model for regulation of the industrial gas industry, allowing it to exploit fully the economies of scale but not to abuse a near-monopoly position.

1.6.3 Gases industry supply chain

The supply chain of many industries can be represented as a supply chain, rather like a food chain in ecology. Just as in nature, the 'food chain' becomes a pyramid when the value of the business at each level is considered: whilst supplying industrial gas plants might be a billion dollar industry, supplying the gases themselves is worth ten billion or more and the applications that use them are themselves worth hundreds of billions.

At the top of the gases industry 'food chain' are the suppliers of gas production plant. Some of these are also operators of the plants that they make; they are 'vertically integrated'. Air Products, Linde, and BOC are examples. Other gas plant makers are independent in the sense that they do not normally operate their own plant and do not offer merchant gas for sale but seek to profit mostly by selling plant, e.g. Costain Petrocarbon and Hitachi.

In parallel with these plant suppliers are the makers of components such as cryogenic storage tanks, compressed gas cylinders, valves and fittings. Again, some gas suppliers also make components, for example AGA and BOC make tanks. In the main, however, components come from specialist companies and specialist divisions of larger firms. Gardner Cryogenics is a large US specialist in cryogenic tanks, for example, and TI Chesterfield of the UK makes huge numbers of hot forged steel cylinders, a business that requires particularly massive and expensive specialised equipment.

Many valves and fittings used in the gases industry are little different from those in general use for flow control in methane, fuel gases and compressed air. However, high pressures in gas cylinders and low temperatures in cryogenic tanks have led to special design components for critical places. Cylinder valves and pressure regulators are completely specialised to compressed gases, for example, as are the long-stemmed valves used for liquid nitrogen and oxygen. Specialised companies, and special divisions of larger general flow control component companies, have grown up to meet these needs.

Below the plant and component makers in the 'food chain' are the gases suppliers themselves and below the gases suppliers are the users of gas. Just as some gases suppliers are vertically integrated with their suppliers, some gases suppliers are downwardly vertically integrated with users of gases. Linde, for example, supplies industrial refrigeration plant which is a large user of freon and other industrial gases, while in the 1980s, the 'King Harry' pizza business, a large user of liquid nitrogen for freezing, was part of BOC.

1.6.4 Geography of gas supply

Industrial gases cover the complete range of commodity types studied by geographers. Some are highly specialist, valuable and shipped to customers around the world from a single producer in small compressed gas cylinders. At the other end of the scale are gases produced by large on-site plant, produced by taking services such as power and cooling water from the customer's plant, and delivered solely to that customer a few yards away through a pipe.

'Tonnage' gases are low-cost gases in large volumes such as the oxygen in steelworks. Tonnage gas is frequently taken from one point of production to one point of consumption by a large diameter pipe, although in areas of high industrial development there are networks of piped gas, especially nitrogen. Silicon Valley, south of San Francisco, and the industrial areas of northern France and Belgium have extensive pipes for tonnage industrial gases. Plants are attached on the pipeline in various places and provide continuity of supply by acting as each other's back-up in the event of plant failure. Customers are attached to the pipeline where necessary. By using inexpensive steel ('black iron') pipes of moderate pressure capability, and with the aid of regional assistance funding in some cases, pipelines have been extended over considerable distances at an economical price. With moderate and fairly constant flow rates in many industrial processes, there has been no need for the dramatic cycling of pressure up to 200 bars or more, seen in methane fuel gas supply pipelines. Similarly, the effective length of the pipe is such that intermediate pumping is not needed.

'Bulk' gas means gas delivered by truck, usually in cryogenic liquid form, and transferred to a bulk tank. Some gases (e.g. CO and H_2) can be delivered in compressed form in trailer loads: an articulated semi-trailer fully loaded with 20 or so tonnes of manifolded cylinders. Bulk gas is transported only regionally; above 100 miles or so the transport costs begin to rise and favour supply from a closer plant. Similarly, the costs of common compressed gases, which are delivered by truck up to a 100 miles or so, are composed mainly of transport costs at larger distances and of handling costs at short distances.

It is notable that there is considerable variation in gas costs geographically. The cost of carbon dioxide, for example, is several times lower in the USA compared to Europe. This large discrepancy probably reflects large investment in the extraction of CO_2 from waste streams in the USA. In Europe there are fewer supply sources.[5] In some parts of Europe, CO_2 is in effect a local monopoly because a single large source within the chemical industry is the only economical local source. (Waste streams in the chemical industry often provide a copious and inexpensive source of CO_2.) There are national and regional differences in electrical power costs and these impact tonnage prices greatly, bulk prices to some extent and compressed gas prices only to a very minor extent.

For gases delivered in cylinders, the costs of labour and transportation completely dominate the business and these vary widely from country to country. For bulk gases, prices vary steeply with distance from the nearest plant, with troughs in price near concentrations of plants and peaks in between. In general, the location of plants reflects the location of users.

[5] In the UK, for example, with few low-cost sources, carbon dioxide is imported by a special cryogenic tanker ship to a large holding tank in London.

However, historical reasons (for example, the movement of the steel industry into fewer sites) mean that gas plants are not ideally located. In the case of gases like H_2 and CO_2, sources of raw gas are rarely ideally placed for consumers. Sources are largely in the older steel and chemical industries, whilst many users are in the newer disciplines of food processing and semiconductors.

Smaller gas companies. In amongst the large multinational gas companies, although these companies dominate the business, there are a number of smaller operations. These fall into a number of categories:

(i) The 'Mom and Pop' operations seen widely in the USA, where a small family business delivers cylinder gases, the business often being about primarily propane and butane for heating. The firms also act as agents for the large gas companies for other gases for welding and other purposes, and try to build this up as it is a less seasonal business than heating fuel gas. This sort of firm is a significant feature of most developed countries' gases market. However, the USA market relies on these small local concerns to the extent that most of the large industrial gas players there reach the majority of their cylinder gas customers via these firms.
(ii) Small special gases businesses typically manufacture only one or two gases themselves, relying on bulk supplies from the larger gas firms and large chemical companies for the other 50 or 100 gases they might typically offer. The bulk of their activities are cylinder filling, mixtures, analysis and purification of ultrapure gases, together with an inevitably large amount of effort on sales and shipping. The smallest of these businesses are typically national in extent and concentrate on only one sector, e.g. say calibration mixtures, medical uses or electronics uses. The larger of these businesses are international and the very largest offer ranges wider than the large integrated industrial gas players.
(iii) Local bulk/tonnage gas firms typically might have a supply scheme taking pipeline gas, with some bulk liquid delivery by road tanker to nearby customers. These firms are becoming increasingly rare as the large gas players take them over. Some of them had historically grown to be large companies, e.g. the Big Three Corporation which was a substantial firm, centred on bulk gas in a small area in southern USA, until its takeover by AIRCO at the end of the 1980s. A more unusual example of this kind of company is the UK firm Hydrogen Supplies which delivers H_2 in tube-trailers in north-west England and surrounding areas.
(iv) Chemical companies' by-products, for example the Distillers CO_2 business of ICI in the UK until its takeover by Hoechst's gases subsidiary Messer Griesheim. Distillers CO_2 took by-product CO_2 from ICI's chemical plants and distributed it in liquid form in road tankers. The fashion for 'focusing on core activities' which swept the UK and US corporate world in

the 1980s led to many of these businesses being sold off either to large gas firms or to the in-place set of managers running the business. (Some of these latter 'management buyouts' seem to have proved very successful. However, some seem to have been a convenient way for a large firm to make large numbers of employees redundant at reduced cost or with reduced impact on their own operations, as the new buyout firms typically end up with a much smaller staff.) To a large extent, this hiving-off of non-core businesses is still happening today (1996) and by-product sales and delivery activities are still regarded as non-core in many industrial sectors.

1.7 Financial features of the gases industry

The capital intensiveness of gas production plants has already been remarked on. With a 'typical' on-site cryogenic air-separation unit (ASU) costing several million dollars, and installations at steelworks ASUs often amounting to $50m and upwards, this might appear to need little justification. It should be remembered, however, that many other industries are now rivalling the chemical and steel industries in their concentration of capital: semiconductor 'wafer fabrication' plants, for example, are now being built with single-site investments of over $1bn.

Projects to produce or apply gas are subjected to financial scrutiny which is similar in general terms to that used in all of industry. Breakeven, discounted cash flow and internal rate of return are all regularly calculated. There are differences, however, between gas projects and the general run of industry. Because the gas production processes tend to be capital intensive, some figures are perhaps debated more intensely in the gas industry than in other industries. There is usually (section 1.7.3) a big incentive to build a few very large plants rather than a larger number of smaller ones, although distribution costs also have to be considered. This is not typically seen in the chemical industry where, with small numbers of very large customers, the distribution costs will typically be low. Capital expenditure on gas production is large and must typically be accompanied by generous levels of guaranteed or almost guaranteed sales, a factor which favours the oligopoly seen in the gases industry.

There is a big difference between the return expected in the application of gases and that expected in the production. An application for using gas might have a payback of less than 2 years and an effective IRR of 30% per annum. (The IRR, internal rate of return, is the effective rate of interest that a company pays on interest it has invested in itself.) However, most gas production projects have long paybacks (5 years or more), high breakeven (often over 80% of capacity) and low IRR – just a few percent above inflation. One result of the low level of maximum returns, the low 'upside', has been that means have arisen to minimise the 'downside' (the risk) of production projects. The 'take-or-pay' clause is a typical example (section 1.7.2).

1.7.1 Product exchange in special gases business

Another result of the capital intensiveness of gas production plant is that special gases are often only manufactured or extracted in a relatively small number of locations.

At one end of the scale, some special gases are simply mixed from gases readily available in bulk from many sources. Many calibration gases based on air gases and hydrocarbons are examples of this. At the other end of the scale are gases used in the semiconductor industry, such as silane or arsine. These are only made in bulk at half a dozen locations worldwide in high purity grades. The more exotic compounds, such as organometallics, are quite frequently only available from one supplier. Even a relatively common gas such as helium is only available from a few dozen locations where its high concentration in a large natural gas field makes extraction economic. European He comes largely from two US plants, for example, supplemented by the Polish plant operated by BOC.

The necessary scarcity of sources, balanced with the huge multiplicity of uses for these gases, means that special gas suppliers must rely on a huge and complex network of product exchange. A gas may quite typically be sold three or four times before reaching its end-user.

1.7.2 Take-or-pay and argon credits

In large tonnage gas contracts there are a number of unique features, relative to standard industrial practice. The take-or-pay contract attempts to match the gas company's commitment to a large and expensive plant to the revenue from the customer. Such a contract stipulates that the customer will pay for a minimum quantity of gas, no matter whether that gas is actually consumed or not. There is a good industrial logic behind take-or-pay since it enables the 'cobbler to stick to his last'. An expert gas supplier can operate a plant for a customer who would not be an expert in operating that plant and will have more time to pursue his own business. Without take-or-pay, the risk for the gas supplier, with a heavy capital expenditure, would be too great and the user would have to run his own plant.

Take-or-pay commitments from the gas user often have corresponding commitments from the gas supplier to 'supply-or-pay'. Gas suppliers then build-in storage tanks and vaporisers so that they have back-up supplies of liquid gas which can be vaporised during periods when the plant is undergoing repairs and maintenance.

'Argon credits' are unique to the gas industry. In the past, industrial gas suppliers operating large air-separation plants used to extract a small fraction of the argon from the oxygen stream, with modest profits from the argon sales resulting. Now, however, there is a much larger demand for argon and argon is efficiently extracted and sold to a large and highly profitable market.

In a typical tonnage contract the gas supplier is operating an air-separation plant supplying gas 'over-the-fence' to one customer. That customer will now typically insist on an 'argon credit'. This means that the customer pays a price for his oxygen or nitrogen reduced by an amount reflecting the quantity of liquid argon that the industrial gas company has been able to extract. The potential value of the argon credit is seen if it is remembered that bulk liquid argon can sell for £500 per tonne or more, whilst oxygen prices run as low as £20 per tonne or less. Even with argon at 5% of the oxygen stream, the argon from an air-separation plant could be worth more than the oxygen. However, most industrial gas companies would baulk at allowing the argon credit to exceed the oxygen price and paying their customers to take the gas.

In addition to specifics such as argon credits, there are other more generic features in gas industry contracts which are unusual, although not unheard of, in other areas of business. These are often associated with the necessarily long-term nature of the gas company's involvement with its customer. For example, it is common to sign-up contracts of 5, 10 or even 15 years. Since the value of money cannot remain stable over this kind of time, it is necessary to include in such contracts formulae for changing the gas price charged. These formulae are based on indices of industrial prices for power, raw materials, capital equipment and labour. Such long-term agreements have to be carefully written to be equitable to both sides, even in the event of unforeseen future economic upsets, and have to avoid falling foul of the 'anti-trust' or 'anti-competitive' laws which exist in most countries.

1.7.3 Scaling-up gas plants

A gas production plant is most efficient at a particular product output and will tend to be less efficient when operating at below that output. It will also be less efficient at above that output, if more output is possible. As a result of this, plants of many different sizes are required to match the many different demand sizes. It is therefore important to be able to estimate the cost of building a plant of any size.

To do this a cost-scaling law is applied as follows

$$\text{new cost} = \text{old cost} \times (\text{new capacity/old capacity})^{\alpha}$$

where α is the scaling power.

In this case, the scaling power applies to the whole plant. However, it may be more accurate to calculate the new cost taking into account different scaling powers applying to different components. In other words, taking C'

Table 1.7 Scaling of gas plant component costs

Component	Scaling power
Pressure vessel	1
Non-pressure vessel	2/3
Pipework	2/3
Controls	0
Analysers	0
Reactor bed	1

as the new cost and $C(C = C_x + C_y + C_z + \ldots)$ as the old, scaling powers $\alpha x, \alpha y$, etc., for subsystems x, y, etc., and the capacity ratio new/old to be R, i.e.

$$C' = C'_x + C'_y + C'_z + \ldots = C_x R^{\alpha x} + C_y R^{\alpha y} + C_z R^{\alpha z} + \ldots$$

For example, the cost of controls has a scaling power of 0. Thus the cost of the new capacity plant controls will be the same as the old capacity plant. (The same controls will be required, more or less irrespective of the size of the plant.)

For a reactor bed, the scaling power is 1, and the cost of the new capacity plant will be increased or decreased exactly in proportion to its size increase or decrease. If twice as much bed material is required, then the cost will be roughly twice as much. (For very small plants, this rule may break down, as bed material suppliers may have higher prices for small quantities.)

The scaling power for a pressure vessel is also 1, because the thickness of the walls of the vessel must increase with the size of the vessel (section 2.4.7). The scaling power for a non-pressure vessel is only two-thirds, since only the area of the tank wall needed will increase as the square of the linear tank dimension, whilst the volume will increase as the cube (Table 1.7).

The approximate scaling rules for the components shown in Table 1.7 are easy to estimate and can be used as a guide in the absence of good price information and to interpolate where price information is scant. Scaling rules for many other components where a scaling rule is not at all obvious can be set up by empirically reviewing prices published and paid for similar types of equipment. Some examples of where this has been done are given in Table 1.8.

Tables of component costs of the sort shown in Table 1.8 are available from some trade associations, and are published in some chemical engineering references (e.g. Perry, 1984).

The scaling laws are distorted from what might be expected because of the existence of subsidiary scaling laws:

- in purchasing, bulk buying reduces costs;

Table 1.8 Scaling of gas plant subsystem costs

Component	Scaling power
Large electric motor	0.77
Centrifugal compressor (without motor)	0.60
Distillation tower	1
150 psi pressure vessel	0.62
Cylindrical tank	0.57
Packaged refrigeration	0.73

- there is always a tendency to 'gold-plate' equipment which normally has a zero scaling law, such as controls or instrumentation: equipment that is desirable but might be considered disproportionately expensive for small plants will tend to be included on a bigger plant;
- the timing of purchases will affect costs: some components, such as bed materials, for example, can be bought after the rest of the plant is completed in a big plant, whilst others must be bought near the beginning of fabrication;
- market distortions often affect the very largest pieces of equipment of a particular type because there will often be fewer suppliers of them.

Finally, there are various discontinuities in the cost of plant:

- when the plant exceeds the size that can be transported by road, more on-site work will be needed and this is more expensive than factory work;
- reliability and back-up considerations;
- when plant exceeds handling equipment maxima: if simple, inexpensive and easily hired cranes can be used this minimises costs.

A combination of these scaling laws, weighted according to their importance in a particular project type, lead to the scaling law for that type of plant. Scaling laws can also be derived directly if enough similar plants exist and prices are known, and this has also been done. For example, Perry (1984) gives an exponent of 0.65 for steel plants with outputs in the range 150–4000 ktonnes per annum. Common examples of chemical plant generally fall in the exponent range 0.6 to 0.7. This arises from combining the scale factors of, for example, heat exchangers or furnaces of 0.8 to 0.9 with scale factors for gas compression and pretreatment of 0.55.

Impact of costs on design choices. Decisions about whether to use gases in compressed, bulk liquid or tonnage are not always straightforward. A tonnage installation, for example, requires a large capital investment in plant or possibly pipeline, whereas an equivalent bulk liquid installation will require only modest capital investment but will have a much higher running cost. A large firm which can borrow money easily at low cost and is able to make long-term commitments will prefer the first approach, whereas a small firm with a rapidly changing market will prefer the second option, even if it means considerably higher overall costs.

Similarly, demand patterns have a considerable impact on supply routes. High peak demand, with low average consumption will point towards bulk storage rather than a production plant, while continuous or near-continuous users will be biased towards production plant. The possibility of operating production plant on low-cost night-time electrical power may also bias users towards production plant rather than bulk gas delivery.

However, as a guideline, the smallest economical cryogenic tanks contain a few tonnes of liquid, and would normally be used for gas usage of around a few tens of tonnes per annum. Below this amount (a few tens of thousands of cubic metres) it is wasteful to use cryogenic tanks. There are many small liquid installations, however, down to 200 l 'liquid cylinders'. This is because the handling of many gas cylinders (5–10 m^3 is the maximum capacity) is tedious and because the cost of gas in compressed cylinders is of the order of five times more than bulk liquid gas.

The scaling laws for cryogenic air-separation plants (and many other gas production plants) mean that the production of less than a few tens of tonnes per day is uneconomic. Over the last decade this breakpoint has been blurred to some extent by the advent of pressure swing absorption (PSA) plant. For users needing only moderate purity nitrogen or oxygen with a steady demand, PSA plant offers a high capital/low running cost alternative to bulk liquid gases from the smallest to the largest bulk liquid user.

The financial value of experience. It is a platitude that experience is 'worth its weight in gold' but few attempt to put an actual monetary value on experience. However, many long-term surveys of industrial production have resulted in the 'experience curve' (Figure 1.12). The experience curve is generally reckoned to be a logarithmic one. The unit cost of producing a gadget is reduced by a multiplying factor A, for example $A = 0.9$, each time the total number produced is, for example, doubled. Mathematically, this is

$$\text{unit cost} = (A^{\log 2(N)})C_0$$

where C_0 is the initial cost. Sometimes the curve is plotted in terms of profit per unit, in which case it is a curve tending asymptotically to a maximum if price is constant.

The experience curve can actually be used to put a value on experience in, for example, assessing the value of a company's technology during a merger. If a company has to start again from square one, then it will have to climb up the experience curve.

Similar curves have been drawn for gas production and similar effects are seen. For many gas production plants the technology is now well-established and the asymptotic efficiency and unit cost is being approached. The

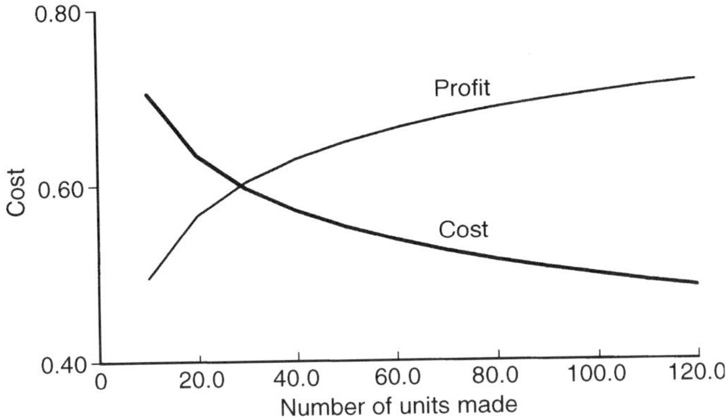

Figure 1.12 The experience curve.

well-developed nature of plant technology is seen in the high 'up-time' figures claimed by the manufacturers. Up-time, the proportion of a gas plant's lifetime that is spent either producing gas or ready to produce gas, is now claimed to be better than 99%, whilst many larger plants with redundant systems and back-up claim 99.9% or better.

1.8 The key role of gases in industry

Industrial gases pervade industry today. In a country of 50 million or so people, like the UK, there will be well over a hundred thousand different locations where gases are delivered. Every day well over a million tonnes of industrial gases, non-fuel gases, are consumed in industry worldwide. The ubiquitous role of gases in industry can be illustrated by looking at a complex consumer product and tracing its components back to their origins to see how gases were used in their manufacture. For example, the gases involved in the manufacture of a pocket calculator are listed in Table 1.9.

Use of gases in the manufacture of a spraycan or 'aerosol'. The role that industrial gases have taken in the production of many modern devices is now considerable. The aerosol or spraycan is a packaging technology, for example, which embodies gases in almost every single component:

Propellants. Propane and isobutane mixtures are the mainstay of aerosol propellants in the 1990s. CFC compounds went out of fashion for all but medical applications and some paints as ozone layer concerns grew. Nitrogen, CO_2, and other compressed 'permanent' gases have been used to propel liquids out of aerosol cans since the 1960s and are increasingly popular.

Table 1.9 Use of gases in the manufacture of a pocket calculator

Component	Gas	Use
Plastic body	Ethylene	In synthesis of polymer
	Liquid nitrogen	In cooling injection moulding
Rubber keys	Propylene/butadiene	In synthesis of polymer
	Liquid nitrogen	In cryotumbling to remove flash
Silicon chip	Argon	In silicon ultrarefining
	Oxygen	In substrate 'field oxide' layer
Silicon chip (surface)	HCFCs, silane (hydrochlorofluorocarbon)	In etching silane (in deposited layers) and depositing silicon layers
Display LEDs	Argon	In Ga/Al/As compound refining
	Trimethylaluminium	In GaAlAs vapour-phase expitaxy
Epoxy/copper board Copper	O_2 Smelting	
Soldering	N_2	Over solder in solder wave machine
Connectors	H_2	In bright annealing of steel plate
Glass faceplate	O_2	In glass melting
Battery, nickel	O_2	In mineral leaching

Formulations. Many aerosol formulations include gases actually in the dispensed mixture or in the manufacture of the ingredients. Materials included in mixtures vary from carbon dioxide or nitrous oxide to hydrocarbons or dimethyl ether. For example, shave gels include a small percentage of isopentane, a compound which exists as a liquid misicible with the shaving gel whilst under pressure inside the aerosol can. However, when the gel is released from pressure and warmed up as it is dispensed onto the hand, the isopentane boils and bubbles, converting the concentrated gel into a lubricious foam ideal for shaving.

Can fabrication. The tinplate aerosol can will have been produced by an oxygen-based steel process, whilst the steel may well have been rolled and heat-treated in a hydrogen or other bright-annealing atmosphere prior to electro-tinning. The aluminium aerosol will have been made from aluminium electrolytically extracted at high temperature with an argon atmosphere.

The plastic and rubber parts of the aerosol's valve and seals may also have been processed with gas-based processes: moulding machines cooled with liquid nitrogen or tumbled in a liquid nitrogen deflashing operation. The plastics themselves may derive from ethylene, propylene or butadiene gases, polymerised on their own or with other compounds.

1.8.1 Smokestack and leapfrog economies

A vital parameter in the design of new cryogenic air separation units is the ratio of nitrogen to oxygen produced. Market studies carried out before such plants are constructed have therefore been very interested not just in the quantities of gases which a given market might need but also in the ratio of oxygen to nitrogen needed. Observers in the gases industry have noted that this ratio actually varies between different countries, depending on the nature of their industrial base.

A newly industrialised country will be carrying out relatively large amounts of steel fabrication (cutting and welding) and will tend not to have the 'luxury' industries of food processing and semiconductors. These 'emergent economies' tend to use a lot of oxygen and much less nitrogen. An example of one of these emergent industrial economies would be a smaller African nation or a poorer Asian nation such as Bangladesh.

'Smokestack' economies, based predominantly on large-scale heavy industry, have an oxygen to nitrogen ratio around 1:1. A smokestack economy will have enormous steel industry applications, requiring nitrogen and argon as well as oxygen (fabrication of oxygen will be only a small proportion of requirements), while other metal processes such as heat treatment will be using nitrogen. In addition, the smokestack economy will have a bulk chemical industry, using nitrogen in fair quantities, so as well as requiring much greater quantities than the emergent economies, the smokestack economy $O_2:N_2$ ratio will be closer to 1:1. In the USA in the early 1960s $O_2:N_2$ was roughly 1:1, while today countries such as India have large steelworks and fabrication industries but a large and growing sector in services and high technology, so they also have a ratio of 1:1

Today the mature industrial economies of the USA and Europe have a $O_2:N_2$ ratio of 1:2, reflecting both the decreasing amount of heavy industry, such as steel, using oxygen and increased use of nitrogen in the service, food and high-technology sectors.

A recent phenomenon are the 'leapfrog' economies. Some newly industrialised countries on the Pacific Rim are jumping straight from an emergent stage into high-technology manufacturing, such as the semiconductor industry, which use large amounts of nitrogen. They have a low $O_2:N_2$ ratio of 1:4, similar to the ratio in air. The gas consumption ratios for different economies are given in Table 1.10.

Table 1.10 Gas consumption ratios in states with differing economies

	Emergent economy	Smokestack economy	Leapfrog economy	Mature industrial economy
Nitrogen	30%	50%	80%	70%
Oxygen	70%	50%	20%	30%

Detailed studies of the local demand pattern for gases are necessarily added to this overall market view before air-separation and other plants are built. Most often, a merchant gas operation will be run as a 'piggy-back' operation on a major pipeline gas supply plant, such as an oxygen plant supplying a steelworks. The scaling laws of cryogenic plant particularly favour this approach.

2 Gas technology

2.1 Separating air gases

2.1.1 Introduction

It may seem surprising that any energy at all is needed to separate mixtures of gases; after all, the molecules of gases are not attracted towards each other so the separation process does not have to exert any force on the molecules. However, the act of mixing gases together increases disorder, or entropy, and energy is always required to reverse entropy. The theoretical minimum energy required to physically separate an equal mixture of two similar gas molecules is about 1.72 kJ mol^{-1}. This is much less, by a factor of a hundred or more, than the energy of typical chemical reactions. This, then, is the fundamental reason why physical separation processes for gases, even rather imperfect physical separations, are favoured over chemical generation.

Thermodynamics of gas separation. Consider a perfect separator, where an input mixture of equal parts of A, B, C and so forth is pressurised into a chamber lined with membranes (Figure 2.1). One membrane perfectly passes A, one B, one C and so on. To produce products at one atmosphere absolute pressure, the feed must be supplied at an absolute pressure equal to the number of products.

For a mixture of two components, such as air, the energy required is thus that needed to compress air under ideal conditions (section 1.3.1) to 2 bara. Hence

$$\text{energy for separating mixture} = RT\ln(P_{out}/P_{in}) = RT\ln(2) = 1.72\ \text{kJ mol}^{-1}$$

The energy needed for non-equal parts of gases is reduced, so that air separation into nitrogen and oxygen requires only 1.3 kJ mol^{-1}. However, if only the 20% oxygen of the separated gas is required, then the actual separation costs of a 50/50 mixture are doubled to 3.4 kJ mol^{-1}, whilst oxygen cost will rise to 6 kJ mol^{-1} (Table 2.1). The relevant equation for W_{min}, the separation energy per mole of product from a mixture of a% A and b% B is

$$W_{min} = -RT/a[a\ln(a) + b\ln(b)]$$

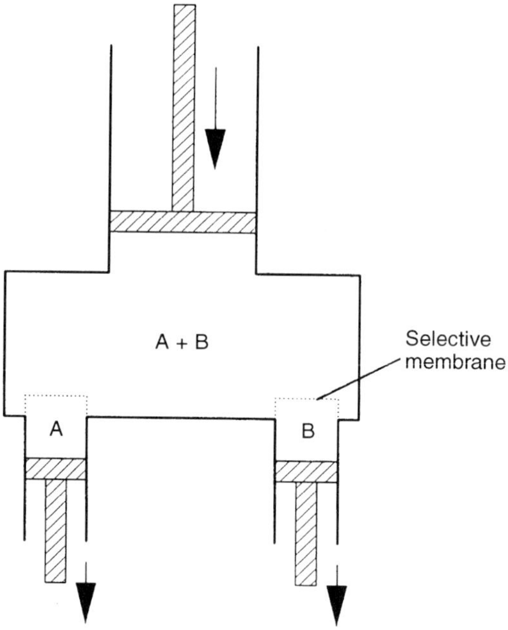

Figure 2.1 Theoretical model of air separation.

Table 2.1 Theoretical minimum energy requirements of gas mixtures

Wanted gas	Notes	Separation energy ($kJ\ mol^{-1}$)
50.000%	50/50	3.4
21.000%	O_2 from air	6.0
0.930%	Ar from air	13.8
0.030% 300 ppm	CO_2 from air	22.2
18 ppm	Ne from air	29.1
5.2 ppm	He from air	32.1
1.5 ppm	CH_4 from air	35.1
1.14 ppm	Kr from air	35.8
0.086 ppm	Xe from air	42.1

It is perhaps surprising that the extraction of even a rare gas such as xenon does not theoretically require a particularly large amount of energy.[6]

Unfortunately, the ideal semipermeable membranes required to carry out the trick shown in Figure 2.1 do not exist, although imperfect membrane

[6] It is interesting to note that in principle energy is available free simply by extracting methane from the air: methane combustion generates around 850 $kJ\ mol^{-1}$, far more than is required to extract it. Also, CO_2 is surprisingly easily extracted from the air, a fact not used in any chemical processes unless atmospheric CO_2 used as a reactant in the setting process of lime-based mortars counts as a chemical use.

separation is possible, as we shall see. Also, the ideal isothermal compressor, needing only the above energy, can only be approached in large sizes.

This, then, is the simple thermodynamic science of gas separation. The achievement of separation is more complex than this simple science might suggest, however. Since the beginnings of efficient gas separation technology by cryogenics in the late 1890s, many different configurations and cycles have been devised. Before going into some of those cycles, however, it will be useful to look at some of the components which will inevitably be part of them. For example, whatever method is chosen to carry out gas separation, a gas compressor is generally used chosen to energise the separation.

2.1.2 *Gas plant building blocks*

Introduction. Generally speaking, specialist firms are used to fabricate the items which are common between the gas industry and other industries. It is still essential, however, to know the basic characteristics of these building blocks in order to be able to design a complete plant most effectively. The manufacture of gases involves the use of equipment which often differs significantly from hardware used in many other chemical and industrial processes. The hardware used can have a bigger impact on the final plant cost and efficiency than the configuration of that hardware in the plant.

A glance at any diagram for gas separation (Figure 2.13) will reveal that vital parts of all cryogenic plants are compressors, heat exchangers and perhaps turbine expanders, as well as the core of the plant, the distillation column, and a host of smaller items such as valves and a nest of high-quality pipework. Gas compressors for cryogenic work are some of the largest units seen and may vary in output pressure from a few bar to a hundred bar or more to satisfy the thermodynamic requirements of the cycle used.

Because the gas from the compressor will be cooled so strongly, the presence of condensibles such as water and oil which would freeze and clog the plant must be reduced to infinitesimal levels. Whilst water is often contained in the feed gas or air and must be specifically separated, most compressors for cryogenic work must also not allow any oil into the outlet stream. This is in contrast to many room temperature applications, where the presence of a trace of oil mist is actually desirable for lubrication of machinery downstream. The consequences of allowing oil into a cryogenic plant are so dire that practically all compressors for this use are oil-free designs. Non-cryogenic plants also have to have efficient compressors. Pressure swing absorption (PSA) plants are occasionally built in very small sizes where convenience and low purchase cost rather than efficiency is the prime requirement but the economics of most PSA plants still depend on efficient compression.

Heat exchangers and insulation for cryogenic service are similarly specialised. In a plant operating above room temperature the efficiency of a heat exchanger is not quite so important as in a cryogenic plant. A heat loss or leak of 1 kW into the cold parts of a cryogenic plant will require 3 kW of electrical power (assuming the highest possible Carnot efficiency for the refrigeration in the plant) to reject it at 77 K, or 70 kW at liquid helium temperature. By contrast, a 1 kW loss or leak from a high-temperature process will require only 1 kW compensation with direct heating (less with a heat pump). Hence cryogenic heat exchangers and thermal insulation have to be more efficient than is normal.

In addition to this emphasis on efficiency, the range of materials out of which cryogenic heat exchangers can be fabricated is limited. Simple and inexpensive carbon steel fabrications are ruled out because steel embrittles, whilst copper is expensive and difficult to weld. In practice aluminium alloys are the mainstay, being strong, weldable and with good thermal conductivity. Stainless steels such as 304 or 316 are also widely used where higher strength is needed and despite their low thermal conductivity they, like mild steel in high-temperature work, can still be made into effective heat exchangers for some services.

Designs also differ for cryogenic use with the need to provide minimum temperature difference between incoming and outgoing streams while still providing low pressure drop. Plate-fin types dominate, providing high efficiency, with the possibility of balancing heat flows from several high- and low-pressure streams of different mass flow rates being easily satisfied in the versatile plate-fin design. Again these differ from normal practice and the large effect on plant performance means that heat exchangers for gas separation deserve special discussion. (Insulation is discussed in section 2.4.)

2.1.3 Compressors

The trapezoid symbol for compressor in a process diagram hides the details and subtleties on which tens of thousands of engineers have worked for a century or more to perfect. Efficient, reliable gas compression with low plant capital cost still requires wise selection on the part of the process designer and skilful engineering on the part of the compressor designer.

A compressor is characterised by its throughput, compression ratio and efficiency. One might add, since compressors are often the least reliable single unit in many gas systems, that the mean time between failures (MTBF) and % up-time should also be considered as vital characteristics. The throughput is usually given as gas volume at normal temperature and pressure (NTP), whilst the compression ratio is simply the ratio of the outlet over inlet absolute pressures. The efficiency measures are less standardised but comparison with an ideal adiabatic compressor is useful.

Some of the fundamental difficulties of gas compressors in the region of 10 bar/20 000 $m^3 h^{-1}$ often used for air separation can be summarised as follows:

- axial compressors cannot achieve high enough pressures except in the largest sizes and with huge numbers of stages (at 10 MW or above axial compressors become competitive)
- centrifugal compressors can only achieve high pressures in large sizes and then only with multiple stages
- reciprocating compressors have many wearing parts and suffer from poor reliability, efficiency and up-time except in small sizes[7]
- gases at high pressures easily leak out of loose seals but tight seals involve wear problems and frictional losses
- many compressors cannot use any seals and use very small clearances and high speed. Slight distortions of rotor or stator in this case can lead to failure
- oxygen at high pressure will encourage ignition of many seal and constructional materials, particularly if friction causes local heating
- all air compressors must efficiently separate out water from their outlets: compressing air causes water to condense out (section 2.1.5)
- screw and reciprocating oil-lubricated compressors must efficiently separate oil, water and air in their outlets.

An interesting variation on the positive displacement types of compressor is the 'water-ring' compressor. This looks at first glance like a centrifugal compressor, being a central set of vanes rotating in a housing. However, the pumping action is provided by the vanes being mounted eccentrically in the housing, whilst a rotating 'ring' (actually a hollow cylinder) of water provides the seal and piston in the design. The water ring is held in place by the high centripetal force. Water-ring pumps can be designed with large clearances between the mechanical parts, yet still provide good compression ratios. Designs such as the water ring are useful for feed gases loaded with particulates, which would cause excessive wear in standard compressors. Water-ring and similar compressors, however, must separate the water or other working fluid from the output even more efficiently than types which have only a lubricating fluid present, which adds another potential complication.

Figure 2.2. shows two compressor types: centrifugal and screw. The large centrifugal compressor from Sulzer is a five-stage compressor with inter-cooling using tube bundles (three of these can be seen in an upright position in the figure).

[7] Reciprocating compressors can, however, last a very long time: one factory the author worked in is still using a large reciprocating compressor made by a Paris firm 90 years earlier for commercial cylinder filling of special gases.

(a)

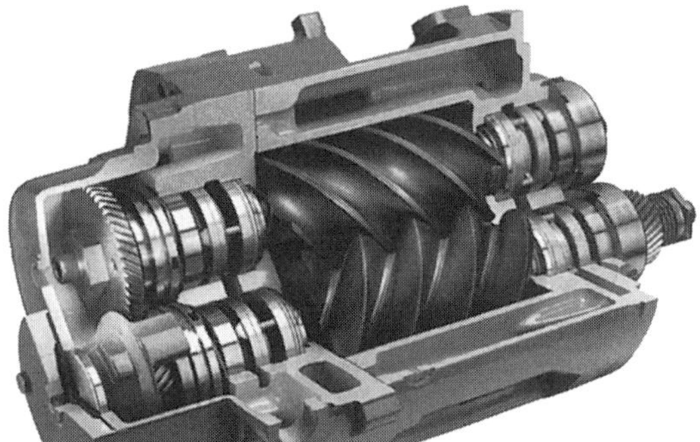

(b)

Figure 2.2 Compressor types: (a) centrifugal (courtesy of Sulzer (UK) Ltd); (b) screw.

Axial and centrifugal compressors. Axial compressors are typically suitable for large-scale compression to moderate pressures, i.e. a few bar. A single stage is capable of only a few hundred millibar or so pressure rise, beyond which blade tip sealing and aerodynamic stalling occur, so that any practical axial compressor requires several or even dozens of stages. The very largest plants may use axial compressors up to high-pressure but require massive installations with many dozens of stages.

Centrifugal compressors. The centrifugal compressor is suitable for medium- to large-scale compression to a few tens of bars. It is efficient and is common at larger air-separation units. The pressure rise achievable is limited by the tip speed and the air density. The maximum pressure rise P_{out} in a centrifugal compressor with tip speed V_{tip} working on air can be estimated as approximately

$$P_{out} = P_{in} \exp(kV_{tip}^2 / 2)$$

where k is a constant for air of about $1.2 \times 10^{-5}\,m^{-2}\,s^2$. This follows simply from considering an element of gas at equilibrium inside the fan while it is whirling around, i.e. with the fan in the stalled condition.

The pressure achieved is a very steep function of the fan tip speed. To give some idea, a tip speed approaching the speed of sound is needed, $300\,m\,s^{-1}$ or so, to give compression ratios of better than 2:1 or so and 3 bara. For large compression ratios, centrifugal compressors are normally cascaded, with coaxially mounted units all driven from one shaft and aerodynamically designed smooth housings or 'volutes' to carry the gas efficiently between stages.

Because of the steep function of tip speed, and limits on the rotational speeds of machinery, very high speed and large size must bc achieved for adequate compression, limiting these compressors to relatively high volume requirements, i.e. above $1000\,m^3\,h^{-1}$.

Rotary compressor seals. One of the problems of all compressors at high compression ratios is that the compressed outlet gas leaks past the shaft seal between the motor and compressor stages. Compressors with a comparatively small compression ratio can in effect simply accept leakage by using a 'labyrinth' seal. This is no more than a set of grooves on shaft and housing which forces the leaking gas to pass though a tortuous path of annuli.

For high compression ratios at least, however, sealing is a problem and much ingenuity is often devoted to its solution. The standard system of the 'stuffing box' seal, where lubricated fibrous ring seals are pressed around the shaft by an adjustable collar, is rarely acceptable. The stuffing seal is more efficient if the collar forces it down hard onto the shaft but then power is lost through friction. Rubber seals are often used, using carbon filler to reduce the coefficient of friction which would otherwise be high in the absence of oil, and elastomer composition chosen for compatibility with gas and lubricant. These are often tensioned with internal metal springs or use a shape which allows operational gas pressure to aid their task. Carbon seals are also possible and have advantages in being self-lubricating. A small leakage is often designed into seals, in effect allowing the seals some gas lubrication.

Yet other compressors avoid problems by using what might loosely be termed 'active seals'. In these designs, a face seal with a pneumatically regulated thrust ring is employed. The seal is closed by a spring pushing the

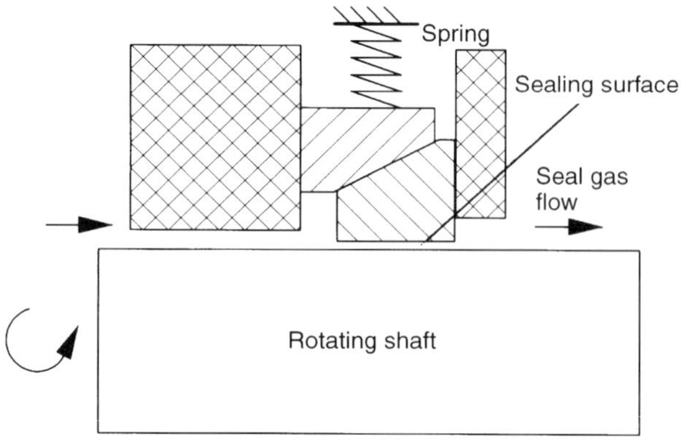

Figure 2.3 Schematic of a pneumatic thrust ring seal with a mechanical self-balancing action.

two faces together. In operation, after start-up, the high pressure gas from the compressed side compresses the spring and opens the seal slightly, allowing gas to drop in pressure down the narrow annulus around the shaft and closing the seal again. The seal takes up an intermediate position with the leak rate (and shaft to seal clearance) regulated by the spring strength and the size of the annulus supplying gas from the high-pressure side.

Some designs of sprung plastic or metal rotating annular seals also achieve a feedback effect, as the two-part annular spring seal is wedged open by the high-pressure side but as soon as the leak through the seal increases, lowering the pressure inside, the spring tightens the seal. Figure 2.3 shows how such a pneumatic ring seal with a mechanical self-balancing action works. By choosing the spring pressure correctly the leak can be regulated to an acceptable level, whilst achieving the very low friction of a non-contact seal. Both these 'active seals' have the feature, often an advantage, that they provide an almost hermetic seal on the shaft when it is stationary or at slow speeds. However, low-speed operation can lead to excessive wear. Similar arrangements are also seen with oil seals but these are rarely seen in air separation unit (ASU) compressors because of the potential for oil contamination.

There is an interesting variation often used on centrifugal compressors which avoids many of these problems with seals: the motor can be operated pressurised with the outlet gas. Yet another way of avoiding the seal problem is to use magnetic coupling. In this the pump is driven by a magnet inside a pressurised diaphragm, while a corresponding magnet outside the diaphragm is driven by the external motor. Clearly, when the drive on the outside is an a.c. electromagnet system, then the magnetic coupling system amounts to the system above.

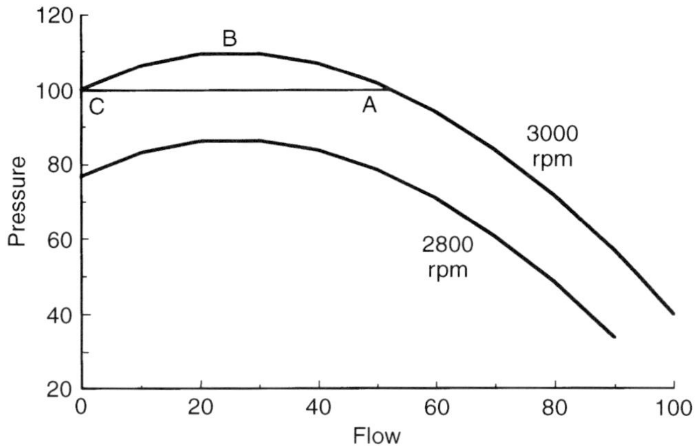

Figure 2.4 Centrifugal compressor flow versus discharge pressure showing surge line.

Compressor surging. Occasionally, typically during start-up or an unforeseen accident, centrifugal and axial compressors can suffer large surges up and down in revolutions, accompanied by large variations in output pressure and vibration. Surging is a result of the interaction between a compressor and the process load it is put under.

Looking at Figure 2.4, it can be seen that if flow is decreased a little from point A, then pressure builds up a little. This is stable, since the increased pressure will tend to increase the flow rate and bring the system back towards to A again. If, however, the flow is decreased past point B as a result of throttling in the downstream process, then a slight decrease in flow will lead to a decrease in pressure, which leads to a further decrease in flow, which leads to a further decrease in flow, and so on. The final effect is that the flow from the compressor will collapse towards C, when the load on the compressor motor falls, so that it can accelerate until the rotational speed has built up sufficiently, perhaps to 3200 rpm to raise the pressure beyond the pressure at B at zero flow. Then flow begins again, slowing down the compressor back to its 3000 rpm norm. This cycle manifests itself in a hunting or surging effect on the compressor speed and is highly undesirable because vibration levels are raised and fatigue damage to critically stressed rotating parts and welds in the static parts is increased.

Surging can be dealt with, particularly in larger systems, by bypassing some of the compressor output if the surge point B is approached. This 'dump valve' assembly increases the flow through the compressor and reduces the pressure achieved, forcing the compressor to go back down its characteristic curve towards A on Figure 2.4. The dump valve can be to a vent line or back to the compressor inlet.

A pressure vessel downstream of the compressor is another way of dealing with the problem of surging. A comparatively small pressure vessel installed downstream of the compressor will suppress surging and also provide a buffer of compressed gas to handle sudden downstream demand changes whilst allowing the compressor time to adjust.

Reciprocating, oil-injected screw and gear pumps. Reciprocating compressors are suitable for extremely high pressures or low volumes at moderate pressures. The simple piston and cylinder linked to a rotating crankshaft with non-return valves on the inlet and outlet is still the commonest design. A single-cylinder compressor, however, has a very unsteady output and needs to be linked to a large pressure vessel for applications requiring continuous output. Multi-cylinder arrangements with suitable valvegear enable fairly steady pressure and flow to be maintained. There are many wearing parts in a reciprocating compressor: piston seals and skirt, cylinder walls, bearings of piston and crankshaft, and non-return valves.

Simplified designs linking a piston directly to a reciprocating drive mechanism, such as an electric solenoid, are available in small sizes. Because they eliminate the crankshaft and several rotating bearings they offer some advantages. There are even small non-contact reciprocating compressors available which mount the piston and drive unit on springs. The springs are arranged to allow a large axial movement but to control radial movement accurately. For low compression ratios, where the necessary clearances are acceptable, the piston can then move in and out of the cylinder without touching the cylinder walls. Such units are used in spacecraft and similar applications where ultimate reliability is needed.

Reciprocating compressors are relatively simple to make and maintain, and are generally cheap to purchase. However, they have drawbacks in the form of high maintenance and reliability problems and are generally avoided in industrial installations today.

Screw compressors are now a commonly used alternative to a reciprocating unit. Screw compressors can achieve a few bar (or a few tens of bars with multiple rotors) in a simple unit which can be made in sizes down to a few tens of cubic metres per hour. They give a steadier output than a reciprocating unit, with fewer moving parts. At their heart are two multi-thread screws with very exact profiles which interlock: a male profile on one and a female profile on the other. The screws or 'rotors' are often sychronised, likes Rootes and Northey blowers, by gears mounted on their shafts.

Screw compressors rely on the carefully machined screws fitting together precisely to fine tolerances, however, and are potentially sensitive to damage from clashing of the rotors in the event of dust contamination or temperature excursion. They are of very high efficiency with only very small losses from leakage from output to input, at least at lower compression ratios. To achieve higher compression ratios (10:1 for air, typically) they are normally operated with continuous injection and recovery of oil. The oil enables the rotors to act as their own synchronizing gears and reduces leakage losses as well as making a more forgiving, lubricated assembly. The injected oil is recovered in a cyclone and a coalescing filter and then cooled prior to reinjection, the oil thus acting as an additional coolant. Special oil must be used which does not easily emulsify as this would lead to accumulation of water from the feed in the recirculated oil. Oil-injected screw compressors can give longer life, better efficiency and improved reliability relative to the reciprocating devices they generally replace.

Older in origin and usage than the screw compressors is the gear pump. This relies on one or more pairs of gears which enmesh gas on the inlet side and carry it to the outlet. The simplest possible gear pump is the Rootes blower, comprising two 'gears' with two 'teeth' on each 'gear', which is thus dumbell or lobe-shaped in outline. The Rootes blower and Northey claw forms of gear pump rely on external synchronising gears. All gear pumps need accurate tooth form and very small clearances to function correctly on gases and are easily contaminated with particulates. However, they can achieve very high compression ratios in small sizes. They are used where very small flow rates are needed and on gases such as propane or butane which are handled as liquids.

Cooling compressors. Compressors do not invariably need cooling. If the subsequent process required high temperature, as well as high-pressure gas, then cooling may not be appropriate. However, compressors, or at least single stages of them, are approximately adiabatic devices. Gas compressed will rise in temperature in accordance with the simple adiabatic rule

$$T_{\text{out}} = T_{\text{in}}(P_{\text{out}} / P_{\text{in}})^{(g-1)/g}$$

thus raising air pressure from 1 bar at temperature 20°C to 10 bar will cause a temperature rise to 293°C or so. This is not what is required for low-temperature cryogenic apparatus and is also quite unsuitable for membrane or PSA separation, which require near room temperature operation. Moreover, the energy expended in raising the temperature can rarely be usefully employed and even if it were it amounts to electrical heating of gas, which is rarely economic. Isothermal compression is the desirable ideal.

Adiabatic temperature rise. The rise in temperature in an adiabatic compression may be visualised as follows. When a piston is forced down on an assembly of gas in a thermally insulated container, the motion of the piston adds to the momentum of gas molecules that hit the piston during the process. This leads to a general increase in the speed of the molecules, which is seen as an increase in temperature.

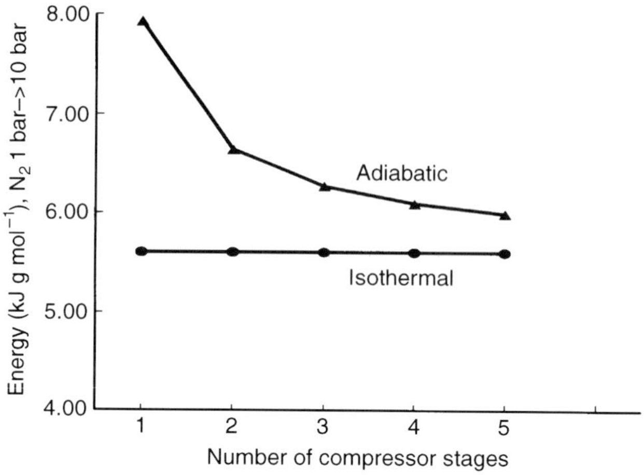

Figure 2.5 Multistage intercooled compressors – approach to isothermal performance.

The simple well-known result for an ideal adiabatic gas compression or expansion is

$$P_o V_o^g = P_i V_i^g$$

where g is the ratio of the specific heats of the gas at constant pressure and constant volume, i.e. C_p/C_v. This in turn leads, when the ideal gas law is added, to the given formula for adiabatic temperature rise.

The optimum gas compressor is an isothermal one. This, as discussed in section 1.3.1, requires an energy of E_{iso} to arrive at pressure P_o from P_i input absolute pressure

$$\begin{aligned} E_{iso} &= RT_i \ln(P_o / P_i) \\ &= 5.6\,\text{kJ mol}^{-1}, \text{ for } P_o = 10\text{ bar}, P_i = 1\text{ bar}. \end{aligned}$$

An adiabatic compressor, were it a single-stage device followed by a cooler, would require more energy

$$\begin{aligned} E_{adi} &= 7/2\ RT_i((P_o / P_i)^{(g-1)/g} - 1) \\ &= 7.94\,\text{kJ mol}^{-1}, P_o = 10\text{ bar}, P_i = 1\text{ bar, for an ideal diatomic gas.} \end{aligned}$$

The energy has gone into heating the gas up, in this case to as much as 293°C.

The optimum approach for compressor efficiency is via multistage compressors with interstage cooling (Figure 2.5). This is the closest approach to isothermal performance possible for practical equipment.

Inlet gas cooling is another option for improving compressor efficiency. If a source of cooling is available, from a waste stream perhaps, then further improvement can be offered, perhaps even exceeding isothermal performance from ambient. Finally, it should be remarked that some larger compressors require careful control of their cooling: sudden changes in feed

Table 2.2 Joule–Thompson coefficients in °C bar^{-1} (nitrogen)

Starting pressure	25°C	100°C
1	0.2217	0.1292
20	0.2060	0.1173
33.5	0.1961	0.1100
60	0.1729	0.0975
100	0.1400	0.0768
140	0.1105	0.0582
200	0.0779	0.0419

air temperature, outlet flow rate or cooling water can cause the compressor to change in temperature too rapidly. When this happens, some components cool or heat more quickly than others and there is the potential for problems of clashing or even seizing occurring, given the close fits used.

Joule–Thompson cooling. Expansion across an orifice or valve in a pipe is the standard versatile source of refrigeration in a cryogenic plant, utilising the Joule–Thompson (J–T) effect. A J–T expansion causes cooling in gases below a certain temperature (the 'inversion temperature') in the following way.

An ideal gas does not cool on expanding across a nozzle: its pressure is reduced but it does no work on any engine and, from considerations of conservation of energy, remains precisely at the same temperature. However, most real gases have a small attractive force between their molecules and they are able to do work against this cohesive force and lose energy, thus cooling, on expanding across a nozzle. The J–T coefficient describes this effect by giving the temperature (in °C) by which the temperature drops on throttling a gas through 1 bar of pressure drop. The J–T coefficient varies with pressure and temperature becoming, in general, negative at higher temperatures (above the 'inversion temperature') and smaller or negative at higher pressures (Table 2.2). The J–T effect is thus not a powerful one: for nitrogen at room temperature compression to 20 bar will yield only a 4°C lowering in temperature.

For nitrogen, the temperature at which the J–T coefficient becomes zero is 348°C. For lighter gases, this inversion temperature is lower, and these gases must be precooled before J–T expansion is useful: hydrogen is at 200 K, whilst helium is 43 K. For gases where data is not to hand, extrapolation of J–T effects can be achieved by noting that the J–T coefficient μ, $(\partial T/\partial P)$ at constant enthalpy, is given by the fundamental thermodynamic formula

$$\mu = (\beta T - 1)V/C_p$$

where β is the thermal expansivity of the gas, i.e.$(1/V)$ $(\partial V/\partial T)$ at constant pressure. As β is simply $1/T$ for an ideal gas, as expected, an ideal gas shows

$\mu = 0$. However, if the van der Waals' coefficients a and b are known for a gas, then an approximate μ can be calculated which at least shows the correct qualitative behaviour and can be used for extrapolation. For example, for a gas at moderate pressures with van der Waals' b coefficient small, the J–T coefficient is given approximately by

$$\mu \simeq 2a/(C_p\ RT)$$

As expected, μ is zero if the van der Waals' a coefficient is zero and increases at lower temperature.

Expansion turbines for adiabatic cooling. The low efficiency of the Joule–Thompson cooling effect and the fact that it is unavailable for many gases near room temperature means that there is a requirement in many cryogenic plants for another source of refrigeration. High-pressure cold gas is often further cooled by an adiabatic expansion. The cooling effect achieved is easily predicted as the converse of the adiabatic heating calculated above for gas heating in compressors, although some corrections will be necessary due to the low temperature of operation leading to non-ideality of the gas. The calculation is more complex where the gas actually condenses partly to liquid on expanding.

Whatever the difficulties of calculation, achieving adiabatic expansion at low temperatures is not straightforward. Some early air-separation plants employed, and some modern helium plants still employ, reciprocating expansion engines, which operated in a very predictable way. However, to achieve necessary throughput of gas such engines rapidly become very large and unwieldy if they are not to have to run at impossible rotation speeds. Add to this the difficulties with sealing the piston to cylinder at such low temperatures, lubricating the whole assembly and operating valves at high speed.

The efficiency of adiabatic expansion turbines in cooling is calculated as follows. If the efficiency of the engine is E_t, there is a perfectly efficient compressor and cooling heat transfer from T_1 to T_2 ($T_2 < T_1$) is required, then Carnot efficiency for the whole cycle, i.e. gas compression + expansion operating as a refrigerator, is given by E_c (= work theoretically needed/ work actually needed) as follows

$$E_c = (T_2 - T_1)E_t\ /(T_2 - T_1E_t)$$

This follows from considering a perfect Carnot cycle heat engine but adding the effects of an expander efficiency of E_t by increasing the heat flow to the hot reservoir by $1/E_t$. Examination of the formula above shows that if the turboexpander has no work output at all, all its work is dissipated as heat in the gas stream itself, then the efficiency of the process sinks to zero and there is no cooling effect at all, it is simply a throttle valve. However, this assumes an ideal gas and for most gases there is a Joule–Thompson cooling effect.

Table 2.3 Comparison of refrigeration cycles

Expansion cycle	Outlet temperature	Power needed (kW Mcal^{-1})
Joule–Thompson (nitrogen)	−40°C	150
Liquid evaporator (propane)	−40°C	0.83
Adiabatic nitrogen expansion	−40°C	6
	153 K	10
	113 K	14

The efficiency of an expansion turbine grows as the output temperature is reduced until, roughly below 100 K or so, it is as efficient to use a turbine as it is to use an evaporating liquid cycle refrigerator. The latter, the standard refrigerator, is most efficient when required to deliver cooling at the temperature of its evaporator and at temperatures close to room temperature. The turbine has no defining optimum temperature of cooling efficiency in the same way and becomes steadily more efficient as temperatures are reduced. Isalski (1989) gives the figures shown in Table 2.3 for power consumption of three different routes to −40°C, using a complete cycle of compressor, heat exchanger and expansion route.

Today, an expansion turbine is the standard means of achieving adiabatic cooling. It necessarily has to run at exceptionally high speed, as it must offer a high pressure drop to give a significant cooling effect. If the turbine is overbraked it will become inefficient with too much gas simply leaking through its blades or around its blade tips at the high differential pressure needed and, as noted above, function simply as a Joule–Thompson valve. The speed requirement makes turbines technically difficult although it has the benefit of making the machine surprisingly small even for large throughputs. Typically single-stage or double-stage centrifugal turbines are used. These run at speeds from 20 000 to 100 000 rpm, the high speeds being used in smaller units. The exotic turbines essential for efficient liquid helium production have to run at much higher speeds, up to nearly 1 million rpm, although the turbines that run at these heroic speeds are very small, only a few millimetres in diameter.

Expansion turbines often employ gas bearings, at least in smaller sizes, to achieve high-speed efficiently. Oiled bearings have the problem, along with ball or roller bearings, that unless insulation is assured (large machines having an advantage here), the low temperatures can cause the oil to freeze or the rollers or balls to fail via fatigue, causing the bearing to seize.

Standard journal bearings suffer from ‘whirl’ at high speed, especially with gas lubrication. This is where the shaft, as well as revolving, orbits inside the bearing. Whirl is suppressed by providing a damping force by arranging gas inlet orifices circumferentially, along with exit orifices which have a small volume above them, around the bearing journal. The small volume gives a damping action thus suppressing whirl. Thrust bearings can also employ an externally fed gas-bearing principle but more often use self-acting spiral groove bearings. These provide a pumping action which creates their own gas film but have the snag of not giving any lubrication when stationary or at low speed. Magnetic bearings have also found application in expansion turbines.

A magnetic bearing can be a passive bearing, that is, a powerful permanent magnet in the bearing repels a powerful permanent magnet on the shaft so that the latter can rotate freely. However, passive bearings are limited. First, they cannot provide a complete suspension system; for example, another type of bearing must stabilise the shaft in the axial direction if the passive bearing provides stability in radial directions. Second, there is a limit to the force that can be practically obtained with the constraint that at least one of the permanent magnets must be on the shaft.

Active magnetic bearings rely on an electronic feedback signal. A shaft, which need not be permanently magnetic, is suspended by attraction to a set of powerful electromagnets, amplifiers varying the current through the electromagnets in order to keep the shaft central. The movement of the shaft in axial or radial directions is detected typically by an electromagnet-based proximity device, providing the input to the suspension magnet amplifiers. Active magnetic bearings can provide full stabilisation so that a shaft need have no physical contact with its bearings or housing.

Expansion turbines sometimes have to deal with a significant amount of 'wet gas' two-phase flow, i.e. partially condensed liquid/gas, and must thermally isolate the cold turbine wheel as effectively as possible from the shaft carrying away the power (Figure 2.6). The power generated by the turbine is often simply dissipated in an oil brake but in large units power is used to generated electrical power or to operate a compressor co-mounted on the shaft.

Ranque–Hilsch vortex tube. The vortex tube achieves, like a turbine, a cooling effect additional to that of the Joule–Thompson expansion but without a conventional expansion engine. In a vortex tube air expands at a nozzle tangentially arranged to one side of the divider in a tube with a central divider (Figure 2.7). Cold air is extracted on the opposite side of the divider from a hole in the middle of it, whilst hot air is collected from the nozzle side of the divider from a annular exit at the end of the tube.

A vortex tube achieves cooling via expansion at the nozzle at the entry to the tube but achieves an overall effect on part of the gas stream better than a J–T valve by separating a cool gas stream in the middle of the tube from the rest of the gas, which exits the tube having slowed down and heated up. In effect, work is done by the cold part of the gas stream to heat up the hot part

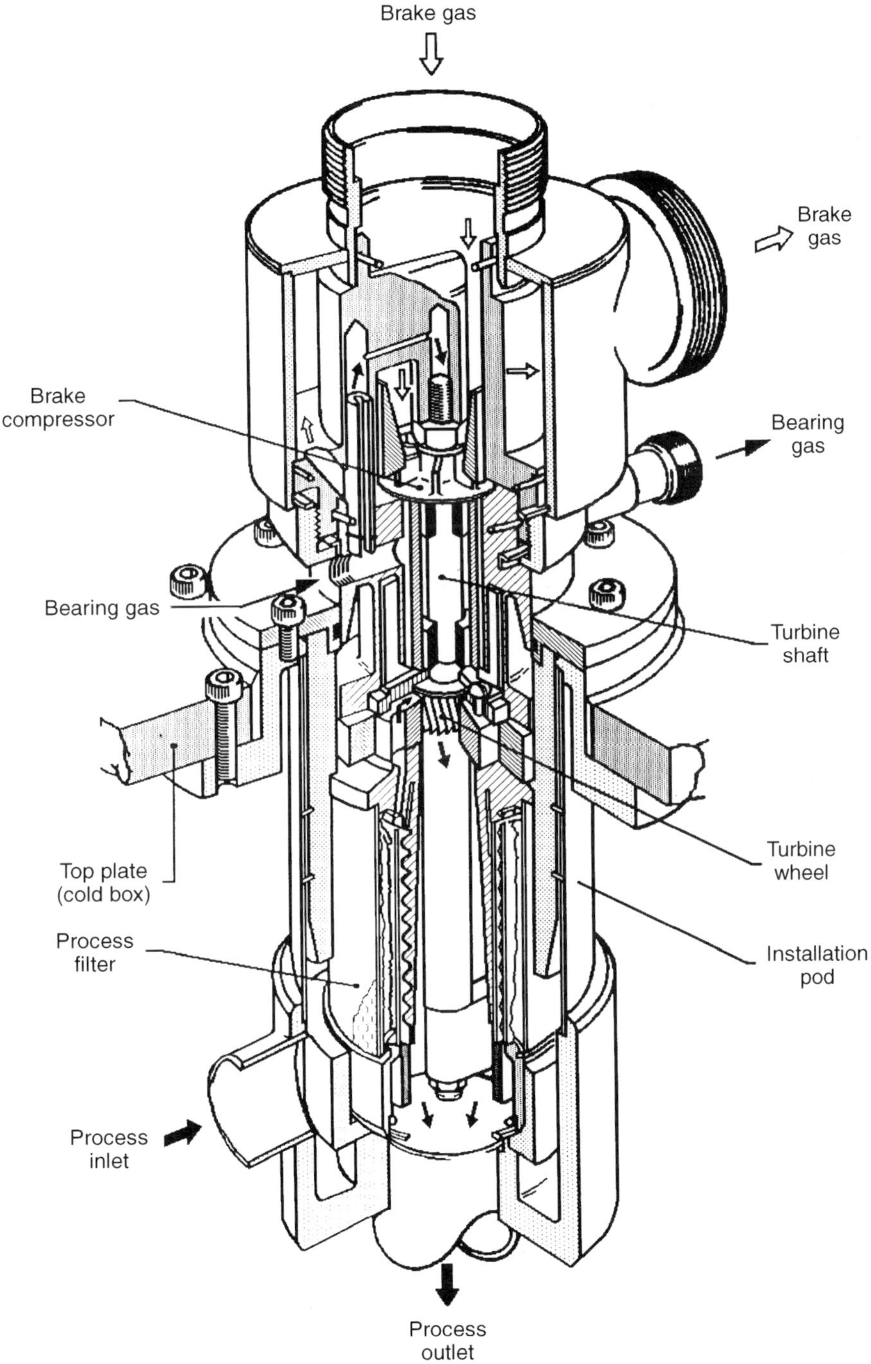

Figure 2.6 Single-stage expansion turbine for small-scale liquid nitrogen use (courtesy of Linde).

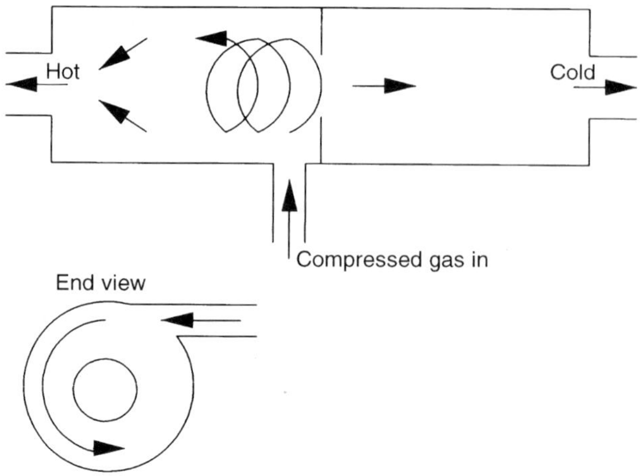

Figure 2.7 Ranque–Hilsch vortex tube.

of the gas stream. The precise theory is incomplete, however, as, because practical vortex tubes operate with a noisy, unsteady flow, this has not yet been modelled. Hilsch's original work used an orifice of 2.2 mm in a 4.6 mm tube supplied with up to 10 barg pressure air and showed temperatures as low as −35°C, from room temperature. Later work improved the achievable temperature drop to 65°C and increased the fraction available as cold air up to 30%. Even with these latter figures, however, efficiency is only 10% Carnot.

2.1.4 Thermal transfer in gases: heat exchangers

Heat exchangers are ubiquitous in gas manufacture and application. In many industrial situations the performance of the heat exchangers is vital to the overall efficiency of a plant. In many applications of gases heat exchangers are necessary to raise or lower the gas to the appropriate temperature of a process or perhaps to stabilise a process against ambient and gas temperature variations. At the most fundamental level, many medium-size gas applications use cryogenic liquefied gas because it is the appropriate form of gas for medium-scale gas supply but actually require room temperature gas. They use simple heat exchangers called vaporisers to vaporise liquid gas and warm it to near room temperature.

A typical situation in industry is where a process is to take place at an elevated temperature but the product gas is needed back at room temperature. If the feed gas at room temperature is warmed up using simple fuel or electric heaters and the product is cooled with tap water, then the process will be very inefficient. The energy used in the warming process is expensive, as is the copious quantity of cooling water needed. Except for the very smallest plant, it will be much more efficient to employ a heat exchanger or heat exchangers, ideally in counter-current mode. In this way hot product exiting the process transfers heat to the incoming cold feed gas, warming it up to near the process temperature. In liquid air distillation plants (section 2.1.5) the process is run at very low temperatures and the emphasis on efficiency is much greater but the principles are the same.

With an ideal counter-current heat exchanger, one that would have a zero temperature difference (ΔT) across it or differential pressure (ΔP) in either process stream, a process could be operated at any temperature with no loss of energy. In practice, ΔT in the region of 1 to 10 K is tolerated as a compromise between capital and running cost. A small ΔT value is in general only achieved by a larger and more expensive heat exchanger, whilst a larger ΔT value wastes energy but can be achieved by a smaller, cheaper exchanger.

Forced convection under turbulent conditions is the most common situation where gases are to be heated or cooled strongly and efficiently. The equations for heat transfer under these circumstances are usually very complex. However, the use of dimensionless groups simplifies equations and allows the setting of operational zones in which reasonably accurate empirical formulae can be used. Whole books are devoted to heat exchanger design for gases, and specialist firms make the units to order. It is interesting nevertheless to consider here some of the possible designs and design considerations.

Exchangers must clearly separate two fluids: the gas being cooled or heated and the working fluid, whether it is liquid cryogen, Freon refrigerant or another gas. The fluids may be at different pressures, as well as different temperatures, and may in generally be potentially corrosive. Nevertheless, the walls of the heat exchanger must be thin so as to maintain the best possible thermal contact.

The choice of high thermal conductivity materials, such as copper or aluminium, is standard although, perhaps surprisingly, inexpensive carbon steel or cast iron can occasionally be used at higher temperature. The strength and chemical resistance of 316L-type stainless steels and the ease of fabrication of small-bore pipes using 316L stainless steel means that these are also popular, despite having thermal conductivity five or ten times worse than aluminium. The decrease in thermal conductivity seen in many metals and alloys works against the designer for heat exchangers needed at liquid air temperatures, with many aluminium and stainless steel alloys almost a factor of two lower in conductivity at 77 K relative to room temperature. It is important to note, however, that different alloys, different batches of alloys and pieces of the same batch differently mechanically worked or heat-treated can show thermal conductivity which will vary by a factor of two or more, with the greatest variations typically below 100 K.

The geometric design of heat exchangers varies enormously. Designs are compromises between providing more intimate contact between fluids via smaller interleaved channels and providing a lower pressure drop via large channels. The straightforward tubes-in-shell design originally seen in steam engines is still popular. For very large units, or those with an arduous service (large ΔT, rapid thermal cycling, high pressure), a development of the shell-tube design is occasionally seen in cryogenics in the form of the helical wound-tube design.

More common in cryogenic practice, however, are welded or vacuum-brazed assemblies of cross-flow or counter-flow plate-fin heat exchangers (Figure 2.8). These are usually more compact and inexpensive for a given service. Plate-fin exchangers, once complete, are large rigid assemblies but have the disadvantage of being more susceptible to unrepairable cracks due to thermal cycling. Differential thermal expansion considerations also limit the maximum possible size of plate-fin units. Aluminium contracts about 0.4% (copper and stainless steel 0.3% for comparison) between room temperature and liquid nitrogen temperature, which is more than the yield strain so distortion of components which are unevenly cooled is likely even where they do not crack. Pipework attachments to heat exchangers and other components running in cryogen must always be designed to allow free thermal movements to avoid problems.

It turns out that the design of a whole heat exchanger can be summed up in the simplest case by the same equation used above (Searl's bar equation) for each small elemental area, dA, of that exchanger by using an effective average temperature difference. This effective average temperature difference in a heat exchanger can be shown by integration, with assumption of constant U value, to be given by the log mean temperature difference

$$\Delta T = (T_{in} - \Delta T_{out})/(\ln(\Delta T_{in}/\Delta T_{out}))$$

The choice of counter-flow in heat exchangers yields a smaller exchanger unit for a given service, as the effective temperature difference is smaller in that arrangement (Figure 2.9). In the idealised example chosen, a certain heat exchanger is compared in performance in counter-current and co-current mode. The achieved final temperature of the hot stream is 223 K in co-current mode and 211 K in counter-current. This reflects the fact that in the co-current case the temperatures in the exchanger vary exponentially so that the small heat transfer at the outlet, where ΔT is small, is not compensated by the large heat transfer at the inlet, where ΔT is large.

In the counter-current case the temperature distribution is still exponential but the ΔT distribution is almost flat, giving a higher average heat transfer. Counter-current mode of operation is the only possible mode where the hot stream must be cooled below the cool stream outlet temperature or where the cool stream must be heated above the hot stream outlet temperature, as shown in Figure 2.9. It is surprising how well large counter-current heat exchangers can work, particularly in larger sizes where

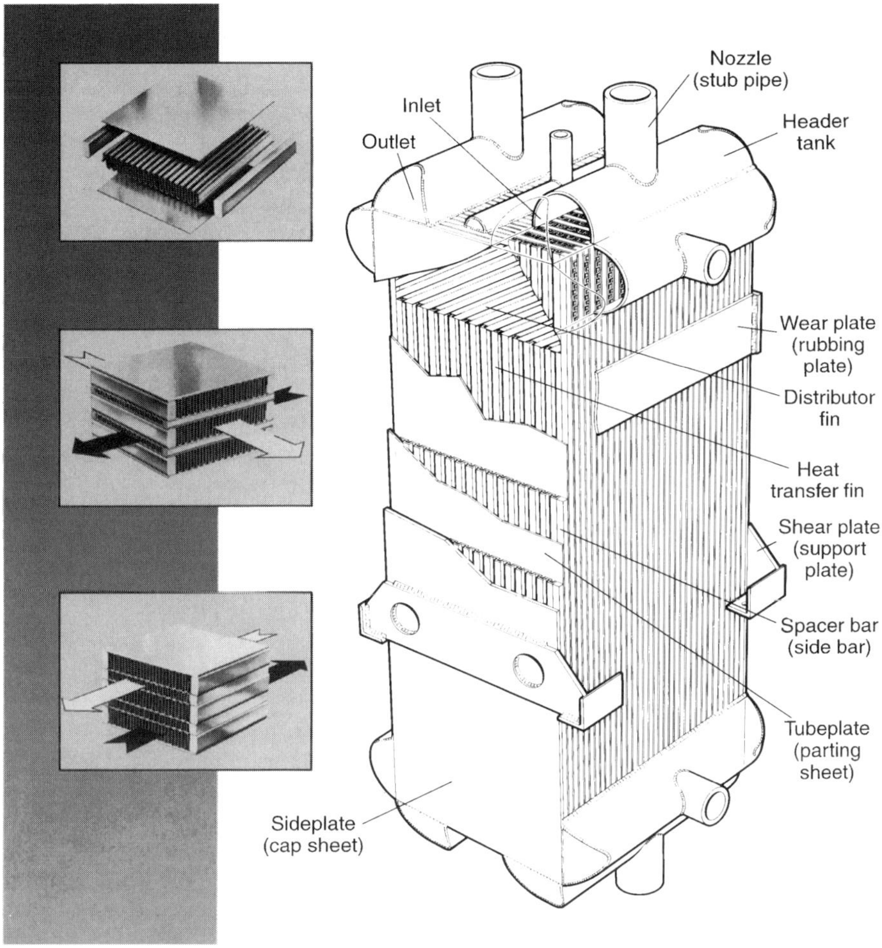

Figure 2.8 Plate-fin heat exchanger (courtesy of IMI Marston Ltd).

heat leaks to the outside are less significant. The rather small amount of cooling from the Joule–Thompson effect can easily overcome the inefficiencies in the heat exchanger system in air liquefaction. If this were not the case, air separation by liquid-air distillation would not have been possible in the early days, using only Joule–Thompson expansion, and even modern plants, with their turbo-expanders giving powerful cooling, would not be so successful commercially.

Passages in a heat exchanger will in general be related to the flow required at reasonably low differential pressures. Liquids will achieve the same mass flow in much smaller channels, for example, so liquid channels will be smaller than gas channels. The need to allow two-phase flow, boiling and condensation further complicates design.

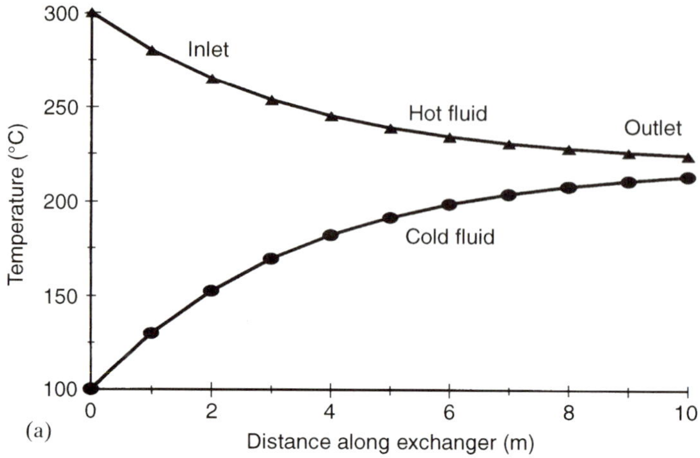

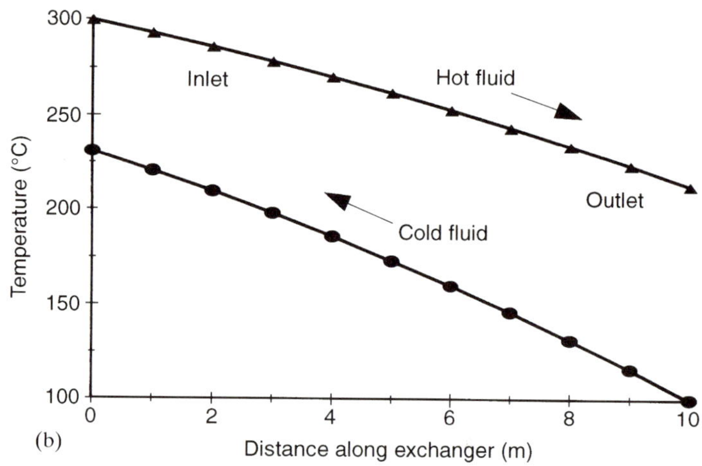

Figure 2.9 Stream temperatures in (a) a co-flow exchanger and (b) a counter-flow exchanger.

A typical calculation for a heat exchanger is

$$Q = UdA(T_2 - T_1)$$

where Q is the heat transferred, dA is a small area of exchanger and $(T_2 - T_1)$ is the temperature difference. The heat transfer coefficient U for a particular design of heat exchanger is usually quoted for different liquids and gases (with different pressures quoted for gases). U can be estimated if this is not the case by adding together the reciprocals of thermal conductance of the exchanger wall and the two film heat-transfer thermal conductances U_f. Under typical forced convection conditions approximate numbers for these can be looked up. For example, for atmospheric air, values of U_f around $100\,W\,m^{-2}\,K^{-1}$ are typical.

The equation

$$1/U = 1/U_{f1} + 1/U_{f2} + t/k$$

accounts for the two film conductances and the wall, of thickness t and thermal conductivity k. The film conductance is so low that in some applications it is unnecessary to use a particularly good thermal conductor for fabricating the exchanger. When the thickness t is not required to be unreasonably thin, and provided the material satisfies thermal expansion considerations, simple steel exchangers can be used. Aluminium is typical, however, and allows for very compact heat exchangers which are physically very strong.

Suppose an air stream of $1500\,m^3\,h^{-1}$ at 300 K and 10 bara is to be counter-current cooled against a $1000\,m^3\,h^{-1}$ nitrogen stream at 100 K and 10 bara. Let us assume that a heat exchanger design is used where U is, say, $100\,W\,m^{-2}\,K^{-1}$ overall. What size (area) of exchanger is needed and what outlet air-stream temperature might be expected? If the simplified assumptions are made that specific heats of the gases are constant and the same, and that U remains constant over the temperature range, then (see Figure 2.9) with counter-current design and an air stream achieving 211 K, the log mean ΔT is 85 K.

$$\text{area needed} = Q/(U\Delta T) \simeq 10\,m^2$$

Clearly, in order to achieve acceptably low ΔT values, large areas of heat exchanger are needed. Hence the need for the convolutions of the plate-fin designs.

2.1.5 Distillation of air

Distillation of cryogenically cooled liquid air at a few tens of bar or less pressure is still, nearly a century after its large-scale use, the technique of choice for preparation of large quantities of oxygen, nitrogen or argon.

The atmosphere: raw material for a \$20 bn industry. The typical composition of the earth's atmosphere at sea level, expressed as fractions of dry air by volume, is as shown in Table 2.4. The amount of water vapour in the atmosphere is frequently forgotten in calculations, but it can be substantial on a warm, humid day.[8]

[8] The saturated vapour pressure of water varies with temperature approximately exponentially (section 1.3.14), being 1 bar at 100°C and 0.023 bar at 20°C. The latter implies that on a day with high, near 100%, humidity, the atmosphere will be about 2.3% water. On compressing to 10 bar the air will still only have the capability of carrying 0.023 bar of moisture. However, 0.023 bar is now 0.23% of the air stream so the other 2.07% will condense in the outlet of the compressor. For every 100 m^3 compressed, in this case 1.8 l of water will be condensed out. This means that a compressor of quite modest capacity, say a few hundred cubic metres per hour, will require a hosepipe to lead away the condensate.

Table 2.4 Composition of the Earth's atmosphere at sea level (expressed as fractions of dry air by volume)

Component	Fraction	Component	Fraction
Nitrogen	78.09%	Krypton	1.14 ppm
Oxygen	20.95%	N_2O	0.5 ppm
(Water	~1–3%)	H_2	0.5 ppm
Argon	0.93%	Ozone	0.4 ppm
CO_2	300 ppm	CO	<5 ppm
Neon	18 ppm	Acetylene	<1 ppm
Helium	5.2 ppm	Xenon	0.086 ppm
Methane	1.5 ppm		

Pre-purification and liquefying of air. Before air can be liquefied it must be free of impurities which would block up the plant. As well as filtering out particulate matter, any gases or vapours which could freeze up and cause a blockage at 100 K will need removing. Also any inflammable impurities need to be removed because most inflammable gases will accumulate in the liquid oxygen and may, when mixed with it, form an explosive mixture. (Several early liquid oxygen plant explosions were caused by the latter problem.[9])

Pre-purifiers may employ PSA or temperature swing absorber systems, or they may use the RHE (reversing heat exchanger) system, which freezes out impurities. Temperature swing absorbers (TSAs) are the norm. Beds of silica gel, zeolite or carbon (section 2.1.6) are selected for their abilities to absorb moisture, carbon dioxide, methane and acetylene at room temperature. This they do for a considerable period before being regenerated by a flow of hot gas. Two beds are employed so that one bed can be regenerated by the heated dry nitrogen stream whilst the other is on-line.

RHEs are often noisy, a fact which is of importance unless the plant is in an isolated location. They function by presenting a cold plate to the cool compressed air passed through them. Moisture and carbon dioxide, the principal impurities, are condensed. After a few minutes, the heat exchanger is purged with a short flow of dry, warm, waste gas, which warms up and drives off the absorbate in seconds, enabling the plate to be used once again. The purge gas, released at large flow rate, is usually noisy, even when it is high up at the top of a vent stack.

Liquefaction of air is still the key technology of modern industrial gases. It is surprisingly simple, once good insulation and the efficient heat exchangers described above can be taken for granted. All that is required is for

[9] Acetylene, unlike methane and other light hydrocarbons often seen in air, is substantially insoluble in liquid oxygen. It is this which leads to the problem because if acetylene is concentrated above its solubility, which is only a few parts per million, it can precipitate, forming a slush with the liquid oxygen which is in effect an explosive. The precipitated acetylene formed in this way is exceptionally dangerous.

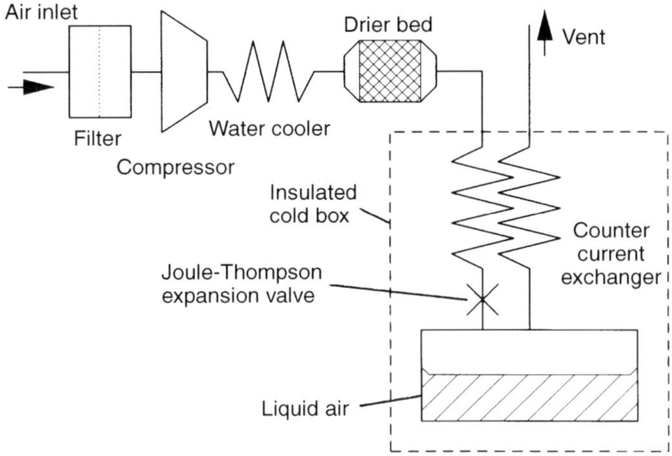

Figure 2.10 Simplified scheme for liquid air production.

compressed air to be cooled against vapours from the liquid air in the process and then expanded. The expansion is most efficiently accomplished by an expansion turbine which allows the cool, compressed air to expand adiabatically to become a mixture of liquid air and cool vapour at lower pressure.

Although efficient in cooling and a potential source of some power, the expansion turbine has to run at around 80 K at very high speed in a two-phase fluid. These are relatively arduous conditions and these turbines are expensive and potentially unreliable. Smaller and simpler plants expand the compressed air through a nozzle or valve using the Joule–Thompson effect. Despite the small temperature drop it gives, with efficient heat exchange the J–T effect is quite sufficient for air liquefaction. A scheme of apparatus using J–T expansion, of the type used around the turn of the century, is shown in Figure 2.10.

Distillation of liquid air. The distillation of any mixture of similar compounds relies on very simple principles. Basically, when the mixture is boiled, the vapour will contain more of the element with the higher vapour pressure. The scale of liquid-air distillation is very large, however. Plants have been constructed which distil up to 10 000 tonnes of air and more per day. A plant of this capacity will have a distillation column 6 m or so in diameter and 35 m tall. The separation of nitrogen and oxygen from air is more difficult than, for example, whisky distillation, however, because the molecules have a close boiling point, a greater purity of product is expected, particularly for nitrogen, and because of the cryogenic temperatures.

Multiple distillation is therefore required and is conceptually performed in a single 'fractionation tower' or 'distillation column'. The column simply

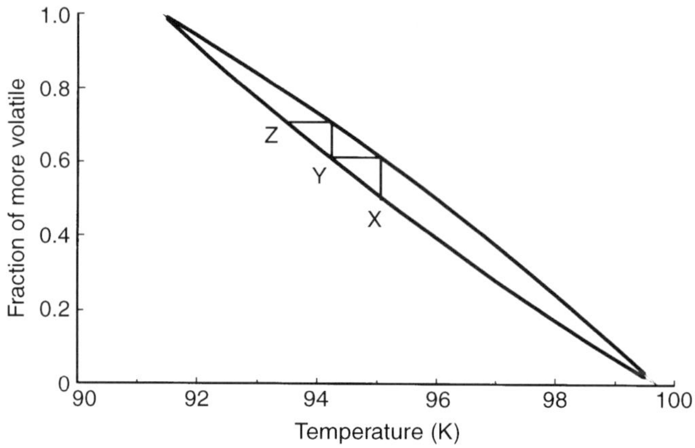

Figure 2.11 Composition diagram for distillation.

integrates a set of retorts and condensers. Liquid air is added to the column centre. The lower boiling nitrogen is concentrated in the stage above, whilst the higher boiling oxygen is concentrated in the stage below. The condenser and reboiler are arranged to provide a 'reflux' flow of liquid and vapour, in addition to that required for product output, in order to ensure that equilibrium is approached in all of the stages. The performance of a distillation column, and the number of stages needed, can be estimated by graphical methods on composition diagrams such as Figure 2.11.

Composition diagram theory. The composition diagram arises in the following way. If Raoult's law is assumed to be followed, then the vapour pressures and liquid mole fractions will be related as follows

$$P_o = P_{a0}\, X_a + P_{b0}\, X_b \text{ at boiling point}$$

where P_o is ambient pressure, P_{a0} and P_{b0} are the saturated vapour pressures of compounds a and b, and the X_a and X_b are the liquid mole fractions ($X_a = 1 - X_b$). If Dalton's law of partial pressures can be used, then

$$X_{a_{vap}} = P_a / P_0 = X_a P_{a0} / P_0$$

and similarly for X_b. Substitution of an exponential type dependence for P_{a0} and P_{b0} and extraction of X_a from the first equation leads to X_a and $X_{a_{vap}}$ as a function of temperature.

It is used by tracking across Figure 2.11 from the bottom line to the top line once for each stage of distillation. A liquid mixture starting at X on stage 1 will distil a vapour of composition X. If this is condensed on stage 2 to liquid and evaporated, a new purer product Y evolves. If Y is condensed on stage 3 to liquid and evaporated, it leads to vapour of composition Z, and so on.

In this way the purity which a certain number of stages will achieve can be judged. Numerical design methods which automate this process are now used.

Note that the simple theory above, what might be dubbed a 'perfect mixture' is only relatively common. Equally common are composition diagrams which show a peak or trough in pressure due to intermolecular forces. On distillation these have a composition(s) which cannot be separated by distillation at a given pressure; the vapour has the same composition as the liquid being boiled. In these cases, products may be only partially separated, pure A from azeotrope or pure B from azeotrope, or the distillation pressure may be changed to avoid azeotropic conditions.

In the simplest arrangement for liquid-air distillation, the column has a liquid-air feed in the middle of the column from which the heavier oxygen-rich vapour is stripped of its lighter N_2 fraction in the descending or 'stripping' section and the lighter nitrogen-rich vapour migrates upwards, becoming purer at each stage, in the upper 'rectification' column.

The progress of each product along the composition diagram lines shown is followed only when little product is withdrawn. In other words, the reflux ratio (liquid mass flow/vapour mass flow) is equal to one. As soon as appreciable product is withdrawn or, equivalently, as less vapour is condensed for reflux and/or less liquid reboiled, the system will follow a somewhat different course.

First consider the rectification section. Here the reflux ratio is less than one and the lighter fraction of the liquid condensed for the next stage should be reduced, reflecting the fact that vapour containing mostly lighter fraction is being withdrawn so that the fraction of the lighter component in the liquid condensed is effectively reduced.

Similarly, in the stripper section, the reflux ratio is greater than one and the heavy fraction of the liquid condensing on the next tray down is not so great as before, reflecting the withdrawal of liquid rich in heavy fraction from the reboiler end of the column.

It can be shown that, under simplifying assumptions, the effect of product feed and withdrawal points can be modelled by a graphical method involving the plot of an equilibrium curve of the fraction of A (the most volatile constituent) in the vapour as a function of the fraction of A in the liquid. This is the McCabe–Thiele approximation and it is well-described in Perry (1984). Similar but more complex constructions can be used for multiple feed and product streams, although numerical simulation is now used for all practical designs.

There is a minimum reflux in the upper column (or maximum reflux in the lower column) beyond which an infinite number of stages would be required to effect separation. Practical plants with finite columns must operate with reflux in excess of this amount, typically by 10–100%. Clearly, operating costs increase with increasing reflux, more work is need to the heat pump

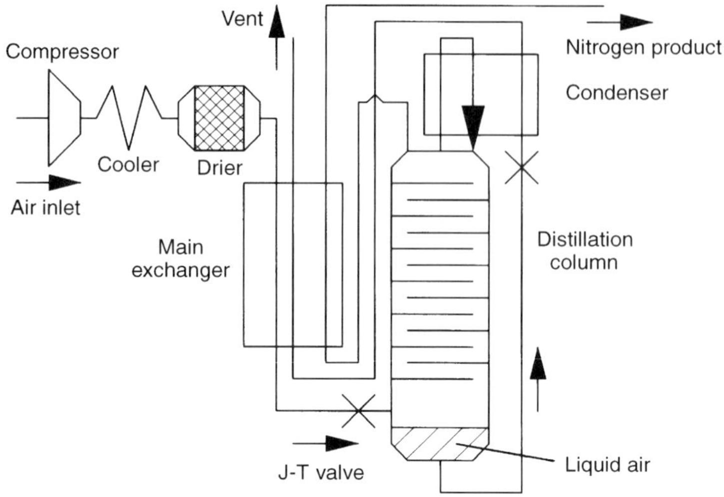

Figure 2.12 Single-column nitrogen separation unit.

system, but capital costs increase with increasing numbers of trays in the distillation column. The best compromise will depend on the relative costs of power and fabrications.

Single column oxygen and nitrogen distillation. The idealised distillation column has a feed of mixture to the middle and two pure products but requires an external heat pump, the refrigerator coil in the condensor and the heater coil in the boiler, to recycle heat from the condenser to the boiler. This heat pump requirement can be very large in the above arrangement; even where a reasonable amount of product can be withdrawn it is inefficient in practice and very expensive on capital installation.

It is very much more efficient to employ the large main air compressors as the energy source to drive the process and provide refrigeration. This can be done by feeding liquid air to the bottom of the column, with some of the liquid from the bottom being taken off to be flashed off to cool the condenser. This results in a pure nitrogen stream at the top and a nitrogen-poor stream at the bottom, i.e. an N_2 rectification system or nitrogen generator system (Figure 2.12).

One important advantage of single-column N_2 units, in contrast to the otherwise more efficient Linde double-column unit, is that the nitrogen stream produced is at a pressure a little less than that of the feed air compressor. A converse trick can be used to make pure oxygen at the bottom and oxygen-depleted air at the top, i.e. an O_2 stripper or oxygen-generator system. In this case, the feed is at the top of the column and the pure O_2 is drawn off at the bottom of the stripper with reflux vapour coming from the feed air boiling the O_2 in the sump. However, although early plants were of this type, O_2 plants of this type are now unusual with O_2 separation being based mostly on double-column units.

The Linde double column. The distillation of liquid air is frequently required to provide two, three or more pure products (nitrogen, oxygen, argon and other noble gases) simultaneously. The outline above would indicate that for each product required at high purity another distillation column is required. The rectification action normally provides a pure stream of A and a stream depleted in A, rather than simply separating A from B and providing two pure products.

It is possible, as in conventional petrochemical distillation, to use a feed point in the middle of a column, simply using the top of the column as the rectifier and the bottom as the stripper. This approach is used in a few plants, for example in the BOC Group's Widnes UK plant, with a single column at high pressure. However, these are exceptional plants where 100% liquid N_2 and liquid O_2 are produced. In these plants the energy required to produce liquefaction make any slight deficiencies in efficiency in separation insignificant. The necessity to operate a heat engine to provide a boiling and condensing service is removed because that heat engine service is already installed for the liquefaction capability.

The single-column O_2 and N_2 plants typically only recover 50 or 60% of their air feed as product as they have considerable amounts of the wanted feed-air component in the waste stream, which is the source of refrigeration for the reflux they need. To achieve a high efficiency and a high recovery a two-column system is required, where the waste stream from one column is employed as feed to the other in some way.

By a happy coincidence it is possible, by choosing pressures correctly, to employ the boiler in a middle-fed column (the main column) to cool a nitrogen-column condenser and to use waste liquid from that higher pressure nitrogen column to flash off and give N_2 reflux at the top of the main column. The thermal linking of two columns operating different separations at different pressures is a quite general opportunity in many chemical separations although it is comparatively rarely used. The ability to condense nitrogen with liquid oxygen depends on the boiling point of liquid oxygen at 5 or 6 bara being 3° lower than the boiling point of liquid nitrogen near atmospheric pressure.

The advantages of this concept were first spotted and engineered by Carl von Linde in his now classic double-column design (Figure 2.13). It has a lower high-pressure rectifier section to which the feed air is taken, whose condenser is the reboiler of the upper low-pressure main column.

The very common double-column design links the two columns also by pipework as the feed air goes to the bottom N_2 rectifier and both liquid

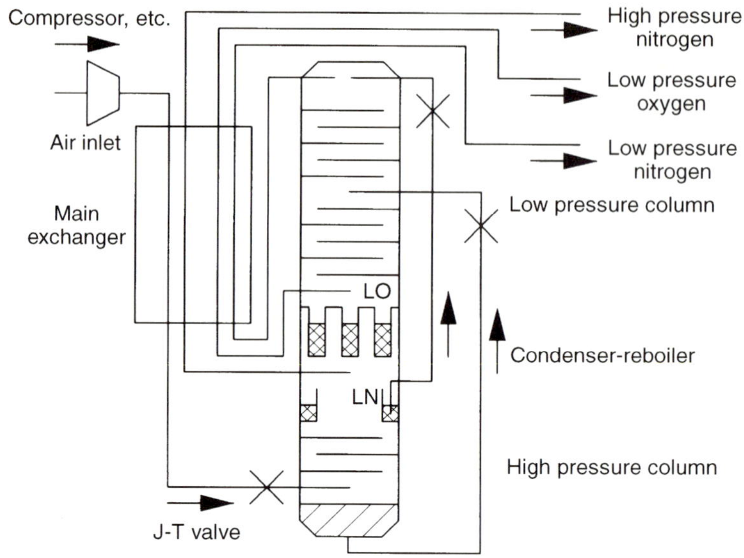

Figure 2.13 Linde double-column air-distillation unit.

nitrogen and rectifier bottoms are fed to the low-pressure upper main column. The feed provides boiling service for the rectifier with condensing service coming from the central reboiler/condenser. The main column feeds the rectifier bottoms, which are of course an O_2/N_2 mixture, to its middle, flashing them off to its lower pressure. For the main column, reboiling service comes from the central reboiler/condenser and condensing service from the flashed off liquid nitrogen fed to the top from the rectifier.

In principle, then, the Linde system provides pure O_2 and pure N_2 without an external heat pump. However, practical designs are made more complex by:

- the need to deal with argon (which, of course, being heavier, ends up in the O_2 stream;
- the accumulation of uncondensible light gases, neon and helium, which must be purged out of the top of the Linde rectifier column;
- the need to provide waste streams which can be used to purge out pre-purifiers or reversing heat exchangers;
- the need for liquid product output, and addition of refrigeration cycles, requiring a turbine at one or more of various possible points to carry out expansion, rather than the inefficient Joule–Thomson expansion valves indicated above;
- complexities associated with the most efficient use of heat exchangers;
- the need to avoid the accumulation of small quantities of hydrocarbons

which may have escaped pre-purification and could end up mixed with liquid oxygen and form an explosive.

Distillation column design. Sieve plates, consisting simply of a sheet with many holes punched in it, are the mainstay of the cryogenic distillation column. With shallow weirs (a few centimetres) sieve trays are made of thin aluminium alloy sheet, often less than a millimetre. Vapour bubbles up through the holes, frothing up the liquid on the tray, whilst liquid overflows over a weir at the edge of the plate and flows down a wide pipe onto the plate below, the weir height controlling the height of the liquid on the sieve plate. Alternate plates have weirs on opposite sides of the column, thus forcing the liquid to zig zag down. The liquid on each tray closely approachs equilibrium with the vapour because of the intimate mixture of the two that occurs in the froth which forms as the vapour bubbles through the liquid.
Distillation columns can have problems, however:

- *Weeping*: liquid streams from a plate to that below effectively 'shorting out' the action of that plate. Weeping gives the lower limit to the column vapour rate before a certain sieve plate design. Sieve plates with small holes to give lower weeping will also have high gas pressure drop, however, so a compromise design has to be adopted.
- *Foaming*: liquid foam rises up and 'shorts out' a plate to the plate above. Foaming gives the upper limit to the column vapour rate. Foam in distillation columns is not like the low-density stable soap/water foam familiar in the kitchen: column foams are of much higher density, 20% liquid or more, and are not stable, breaking back to liquid and gas in seconds. They are nevertheless troublesome.
- *Sieve tray deflection*: if gas flow is excessive, then trays will bulge upwards. If design thickness is inadequate for liquid loading, then sagging will occur. Also at high gas flows vibrations can be set up which can cause excessive weeping or damage to the column in the long term. These vibrations have been linked to synchronised bubbling of vapour through the sieve holes, *inter alia*.
- *Channelling*: liquid and vapour are not evenly distributed across the diameter of the column. It is dealt with in packed columns by adding distributor pipe assemblies every few pipe diameters. In plated columns correct design of feed of liquid to the column at the top, the liquid path from plate to plate, and strictly flat and horizontal plates usually avoid any problems.

A measure of the efficiency of a column is the number of theoretical plates in a certain size of column or the ratio between this and the actual number of plates, i.e. the plate efficiency. The number of theoretical plates needed for a particular separation is dictated by the composition diagram and the purity of product desired but is generally quite small. The small losses of efficiency due to pressure drops bigger than necessary which are entailed by installing a surfeit of plates are more than compensated for by the additional freedom of operational range allowed by a longer column. In particular, it is often convenient to operate columns with a relatively small reflux flow ratio, which means that extra plates are needed.

Packed columns, generally 'structured packing' comprising regular rows of crimped foils rather than randomly packed columns of irregular mineral or the manufactured 'Raschig rings', are an alternative to fabricated plate columns. Height of equivalent theoretical plate (HETP) is the efficiency measure used. HETP values are normally very small, a few centimetres even with quite coarse packing, and this is why packed columns are widely researched. Not only are packed columns smaller than trayed columns, leading to lower capital costs, but there are potential running cost requirements as well: lower cold box heat inleak losses and lower hydraulic pumping losses due to the need to pump liquid up a tall column. Pressure losses, due to having to pump vapour through the liquid and the energy expended in creating the froth, are also avoided in packed columns, where the essential gas/liquid contact is achieved by the vapour simply passing the large area of thin film on the packing.

Packed columns are increasingly being used instead of trayed columns, with companies like the cryogenic plant divisions of L'Air Liquid and Linde leading the way to plant entirely based on packed columns. They are not invariably the correct solution, however, because problems with achieving high turndown are common and packed columns generally have to be operated within comparatively narrow limits of liquid and vapour flow rate. In the opposite direction of performance is the bubble-cap tray. Bubble-cap trayed columns have the advantage of high turndown and generally wide operating limits, and high plate efficiency. However, in cryogenic service they are generally regarded as having pressure losses which are too high and large plate heights, i.e. much larger than the few inches necessary in trayed columns.

A 'halfway-house' between packed columns and trays in some respects is the use of trays made of a porous material. These have been claimed to give good gas flow while giving relative freedom from weeping and foaming. However, no commercial plants are using this concept to date.

Finally, just becoming popular are 'refluxing exchangers'. These are heat exchangers designed for distillation service. The types being extensively trialled now are mostly heat exchangers of the plate-fin type mounted in a vertical plane which integrate the action of heat exchange and distillation. Typically distillation takes place in vertical fin passages whilst horizontal passages are connected to provide condensing and boiling service, and maintain appropriate temperature distribution. Clearly, distillation based on these units will have restricted regions of operational liquid and vapour flow but within those limits they have the potential for very high efficiency.

ASU detailed design. Once the basic cryogenic design has been complete, there are a myriad of details to be determined on a practical air-separation unit.

Instrumentation is a fundamental part of all ASUs, varying in complexity from the humble thermocouple costing a few dollars to $50k or more for multigas analysers such as process chromatographs. With all the cryogenic part of the ASU hidden away deep inside its cold box, it is necessary to have more instrumentation than a standard chemical plant. Levels at various points in the equipment, as well as in the storage tanks, must be recorded and the mainstay for level monitoring is the differential pressure (DP) gauge of the type used in storage tanks (section 2.3.1). Several points going up the column will be monitored for temperature, as will the different pressures in columns and coming from compressors. Flow rate measurement of cryogenic liquids is difficult to carry out accurately (section 2.3.1) and often avoided. Gas flows, although also somewhat difficult to gauge accurately, are useful and accurate venturi or turbine meter flow measurement is necessary with a pipeline supply in order to be able to charge the user for the gas used.

The compressor is a significant (and expensive) machine in its own right and is normally monitored carefully for electrical input versus flow and pressure output, giving approximate absolute efficiency and a good measurement of any changes of efficiency. Parameters such as bearing oil feed pressure and temperatures may be monitored in addition to the basic input/output monitoring.

Early ASUs occasionally exploded when hydrocarbons accumulated to dangerous levels in liquid oxygen somewhere in the system and the mixture ignited. Although these incidents were rare and often of small scale, modern plants take pains to avoid these problems by using additional absorber units based on silica gel in key points in the design. These are occasionally regenerated by being warmed up with a stream of purge gas. As hydrocarbons accumulate in liquid oxygen, two absorbers are often used, connected to the sumps of the lower column and upper column, absorbing acetylene and methane, for example, from oxygen-rich and pure oxygen liquids respectively. The gel absorbers work at cryogenic temperatures with good efficiency.

Gas analytical instrumentation is essential both for monitoring the plant status and also for checking product purity. For ASUs with argon production, for example, it is useful to measure Ar content in the argon side stream, which amounts to plant status information. A process gas chromatograph using zeolite molecular sieve columns is used. For a semiconductor industry N_2 generator, output moisture and O_2 content are critical and analysers much more sensitive than the norm will invariably be used to check output. Occasionally, gas analysers can also have a safety aspect, for example detecting the presence of traces of methane, acetylene and other

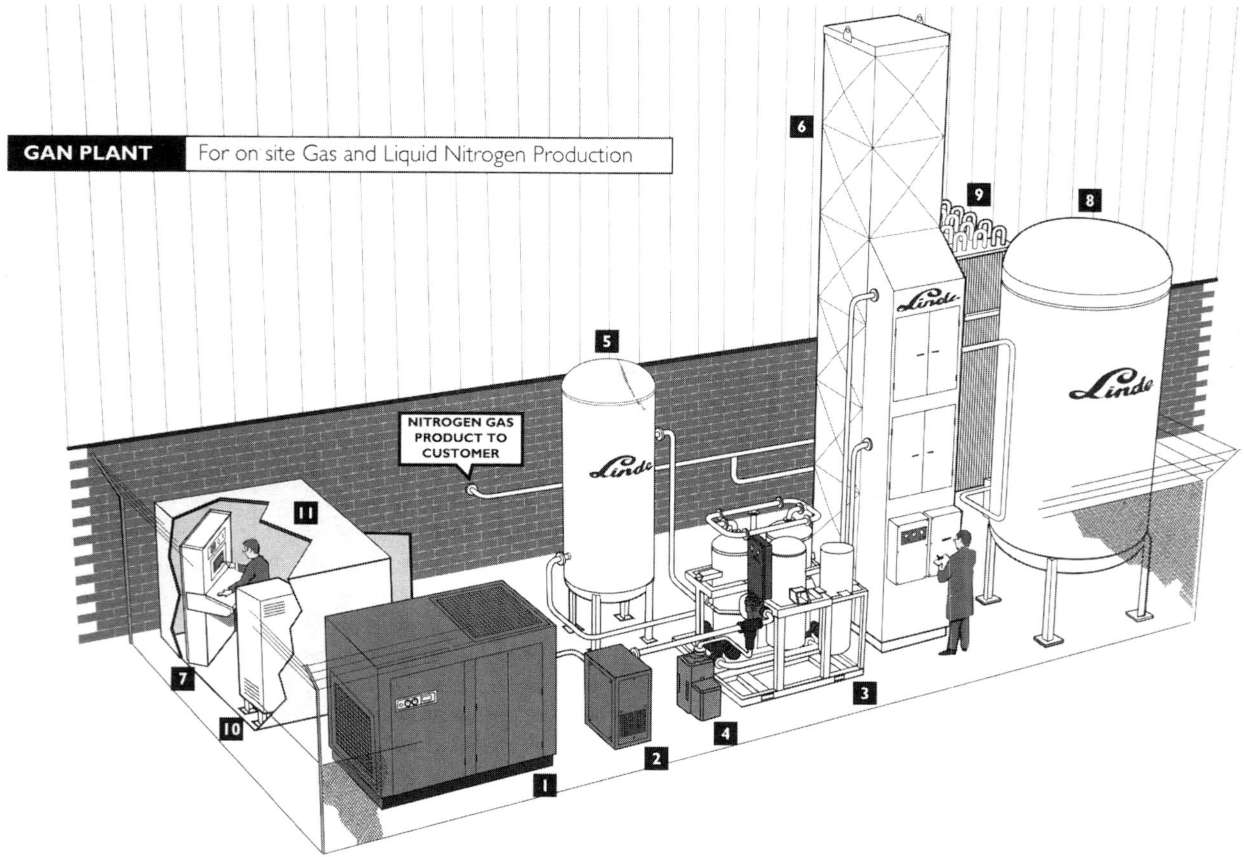

Figure 2.14 Small-scale cryogenic air-separation plant for nitrogen showing typical arrangement (courtesy of Linde). 1, air compressor; 2, after cooler; 3, air purification unit; 4, oil/water separator; 5, air receiver; 6, cold box module; 7, control unit; 8, liquid nitrogen storage tank; 9, ambient air vaporiser bank; 10, power distribution unit; 11, control cabin.

hydrocarbons in the feed air or liquid oxygen may help avoid explosive mixtures building up. Analysers employed for output monitoring can also be used to give plant status with respect to some conditions.[10]

Modern ASUs are fully instrumented in all but the smallest sizes with numerous thermocouples, pressure, flow and level gauges and gas analysis equipment as described. All the data from these sensors are led back, typically via a number of programmable logic controllers and set-point PID (proportional-integral-differential) controllers, to a central computer in the control room of the site. This displays mimic diagrams of the whole plant, indicating the gauge readings and positions of valves. Programs will be set up on this computer for optimum performance in various modes. One typical mode would be maximum liquid nitrogen production. During the night, when electrical power is cheap, liquid nitrogen production is maximised. During daylight, higher power cost may dictate a maximum gaseous oxygen efficiency mode or even a minimum energy consumption 'stand-by' mode. At other times maximum argon may be demanded or a special nitrogen purity may be required. Figure 2.14 shows a tonnage oxygen ASU with tank, column in cold box and compressor.

Selection of distillation designs. In industrial practice there are many different designs of cryogenic air separation units in use. Convergent design, the mark of an efficient, maturing industry, is not always obvious in ASU equipment. This is partly because the advantages of scaling-up plant are such that standard off-the-shelf units are unattractive because the plant needs to match quantities required precisely. However, the wide variety of plants is also a function of the form and relative quantities of product gases required. There are advantages to operating plant designed as modular units, however, and this approach is seen occasionally in plants designed for gaseous nitrogen which have been retrofitted with a separate bolt-on cryogenic liquefier to produce liquid product.

Designs must be thought through if the correct decisions on plant design and operating are to be made. If nitrogen is required at high pressure, 100 bar for example, then the most efficient plant may well be one operated at (paradoxically) low pressure. The low-pressure product is compressed for the downstream process, which means that large amounts of energy needed for compression to 100 bar are not wasted on the waste streams of nitrogen and oxygen.

Separating argon. Argon used to be regarded as a nuisance, contaminating

[10] An example of the latter is the case of a plant where purity of the output liquid nitrogen was being compromised by small amounts of O_2. By comparing the output analyser with results from an intermediate analyser it was possible to show that the distillation process was fine but that oxygen was somehow ending up in the product after distillation. Suspicion fell on a heat exchanger which subsequently proved to contain minute cracks.

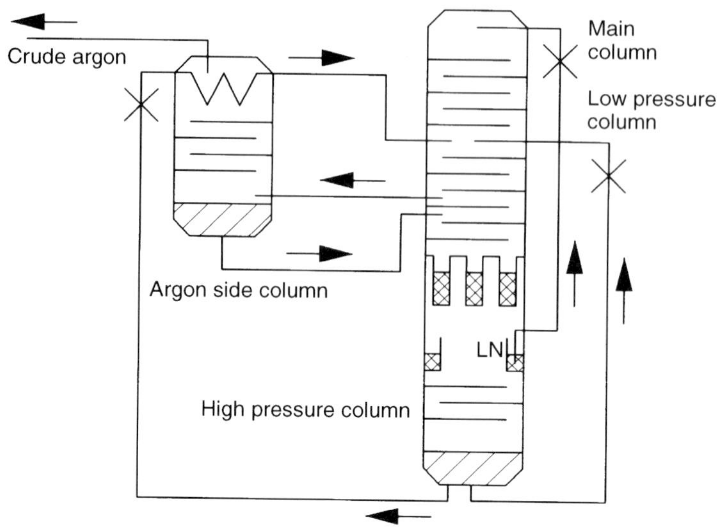

Figure 2.15 Separation of argon using a side column.

oxygen and making it less effective in oxy-acetylene cutting, for example, although its inertness means that it rarely caused any real trouble. With the rise in uses for argon, however, from sputtering of coatings to welding of 100 mm steel plates for chemical plant vessels, argon is now separated out in all cryogenic air-separation units of substantial size.

With its boiling point intermediate between liquid oxygen (LO) and liquid nitrogen (LN), liquid argon (LAr) concentrations are highest somewhere below the middle of the upper column in a double-column ASU when the ASU is producing pure (99.5% or more) oxygen. If argon is not withdrawn, it will accumulate in the distillation column until its concentration is such as to contaminate the oxygen output. In plants producing low-purity oxygen, argon is mostly lost in that oxygen and recovery may not be worthwhile.

In a typical ASU, a liquid product containing mainly oxygen, the composition being typically 100 ppm LN, 10% LAr and 90% LO, can be withdrawn. This can be further distilled in a separate column, which is often placed alongside the main distillation column to avoid having to pump the cryogens up or downhill and operated at a similar pressure to the main column; this is the 'argon side column' system (Figure 2.15). The side column upgrades the argon content up to 95% or 98% or so Ar, the produced argon being the top product. The bottom product, almost pure oxygen, is fed back to the main column a little further down below the take-off point. The condensing service for the column is provided by expanding a small amount of liquid air from the bottom high-pressure column across a valve and passing the cool air through a condensing heat exchanger at the top of the side column. The expanded air is wasted via a heat exchanger or fed back into the low-pressure column nearer to the nitrogen end.

The 'industrial' or 'crude' argon produced by the side column is saleable to some users, for example those welding shield gases where some oxygen is allowed in the inert argon mix. For many crude argon applications low nitrogen content is more important than the few per cent residual oxygen. For most users, however, a few per cent oxygen is completely unacceptable and a true inert gas is needed with only parts per million of reactive gases such as oxygen or even nitrogen. This 'pure argon' is often produced by a standard sequence of a catalytic reaction with excess hydrogen to remove oxygen, followed by an additional distillation step.

The distillation step is preceded by a condenser and a temperature-swing drier, where water formed in the catalytic step by hydrogen/oxygen combination is removed. In the additional stand-alone distillation 'column' (it is little more than a flash-off tank), excess hydrogen and residual nitrogen are removed as lights and liquid argon removed as bottoms. The catalytic reactor, because of the high degree of purity of the gases used, can be of the precious metal/inert support type without fear of loss of activity from poisoning.

Separating neon, krypton, and xenon. The inert gases He, Ne, Kr and Xe are exceptionally difficult to detect in the air and their separation from it is no less difficult.

Helium and neon are left dissolved in liquid nitrogen separated in a Linde double column-system. By expanding the liquid in a phase separator prior to storage, a top gas can be obtained with several per cent neon and helium together. (Helium is inexpensively obtained by processing of natural gas (section 2.2.8), so neon is the main product of the system.) However, commercial production of neon is carried out in a minority of large air-separation plants by drawing a neon-enriched vapour stream from the top of the bottom (high-pressure) column of the Linde double-column unit. The vapour is taken to a neon side column where it provides boil-up vapour, the condensing service being a liquid-nitrogen-fed heat exchanger condenser at the top. The crude neon can be pulled off from the top of this column at 50% efficiency and at 50% concentration, the balance being nitrogen and helium with minor amounts of hydrogen.

Neon is further purified by adsorption onto porous beds, typically activated carbon. First, hydrogen can be removed if necessary by catalytic combustion. Next come carbon adsorption units. The final bed in a multi-stage, multi-bed system is often a LN-cooled carbon adsorption temperature-swing regenerated unit, while the first stage is usually a pressure-swing regenerated unit at room temperature.

The final stage of purifying neon is to liquefy it either by a mechanical expansion cycle or by cooling with a liquid hydrogen or gaseous helium cooling circuit. After liquefaction, remaining helium is lost by flashing off the liquid neon to atmospheric pressure, recycling the top-gas. Neon is now supplied in relatively large quantities. It is packaged both in compressed cylinders, on tube-trailers of cylinders and in liquid form in small cryogenic dewars, almost identical to those used for LN. (Neon production was greatly expanded in the USA for a while in the 1980s as a result of US Department of Defense needs.)

Krypton and xenon are only present to the extent of 1 and 0.08 ppm in the air, making them exceptionally difficult to separate. In practice separation is carried out in several steps:

1. Withdraw an LO stream: this, in principle, will have Kr/Xe at 5 and 0.4 ppm.
2. Boil up LO in a distillation side column, use N in the condenser, producing a concentrate containing 0.2% Kr and 0.02% Xe in oxygen.
3. 'Burn-out' residual hydrocarbons in a catalytic combuster (the excess oxygen providing oxidation service).
4. Absorb CO_2/H_2O.
5. Distil cryogenically again, the bottoms containing up to 80% Kr and 6% Xe (oxygen being a top product).
6. 'Burn-out' residual hydrocarbons again.
7. Absorb CO_2/H_2O again.
8. Bottle the krypton/xenon mixture for sale *or* accumulate enough Kr/Xe to liquefy against LN and distil to pure Kr and Xe.
9. Bottle Kr and Xe.

2.1.6 Pressure-swing absorption

PSA adsorbers and molecular sieves. It is usual in chemical engineering to distinguish between absorption processes, where a gas is absorbed by being dissolved in the bulk of a liquid, and adsorption processes, where a gas is adsorbed in the surface of microporous solid. A biological example of the former is the absorption of oxygen by haemoglobin in blood whilst in industry liquid absorption is used to extract carbon dioxide and hydrogen sulphide from gas streams. Absorption of acid fume and toxic reactive gases by water spray scrubbers is another example. Adsorption by solids is met most commonly in the form of activated carbon granules, which are used to remove impurities in air in many applications. In the latter two cases, the process is often not reversed; the water and activated carbon are simply dumped when their capacity is exhausted. For separation of gases, as opposed to removal of trace impurities, a cyclic process with regeneration of the adsorber is necessary. Pressure-swing adsorption (PSA) is an example of such a reversible adsorption process.

PSA units for air or gas separation mostly rely on 'molecular sieves'. These are not sheets of material with very small holes through which a gas is forced leaving the big molecules behind (this is the principle of membrane separation described in section 2.1.7). They are in fact microporous solids, with molecular-size pores, which can be used to adsorb and desorb gases. Mostly because of the different forces of attraction between different gas molecules and the micropores, but also because of the different size of gas molecules, the adsorption differs between gases and can be used as the basis for separation.

Almost all microporous materials will show separation effects, at least for some mixtures of gases, but few satisfy more stringent criteria for a useful absorbent:

- reasonable crush strength so that they can be compacted into tall vertical beds without blocking;
- little tendency to 'dusting', i.e. converting from granular solids into fine dust;
- controllable microporosity so homogeneous beds of known capacity can be prepared;
- not too strong a binding to any one gas because desorption will then be difficult;
- small macropore/interpellet void volume: materials with a large volume which is not in the active micropores require larger purge volumes to achieve high purity products and high recovery;
- macropore structure which allows good access for gas flow through the bed and to the micropores.

These criteria have all but eliminated most natural products and many less reproducible synthetics such as calcined clays. Early pressure-swing adsorption was based around activated carbon or the porous solids well-known as easily-regenerable drying agents. Silica gel, made by neutralising strong solutions of sodium silicate acid (water glass) with acid, and hydrated alumina, made by dehydrating hydrated alumina, are examples of the latter. Silica gel and activated alumina both retain water molecules strongly, which makes them good materials for use in a gas drier unit but less useful in gas separation where the gas mixture may also contain moisture. However, the moisture in these materials is loosely bound and can be removed by heating to 150°C or by pressure reduction as part of a PSA cycle. Hence the first part of the gas-separation bed using alumina or silica gel in fact functions as a built-in drier unit.

Activated carbon has been a big industry since the extensive nineteenth century investigations into the many interesting properties of carbon. Carefully prepared pellets of carbon with volatile binders are first heated to 500°C or so to eliminate the binder and other hydrogenous material, followed by stronger heating in special atmosphere, typically containing mild oxidants such as CO_2 or steam, causing the slow development of microporosity throughout the bulk of the material. Activated carbon is particularly useful in separating gases which may contain moisture because its surface is almost non-polar and will not attract and retain water molecules.

Two types of microporous solids are now of importance in gas separation by pressure swing: carbon molecular sieves and zeolite (aluminosilicate) molecular sieves. Molecular sieve materials are the only materials capable of separating air gases: even on typical molecular sieves the ratio of O_2 to N_2 adsorption is only 2 or 4 (with alumina or ordinary activated carbon the ratio is close to 1).

Molecular sieve carbon (MSC) was developed by applying activated carbon treatments to particular carbon sources or by coating the pores of conventionally made activated carbon with more carbon. One method uses a thermosetting polymer which is then cooked on to form an additional layer of carbon or carbonised material. Heating MSC materials in hydrocarbon gases can deposit carbon in the micropores, adjusting the pore size. MSC has, in general, more micropores within the active size range of importance in gas separation (5–15 Å) than the microporous solids above, which have little porosity below 20 Å. Zeolites also have microporosity below 10 Å, which is intrinsic to the crystallography of aluminosilicates. The synthetic zeolites used are made by crystallising a sodium/aluminate/silicate water gel by heating under pressure. The resulting crystals are in effect 'clathrates': a continuous structure with cages containing water molecules. On calcining above 500°C, the water is eliminated to yield the zeolite. Zeolites absorb water strongly and this can only be removed completely by heating to 300°C or so, meaning that gas mixtures to be separated should in general be dried before an adsorbtion separation is attempted.

MSC and zeolite both have structure at several levels: macropores which allow bulk flow of the input mixture and purge gases, pores which are microscopic but still large compared to gas molecules, and finally the micropores, which are the part of the structure that shows appreciable differences in absorption of gases. In the case of carbon molecular sieves there is a continuous spectrum of pore size whilst in zeolites there is a bimodal distribution with micropores of a few Å and macropores between crystallites of a few microns.

Some molecular sieves have been 'doped' with a chemical agent to make them more selective; an example is the BP carbon monoxide adsorber which adds copper ions to a zeolite molecular sieve. In others the proportion of cations in the zeolite aluminosilicate lattice of a zeolite is modified to change the pore size and substitution of sodium and potassium atoms yields the now standard 4, 5 and 13 Å pore sizes. PRAXAIR patents also name lithium as a useful cation substitution in its zeolites. (PRAXAIR, formerly the Union

Carbide gases business, carried out much of the early work on synthetic zeolite molecular sieves and developed the market for the basic 4 Å, 5 Å, 13 Å, etc., sieve types.)

PSA/VSA concepts. Pressure-swing adsorption is a simple concept in which gas is admitted to an adsorber bed at high pressure then desorbed at a lower pressure. There is either a difference in adsorption rate or a difference in static amount of adsorption. Hence the two gases present in the input gas are differentially adsorbed; one is concentrated in the bed whilst the other is released in the depressurisation stream. After the depressurisation stream has been flowing for a preset time to the outlet, the depressurisation is diverted to waste until it is at near-atmospheric pressure, releasing the adsorbate, perhaps with the aid of a purge stream.

In a so-called vacuum-swing adsorption (VSA) unit, the depressurisation is continued past atmospheric pressure. In either case, this part of the cycle ensures that a minimum of the adsorbed gas is left so that the gas mixture is not distorted when the next cycle of raw gas mixture admission begins. The trade off is, of course, the cost of installing and running the necessary vacuum pump. The size of vacuum pump needed climbs very quickly with required throughput since vacuum pumps have to handle large volumes to achieve reasonable mass flow rates. However, the use of the vacuum swing enables more efficient use of molecular sieve. Typical absorption isotherms do not favour PSA operation with a high delivery pressure so most PSAs are best operated with venting down to near 1 bara, followed by a compressor to step up pressure if necessary. Vacuum-swing absorbtion units for oxygen have shown greater economy than PSAs even when the extra cost of buying a large vacuum pump is taken into account. The vacuum-swing cycle makes more effective use of the isotherm and power efficiencies can be increased by 75%. VSA also usually allows a simple, low-cost two-bed design where a PSA system might require multiple beds.

More exotic pressure-swing systems are continually being proposed. Systems with large numbers of beds, all undergoing different processes at different pressures simultaneously, are typical. One ideal system to which some have aspired is a moving bed system in which gas coming in is continuously met by fresh absorber, which is slowly removed and regenerated in such a way that the whole system operates in a dynamic steady state. Systems have been proposed which simulate the pseudo-moving bed SORBEX organic solvent separation process. SORBEX swaps process streams between dozens of inlet pipes to a single large fixed adsorber column to simulate the moving bed. One more recent exotic system proposed is pressure parametric pumping (PPP). PPP involves pumping gas through a column in one direction at high pressure and in the reverse direction at low pressure, using a specially designed pump. The high-pressure outlet end concentrates the less absorbed component even when the ‘cycle time’, the pump rotation time, is a few seconds because the gases migrate down the column a little with each pump cycle.

Isotherms and adsorber bed dynamics. In order to use the adsorber beds and the compressor energy efficiently, it is necessary to make the depressurisation to product stream as long as possible; the longer this goes on, the greater the output of desired gas for a given plant and energy consumption. However, the longer the desorption is carried out to the product stream, the more the product will be polluted by the other elements of the mixture breaking through.

There are some tricks which can improve matters, however. Absorber bed materials which have a 'favourable isotherm' will have a sharp breakthrough characteristic. This means that nearly all the bed capacity can be exhausted before breakthrough occurs if flow patterns through to the active micropores permit. Nearly all practical bed materials have this characteristic.

The 'isotherm' of an absorber is the graph showing the amount, Q, of a gas absorbed at equilibrium versus concentration, C, of that gas at a particular temperature. A favourable isotherm is one which does not show a linear relation of Q to C but shows a shape closer to that of an irreversible reaction, i.e. Q grows rapidly at small C but then saturates at higher values of C. Such an isotherm will tend to sharpen the concentration front passing through an absorber bed, as high-concentration fronts move more rapidly than low-concentration fronts, so that an initial sloping front will tend to grow into a very sharp 'shock wave' and approach a step change in concentration.

A commonly used isotherm is that named after Langmuir. Langmuir based his simple theory on an equilibrium between adsorption and desorption in the following way. Kinetic theory gives $v = P/((2\pi mkT)^{0.5})$ as the collision frequency of a gas. Rate of desorption is a thermally activated process with an activation energy given by the heat of adsorption (ΔH) of the gas. Hence rate of desorption is proportional (with constant b) to fractional coverage of adsorbate (f) on the material surface multiplied by a Boltzmann factor $e^{-\Delta H/kT}$. Thus, if a is the sticking probability of a gas molecule hitting the surface

$$av(1 - f) = bfe^{-\Delta H/kT}$$

or

$$f = BP/(1 + BP)$$

where

$$B = \text{Langmuir constant} = a\exp(\Delta H/kT)/(b\surd(2\pi mkT))$$

and m is the molecular mass. Figure 2.16 plots Langmuir isotherms for a typical molecular sieve absorber at two different temperatures.

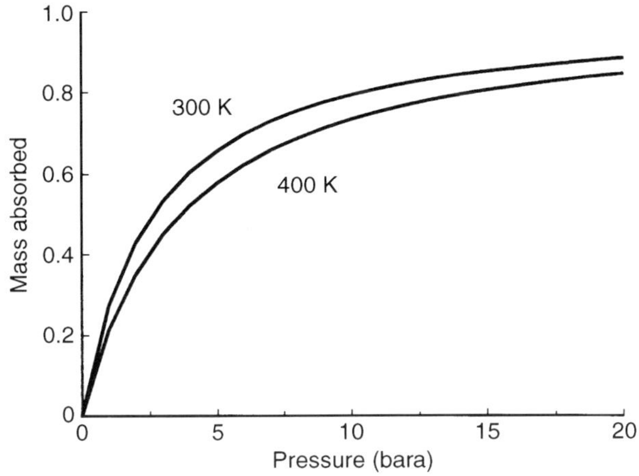

Figure 2.16 Langmuir isotherm plots.

With an isotherm of this shape, clearly higher concentrations will be less absorbed relative to lower concentrations, as required to sharpen the concentration front. In practice, absorber beds do not show the sharp fronts expected from this simple theory because

- flow into the micropores of the absorber is not instantaneous;
- absorbers are prepared in rather large pellets, with large voids between them, to avoid pressure drop;
- the bed will tend to heat up as adsorption takes place;
- zeolites have two levels of microporosity;
- pellets will not in general be identical spheres (the usual assumption in modelling).

One conceptual model, which approaches more nearly the practical situation, considers the absorber bed to be a set of well-mixed tanks connected by pipes. In this way a simple numerical model can be built which can be tuned to match empirical measurements in the laboratory or on small units, the model being then usable to help in the design of full-scale units.

Many absorber beds in practice show S-shaped concentration fronts (Figure 2.17). These quite typically show some broadening as the front progresses through the bed, despite the sharpening effect of the typical favourable isotherm. In long beds, however, a constant MTZ (mass transfer zone) develops, representing an equilibrium between the front-broadening due to mass transfer within the pellets of adsorber and the front-sharpening due to the isotherm. In this case, a constant S-shaped curve is seen progressing through the bed. Often an effective MTZ is assumed in order to carry out approximate calculations.

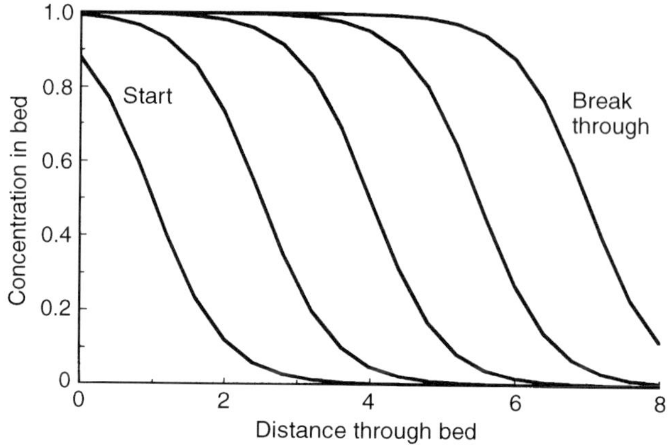

Figure 2.17 Concentration front curves progressing through a PSA bed.

PSA design considerations. A PSA must regenerate its adsorber material, hence, in addition to knowing when impurities will break through an adsorber bed, it is also necessary to know when desorption of those impurities by vacuum, purge gas pressurisation or flow is complete. An essential part of the economics of PSA is having a sufficiently short regeneration time for the adsorber bed whilst not wasting excessive amounts of compressor energy or feed gas.

The heat of adsorption means that a PSA bed is not isothermal. A typical adsorber may rise in temperature by 100°C or more during adsorbtion and fall by a similar amount during desorbtion. (It has even been proposed to use the cooling effect in a PSA unit as a kind of refrigerator; one having the advantage of using no CFCs, of course.) Also, the compressed supply may be not cooled completely to room temperature. To a large extent, the non-isothermal nature of the bed can be controlled simply by increasing the size of the beds relative to their total possible capacity or, equivalently, choosing a much shorter cycle time than would be indicated. This is the approach generally taken.

Other approaches to non-isothermality are possible, however. Relatively small heaters installed near the feed end of beds help as this part of the bed tends to cool below ambient, limiting desorption efficiency during depressurisation. Another approach is to mix the adsorbent with an inert substance with a high heat capacity. Candidates are those which are conveniently stable and inexpensive, and have not too low a thermal conductivity. Iron has been used but non-microporous alumina, magnesia or aluminium would appear to be candidates, with alumina and magnesia offering safety advantages over iron with oxygen-containing gases, and aluminum offering better thermal conductivity.

Uneven distribution of flow can seriously degrade a PSAs bed concentration profile, even in the presence of a favourable isotherm. The packing of the bed must be done very carefully not only so that it gives even flow but also so that it does not move, even during the pulsed gas flows intrinsic to PSA operation. Interpellet voids are unavoidable; they result in a sizeable dead-space volume with a typical absorber. It has occasionally been proposed to include, along with the large pellets, smaller pellets to fill in between the larger pellets. With careful control of the relative numbers of larger and smaller ones, it would be possible to decrease dead-volume substantially while still leaving good flow characteristics. (It is well known in road construction that a combination of coarse and fine aggregate roadstone will settle to a much higher density than either on their own whilst still maintaining porosity.)

Additional small purge flows after completion of depressurisation help to remove residual impurities and improve purity in PSA units. The slight loss of product recovery is more than compensated for by the increased purity of product.

The PSA unit as described needs an intermittent supply from the compressor and itself provides an intermittent stream of product. In piston engines the intermittency of output can be dealt with by linking several cylinders, which provide a sequence of closely spaced power pulses, and by a flywheel to store some energy from one power pulse until the next pulse. PSA units use both these approaches. They may employ an air receiver vessel to store compressed gas from the compressor at the inlet and a product receiver vessel to store product. Almost all practical units use at least two adsorber vessels; one which is providing product whilst the other(s) is being repressurised or purged. The two-bed system can be regarded the mainstay of PSA plant and the standard by which other plants are judged (Figure 2.18).

The sequence of operations in a PSA plant is an important part of its optimisation. Despite the apparent complexity of many large PSA sequences, however, there are really only four essential steps going on:

1. pressurisation/production (on-line bed)
2. vent
3. purge to lower partial pressure of adsorbed species at 1 bara total pressure (or use vacuum)
4. backfill to push remaining unwanted gas out

(5. pressure equalisation as an energy conservation measure makes five steps).

Large improvements in efficiency can be obtained for gas separations by increasing the number of beds and suitable switching arrangements between them.

With more beds there is less variation in load on the compressor and less

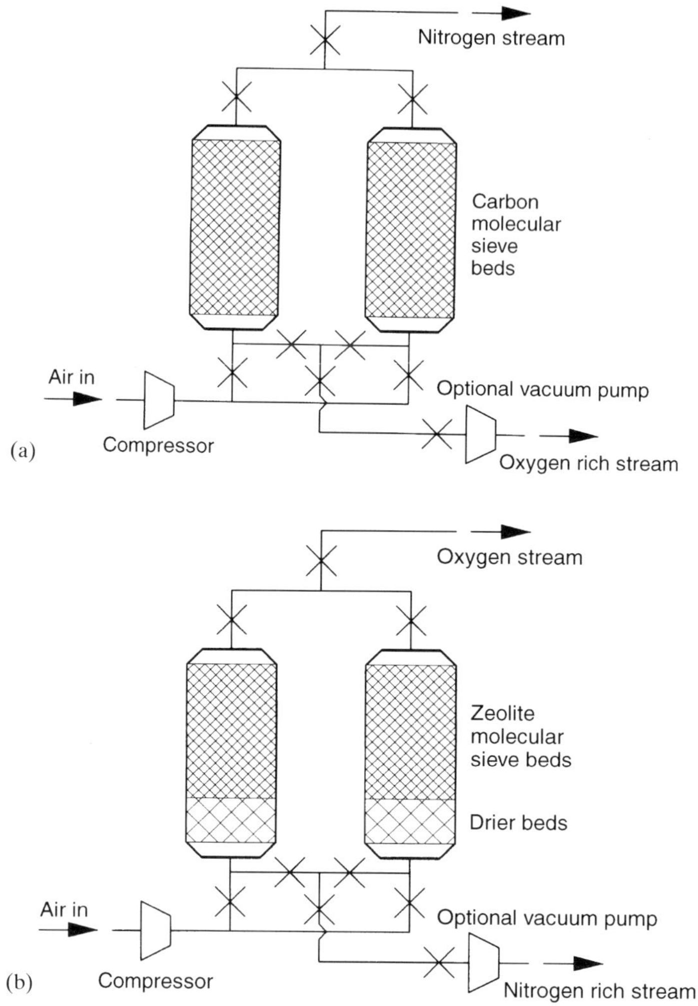

Figure 2.18 Simplified two-bed pressure swing absorber air separator. (a) N_2; (b) O_2.

variation in product supply pressure, obviating the need for surge vessels. There is also the possibility of including multiple pressure equalisation steps in the PSA cycle. Pressure equalisation economises on compressor power by raising and lowering bed pressure in steps by connecting beds to each other instead of only to the inlet, waste and outlet streams. In a four-bed system (Table 2.5) each bed is both lowered and raised in pressure in two steps. Both energy requirements and product recovery are enhanced relative to an equivalent unit with two beds. The steps 1 to 9 are of different lengths and

Table 2.5 Cycle of operations of a four-bed PSA gas absorption system (Yang, 1987)

1. adsorb in co-current direction (bed on line) (pressure in bed is at line pressure, maximum)
2. equalisation in co-current direction (pressure falls)
3. depressurise in co-current direction (pressure falls)
4. equalisation in co-current direction (pressure falls)
5. depressurise in counter-current direction (pressure falls to minimum)
6. purge in counter-current direction (pressure in bed is at a minimum)
7. equalisation in co-current direction (pressure increases)
8. equalisation in co-current direction (pressure further increases)
9. repressurise in counter-current direction from on-line bed (pressure increases up to line pressure)

are phased by microprocessor sequence controllers so that one bed is always on line whilst the other beds are purging or equalising.

Hydrogen PSA. Hydrogen extraction was the first major process, after the now ubiquitous PSA drier (considered in section 2.2.11), to switch to using a PSA process to a significant extent. Hydrogen is particularly easy to separate using PSA because of its low molecular weight. The hydrocarbon gases and CO_x mixtures it is found in are all easily absorbed by molecular sieves, leaving hydrogen in the gas stream. Because hydrogen is typically extracted in very large units (e.g. steam reformer plants, catalytic crackers in oil refineries and ethylene plant by-product streams which are all often in the range of 1000s of tonnes per day), these are large and sophisticated multi-bed units.

The PRAXAIR hydrogen PSA process uses beds packed in the first section with activated carbon for dealing with the highly absorbed species (carbon dioxide and water) and the second section with 5 Å zeolite for dealing with the less highly absorbed species (such as methane and carbon monoxide). Product purities can exceed 99.999% on 70% hydrogen feedstock and units with a million cubic metres per day (100 tonnes per day) capacity are operating.

Nitrogen and dynamic adsorption PSA. Nitrogen PSA separation relies on carbon molecular sieves whose adsorption rate for nitrogen is different from that for oxygen. It is important to note that carbon molecular sieves have the same equilibrium absorption properties for nitrogen as they do for oxygen. In order to separate efficiently nitrogen PSAs operate quickly relative to the gas absorption rate and never approach the equilibrium state where their selectivity factor reduces back to zero.

A typical PSA plant of a few tens or hundreds of cubic metres per hour employs an oil-inject screw compressor as its motive power. The use of oil-injected screw compressors and other 'wet' compressor types carries some risk with it because oil can be carried over into the PSA carbon bed where it soon poisons the adsorber. However, these compressors are most efficient in smaller sizes; precautions such as additional preventative maintenance measures, additional oil traps and cut-out devices are therefore indicated.

Small-scale generation, a few litres per minute of nitrogen, is possible with PSA, although no such systems have yet become at all common. One application which is a candidate for such micro-PSA systems is the production of nitrogen for the dispensing of beer (section 3.4.3). Another is the production of a small, steady stream of nitrogen for gas chromatography (particularly valuable for GCs which are used for continuous monitoring since it avoids tedious cylinder changing). By accepting a very low yield relative to the input compressed air, generous sizing of beds (relative to the very low flow rate of less than a litre per minute) and other optimisation, it is possible to produce a 1 ppm oxygen level nitrogen.

Oxygen and equilibrium adsorption PSA and VSA. Oxygen PSA relies on artificial zeolite molecular sieve materials which have, at equilibrium, a difference in their retention of oxygen and nitrogen. Oxygen PSA systems have a cycle time which is long relative to the gas absorption time constant, so that equilibrium is almost reached, maximising capacity and relying on the fact that their selectivity at equilibrium is still good. As mentioned above, oxygen systems are often simplified and rendered more efficient by using a vacuum pump to vent the beds after adsorption. These systems are referred to as VSAs.

Compressed air for use in a zeolite PSA must be reasonably dry, as all zeolites will preferentially adsorb H_2O until saturated and fail to function as oxygen separators. Fortunately, the compressor will in any case intrinsically remove the bulk of the water (see footnote in section 2.1.5) but further removal is still needed. Zeolites are, in fact, themselves frequently used in gas drying systems because of their strong but reversible adsorption of water. Zeolites are best heated to remove moisture effectively and heated bed driers (temperature swing absorption, TSA) are possible. The problem is solved if a twin-bed PSA drier, perhaps using silica gel or alumina, can be used in front of the oxygen PSA. A more common solution is to use drier materials such as activated alumina in the first part of the adsorber bed. The activated alumina has a more easily reversible binding to the water molecules than zeolite and therefore water is more easily desorbed during the blowdown and purge steps.

Many tens of thousands of oxygen PSA systems have been sold worldwide and probably 99% of these have been small systems for one special application; supplying a few litres per minutes of oxygen-enriched air for medical purposes. These systems are about the size of a small suitcase and require a power supply of 500 W or so, making them a thousand times smaller than a typical industrial oxygen system. Tiny systems such as this

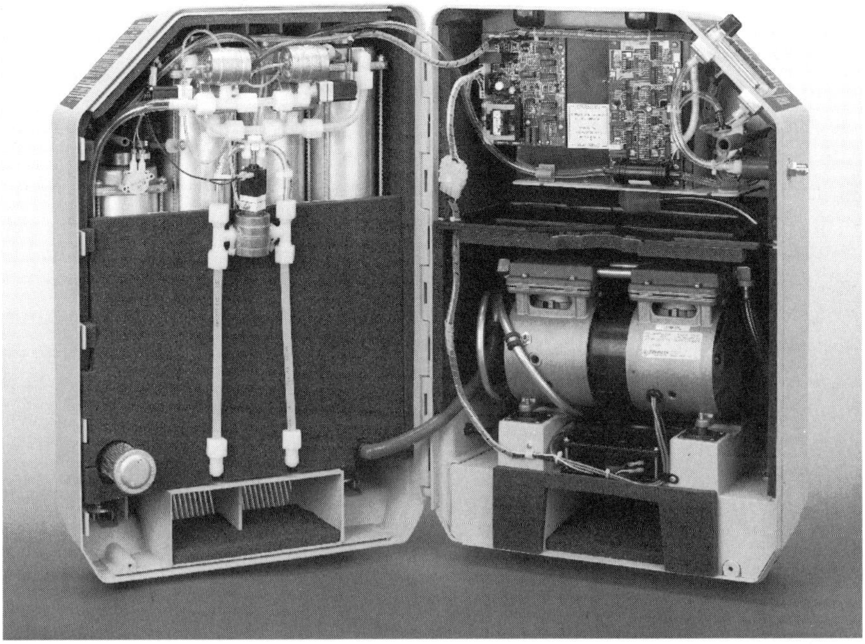

Figure 2.19 Small-scale PSA oxygen generator for medical use. The unit hinges apart as shown with, on the left, four large aluminium canisters containing molecular sieve adsorber and, on the right, the compressor, gas analyser and control units. Pneumatically driven poppet valves (five are visible on the left at the top) are used (courtesy of Puritan-Bennett).

must minimise complexity and hence cost; the whole system is typically \$2000 or so. A very simple implementation of the straightforward two-bed PSA cycle is used in many units, with two beds of zeolite protected by silica gel dessicant. Their performance is relatively poor compared to industrial units, with only 30% or so O_2 recovery, for example. A difficulty is the compressor, which must run continuously and very quietly, and yet supply 100 standard litres per minute of air. To minimise compressor cost and noise, low pressures are used, less than 2 barg in some designs, and the compressor is housed in an elaborately padded box. The purity achieved varies from only 50% to 95% or so, depending on the design, but even the lower figure is entirely adequate for the purpose. Figure 2.19 shows a typical medical oxygen concentrator.

As mentioned above, the most power-efficient oxygen systems employ a vacuum cycle purge. Compression to a low pressure, less than 1 barg and sometimes as low as 0.25 barg, is used to produce oxygen from a freshly regenerated bed at just above atmospheric pressure. This low pressure feed is acceptable only because purging/desorbtion will be at vacuum pressures.

The gas is then compressed if necessary. This is efficient as only the product (20% of the overall system gas flow) is compressed. The vacuum cycle removes nitrogen more efficiently than an atmospheric purge cycle (section 2.5.10) because of the large ratio between absorbing pressure and desorbing pressure, and because it is a better match to typical zeolite isotherms. Two-bed systems are common in VSA designs, despite their relatively large scale (up to 100 tons/day or so), since the improvements in power efficiency do not justify the additional capital expenditure on a more complex system.

To minimise the complexity and hence cost of very small systems, the Union Carbide Corporation, now PRAXAIR, developed earlier work which used a single bed. With surge vessels smoothing the varying load on the compressor and variations in the supply rate, an inexpensive system can be devised. The size of the whole system, bed and surge vessels, is minimised by using the fastest possible cycle time, i.e. just a few seconds. Gas throughput is very high for the unit size. The valve layout with one bed is very simple, with just two valves on the inlet side of the bed governing whether the bed is pressurising or depressurising/backpurging to waste, and the outlet surge vessel accumulating the less highly absorbed oxygen. The single bed must use a finer grade of pelleted zeolite; the 1 to 3 mm pellets normally used are too large. The fine pellets mean that the bed has a high pressure drop across it, which is required for the system to work. The single-bed system would appear to be well-matched to applications such as medical oxygen generation.

Practical PSA considerations. The control of a PSA unit can be as simple as a timer which switches a spool valve from one position to the other at fixed intervals. The use of a single complex spool valve appears to have advantages in PSA designs because such a single valve can provide all the gas switching with, as well, low-pressure drops. For many high-production rate, low-purity plants, this may well be the correct choice. However, spool valves are probably less reliable in general than poppet valves. Spool valves have large numbers of sliding seals and are made in small numbers to special designs; the converse of poppet valves. Leaking valves are only important, of course, where the purity of the product is important but purity is often important for nitrogen units. The use of poppet valves also enables the maximum advantage to be gained from modern programmable logic controllers. A PLC can drive many valves at different timings, optimising the PSA efficiency, and can even be used to change timings dependent on product stream needs.

PSA beds are intrinsically subject to large transient flows of gas as beds switch, although the multibed systems with one or more pressure equalisation steps are superior. During these brief pulses of flow, the adsorber materials can move slightly. Once this starts to happen, the pellets tend to abrade and generate fines, and become loose so that they can abrade more.

The gas inrush causes fluidisation eventually, the whole bed starts to abrade and, in the long term, to break down. Furthermore, channels may form in the bed causing inefficient distribution of feed gas to the adsorber. To restrain these undesirable effects sieve materials must be carefully chosen for good crush strength and packed with some compressive stress built in.

There is a trade-off between purity and production rate in PSA systems. If some of the adsorbate is discarded, then purity can be improved. However, as adsorber materials improve, then the trade-off shifts to higher purity or higher efficiencies. It is important to recognise that economically effective PSA systems are only currently useful for relatively low-purity nitrogen and oxygen. (Although, as noted in section 2.4, they can achieve much higher levels of purity on hydrogen, and low yield systems may achieve high purity.) Oxygen of purity better than about 95% or nitrogen of greater than about 99.9% are currently probably best derived from delivered cryogenic liquids. Alternatively, the additional costs and complexity of a downstream gas purifier unit must be accepted.

Today, much computer modelling can be done to guide PSA design and operation; both at the micro-scale of absorbers and gas passages within them and at the macro-scale of sequencing of valve operation and controlling flow rates during the sequence. The importance of modelling is that it allows the scaling of plant; custom-made PSAs for nitrogen and oxygen are now being made at the 100–200 tonnes per day level and this maximum size has grown by a factor of two every four or five years. PSA units for lower purity gases from air separation are up to twice as efficient in energy terms as cryogenic air separation, so there is a strong driving force behind this size growth. The early role of sieve supplier companies in computer modelling and experimental work, like Bergbau Forschung in Germany, will most likely be overtaken by the gas companies as PSA grows to be more and more a core technology for the latter.

2.1.7 Membrane separation of gases

The membrane is the simplest conceivable device for gas separation. The ubiquity of filters to remove dust particles from gas leads naturally to the assumption that a similar separation can be achieved for gas molecules simply by decreasing the size of the filter holes. However, it turns out that using filter membranes for gas separation is much more difficult. Gas molecules are only fractionally larger than the atoms of which the membranes must be made. Sieving membrane filters are impossible because of the impossible thinness of the membrane required, so membranes can at best act as deep filters. However, one gas molecule is very much the same size as another, whereas a dust particle of 0.1 μm diameter is hundreds of millions

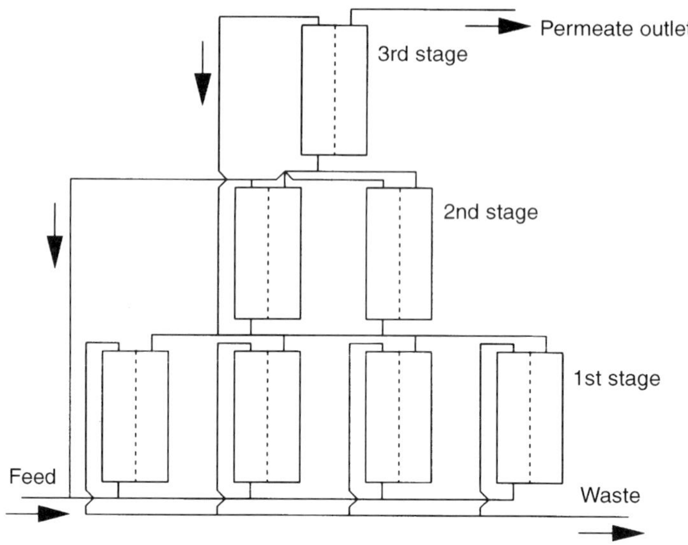

Figure 2.20 Cascaded gas separation membranes.

of times as big as the gas from which it is filtered, so filtration mechanisms are quite different.

Uranium and membranes. Membranes have a long history of experimentation but very little history of actual practical application. One of the most famous examples of separation, and the first major application for membranes in gas separation, was the giant Manhattan Project of the Second World War. This employed dynamic separation across semipermeable membranes to prepare one isotope of uranium, U^{235} for the first atomic bomb. The separation factor was very small indeed; just that due to the difference in speed of the two uranium hexafluoride molecules, which is proportional to the square root of the ratio of their masses, 1.00429.

To obtain useful separation, and to avoid throwing away most of the U^{235}, a cascade of membranes was used, with the permeate and retentate of the first unit being led away to two subsequent units whose permeate and retentate were in turn led away (Figure 2.20). Even so, the separation achieved was highly incomplete.

Fortunately, only partial separation of the isotopes was needed and a large amount of money was available for the job. The resulting plant cost billions of dollars in today's money, covered square miles and required hundreds of megawatts to run it. A modern successor to the Manhattan Project is the French plant at Pierrelatte. This leviathon amongst chemical plants actually requires the power of four nuclear power stations to run it, whilst each of the pumps propelling the UF_6 through each membrane is of megawatt size.

Nitrogen from membranes. Fortunately, the production of nitrogen via a membrane method is considerably easier than the separation of uranium. Rather than using a semipermeable membrane, air separation uses completely sealed but very thin polymer membranes which are much more efficient. However, nitrogen has a very low market price compared to enriched uranium and it is only very recently that economic plants have become available.

The breakthrough that has enabled cost-effective nitrogen production is the advent of chemically selective membranes. Rather than relying on the small difference in molecular weight or other physical properties, as the uranium separation does, selective membranes achieve orders of magnitude better separation in a single stage by use of a chemically active membrane.

A semipermeable membrane relies on 'microporous' structure. However, the micropores are in fact much larger than the gas molecules being separated. They simply ensure that no bulk flow of gas takes place through the membrane. Then the transit of gas across the membrane simply depends on molecular speed. An active membrane works by allowing the gas to dissolve, diffusion then causing the gas to migrate through it. Separation is both by means of the differential solubility of the gases and by their different diffusivities.

Gas flow, rate F, across a membrane is given by Fick's law

$$F = P_e A(P_i - P_f)/t$$

where P_e is the membrane permeability, t its thickness, A its area and $P_{i/f}$ the initial and final pressures. Hence membranes of the maximum area and smallest thickness are needed, whilst large pressure drops are inevitable.

A typical membrane separator has a thin (1000 Å) active but not very permeable membrane supported on an easily permeable non-selective (microporous) membrane. This kind of structure is known as an 'asymmetric membrane' and was first developed in the 1960s using dense (selective) cellulose acetate on low-density microporous cellulose acetate (non-selective). The asymmetric structure enables the use of the thinnest possible active membrane and enables the use of mechanically weak selective membrane polymer. Membrane areas are claimed for fibre modules which have 10 000 m^2 of membrane for each cubic metre of module volume. Membrane chemistry is shrouded in commercial secrecy but many membranes are simply polymers such as unplasticised PVC, cellulose acetate, polysulphone or silicone. A few results are now published on metal–organic complex compounds. Polysulphone and acetate membranes give oxygen permeabilities four or five times that of N_2 and this is typical of the simpler commercial membranes, although these have greater permeabilities for both components. Permeabilities increase with temperature according to an Arrhenius activation relation, with activation energies in the range of 1 to 3 kJ gmol^{-1}

$$P_e = P_{eo} \exp(-E_a/kT)$$

where P_e is permeability, P_{eo} is permeability at a reference temperature, E_a is the activation energy for the process, k is Boltzmann's constant and T is absolute temperature.

There are thus considerable increases in permeabilities due to temperature and increasing temperature of operation from 20 to 65°C, for example, would give, with an activation energy of 40 kJ mol^{-1}, a ten times greater throughput. Typical relative permeabilities for plastics are

H_2	N_2	O_2	CH_4	CO_2
10	0.2	1 (reference)	0.2	6

Within these relative permeabilities, which vary surprisingly little for most plastics, there is a millionfold variation in absolute value of permeability; specially made 'barrier' films, such as polyvinylidene chloride, have permeabilities which are 1/2000 times that of polyethylene polystyrene or Teflon, whilst dimethyl silicone rubber has a permeability which is a hundred or so times greater than these. Silicone and natural rubber are anomalous in that they do not show a narrower range of relative permeabilities than the ratios shown above. There is an unhelpful trend that high relative permeability ratios are seen only in low permeability polymers. Hydrogen, for example, permeates only a factor of two faster than oxygen on rubber. Some oxygen permeabilities for commercial polymer films (in 10^{-10} m^2 day^{-1} bar^{-1}) are:

Rubber	Low density polyethylene (LDPE)	High density polyethylene (HDPE)	Nylon 6,6	Poly-vinylidene chloride (PVDC)	Polyacrylo-nitrile (PAN)
18000	1900	430	30	5	0.16

Commercially, nylon, polyvinylidene dichloride (PVDC) and polyacrylonitrile (PAN) are regarded as barrier polymers. Permeabilities are sometimes measured in units of cm^3 cm^{-1} cm^{-2} s^{-1} cmHg $\times$ 10^{-10} at standard temperature and pressure. This conveniently gives oxygen permeabilities a number little bigger than unity for the common plastics and only a factor of few hundred for silicones. The presence of the factor 10^{-10} is a reminder that permeation is not an easy process.

These figures also have relevance for the design of many other artefacts. In food packaging (section 3.4.3), it is clear that carbon dioxide will be lost quickly from packaging whilst nitrogen can largely be retained. For food packaging membranes permeability of oxygen is often quoted and labelled 'oxtran' (oxygen transmission).

Membrane separation is highly effective for the separation of heavier organic or toxic components from air; a fact that is occasionally employed in instruments for the monitoring of toxic gases e.g. mass spectrometric analysis on molecules passing through a silicone membrane. One such instrument (VG Instruments, Middlewich, UK, model PETRA) claims thousandfold or more enhancements of toxic trace gas component peaks using the technique.

The figures also make it clear that membrane separation of H_2 from air or methane is straightforward. This separation was the subject of the first commercially successful selective membrane process, the Monsanto PRISM process, based on a polysulphone selective membrane. The PRISM membrane is an asymmetric membrane moreover, although a crude separation could be achieved using a simple membrane. Monsanto membrane units are already in service carrying out H_2 reclaim from ammonia plant gases and more recently for He recovery from diver's breathing mixtures used in offshore oil production work. It is also clear that the removal of CO_2 from, for example, fuel gases, to upgrade their calorific value, can be carried out easily given the very high permeation rate of CO_2.

There is, as in PSA, a fundamental compromise which has to be made in membrane separation: to achieve high purities the membrane must be absolutely pinhole free and this means that it must be thick. However, to maximise throughput of gas permeate the membrane must be as thin as possible. Only recently have selective membranes become available that satisfy these requirements to a sufficient degree. It remains, however, a difficult technology; a typical gas-cylinder size separator has up to 1 million fibre strands, any of which could leak, reducing the assembly performance.

A typical membrane separation unit (Figure 2.2.1) comprises bundles of hollow membrane tubes or 'fibres', which can be from 40 μm to 0.2 mm in diameter, none too big to pass through the eye of a small sewing needle, mounted in end blocks inside a cylinder. The fibre manufacture is largely shrouded in commercial secrecy but available fibres, such as those from DOW, are based on well-known principles of the asymmetric membrane. Dealing with pinholes in the selective membrane is the subject of some secret commercial activity but, to judge from recent claims, has been largely overcome.

A membrane unit in general works as follows. High pressure raw gas is continuously admitted to the outside of the fibres at one end of the cylinder whilst at the other end the depleted waste stream is exhausted from the fibres. The product is drawn off from the fibre mounting block. The hollow fibres consist of highly permeable non-selective polymer fibres coated with a thin coating of selective membrane. Typical inlet pressures vary from a few bars to a few tens of bars.

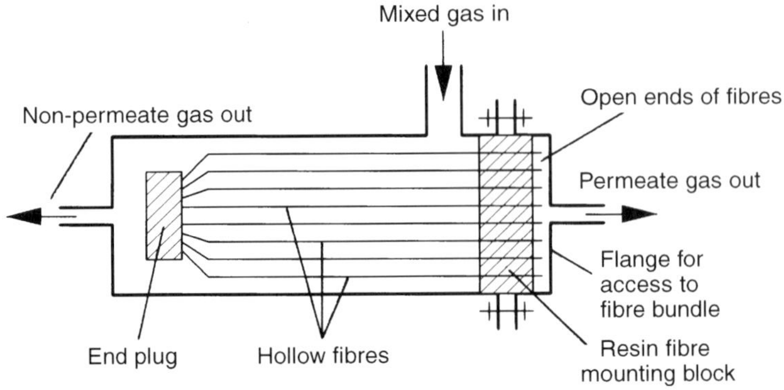

Figure 2.21 Typical membrane gas separation unit.

In the case of nitrogen generation from air, nitrogen is less permeating than oxygen. Hence oxygen is the 'product' in the above description and nitrogen is the 'waste stream'. Clearly, by selection of flow rates, and perhaps by the addition of purge streams (for greater nitrogen purity, a purge stream of air down the fibres carrying oxygen permeate away helps), the process can be optimised. The use of counter-current can also offer some improvements in performance.

A single membrane separator is often followed by a further membrane separator in series, to improve purity or decrease product lost, as current selection factors for single stages are insufficient in many applications. A cascade of membranes can be used, as in uranium separation, if necessary. Standard membrane separators are available for nitrogen, with a membrane selectivity of 6:1 or better. As with PSA, recovery, percentage of the feed air output as gas, depends on the purity of output required. For example, with membrane selectivity of 5, and 99% N_2 purity, a recovery of 30% is possible, or with 90% N_2, recovery of 70%. For a selectivity of 10, which may be available in the future, recovery would rise to 50% on 99% N_2.

For low purities, membranes are showing great promise with comparable recovery to PSA. Prices for membrane units are still high but are expected to fall with time. With lower cost membranes in the future, membrane systems will become more popular, particularly in small applications where their simplicity is particularly appealing. Isalski (1989) suggests various schemes for combining membranes with cryogenic separation, using the membrane to carry out a coarse bulk separation on the feed gas, followed by a cryogenic unit to give high purity output gases. These more complex schemes will depend on membrane units reducing in cost considerably, as they are only make sense for large-scale systems. The Fluor company promotes a fuel gas + ethane separation system, using membranes to reject CO_2 along with cryogenic distillation, however, so at least some schemes are economic even today.

Gas separation by centrifuge. Membranes are unlikely to be used for separation of uranium isotopes in the future as more efficient methods based on centrifugal separation are available. UF_6 is put into a set of devices looking remarkably like spin-dryers except that they are several metres across and revolve at speeds which would cause ordinary materials to fly apart. At the many hundreds of g achieved in these machines stage separation factors several times larger than the membrane method are available, with the additional advantage of lower losses per stage. Practical gas centrifuges employ an axial counter-current flow, achieving the equivalent of many stages in each rotating cylinder; in effect the system can be considered to be a set of disc-shaped elements connected rather like a distillation column. The feed gas is sent to an appropriate place in the middle of the cylinder and streams enriched in light and heavy components extracted from the periphery and the axis of the cylinder at opposite ends.

There seems little chance that a centrifuge method, conceptually simple as it is, would have anything to offer in, for example, air separation, because the separation factors with inexpensive centrifuges are too close to unity to be useful. A simple analysis shows that the separation factor is proportional to the difference between molecular weights, so separating N_2 and O_2, which differ by four units, is barely easier than U^{235} and U^{238}, differing by three. Stable isotope separation of gases may be a more practical application for the technology.

There have been proposals, however, for carrying out distillation in a 'column' inside a spinning centrifuge. It is easy to show that such a column could be made a very small fraction of the size of a conventional gravity column. US military work pursued a project to make such a compact centrifugal column and this was more or less technically successful. However, the complexity of feeding liquid cryogens in and out of a spinning column, and controlling the process, makes it unattractive except for specialised military applications.

2.2 Making other gases

In many cases, the manufacture of non-air gases is very similar to that of air gases, except that separation is from an industrial waste gas stream rather than the atmosphere. Cryogenic distillation of waste streams is rare, however, because waste streams rarely consist of low-boiling gases like nitrogen and oxygen. However, PSA and membrane separation methods are common. In mixtures of hydrogen with other gases, for example, both PSA and membrane methods are particularly simple and efficient, because the H_2 is so far from the other components, such as hydrocarbons and CO_x, in its properties.

2.2.1 Gases from chemicals

Many schoolchildren finish their studies in chemistry under the impression that gases are mostly made by chemical means. In fact, chemical generation is not a standard means of production for most gases in industry. It is nevertheless important.

Chemical generation of oxygen. Chemical reaction-based schemes for separating oxygen from the air are still occasionally considered today for special purposes, for example in purifier use. As well as the barium process, there are possible cycles using high temperature to release oxygen absorbed at lower temperatures based on nitrate/nitrite changes and manganese oxide/dioxide changes. MOLTOX is an example of this kind of process, using alkali nitrate and nitrite transitions at 500 to 650°C (Dunbobbin, 1987), although it is still under development. MOLTOX might be very efficient if integrated with a process (a steelworks, perhaps) which required operation at elevated temperature.

Less energetic reactions with oxygen have been sought in order to lower the regeneration temperature in a process. Complexes involving transition metals (of which, of course, the haemoglobin used in nature is one) are all possible, including those based on copper (cuprammonium), iron and manganese. Organo-metal complexes are possible, as are compounds based on precious metals. Any compounds which form a very weak bond with oxygen are candidates for a reversible chemical separation process, although the strong oxygen–oxygen bond must be broken and remade, causing problems for many processes.

Oxygen is occasionally supplied for emergency breathing use via chemical generation, for example from chemicals such as certain perchlorates or peroxides which react with water to release oxygen, although the purity and smell of chemically released oxygen is difficult to control and the gas may need cooling from some mixtures before it can be breathed. In confined spaces the release of carbon dioxide, as well as the lack of oxygen, is a danger to health. Sodium peroxide used to be supplied for use in submarines, for example, as 'oxygen candles', as it has the additional advantage of absorbing carbon dioxide as well as releasing oxygen.

Nitrogen from inert gas generators. The term 'inert gas generator' is usually taken to mean a system for burning a clean fuel in a precisely metered supply of air, followed, if necessary, by condensing of water vapour. The inert gas produced by such an arrangement will vary in its carbon dioxide content with the carbon in the fuel and will contain impurities reflecting those in the fuel. However, with good control of the burner, the composition should be 90% nitrogen with 10% carbon dioxide. The production process is relatively efficient because around nine volumes of inert gas are produced for every volume of fuel gas.

Inert gas generators are most commonly used with careful stoichiometric control of the oxygen supply so that the oxygen content is minimised, whilst ensuring that little fuel remains in the gas stream. The use of very clean fuels, methane is most common, improves the quality of inert gas generator output over that obtained by combusting heavy oils of variable composition and high sulphur content. Residual fuel, nitrogen oxides, carbon dioxide and sulphur compounds will all, however, still be present in trace quantities and may cause process problems or corrode the pipework.

Inert gas generators are only nowadays used in processes with minimal requirements on gas quality and a steady demand pattern, and then only when capital and running requirements must be minimised and the size of gas stream needed is not too large. In the absence of many favouring factors, the use of cryogenic nitrogen, either on-site or in bulk liquid tanks, is preferred. In applications requiring a reducing gas, an inert gas generator can be run with excess fuel. This system is often dubbed an ‘endothermic’ generator and is typically used in heat treatment of metals (section 3.2.2).

2.2.2 Carbon dioxide

Smokestack extraction. In some ways the most obvious source of CO_2 is exhaust stack gas. Extraction from exhaust stack gases in which, after N_2, CO_2 is the biggest component is certainly superficially attractive. However, there are quite considerable problems. The stack gases are generally hot, dusty, humid, oxidising and highly acidic. Small amounts of sulphur in feedstocks for most burners result in sulphur dioxide and trioxide, and sulphuric acid. Some processes produce particularly clean exhaust streams, some refinery combustion products, for example, and these are generally preferred over extraction from general furnace and boiler exhaust streams.

Increasingly today, in large stacks, many of these objectionable impurities are filtered or scrubbed out as an anti-pollution measure. With a cleaner gas stream CO_2 extraction is much more economical. However, the way in which scrubbing is done affects the CO_2 content of the stack gas.

After prewashing, compression and water separation, a waste stream is turned into pure CO_2 via a sequence of processes which varies according to source and use purity needs. Charcoal beds are commonly used to absorb remaining contaminants followed, if necessary, by cooling to liquid and distillation. The condensation to liquid eliminates the large amounts of permanent gas contamination (nitrogen, for example) and crude distillation can remove residual unburnt fuel components. The resultant liquid carbon dioxide is stored under 20 bar or so pressure and at a reduced temperature (−20°C) in insulated tanks.

Ammonia plant extraction. Ammonia plants are an important source of relatively pure carbon dioxide for the industrial gases business. Ammonia plants need vast quantities of nitrogen and hydrogen in the stoichiometric ratio of 1 part nitrogen to 3 parts hydrogen. Such large volumes of hydrogen are a problem. The only economic method is generation via steam/methane reaction unless a nearby petroleum refinery has hydrogen supplies available.

The reaction produces a mixture of H_2, CO and CO_2 from which fairly pure H_2 must be extracted for the ammonia synthesis. The reaction can be biased towards H_2 and away from CO production by high steam–methane ratios. Alternatively, the nitrogen required can be generated by partial oxidation of the methane with air, which can be arranged to leave a stream which contains nitrogen and hydrogen in 1:3 ratio with mainly CO_2 by-product.

The CO and CO_2 rejection which has to be done provides a natural by-product stream which, with little further processing, can be liquefied and stored. The CO_2 extraction is most often performed by an acid gas wet scrubber using simple aqueous ethanolamines or specially designed molecules such as sulpholane. (In gas streams with hydrogen sulphide, methyldiethanolamine and other tertiary amines are used now as selective absorbers for H_2S which leave the CO_2 in the stream; this can be convenient, not least because H_2S solutions are highly corrosive to steel plant equipment, forming iron sulphide.) The feed gas is passed up a tower with counter-current absorber where the CO_2 is stripped into the liquid. The liquid is then heated, typically with steam, to remove the CO_2 and the liquid recycled. This simple description ignores, of course, difficult details such as the effect of other sulphur compounds, corrosion, foaming, and degradation of the scrubbing liquid.

2.2.3 Acetylene, ethylene and unsaturated hydrocarbons

Acetylene. Acetylene is still largely made by the addition of water to calcium carbide. (There are alternative methods from petrochemical sources.) The carbide is produced in electrical arc furnaces where calcium oxide (lime) and coke or anthracite (carbon) are heated to 1850°C by arcs between three carbon electrodes connected to a powerful (multiple tens of megawatts) three-phase source. The furnace is a metal shell lined with firebrick and, at the bottom, carbon to resist the red-hot molten carbide. The furnace can be run almost continuously, with further additions of lime and coke, while molten carbide can be run off into moulds. The resultant impure calcium carbide is broken up when cool into the once-familiar carbide chips. Carbide furnaces emit substantial amounts of noxious fumes, smoke and noise, and require low-cost electric power, typically hydroelectric. Today, more effort is made to recycle some of the carbon monoxide evolved when the reaction occurs

$$CaO + 3C \Rightarrow CaC_2 + CO$$

The carbide/water reaction can be carried out with a large excess of water, in which case the reaction occurs smoothly and safely at low temperature. The by-product, however, is a large volume of unpleasant lime sludge. Some plants mix a roughly stoichiometric amount of water with the carbide and apply cooling and agitation to avoid overheating; the pay-off being that only a small volume of fairly usable impure lime is produced which can be used once again in the carbide furnace, for example. Various impurities, including such dangerous gases as arsine, phosphine, silane and hydrogen sulphide are also produced by the carbide/water reaction, deriving from impurities in the original materials to the carbide furnace, and must be absorbed, typically by an acid wash tower to absorb hydrides and basic impurities, and then an alkali tower to absorb acid impurities such as hydrogen sulphide. Finally, the acetylene is ready to be pumped into special cylinders.

$$CaC_2 + H_2O \Rightarrow CaO + C_2H_2$$

The acetylene process has been associated with pollution in a number of forms and even today is still a potential source of environmental problems. First, the carbide furnaces are a source of fume and smoke, which are difficult to suppress. Add to this that carbide furnaces are usually placed near to cheap hydroelectric power sources, which tend to be in areas of natural beauty. Finally, the typical by-product of acetylene production from carbide is a lime sludge, which is difficult to dispose of because it has a high water content (and thus low lime content) and is often foul smelling as a result of impurities in the feedstock.

Alternative to the carbide route to acetylene are processes that aim to convert hydrocarbon feedstock. There is an electric arc process producing acetylene and hydrogen by heating methane to over 1500°C in a multi-megawatt electric arc for a few milliseconds followed by quenching. A more typical alternative process adds preheated (600°C) methane and oxygen and burns them, following the flame by a water quench system. With the correct very rich flame, the reaction products are 50% H_2, 25% CO, unreacted and fully combusted products, and less than 10% C_2H_2. This can be washed out by absorbing it in dimethylformamide in a counter-current scrubbing tower, leaving the unwanted gases to be used as intermediates or burnt as fuel gas. The acetylene is flashed off from the dimethylformamide by steam heat.

Absorbers for acetylene in cylinders. Unfortunately, acetylene is unstable under pressure of more than a few tens of bar, or in liquid or solid form, and can detonate explosively and disastrously[11] (section 3.2.3). Acetylene dissolved in an organic solvent under pressure is stable, however. Typically the gas is dissolved in acetone, which is in turn held in a porous spongy solid. Organic solvents, such as acetone and tetrahydrofuran, are well known for their ability to contain large volumes of dissolved gas. The large volume of acetylene, $0.2\,kg^{-1}$ at 18 barg, which can be dissolved in acetone was a key factor in allowing the convenient and widespread use of acetylene in welding and cutting.[12]

The stability of the solution is enhanced if the solution is absorbed onto a microporous spongy substance such as kapok plant fibres, asbestos mineral fibres, a porous plaster mix or a silica such as kieselguhr. There is a further complication in that, in order that the acetylene will bubble out of solution steadily during gas usage, nucleation centres may need to be provided for bubbles to form. These might be small pieces of charcoal. The porous mass should also have larger passages in its structure to allow the flow of gas from remote parts of the cylinder to the valve. A medium with all the ideal properties is not available so a compromise choice has to be made, perhaps explaining why there are a number of different alternative materials in service.

Nevertheless, the search goes on for safer or more efficient dissolved gas packaging for acetylene (and potentially for other gases). Acetone has a low boiling point (56°C) and a very low flash point, making it a dangerously inflammable substance to handle. Dimethylformamide has been tried as an acetylene absorber and, although much more expensive, has a good solubility for acetylene and is safer. Its high flash point and high boiling point (150°C) make it fairly safe to handle, and its low vapour pressure means that less solvent is lost by carryover of vapour with the acetylene.

Solid adsorbers are a possibility for acetylene storage although none has yet achieved any commercial success. Clathrates, specially synthesised cage-shaped molecules such as the silicate-based molecular sieve materials or the organic-based hydroquinone compounds, have often been proposed

[11] 'Friable buildings' are commonly constructed around acetylene plant. The idea is that in the very unlikely event of an explosion the blast will be absorbed by the disintegration of the building. Furthermore, often the roof of such a building is designed to offer little resistance to blast, which thus tends to be dissipated upwards, whilst heavy walls deter lateral blast effects.

[12] Acetylene used to be readily available without such specialised gas cylinders; instead calcium carbide chips were used. This most interesting, if dangerous, compound was sold for putting in acetylene generators. These consisted of a water tank with spring-loaded diaphragm underneath a hopper on the top for the calcium carbide chips. When the outlet was opened and the pressure fell, the spring-loaded diaphragm relaxed, the hopper allowing carbide chips to fall into the water. If the pressure increased, the diaphragm closed the hopper once again in a crude regulation of pressure. A metal version of the laboratory Kipp's apparatus would not have worked too well; the carbide in the middle chamber would have heated up, generating steam, and pressure control would have been even worse.

as absorbers. However, the cost of these materials relative to their technical advantage has always so far prevented their adoption. More recently, the use of the carbon pentagon/hexagon spheroidal carbon skeletons named after Buckminster-Fuller (fullerenes) has been proposed. Fullerenes, although only recently recognised, are in fact fairly easily produced, albeit inefficiently, by dissolving soot from carefully controlled flames in benzene. The resultant red solution can be evaporated to give a mixture of fullerenes of around a hundred or so carbon atoms or less. These have some remarkable properties (for example, superconductivity when suitably doped with metal ions) and are also remarkable absorbers. Although currently expensive, fullerenes are composed of carbon and so could become cheap if efficient production methods become available.

Ethylene. Ethylene is the cheapest unsaturated hydrocarbon. It is extracted as a by-product from petroleum processing streams, particularly cracking plants, and is also made by cracking of ethane.

Ethylene is used on such a huge scale that massive plants have been built to create ethylene from almost any hydrocarbon feedstock. A simple ethane cracker might simply pass the gas through a counter-current heat exchanger and then rapidly through a 900°C furnace for a fraction of a second. The output gas stream is then passed back up the heat exchanger to heat more inlet gas. The exhaust gas will contain ethylene and also methane, acetylene and other products. It is compressed, when heavier fractions condense, and scrubbed to remove CO_2 and other components which might freeze. A cryogenic separation process at below ethylene boiling point is then used and non-ethylene components recycled.

Ethylene is occasionally subject to runaway polymerisation in the gas cylinder, evolving large quantities of heat, or in processes at high pressure; it is, after all, the monomer of that most common of plastics, polyethylene (PE), which is formed by applying high pressure. Dry ethylene in a dessicator in the presence of acid as a catalyst is particularly subject to this problem. In large quantities, ethylene is delivered in liquid form at low pressure. However, with precautions, cylinder ethylene is not regarded as a particular hazard.

Methylacetylene and propylene. There are a number of substituted acetylenes and more complex unsaturated gaseous hydrocarbons which have from time to time found applications. They have lower performance flames than acetylene but can typically be stored in liquid form and are therefore more conveniently handled. Petrochemical sources are normal.

Propylene is, like ethylene, an important intermediate in the petrochemical industry. It is sourced as a by-product from ethylene manufacture. At high pressures, dry propylene can, like ethylene, easily be subject to polymerisation. With polypropylene explosive polymerisation can take place with acidic catalysts but, subject to precautions, it is perfectly safe under its own vapour pressure at room temperature. It is therefore normally delivered in liquid form at 10 barg.

2.2.4 Fuel gases: methane, propane and butane

Methane is familiar as domestic piped gas in the more urban parts of the world. Propane is most familiar in liquid form in low-pressure cylinders for domestic use, while butane is used in mixtures with propane as the propellant for spray cans and as the fuel for small domestic burners such as cigarette lighters.

These gases are all extracted from natural gas or petrochemical streams. These streams contain carbon dioxide and hydrogen sulphide, which are typically removed by scrubbing with absorber liquids. This done, the streams normally consist of a mixture of light hydrocarbons, which require separation by cryogenic or high-pressure distillation. Ethane is normally removed, as are ethylene and other unsaturates, as they are more valuable as chemical intermediates than as fuels. The lightest fraction is mainly methane and is either piped direct or liquefied for transport by ship to consumers. The fraction heavier than ethane can be separated into fairly pure propane, isobutane and butane (with a small admixture of ethane and pentanes) or can be sold as mixtures. The mixtures are sold as having boiling points in certain ranges or, which amounts to much the same thing, as having vapour pressure between certain limits at room temperature. A stenchant such as a powerful smelling mercaptan or other vapour based on sulphur is usually added to commercial fuel gases as an aid to leak detection. For applications requiring complete freedom from impurities such as sulphur (many chemical processes) or low odour (aerosol can propellant) the gases are sometimes available without stenchant or else an regenerable adsorber cartridge 'purifier' can be used to remove the mercaptan compounds.

Methane has the unusual distinction of forming stable solids with water at comparatively low pressures and comparatively high temperatures (Melvin, 1988). Hydrates such as $(CH_4)_8 (H_2O)_{46}$ are formed at 70 bar and 10°C, for example. Methane hydrates cause blockage in high pressure pipelines in winter and pipelines are often cleaned out with methanol (which decomposes hydrates). It is thought that huge reserves of methane gas, perhaps as much as in current gaseous reserves, are stored in the earth as solid hydrates but it is very difficult to extract the gas from them.

The fuel gases are nearly always simply burned in air as a source of heat and are, as a result, largely interchangeable with each other, with other fuels, from petroleum to coal, and sometimes interchangeable with electrical or other heat sources. Because of this interchangeability, and because they are often used in domestic applications, these gases are not 'industrial gases' in the same sense as the gases such as nitrogen, oxygen or acetylene. It is for this reason that they are dealt with only cursorily in this book. There are some significant non-fuel uses for fuel gases, in heat treatment of metals in furnaces, for example (section 3.2.2). Propane, isobutane and butane are used in propellant mixtures, mixed with liquid formulations such as hair spray or paint, in aerosol cans (section 3.8.12).

2.2.5 Hydrogen

Coke oven gas and reformer gas. Hydrogen is a by-product of coke oven operation. Coke ovens are still widely operated to produce high-quality carbon for the reduction of iron ore to iron in blast furnaces, although changes in blast furnace practice, such as the use of large quantities of oil injection, have reduced coke oven use.

Coke oven gas consists of CO, H_2, CO_2, steam, ammonia and smaller amounts of many other gases and vapours. After crude washing to remove the more reactive contaminants PSA units are used to extract the hydrogen in the stream.

The reaction of methane from natural gas with water, producing the CO/H_2 mixture known as reformer gas, has been in use ever since methane became readily available. The cleanness of the feedstock simplifies plant design and stand-alone plants for hydrogen production are common in large chemical complexes. The clean feedstock also enables the use of more advanced separation technology, membrane separation, to separate the H_2. Reformer conditions can be adjusted to maximise hydrogen yield from the reactions

$$CH_4 + H_2O \Rightarrow CO + 3H_2$$

$$CO + H_2O \Rightarrow CO_2 + H_2$$

The reforming of other hydrocarbons also yields hydrogen, along with much more CO. Cracking of hydrocarbons also yields hydrogen and this is sometimes separated in large petrochemical plants.

Hydrogen from methanol and ammonia. The above methods cannot be carried out effectively on a small scale and smaller industrial plants generally carry out 'cracking' of methanol or ammonia.

Methanol can be cracked by suitable catalysts into a mixture of carbon oxides and hydrogen, again followed by H_2 separation on membranes or by PSA, but the latter processes mean that this is only economical at the larger end of 'small scale'. Similarly, ammonia can be heated to yield a mixture of nitrogen and hydrogen ('cracker' or 'forming' gas). This is generally used directly in processes where the 30% N_2 content is unimportant, such as in bright annealing of steel. Forming gas contains only a few other impurities; for processes tolerant of nitrogen and those impurities requiring moderate volumes of hydrogen, cracking units can be a good solution. Forming gas generators are simple, needing only a controlled temperature hot chamber, valving systems and some precautions during start-up. Furthermore, forming gas can often be formed by drip feed of ammonia into a furnace which is sufficiently hot to ensure complete cracking, again with precautions during start-up. Bulk ammonia storage and delivery is hazardous, however, and the relatively volatile bulk ammonia price affects the cost-effectiveness of forming gas.

Solvay cell and membrane electrolysis for hydrogen and chlorine. Electrolysis of brine yields hydrogen and chlorine, whilst electrolysis of alkali yields hydrogen and oxygen.

Where electricity costs are low, for example, near large hydro-electric schemes, and where chlorine must be generated, it is economic to generate hydrogen on a large scale by electrolysis. Occasionally smaller alkali electrolysers are used to supply pure hydrogen which would otherwise be delivered in manifolded compressed gas cylinders, perhaps because cylinder H_2 is locally expensive or because a strong security of supply is demanded; some semiconductor facilities have electrolysis equipment, although liquid hydrogen is now preferred.

Solvay cells electrolyse brine between a flowing liquid mercury cathode and a carbon anode. Chlorine is produced at the anode, whilst sodium ions are discharged at the mercury cathode and form a sodium/mercury alloy or amalgam. The amalgam is then pumped away to a separate cell where it is contacted with water and graphite, evolving hydrogen from the graphite surface (which reduces the H_2 production overpotential) and producing concentrated (50%) aqueous sodium hydroxide. The resulting hydrogen is substantially pure apart from water and a trace of mercury vapour. Mercury contamination must be dealt with by passing over iodised carbon filters, although much of it is removed by compression of the product. Because of fears of mercury contamination, Solvay cells are now unusual in new plant, although with suitable monitoring and controls new construction is possible.

The membrane or diaphragm electrolysis cell employs straightforward carbon electrodes, with a semipermeable membrane, based typically on asbestos fibre but sometimes a special polymer, between the anode and cathode. Membrane cells can be used for brine and alkali electrolysis. Brine electrolysis in these cells has the problem that the concentration of sodium hydroxide that can be achieved is rather low, little better than 10% and the alkali is contaminated with unreacted brine and sodium chlorate (from chlorine) because of leaking across the membrane. Much power is then needed to extract water from the deliquescent alkali to bring it up to commercial (50%) strength. Polymer membranes made from cation exchange resins can improve this performance dramatically, producing a purer alkali with less chlorate and chloride contamination.

For simultaneous H_2/O_2 gas production a cell using sodium or potassium hydroxide electrolyte is normal.[13] Nuclear submarines employ electrolysis cells using KOH to generate breathing oxygen during long underwater voyages. The output is carefully monitored with multiple gas analysers and with a mass spectrometer to ensure the safety of the atmosphere for the sailors. The latest systems use membranes made of advanced polymers, cation exchange resins, instead of asbestos, which allow only sodium (or potassium) ions to pass, and offer purer products and a more efficient system.

A surprisingly large amount of electricity is needed to decompose a small volume of water: about 4 kWHr per kilo of water. The low rate of production is clear from laboratory electrolysers, where a substantial 10 Å or so power supply yields a weak stream of bubbles. The power is needed at an inconveniently low 2 or 3 volts d.c. or so. As a result electrolysers are usually designed in series connected banks so that their operating voltages come up to levels at which transformers and rectifiers are efficient. The current electrodes have current densities of from hundreds to thousands Å m^{-2}. With suitable containment vessels electrolysers can easily produce gas under pressure and in fact efficiencies can be higher at high pressure with less blinding of the electrodes by bubbles; compact KOH systems in submarines take advantage of this.

Electrolysis hydrogen, although very pure otherwise, normally contains O_2 or Cl_2 (and the oxygen or chlorine will contain H_2), as well as water vapour. Downstream processing of the gases aims first to remove O_2 or H_2 contamination by passing over a heated precious-metal combustion catalyst (deoxo), then to remove moisture and then residual contamination.

Oxygen from electrolysers is valuable because it is entirely free of carbon compounds, argon and other inert impurities that oxygen from air separation normally contains. Some applications, such as in the nuclear reactor service, which require small quantities of oxygen free from argon, specify oxygen supplied by the electrolytic route.

Liquefaction of hydrogen. Hydrogen is normally liquefied only on a large scale for rocket launching, as liquefaction is an expensive process for this low-cost commodity. Rocket testing and launching continues to drive large liquid hydrogen (LH) installations, although necessarily long-haul supply routes in some parts of the North America justify the liquefaction of hydrogen for transport purposes. One Canadian plant, for example, with plentiful low-cost energy but far from H_2 uses, incorporates a liquefier and exports the gas to the USA.

Liquid hydrogen is an exceptionally pure product as it will dissolve only

[13] Why school demonstrations use dilute sulphuric acid for demonstrations of electrolysis is unclear; both on a small and on a large scale, NaOH/KOH seems superior.

ppb level impurities at its 20 K boiling point. As a result many semiconductor manufacturers now specify LH for their hydrogen needs. Most industrialised countries therefore now require at least modest supplies of liquid hydrogen to be available. Plants for the liquefaction of hydrogen are, like many cryogenic plants, subject to scaling rules which strongly favour large installations. There is therefore today a considerable trade in LH to efficiently use the output from the few large plants, distributing the LH in cryogenic tanks mounted in standard ISO shipping containers.

Liquid hydrogen is produced by first further purifying a substantially pure H_2 product to remove air gases and moisture, then cooling to 20 K, when liquefaction takes place, and finally converting the majority (75%) *ortho* form to the *para* form. Purification via conventional temperature swing absorption (TSA) systems (section 2.1.5) can guarantee low levels of CO_2 and moisture but further removal of air gases may well be necessary, often with an LN-cooled activated-carbon bed and followed in some cases by an LH bed or filter just prior to the liquefier.

The liquefaction of H_2, using Carnot ideal heat engines, requires 12 200 kJ kg^{-1}, corresponding to 434 kJ kg^{-1} specific heat at 20 K (6100 kJ kg^{-1} work needed) plus a further 6100 kJ kg^{-1} cooling from room temperature. LH thus costs 15 times more than LN to produce, even with perfectly efficient machinery. The insulation for all liquid H_2 equipment needs to be of a very high standard; similar to that used for liquid He, with vacuum insulation frequently being demanded. Safety considerations must be taken into account with LH plant; roof vents are employed with detectors to avoid accumulation of H_2 there, along with intrinsically safe electrical equipment where possible.

Small liquefiers, such as those used in research, work in a simple way, often using a Joule–Thompson valve to expand 100 bar LN-cooled hydrogen and recycling most of the H_2 back to the compressor. Larger, more efficient liquefaction schemes are used above about 1 tonne per day, with 50 tonne per day units a typical size. They involve an expansion turbine rather than LN input to provide cooling to LN temperatures and an *ortho–para* conversion at an intermediate temperature such as 77 K as well as at 20 K. They may also incorporate an expansion turbine for cooling the H_2 below 77 K instead of or in addition to the final J–T expansion valve.

Molecular hydrogen exists in two forms with the spin 1/2 nuclei (protons) of the two H_1 atoms parallel or opposite, giving total spin 1or spin 0. (The H_1 isotope is 99.99% of natural H.) With most molecules, the nuclear spin energies are undetectably close and no consequences are seen. However, with H_2, the nuclear spin alignment in the molecular electric field corresponds to energies of 710 kJ kg^{-1}, comparable with specific heat capacities. At room temperatures the spins are randomly assigned, giving 25% *para*-H_2 and 75% *ortho*-H_2. This is the ratio of the number of same-energy *para* (nuclear spin 1) and *ortho* (nuclear spin 0) states that are possible. At cryogenic temperatures the stable form is the lower energy *para*-H_2, and *ortho*-H_2 in freshly made LH slowly converts to the *para*-form, evolving the 710 kJ kg^{-1} heat as it does so. The latter heat causes boil-off and loss of (ultimately) nearly 50% of the LH over the several weeks that the transition takes. For transport in tanks, and for many applications, this rate of boil-off is unacceptable for economic or safety reasons and 100% *para*-H_2 is required.

Para-H_2 can be produced simply by allowing LH to sit in tanks for a while and recycling the boil-off to the liquefier. A better method is to speed up the process to a few seconds by employing a catalytic *ortho–para* converter, usually transition metal salts or oxides. The catalysts can be used near the boiling point of LH but are more efficiently used in stages, converting a little more of the stream to *para*-form at each temperature stage drop in the liquefier, and hence removing some of the conversion heat at a higher temperature than 20 K.

2.2.6 Chlorine and fluorine

Chlorine and alkali from brine electrolysers are major components of the chemical industry. A typical production site is that of ICI in Bromborough, UK, where hundreds of giant cells are wired in series (to minimise ohmic losses in the cables) to one of the largest d.c. power stations in the world. The chlorine produced is purified by drying and then bubbling through liquid chlorine; this eliminates small amounts of organochlorine compounds formed during electrolysis. It can then be stored and delivered in low-pressure (vapour pressure is 6 bara at room temperature) steel or stainless steel containers as liquefied gas (Figure 2.22).

Chloride and fluoride gases. Chlorine gas is often used to oxidise hydrogen to make HCl gas, which is sold in small amounts in anhydrous form in gas cylinders but is mostly dissolved in water and sold as hydrochloric acid. The combuster is usually a straightforward enclosed flame design, with water cooling, although materials problems loom large; hot and slightly moist HCl gas is very aggressive, even on high chromium stainless steel, so carbon and expensive lined metals are necessary. Hydrogen and chlorine in 1:1 ratio are fed in and electrically ignited, and the resultant HCl dried, initially by cooling well below freezing point, to remove moisture. The HCl made in this direct combustion is favoured for those applications needing high-purity product. Other chloride gases can also be made by combustion of elements in chlorine. One example is BCl_3. Solid lumps of boron are burnt by simply heating to initiate and then regulating chlorine flow to control the reaction and condensing the BCl_3 (which boils at 12°C) produced.

Although known in compounds since Scheeles work in 1771, elemental fluorine was not produced until 1886 because it is so strongly bound in its compounds. The element is so reactive and dangerous to handle that it is nowadays rarely used; one of the few applications encountered in the gas

Figure 2.22 Industrial brine electrolysis cells for chlorine (courtesy of ICI).

industry is the fluorine passivation of stainless steel pipework and vessels. However, compounds of fluorine such as HF, SF_6 and perfluorinated gases such as perfluoropropane are now relatively well-used cylinder gases, and the CFCs, HCFCs and HFCs used in refrigeration are still gases of major significance.

Fluorine can be prepared by electrolysis of potassium hydrogen fluoride solutions in liquid HF. The preparation of HF by dissolving fluorspar in acid is more important industrially, however, as elemental fluorine is rarely required. Anhydrous HF made in this way is then purified and liquefied by refrigeration. Many fluoro gases can be made by using HF in a substitution reaction with a chloride, using catalysts such as antimony pentachloride

$$CHCl_3 + HF \Rightarrow CHClF_2 + 2HCl$$

2.2.7 Ozone generators

Cryogenic liquid ozone (O_3) is simply a laboratory curiosity. A deep blue in colour, it is dangerously unstable. Likewise, ozone mixtures in air or oxygen are unstable, reverting over a day or so back to oxygen. Ozone can be dissolved in water, it is more soluble than O_2, but only in relatively dilute solutions and it is even then typically slightly unstable. Ozone can be stored stably in solution in chlorofluorocarbons. However, although this is convenient for laboratory experiments, it is too expensive for any commercial application and in any case CFCs are now regarded as too serious an environmental problem to be entertained seriously for any new industrial process. Perhaps studies on the solubility of ozone in more environmentally friendly HFCs or perfluorinated compounds are indicated.

Because of these storage problems, ozone is therefore prepared commercially by on-site generators. These vary in size depending on application from less than 1 kg per day to many tonnes per day. Most generators rely on electrical corona discharge in air or oxygen at pressures from 1 to 2 bara. The corona discharge is a current-limited high-voltage discharge, typically using 10–20 kV a.c. The electrodes are straightforward but must be resistant to oxidation and need liquid cooling in all but the smallest generators. Another typical feature of the ozone generators is the requirement that one of the electrodes be covered with insulator, typically glass. This largely avoids arcing and consequent damage, although electrodes still fail occasionally. The alternating voltage is somtimes a straightforward 50 or 60 Hz mains, although there are some advantages to using higher (audio) frequencies, which are therefore used by many units. Figure 2.23 shows the principle used by a corona discharge ozone generator unit.

The discharge forms oxygen atoms by bombardment of the oxygen molecules by electrons in the discharge. The principal reaction of the oxygen atoms is to add to an oxygen molecule forming O_3. An undesired side reaction is the formation of oxygen ions, which hinder the electron bombardment process. Many units use a specially shaped waveform with millisecond pauses at zero voltage to allow oxygen ion neutralisation, minimising this side reaction. Once made the ozone will decompose at a rate which increases with concentration and with temperature. Approximately, the half-life, $T_{1/2}$ behaves as

$$T_{1/2} \simeq T_{1/2}\text{o}\ \exp(Ea/kT)/\%O_3$$

where $T_{1/2}\text{o}$ = a constant. A typical ozone half-life would be an hour at 100°C and a day at 20°C, although ozone dissolved in water, or ozone in the presence of catalytic substances, may decompose much more rapidly. The output ozone concentration is limited; with pure oxygen supply, 8% ozone is possible (highly concentrated ozone can detonate, so is in any case undesirable) while with air supply, 1–2% is normal. The feed gas, whether air or oxygen, must be dry to avoid the formation of corrosive nitric acid.

Electrical efficiencies depend on size of unit and gas supply, with oxygen units typically twice as efficient as air units. The highest efficiencies are only around 5 kWHr kg^{-1}, which is low compared to the theoretical value of 0.84 kWHr kg^{-1}. This inefficiency manifests itself as heating of the gas input, which is undesirable as the discharge is less efficient at higher temperatures

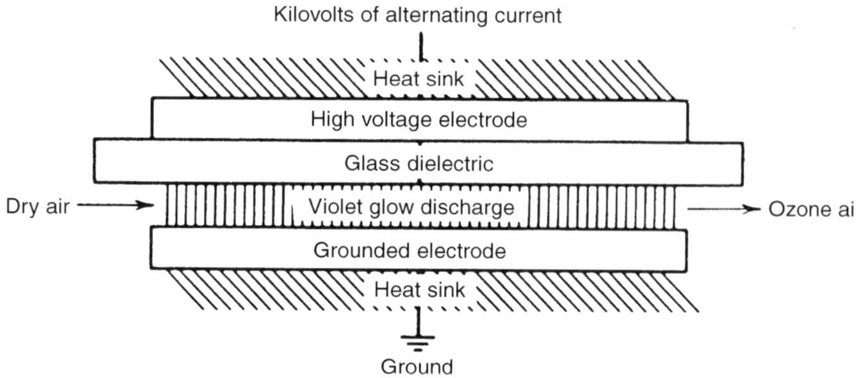

Figure 2.23 Corona discharge ozone generator unit.

and the ozone formed decays more rapidly. Where an ozone generator is being fed from a liquid oxygen supply, it is open to the designer to only partially heat the oxygen gas whose sensible heat, in being warmed from 100 K or so, can therefore be employed to cool the process.[14]

2.2.8 Helium

Helium is usually found in natural gases but is usually only economically extracted in a few parts of the world; for example, in the USA and Poland where the concentration is particularly high.

Separation of helium from natural gas could in principle be done with a PSA process, as used for separation of hydrogen from by-product gas streams. However, in practice, the plants employed hitherto have employed variations on a cryogenic route. The precise design chosen depends on the nitrogen and other contents of the natural gas stream.

A typical natural gas contains methane, along with a few per cent of other hydrocarbons, and up to a few per cent He, up to 50% nitrogen, and moisture, H_2S and CO_2. The prime aim of the cryogenic process is often to remove the nitrogen, heavier hydrocarbons and acid gases to make the natural gas saleable as a fuel gas. In cases where the removed waste stream contains more than 1% or so of He, it makes sense to extract He as well but this is the case in only a few places.

The raw methane is washed with an amine solution to remove H_2S and CO_2, followed by drying over a molecular sieve. It is then admitted, often under natural pressure but sometimes compressed, to the nitrogen separator. This cools it and removes the heavier hydrocarbons as liquid, then

[14] There is the question of whether ozone can be formed in a discharge in the liquid oxygen directly and whether this would show improvements in efficiency due to the cooler environment.

expands it though a J–T valve, yielding a mixture which can be distilled to give methane and an He-containing purge gas. A nitrogen-cycle refrigerator using a turbo-expander provides cooling for the process.

Whereas hydrogen is only liquefied for special purposes, helium is normally liquefied because of the intercontinental distances over which it is transported. There are few sources but many small uses to which it is put. In addition, many research applications need He in liquid form.

The crude He extracted by natural gas processing must be purified to a very high standard before liquefaction as absolutely all impurities will freeze and be potential causes of blockage at liquid helium temperatures. Hydrogen may be removed in a similar way to Ar processing (section 2.1.5) by heating in a catalytic combustor after adding excess oxygen. Most impurities are removed by a stripping process using liquid nitrogen. Use of LN at low pressure ensures that all N_2 is condensed.

A helium liquefier does not have the *ortho–para* conversion difficulty that a hydrogen liquefier has but has the much more serious problem that it must reach 4.2 K, which is close to absolute zero and is the coldest temperature at which any industrial process runs. At these temperatures insulation must be of the very highest order and radiation cold losses can be very serious, so foil heat shields are necessary where cold parts could 'see' room temperature parts. The last stage of purification of the helium leaves the gas already pre-cooled below 77 K at a few bar or tens of bar. Counter-current cooling in heat exchangers is then used to reach a few degrees above boiling point, when a Joule–Thompson expansion forms liquid He and gas. Some of the gas is expanded in two or three expansion turbines; these are small, ultra high speed devices, with the cool gas cycled back through the heat exchangers to provide most of the formidable cooling power needed. Figure 2.24 shows the typical scheme used to liquefy He today – a cascade of expanders.

2.2.9 Nitrous oxide

Nitrous oxide (N_2O) is made by thermal decomposition of ammonium nitrate. The latter requires some handling precautions, as it is potentially a dangerously explosive compound: often used by terrorists, for example, in home-made bombs. Inadvertent mixtures with organic or other finely divided oxidisable materials are extremely dangerous. The nitrate is gently melted at 170°C then pumped into a reactor vessel where is it more strongly heated to around 250°C where it decomposes to steam and nitrous oxide

$$NH_4NO_3 \Rightarrow N_2O + 2H_2O$$

Subsequently, a condenser separates out the water and cools the gas. Further treatment is then necessary to remove acidic NO and NO_2, and basic impurities such as NH_3. This is done by successive wash bottles containing water, alkali and acid. The gas is highly soluble (more than 1 volume per

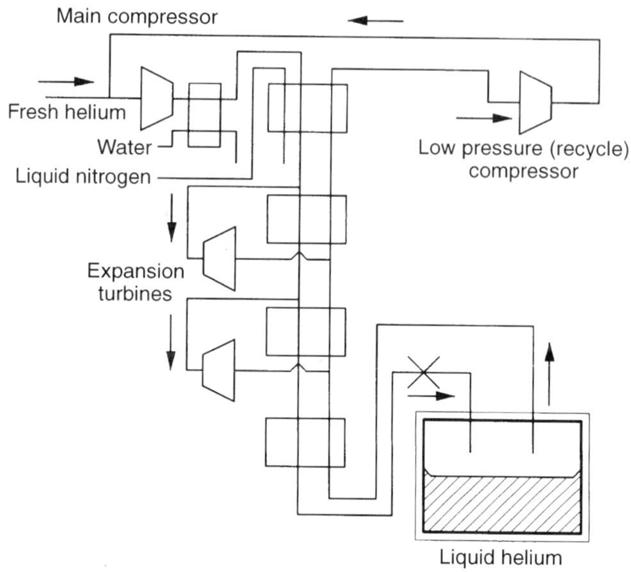

Figure 2.24 Cascade of expanders to liquefy He.

volume soluble in water), so the wash liquids must be allowed to become saturated with gas before full output is obtained from the plant. The wet gas is then dried and stored as a liquid by cooling and compressing either at room temperature under pressure (50 barg) or refrigerated at lower pressure.

2.2.10 Production of special gases

The myriad of 'special gases' in use in industry means that it is impossible to describe the manufacture of all of these. A special gas is supplied in small quantities. For all special gases, therefore, suitably compatible compressed gas cylinders must be evacuated or purged prior to filling from a bulk source. For many purposes special purity is needed, above that of available bulk sources, so small-scale purification is often used. For still other special gases bulk sources are not available. These gases are then made in small-scale purpose-built plants. For the more unusual gases it is not economic to run more than one or two plants and cylinders from these plants are exported worldwide. Examples of special gases in this category include silane, nitrogen trifluoride, boron trichloride and phosphine, which are all produced for the semiconductor industry. Boron trichloride is produced by burning solid boron in chlorine gas, followed by purification by absorption and distillation (Figure 2.25).

Some hazardous special gases can be manufactured in-situ, with some potential advantages in safety, although this is highly exceptional. One US

Figure 2.25 Production of BCl_3. After burning solid boron in a stream of chlorine gas, the BCl_3 is purified in these absorption columns (courtesy of Epichem Ltd).

company, for example, installed 100 cc min^{-1} gas plants for generating arsine gas (AsH_3) in a mixture with hydrogen. The process is not described in detail by the firm but presumably involves electrolysis at pure arsenic electrodes in an aqueous solution, when arsine is mostly evolved along with hydrogen, followed by removal of moisture.

Making silane. Pure silane for use in the semiconductor industry is often made by reduction of a chlorosilane with lithium aluminium hydride, the process being carried out in an organic solvent such as tetrahydrofuran. The reagents are typically stirred together in a reactor with careful temperature control, with the whole assembly being located in a deliberately flimsy room, a 'friable building', with the operators behind a blast wall. Silane evolved from the mixture is first cooled, the condensate of chlorosilanes being recycled, and then cooled further for redistillation to remove solvent and unreacted or partially reacted chlorosilanes. A typical scale of plant might make a few tens of tonnes per annum.[15]

In the last few years the economics of silane production have been drastically changed with the arrival of one or two very large plants (hundreds of tonnes per annum) such as that of DENAL in Japan or Advanced Silicon Materials, Inc. in the USA. Product from the DENAL and Advanced Silicon plants is so pure and so cheap that there is now less justification for small silane plants to produce raw silane. Most gas suppliers now concentrate on purifying and certifying silane from these large plants. They use large 'tube trailers' of high-grade cylinders, other transfilling equipment, gas analysers and purifiers, so that they can deliver silane of the required grade from these bulk sources efficiently. The days of small silane plants would appear to be numbered and silane production will become an oligopoly similar to the more specialist gases such as arsine, phosphine or nitrogen trifluoride.

Stable and radioactive isotopes. Gaseous radioactive isotopes are highly dangerous because they can easily be ingested and cause genetic damage in humans and animals. At sufficiently low concentrations these hazards are negligible, however, and gaseous radioisotopes are potentially very useful tracer gases. The simple apparatus needed for the detection of radioisotopes, even at exceedingly minute concentrations, commends their use in applications where immense dilution factors must be dealt with. Similarly, applications with a need for many inexpensive detectors may find them useful.

There are only a few radioisotopes of the common gases with convenient properties. Tritium is available at low cost from nuclear reactors, where neutron absorption by lithium produces tritium and helium. There has been much research on tritium in the worldwide nuclear fusion development programme because its reaction with deuterium is the lowest-temperature nuclear fusion reaction known. Tritium in industry is used as a source of ionisation in low-pressure lighting tubes, which glow without an external power source.

Krypton-85 is also used for lamps and also in vacuum tube devices which require an internal source of ions. Radon-220 is produced by purging a thorium oxide container with carrier gas and can be used for high speed tracer work where its exceptionally short half-life is not a problem. Some radioisotopes used in gas tracer work are shown in Table 2.6.

[15] Even more specialist than silane is the compound disilane (SiH_3–SiH_3), which is made only in kilogramme quantities. L'Air Liquide have a patented process for producting disilane from silane in an electric discharge.

Table 2.6 Radiosotopes used in gas tracer work

Radioisotope	Half-life (s)	Remarks
Hydrogen-3 (tritium)	3.87×10^8	Very low energy electron emitted
Argon-42	1.1×10^8	Availability?
Krypton-85	3.4×10^9	Good availability
Xenon-133	4.55×10^5	Availability?
Radon-220	56	Short but can be made on site by purging thorium
Sulphur-35	7.49×10^7	Used as SF_6, but molecular weight is high
Nitrogen-13	600	Life too short
Carbon-14	1.81×10^{11}	Life too long, activity low

Stable isotopes. Stable isotopes of the common gases are safe to use and relatively inexpensive. They have a number of uses but tracer applications probably predominate. By using a mass spectrometric detector, which can be made very sensitive, many chemical and biochemical studies can be carried out.

Deuterium extraction is often carried out on large hydrogen-producing plants with various techniques varying from electrolysis to distillation of LH being used. Deuterium is the least expensive of the stable isotopes, costing only a few hundred dollars per kilo (expressed as heavy water), reflecting the ease with which it is separated. Important in chemical research work, deuterium is often used as heavy water as a 'moderator' for slowing down neutrons in nuclear reactors, particularly those used for research. Deuterated compounds often show very different reaction rates from normal hydrogenated compounds. A recent study, for example, showed that vaccines prepared in a heavy water (D_2O) solution were stable for much longer than vaccines of normal formulation, avoiding the need for refrigeration.

Carbon-13 and oxygen-18 are often made by distillation, e.g. distillation of carbon monoxide, but are much more expensive than deuterium because their atomic weights differ by only 10% or less from the majority isotopes. Carbon-13 as carbon dioxide is readily detected in mass spectrometers and is useful for examining the metabolism of plants, while oxygen-18 labelling can be used in human studies. Boron-10 gases are used to generate boron-10 beams for ion implantation in the semiconductor industry. Stable isotopes useful in gas tracer work are given in Table 2.7.

2.2.11 *Purifiers*

'Purification' processes may be differentiated from 'separation' processes because the former are designed to remove only small amounts of unwanted

Table 2.7 Stable isotopes useful in gas tracer work

Stable isotope	Abundance (%)	Remarks
Hydrogen-2 (deuterium)	0.015	Cheap, twice as heavy as normal hydrogen
Helium-3	0.0001	Used in low-temperature research
Boron-10	19.8	Used in ion implanters
Carbon-13	1.1	Used in biochemical research
Nitrogen-15	0.37	Used in biochemical research
Oxygen-18	0.204	Used in biochemical research
Sulphur-34	4.22	

gas from a mixed gas stream and are not required to provide the removed components as an output. Absorbers, both liquid and solid, cryogenic adsorbtion, reactive metal getters and reactive organic getters are possible principles in purification.

Absorber purifiers rely on washing the product gas with a liquid which preferentially dissolves the undesired impurities. The most common washing liquid is, of course, water and it is used in most wet gas scrubbers for controlling effluent gas (section 2.6.14). Scrubber purifiers generally use a packed or trayed column to contact the liquid intimately with the gas being processed. The gas, after passage through the contactor, must be cleaned of absorber and dried. The absorber liquid, once it has been used for scrubbing the gas, is usually recycled by passage through another contactor tower, the stripping unit. In bulk chemical production wet purifier processes are ubiquitous. Industrial gases manufacture, however, largely avoids the use of wet purifiers. Generally, the high levels of purity to which the industrial gases industry works does not allow the use of wet processes. Reference works such as Kohl and Riesenfeld (1960) provide a wealth of detail of absorber systems.

Dry purifier principles. Dry purifiers of various kinds are the workhorses of the gases business. Although designs differ considerably in detail, all have common features. First, all must contact a solid surface with the gas. The gas must transfer from the main gas stream to the surface, typically diffusing through micropores in the reactor bed, then react on the active surface, then diffuse back through the micropores to the bulk gas stream once again. The rate at which the purification reaction will take place depends on temperature, pressure, the area of the active surface and the efficiency of the diffusion path through the microstructure.

Most purifier materials as well as having a microstructure also have a macrostructure: the bed material is typically prepared in pellets or granules, varying in size but most often of the order of millimetres. Sometimes the pellets are shaped into rings, giving a large amount of free volume for good flow. Whatever their shape, the pellets or granules are designed to allow the bulk flow of gas easily and evenly throughout the bed but are small enough to ensure that the gas can diffuse through to the centre of the granules in a reasonably short time.

The macrostructure is often complemented by an engineered microstructure. Instead of trying to prepare a pure form of catalyst in microporous form, the catalyst is coated onto the pores of a naturally microporous material. An example would be the use of platinum to coat an inert porous ceramic such as alumina. The bed materials will often require heating, either for reaction or for regeneration, and it is helpful to choose support materials with relatively high thermal conductivities, such as alumina or magnesia. The active material is coated onto the support typically by soaking in a solution of a compound of the active material. The ensemble is then activated by heating to drive off the solvent and then heated, perhaps in a reducing gas, to react or calcine the active material.

Adsorption purifiers. Dry adsorbtion purifiers rely on the formidable capacity of substances such as activated carbon (section 2.1.6) to adsorb small amounts of more reactive gases and vapours from low boiling point gases and vapours. Cryogenic purifiers rely on the principle that most impurities have much higher boiling points than the wanted product. By cooling the product to dry ice or even LN temperatures, many impurities are simply condensed out. The effect can be further magnified by the addition of absorbers like activated carbon, which function at much greater efficiency than at room temperature.

Reactive purifiers contain active reactive chemicals which combine with impurities but not with the product gas. Small-scale purifiers have a very simple design, comprising a single vessel loaded with a solid reactive or adsorbtive material. No provision is made for reuse of the purifier and it is either jettisoned or returned to the supplier after use. Some of these purifiers with straightforward active principles can be repacked by the user with fresh material. Simple cylinders packed with oven-baked magnesium perchlorate or 13X zeolite granules, for example, can be used to remove moisture from air or inert gases.

Larger scale adsorber purifiers all employ adsorber materials which are regenerated in-situ. The 'two-bed' concept is often used for regenerable solid reactor bed purifiers: one purifier bed is on stream purifying gas whilst the other is being regenerated by heat and purge gas and/or a regenerating gas mixture. The regeneration process is preferably a physical process of desorbtion as this makes for the lowest running costs. Temperature, pressure or vacuum swings can be used with a product or inert gas purge and a pressure swing adbsorber purifier capable of the quickest regeneration time (section 2.1.6). The purge gas is generally run in the counter-current direction, so avoiding any possibility of polluting the outlet end of the bed with the impurity.

The simplest industrial adsorber purifiers are the pressure-swing drier units marketed for drying compressed air or inert gases using the supply gas itself as purge and 'power supply' (section 2.1.6). After minutes on-line, the bed (activated alumina is common) is depressurised to atmospheric pressure, purged briefly and then put on-line once again. These purifers are commonly operated by a pneumatic timer unit, thus they conveniently need only gas inlet and outlet connections and a vent; no power or heating inputs are required.

Temperature-swing adsorption operation is more typical of most purifiers and usually involves the use of a hot purge gas, perhaps assisted by heaters. Heating can take a very long time with a large absorber volume so TSA systems operate much longer cycle times than PSA purifiers; hours rather than minutes. In addition to the desorbtion process, it is also necessary to include time for cooling so that the regenerated adsorber is immediately effective when switched on line.

The temperature needed for regeneration is always kept as low as possible. First, the rise in temperature costs energy but, more importantly, the temperature is limited by the adsorber materials. Many adsorber materials suffer mechanical breakdown (producing fines or 'dusting') and loss of capacity as a function of the number of times they have been thermally cycled; the number of cycles before deterioration is a steeply increasing function of temperature.

A three-bed system is sometimes used for large purifiers to increase utilisation of the bed material. This involves having bed 1 in adsorption, followed by bed 2 in series performing a 'polishing' or additional purification, with the remaining bed 3 being regenerated. After the regeneration cycle is completed, bed 1 becomes bed 3 (regeneration), bed 2 becomes bed 1 (on-line) and bed 3 becomes Bed 2 (polisher), and so forth.

Catalytic purifiers and deoxos. Catalytic purifiers are now a very common technology; almost every petrol motor vehicle has one to reduce emission of CO and NO_x. Catalytic purifiers in the gas industry are typified by the argon purifier units (APUs) used to remove the last vestiges of oxygen from argon derived from air-separation units. Such catalytic units to remove oxygen are often dubbed 'deoxos', although reactive getter units which remove O_2 are, confusingly perhaps, also often called deoxos.

The requirement to remove oxygen from inert gases is very common. Argon from air distillation, for example, always contains some residual oxygen because the boiling points of the two gases are so close. Argon is usually used in an application only because it is completely inert; it must therefore be at high purity (welding argon is an exception in this respect). A classic design is a heated precious metal catalyst supported on a support, along with a hydrogen dosing system. After passing over the catalyst, which might be 0.5% palladium on pelleted activated alumina, the argon will contain only water vapour, excess hydrogen and trace nitrogen present in the original gas stream. The system will actually operate at room temperature, although heating is usual as it avoids certain poisoning effects.

Although generally very satisfactory, catalytic purifiers are subject to various mechanisms of 'poisoning' by components in the inlet gas stream. First, particulates or gases which decompose to solids can coat or block the pores or coat some of the active surface. Second, over time, the catalyst may change in physical form, perhaps crumbling into fine dust or sintering together into large lumps.

Reactive getter purifiers. The reaction of a reactive metal with oxygen in a gas stream is an important means of industrial gas purification. For small gas streams containing small amounts of impurity, the getter material is a finely divided alloy of reactive metals. The alloy is chosen to be very brittle. After preparation in a vacuum furnace, the alloy is cooled and comminuted in a vacuum or gas inerted grinder then sintered just below its melting point to form pellets or larger cylindrical 'pills' in another vacuum furnace. The pellets are then loaded into a cartridge which is inert purged and shipped to the user.

After installation, the user simply heats the unit with a simple heater-tape-type heater (although larger purifier units have an internal heater) for an hour or so, which activates the getter material. On exposure to air, the getter material reacts until it has formed an impermeable oxide layer and is then stable. The heating process activates the getter by allowing the oxygen to diffuse out of the surface into the bulk of the getter alloy, thus presenting active surface for deoxygenation action. The purifier can be run hot or periodically heated to present fresh reactive surface. The capacity of a getter material for absorbing an impurity, defined with respect to breakthrough of an impurity at, say, the ppb level, will vary from gas to gas, depending on the heat of reaction of the gas with the getter. Oxygen is strongly bound and an oxygen capacity of 20% is typical, whilst for gases such as CO or hydrogen, which are weakly and reversibly absorbed, the capacity may be as low as 0.6% or 0.4% by weight.

Simple once-through getter purifiers are available for inert gases (He, Ar and Ne); an alloy of Ti or Zr is typical. The SAES company of Milan, Italy, are leaders in this field. They have developed special alloys such as Zr/Fe/V which give improved performance relative to titanium, and removal to ppb levels, and good capacity for nitrogen, hydrocarbon and CO_x as well as oxygen. The purification of nitrogen with reactive metals is more difficult than that of inert gases because the most reactive metals also react with nitrogen and so are unusable. The SAES company has shown that ppb levels of purity can also be achieved on N_2 using a zirconium/iron-based alloy. The alloy has a composition of 75/25 Zr/Fe alloy, which although it slowly adsorbs nitrogen to the extent of a few per cent by weight, it remains active towards impurities. Running at 350°C it will remove CO_x and hydrocarbons to 10 ppb or so, as well as removing moisture and oxygen to undetectable levels.

Once-through getter purifiers are now very common in semiconductor manufacturing plants, being used as 'point-of-use' purifiers. A point-of-use purifier is largely used as an insurance policy: it ensures that gas fed at nominally high purity from a gas distribution system or down a long pipe from a cylinder placed outside will still be guaranteed to be of high purity. Changing of filters, reverse flow and changing of gas cylinders can introduce small amounts of air and moisture, even where the gas supply itself is of the best quality. Point-of-use purifiers protect against this, as well as upgrading the purity of the gas.

Once-through getter purifiers are too expensive for high flow rates and for the purification of gas streams which start off at a lower level of purity. A typical larger scale reactive purifier therefore has to be regenerable, for example the supported copper type occasionally used to remove oxygen from inert gas streams. The copper, spread out on the surface of the inert porous mineral support, is first reduced fully to copper by a heated hydrogen/nitrogen mixture. Then oxygen in any gas stream, up to a fraction of 1%, passing through the purifier will be absorbed in the formation of copper oxide. After partial exhaustion of the bed over a few hours or days, it can be regenerated in a few hours by heating with hydrogen. A similar purifier system that might in the future become popular with semiconductor plant users is based on a supported nickel bed. The nickel-bed purifier, originally developed by AIRCO but now marketed by the SAES company, is aimed at further raising the purity of already good purity cryogenic nitrogen from contaminant levels of a few or tens of ppm to ppb levels.

Membrane diffusion purifiers. Palladium diffusers are an unusual type of purifier usable only for hydrogen. The diffuser purifier allows impure hydrogen under pressure on one side of a heated thin palladium/silver membrane. Hydrogen dissolves and diffuses through the membrane leaving impurities behind. In effect, the hydrogen reacts to form an interstitial hydride with the Pd/Ag alloy, diffuses through the heated bulk metal and then reverses the reaction on the low-pressure side of the membrane, forming H_2 gas once again.

In principle, the membrane method is similar to the air-separation membranes described above. Just as with the polymer membranes above, however, there are problems with leaks. This is particularly the case because the Pd/Ag membrane undergoes some metallurgical changes with high-pressure hot hydrogen becoming, like many metals treated with H_2, brittle and subject to cracks in operation. Also, as in other membrane processes, the membrane must be of the largest possible area and needs to be exceptionally thin to function effectively, and the maximum possible differential pressure must be placed across it, giving ideal conditions for leaks through cracks and pinholes.

Channelling problems in purifiers. The path of gas through a packed purifier unit is never completely uniform. Whether the packing is a support for a solid or a liquid absorber or getter, if that packing is uneven, then the gas will follow the path of least resistance: that which has no packing in the way. In this way, a badly packed purifier may be found to be useless at all but the lowest flow. Just as in distillation column design, fabricated mechanical redistributor plates or additional random structure packing may be useful here. However, an elegant solution is simply to ensure a very even packing of the column with a support material which is not too wide in its particle size distribution. ('Fines' often tend to build up on larger support material and block it up in an uneven way.) Channelling effects tend to get worse with size, giving a potential for difficulties when scale-up of purifier vessels is attempted.

Overheating of purifiers. Another problem that gets worse with size in purifiers is overheating of the bed inside the reactor. It is often forgotten that purification absorbtion reactions are usually exothermic, as are regeneration reactions.

Models of heat lost from and created in a purifier can be very complex. However, simple assumptions on the reactor will lead to a parabolic curve governing the overheat in the middle of cylindrical vessels which is often similar to that seen in practice

$$T \simeq T_w + (R^2 - r^2)\, Q_v/2S$$

where T is the temperature between the walls, T_w is the temperature at the wall, Q_v is the rate of heat generation per unit volume, S is the thermal conductivity of the packing, r is the radial position and R is the wall radius.

This means that for small Q_v or large S the temperature rise will be small. However, if Q_v is large and S is small (as it is for many mineral support packings), then the temperature will rise to a maximum along the axis, with squatter cylindrical or larger reactors considerably worse because of the R^2 factor.[16] The use of higher conductivity packing can be indicated, or the use

[16] A new design of large purifier that the author once set up was subject to overheating problems which only surfaced after an extended period. The unit concerned was of the dry-getter type, where the getter was regenerated with a gaseous mixture. In normal service, the unit ran for some hours purifying, then switched for a shorter period to regenerate. All went well until one day the source gas deteriorated badly and was fed to the purifier rather than being vented. The purifier worked hard to purify the off-spec gas and then switched over to regenerate. Then the regenerate gas began to flow and added its exothermic reaction heat to that of the usual heaters. The temperature rose and continued to rise, at least in the middle of the reactor bed, despite the thermostat cutting off the heaters. The getter was finally overheated and sintered together. The purifier was ruined, although the only external sign of this was a small plastic control pneumatic signal which had melted in the radiant heat from the reactor vessel.

of multiple parallel beds, where temperature rises are thought likely to be troublesome. However, a preferrable solution in computer-controlled units may well be modification of the software. By allowing a slow start to regeneration, perhaps using a diluted regeneration gas, for example, overheating may be avoided in that part of the cycle, whilst overheating in operation might be avoided by sensing impurity levels and reducing mass flow through the unit when levels are high.

2.2.12 Preparation of gas mixtures

With so much effort going into separating gas molecules, it may be a surprise that a highly profitable, though small, part of the gas industry is involved in mixing gases together again. In larger quantities, a gas mixture is most often produced at point-of-use by mixing pure gases with a mass flow controller or pressure regulator-based rig. Such a rig is not quite as straightforward as might be imagined, however, for example a turbulent mixing vessel may be necessary at the injection point of one gas into another. Also, accurate, calibrated flow controls are needed, which function to give an accurate ratio of gases at different flow rates, and there may be problems during start-up or shut-down, with off-spec mixture being briefly produced. Because of these problems, good mixing rigs are relatively expensive for applications needing small quantities. Cylinders of compressed gas mixtures are the best solution where modest supplies are needed, offering convenience, reliability and low capital cost.

In principle the preparation of cylinder gas mixtures is very straightforward. However, some precautions are needed to ensure that the mixture is safe, is actually mixed (and will not condense and unmix) and is accurate.

Safe gas mixtures. Clearly some mixes are completely impossible; a request for a mixture of ammonia with hydrogen chloride, for example, will result in a cylinder containing only a few parts per million of the mixture (at equilibrium) and a large lump of solid ammonium chloride, as the gases spontaneously react.

Some mixes, typically those which contain an oxidiser and a fuel within their flammable limits, are even more dangerous because they are perfectly stable until ignited, when a dangerous explosion will ensue. The prediction of flammability at high pressures is not straightforward. Oxygen mixes, for example, are often more dangerous at high pressure, with much wider limits of flammability.

Other mixtures, those containing reactive gases at low levels diluted in inert gases, will require special cylinder preparation. Over a period of months or less, components like NO or HCl in a gas may react with the cylinder walls or contaminants on them and may disappear altogether. Although aluminium cylinders are intrinsically lower in reactivity with most gases, even Al cylinders need special preparation; drying by heating during evacuation, for example, prior to filling. For ultimately stable mixtures, aluminium cylinders which have been polished followed by addition of a controlled oxide passivating layer (to replace that formed spontaneously in air) offer one of the best solutions. Anodising can be used to provide oxide passivation of aluminium cylinders. Proprietary (some nickel-based) coatings are available for improving the performance of steel cylinders. Cylinders can also be passivated against reaction with fluorine-containing mixtures and other highly reactive mixtures by a controlled exposure to diluted fluorine. Using passivated cylinders, gas mixtures such as 50 ppm H_2S in nitrogen, which are proven stable after five years, can now be supplied.

Filling procedures for mixtures. The simplest way to fill an accurate gas mixture with a mixture of non-condensing gases is to fill by pressure. The smallest component by volume is added first, then the next smallest constituent, then the balance gas.

Often greater accuracy is needed, however, and this is where accurate gravimetric filling is called for. A fundamental difficulty here is the small relative weight of gas which must be measured. A 100 kg cylinder will contain only around 10 kg of gas at typical molecular weights and only 0.5 kg hydrogen, even at full cylinder pressure. However, a 100 ppm component may need to be measured to 1%, which means that 10 mg must be measured on a 100 kg cylinder, requiring a very accurate balance. A further fundamental difficulty here is the effect of the gas connection on the scale reading. A flexible 'pig-tail' of pipe is necessary, and care must be taken so that the pig-tail cannot cause inaccuracy. After filling, a more accurate gravimetric check measurement can be made with the pig-tail disconnected. The gravimetric method can be further extended to low-level mixtures by further gravimetric dilution of a premixture of (say) a gravimetric 1000 ppm mixture; low ppm mixtures can be reached with gravimetric accuracy with a relatively simple weighscale.

Standards laboratories, such as the National Physical Laboratory (NPL) in the UK and the National Institute of Standards and Technology (NIST) in the USA, produce gas mixtures to internationally agreed best procedures known as standard reference materials (SRMs). Gas mixture manufacturers may offer traceability of their gas mixtures to the SRMs as an alternative to cylinders prepared to traceable weight standards via a gravimetric production route.

Low-level mixtures of vapour in gas are most conveniently prepared by injecting the vapour in liquid form with a calibrated syringe into the gas stream going to the cylinder.

Ensuring homogeneous mixtures. The simplest means of mixing cylinders is simply to lie the cylinder down (section 1.3.10). However, for speed of

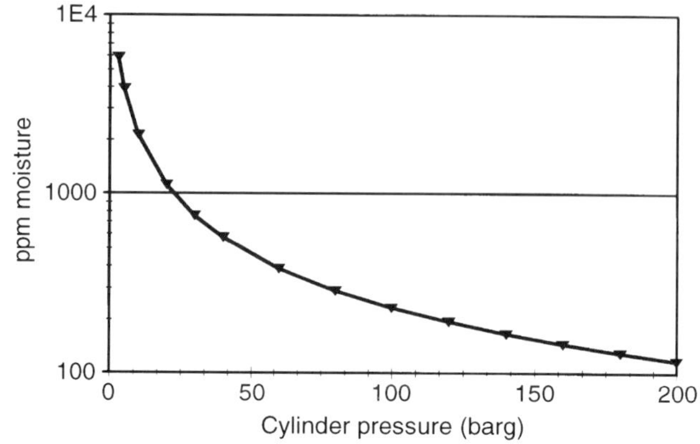

Figure 2.26 Theoretical worst case moisture in a cylinder gas versus cylinder pressure.

operation and to ensure mixing of very heavy and very light gases, 'rolling' is often specified. This involves placing the filled gas cylinder horizontally on a pair of electrically driven parallel rollers. After only a few minutes even a mixture of a very heavy gas with a very light gas, such as SF_6 with hydrogen, will be thoroughly mixed. The horizontal attitude in any case ensures diffusional mixing will take only hours in a normal-sized cylinder and rotation at 60 rpm or so invariably brings the mixing time down to well under an hour.

For large quantities of a gas mix, a 5 m long 'torpedo' or 'tube' cylinder with a capacity of around 110 Nm^3 at 175 bar pressure may be used. This can be filled, mixed by rotation and then decanted into ten or so standard-sized cylinders for delivery, with the large cylinder used for mixing never leaving its horizontal rollers. For mixtures such as those for welding, which do not need to be very accurate and are used in large quantities, filling and handling in the vertical position is convenient. For this, it may be an advantage to fit dip-tubes to the cylinders. A dip-tube tends to give a more effective circulation in the gas cylinder, particularly if light gas is injected into the heavy gas. Attempts have been made to devise special designs of dip-tube which further enhance mixing in the upright gas cylinder, for greater convenience, such as the patent by Jolley (1990) describing a 'venturi' attachment to a half-length dip-tube.

Moisture in gas cylinders. A frequent source of complaint about cylinders of gas mixtures concerns varying levels of moisture as the cylinder is consumed. It is always worth remembering that in a typical gas cylinder a small amount of moisture is always present and this moisture will tend to become concentrated at low pressures as the vapour pressure of the water becomes a larger fraction of the total gas pressure. The (hopefully rare) example of a cylinder that has accidentally been partially filled with water prior to filling with compressed permanent gas is shown in Figure 2.26. This illustrates the moisture content at the outlet, as a function of cylinder pressure, assuming the water vapour stays roughly in equilibrium.

2.3 Gas measurement and analysis

Many gases must be measured in quantity and quality for their suitability for a process. In addition, most articles of commerce have to be measured or counted so that they can be correctly accounted and charged for. Today, also, materials which may end up in the environment after use, or as a result of accidental release, must be monitored before discharge and in the environment. Gases are no exception to these requirements but they are not so easy to measure. The fundamentally intractable properties of gases, invisibility and intangibility, mean that human senses are not useful. Instrumentation of a more or less sophisticated level is essential to work with gases in all but the simplest applications.

2.3.1 Gas flow measurement

Simple measurements are often not difficult with gases. Gases can, like most substances, be weighed in containers although, as noted in section 2.4.7, the container weight is high, which causes accuracy problems. Pressure measurements are also fairly straightforward. If access to the gas container is available, then gas volume can be often be more accurately measured there, via pressure and/or weight measurements, rather than by instruments in the pipeline.

For compressed permanent gases, unless accurate scales are available, it is simpler and reasonably accurate to measure pressure and use the ideal gas law (with a table of compressibility Z values if available). Volumes of gas containers are known or can be measured by weighing empty and full of water. If accurate measurements are made of pressure, the contents of compressed gas cylinders can be accurately given by pressure, and it is often forgotten that observation of the fall in cylinder pressure is capable of giving an accurate average flow rate in many applications. The advent of accurate, relatively drift-free electronic pressure gauges means that flow rate via display of electronic rate of change of pressure can be considered; similarly, flow rate via electronic rate of change of weight is perfectly possible, although apparently rarely done.

However, access to the gas container is often not available, and in any case a more direct measurement of instantaneous flow is usually needed. Flow measurement of gases at an approximate level is not difficult, as the simple rotameter shows. However, accurate flow measurement is much more

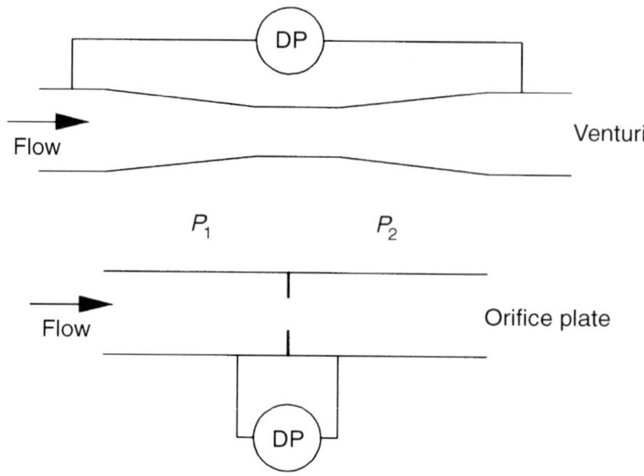

Figure 2.27 Venturi and orifice plate meter systems.

difficult and few techniques are capable of much better than 1% accuracy unless elaborate precautions are taken, as in laboratory tests.

Venturi and orifice meters. Probably the simplest way to measure a gas flow is a to include a venturi or an orifice plate (Figure 2.27). The flow is related to the square root of the pressure drop across the device. The venturi can be regarded simply as a low-loss orifice and both can be treated by similar theory.

This arises simply as a consequence of Bernoulli's equation (section 1.3.5). First, assume that the venturi or orifice is small compared to the general pipe diameter. Then V_1, the velocity in the throat of the device, will be considerably larger than V_2 and $V_1^2 \gg V_2^2$, therefore V_2^2 may be neglected

$$P_1 + \rho_1 V_1^2/2 = P_2 + \rho_2 V_2^2/2$$

Taking a small pressure drop, with density ρ approximately constant and $\Delta P = P_1 - P_2$

$$\Delta P = \rho V_1^2/2$$

Now noting that Q, the volume flow, is given by AV_1

$$Q = AC_d \sqrt{(2\Delta P/\rho)}$$

Note that the Q value obtained also depends on the absolute pressure of the venturi/orifice via the density. Note also that Q will vary from gas to gas, depending on the gas density. The constant C_d is a correction factor for the particular geometry of venturi and is close to 1 for good venturi devices. C_d is less than 1, typically 0.65 but varying with Reynolds number, for orifice plates.

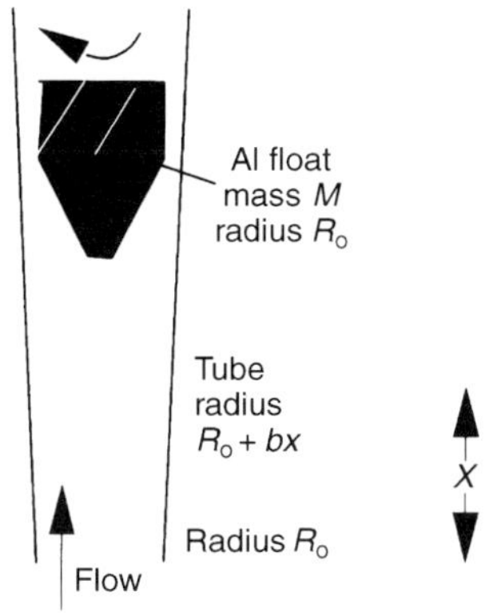

Figure 2.28 Rotameter.

Venturi meters are preferred when accurate measurements are made, or when a low net pressure drop is required; a good venturi will have very little energy loss as the pressure lost in the constriction is nearly all replaced in the diverging section downstream. Orifice plates are used for convenience; they can be inserted into a flanged joint, for example, quickly and at minimal cost. They have the disadvantage that they induce a high pressure drop and are not in general so accurate.

Rotameters. The rotameter or floating ball gauge is very common, despite the increasing usage of electronic flowmetering. The word 'rotameter' includes the 'rota' prefix to indicate that it is a meter with a rotary principle. Accurate rotameters employ a bullet-shaped float with small angled flutes. If flow is present, the float rotates, showing that the meter is working and not jammed.

The basic concept of a rotameter is very simple: if gas flows, it forces the float up the slightly conical inside of the sight tube until the annular gap past the float sustains a differential pressure equal to the weight of the float divided by its area.

A simple rotameter can show a roughly linear relation between float position and volume flow rate (Figure 2.28). To see how this can arise, the formula above for the venturi/orifice meter is first noted, using the annulus area between float and tube as the effective orifice

$$Q \simeq \text{orifice area} \times \surd(2\Delta P/\rho)$$

For a float of mass M, radius R_o

$$\Delta P \pi R_o^2 = Mg$$

or

$$\Delta P = \text{constant} = Mg/(\pi R_o^2)$$

For a tube with a conical shape, $R = R_o + bx$, where the position of the float is x. Hence

$$Q \simeq 2\pi R_o bx\sqrt{(2Mg/(\pi R_o^2\rho))}$$

Note that the dependency on the gas density ρ means that rotameters must be calibrated for different gases.

Although simple and popular, rotameters have a number of problems in practice. The sight tube must be accurately made and has to be glass to avoid the wear and thermal expansion problems inherent in other transparent materials. A glass tube with high pressure inside is potentially very dangerous so a rotameter should generally have a safety tube of clear, strong plastic, such as polycarbonate or polyester, surrounding it. The glass tube needs to be sealed into the pipework system with O-rings, with precautions to avoid leaks. The glass tube is itself not objectionable from a chemical point of view in nearly all gases. However, the polymer O-ring seals can be attacked by reactive gases and the possibility of the glass tube leaking a hazardous gas after breakage is also a problem.

Common designs are limited to 10 bar or so pressure. Some of the problems can be avoided in designs where the position of the rotameter float does not have to be sighted optically. One common design has a magnetic float, which moves an external indicator needle and can operate up to 200 or 300 bar. Another limitation of rotameters is the limited flow range over which they are reasonably accurate. A typical unit can only read to a few per cent over a flow range from 20 to 100% of full-scale. Some rotameters are fitted with two floats of different materials. At the bottom is a high-density steel float whilst above it is a lighter glass or aluminium float, which extends the range available by a factor of the square root of the float density (for identical float dimensions).

Positive displacement meters. If a gas is at a known pressure, then pushing it (gently, so that the known pressure is not changed) through a cylinder/piston or similar positive displacement machine will measure its flow; each revolution of the machine will meter out a known volume of gas.

This is the principle of domestic fuel gas meters and is also widely used in industry where it is practicable as it is both reasonably low in cost in small sizes and accurate. Bellows seals are used, to avoid possible wear and gas leaks via sliding seals, and these meters can be made at surprisingly low cost.

However, this kind of meter has quite a number of problems:

- high mechanical wear and low accuracy when used at high flow rates;
- large sizes difficult and expensive;
- the bellows used are subject to damage with pressure differentials;
- normally only usable at low pressures;
- normally only usable on gas near to room temperature.

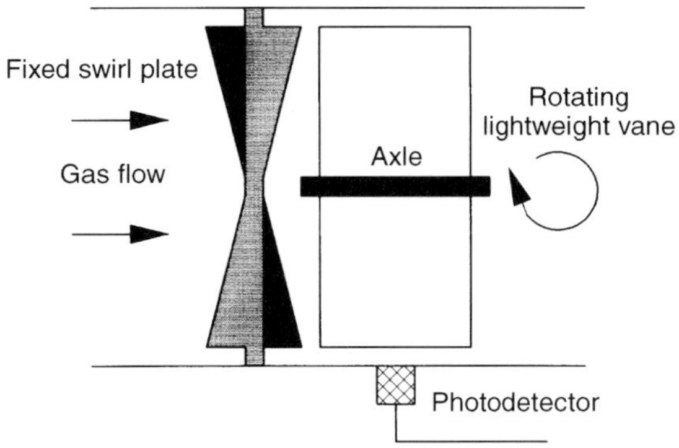

Figure 2.29 Swirl plate turbine flow meter.

Wet versions of these types of meters, for instance those using vanes sealed with water, are used in laboratories and can be made very accurate. However, adding water or oil vapour to the gas, which may then need to be removed, is undesirable in industrial use. Sliding vane meters as used in liquid metering, where the vanes are lightly spring-loaded so as to present a minimum resistance to flow, are also used and can be made accurate.

Turbine meters. In larger industrial installations the turbine meter is very important. A typical turbine meter consists simply of a rotor, often somewhat like a high-pitch marine propellor in form, rotating on an axis down the middle of a special pipe section. The rotor may simply operate a meter display via reduction gears. Alternatively, the rotor may contain a magnet which gives electrical pulses as it passes a pick-up coil mounted on the pipe section wall; this arrangement minimises possible mechanical problems. The bearing used for the rotor is crucial as it must remain absolutely free-running, even in the prescence of reactive gases and adventitious impurities, such as water, which may exacerbate the corrosivity of the gas being monitored. Tungsten carbide or other ceramic ball-bearings are sometimes used.

The 'swirl plate' turbine is a particularly simple form of the turbine meter and can be made highly accurate (Figure 2.29). It comprises a fixed stator which imparts a rotation to the gas. This rotation is then sensed by a thin flat vane rotating about a central axis. The low inertia of the vane relative to a standard turbine impellor means that the swirl plate turbine responds more quickly to changes in flow rate.

This principle has been used in an electronic spirometer for human breathing measurement. In this, the vane is an ultralightweight plastic type, metal coated. Its rotation is sensed by a photodiode which is illuminated by an infrared beam reflected off the vane from a light-emitting diode.

Vortex shedding meters. Vortex meters work by placing an ingeniously unstreamlined body in a gas stream. As the gas flows, it generates a regular series (a 'von Karman street') of eddies; vortices which are carried along with the stream behind the body. It so happens that, provided the stream velocity lies between certain limits, the number of eddies is precisely proportional to the volume of gas that has passed. The eddies are normally detected with a small piezoelectric microphone or with small anemometer hot wire sensors. Although technically interesting, they are rarely used in the gas industry.

Correlation flow meters. In a sense related to the vortex shedding meter above, in that they rely on turbulence in the flow, correlation meters measure a gas parameter, such as instantaneous pressure $P(t)$, at two points x and y, and correlate the signal at the two points, one upstream of the other by a short distance. The correlation is performed by a small microprocessor, based on a simple algorithm, such as

choose ΔT
calculate correlation sum: $[\Sigma P_x(t+\Delta T)P_y(t)]$ (summed over a sequence of 100 or so t values)
choose different values of ΔT
find maximum C_{sum}
output ΔT (corresponding to maximum C_{sum})

The correlation summmation has a maximum when the offset time, ΔT, is such that the two signals have a similar time sequence. ΔT essentially gives the transit time of flow noise from x to y, and, knowing the distance x to y and the pipeline flow characteristics and cross-section, can be used estimate volume flow.

Thermal mass flow meters. The thermal mass flowmeter, often abbreviated to MFM, is now one of the standard forms of flowmeter used in the gases industry and in some user industries (Figure 2.30). The MFM is often used as the sensing element in a mass flow controller (MFC). The thermal mass flowmetering element is followed by a solenoid control valve which is operated by an electronic feedback circuit such that the flow through the valve is regulated to be equal to the set flow required. MFCs are now the standard instrument for the regulation of relatively small gas flows in gas applications in semiconductor manufacture.

Most MFMs rely on two resistive windings around an insulated small bore stainless steel tube. Essentially, the windings heat the tube up. In the absence of flow, the two windings on either side will be at the same temperature and therefore the same resistance. These windings are connected into a bridge circuit. Once flow begins, the temperature of the downstream resistance winding will increase whilst that of the upstream

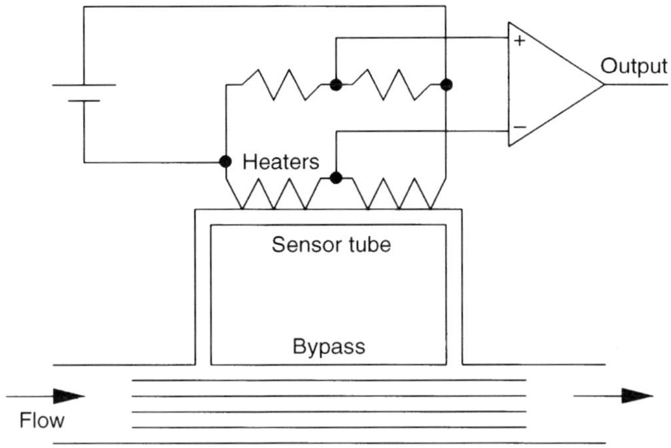

Figure 2.30 Thermal mass flow meter construction.

winding will decrease, unbalancing the bridge circuit. The unbalanced voltage across the bridge, if the construction of the sensor is correct, is approximately linear with the thermal mass of gas flowing. Gases of different thermal capacity will give different results but this is easily allowed for.

For flow rates which are large, above about 50 $cm^3 min^{-1}$, the simple tube sensor becomes nonlinear and inaccurate and presents too large a pressure drop. This problem can be handled by arranging a bypass flow path so that a small but constant fraction of the flow bypasses the sensor tube. The approach generally taken is to operate both the sensor and the bypass in the linear flow regime by using very small effective tube diameters (and many tubes in the bypass circuit).

The thermal MFM has wide applicability: its wholly stainless steel bore is inert to most gases, its output is independent of the supply and outlet pressure, its construction is such as to give good long-term stability, and it can be built to have a fast (<1 s) response time.

Thermistor/hot-wire anemometers. The thermistor or hot-wire anemometer is frequently used to measure gas velocity but it can also be installed in a pipe section to act as a flowmeter (Figure 2.31). It is also a convenient instrument with which to probe gas flows in reactor vessels, for example. Its advantages are that it gives gas flow speed irrespective of orientation of flow and over a wide dynamic range. By using tiny pieces of tungsten or platinum wire, and by using miniature bead thermistors (sizes down to about 1 mm square are readily available), a high speed response can be obtained; as little as milliseconds. Also, by using an array of these tiny sensors, or by scanning them mechanically, gas velocity can be probed on a scale of millimetres.

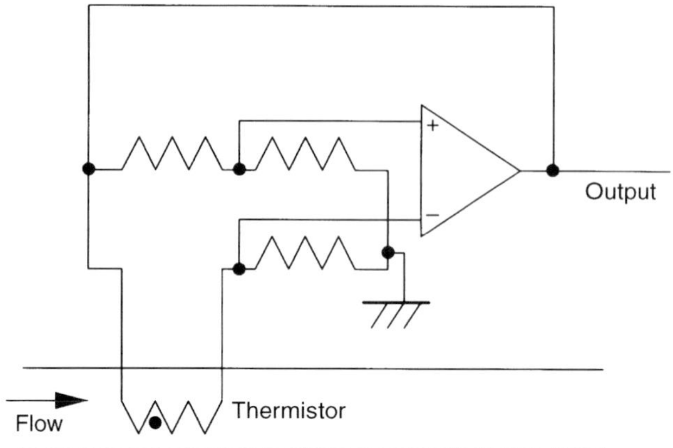

Figure 2.31 Thermistor anemometer principle and circuit.

The most common form of thermal anemometer utilises a feedback electronic circuit to maintain the sensor element at constant resistance (and therefore temperature). The parameter measured is the power needed to maintain this set-point elevated temperature. The elevated temperature can be anything from a few °C above room temperature for a thermistor to hundreds of °C for a hot-wire sensor.

At extremely low flow rates, there is some distortion of the response due to convection currents from the sensor. However, at even small flow rates the forced convection response is seen; it is nonlinear but can easily be calibrated. Various forms of equation can be used for the heat loss from the wire or thermistor but they all give approximately the same fourth power in voltage applied result for the flow velocity

$$\text{rate of heat loss, } \mathrm{d}Q/\mathrm{d}t \simeq A(U\rho + B)^{0.5}$$

where A and B are constants, U is the gas velocity and ρ is the gas density. B is fairly small in normal operation. Noting that

$$\mathrm{d}Q/\mathrm{d}t = V^2/R$$

where V is the voltage applied and R is the (constant) resistance of the sensor element, we can convert the equation to give the electrical response to flow, i.e.

$$U = (V^4/A^2R^2 - B)/\rho$$

Temperature compensation is necessary, particularly for the lower temperature thermistor-based anemometers. A change in the gas temperature will change the values of A and B. The selection of wire or thermistor element is made on criteria of high-temperature coefficient of resistance (this increases the accuracy of the temperature regulating circuit), corrosion resistance and resistance value to match the supply voltage chosen.

Ultrasonic flowmeters. The blood flow indicator used in medicine is an example of a Doppler ultrasound meter. It uses 2 MHz soundwaves reflected from particles in human blood to give a Doppler shift to the reflected signal, which gives a measure of the blood velocity. Ultrasonic meters for gases cannot function in this way because gases contain little or no particulates in suspension and must rely on the intuitively straightforward time-of-flight technique or an equivalent phase shift. Meters have been made that will work in pipework down to 10 mm or so using 100 kHz signals and phase shift detection. However, the technique is still only applied to gases rarely and costs are high relative to more traditional flowmetering methods.

Two tranducers are needed, typically mounted in a special pipe section which allows smooth flow of gas while allowing ultrasonic pulses or modulated waves to cross the pipe diagonally. The transit time in both upstream, T_u, and downstream, T_d, directions is taken together with the distance apart, D, and the resultant velocity, $V\ (=2D/(T_u - T_d))$, calculated is very nearly the average gas velocity, with the effects of gas composition (speed of sound varies as $1/(\sqrt{\text{molecular weight}})$) largely eliminated. There are quite a few problems with such a system in the signal processing. First of all, a pulse applied to an ultrasonic transducer produces a rather messy signal because of resonances in the transducer and its mounting. Secondly, the received pulse is even more messy due to multiple reflections and conduction along the pipe walls (the speed of sound in the steel of the pipe is much higher), and it is difficult to define arrival time precisely. However, the arrival time changes only slightly with normal gas velocities and it must be measured precisely. Inexpensive digital signal processing hardware allows sophisticated processing to alleviate these problems. Gated-phase measurements on the transmitted and received pulses, for example, can be used; the measurement uses phase-shift on a portion of the received waveform which is large enough to be clear of small contaminating signals but early enough to avoid the larger multiple-reflection signals.

The use of ultrasonic devices on larger pipes is now well-established: larger pipes are technically easier to measure. On larger pipes, too, measurements over four or more paths across the gas stream can be made by using three or four transducers. This allows corrections for changes in radial flow velocity profile. However, with the increasingly lower cost of processing capabilities, it is to be expected that ultrasonic flowmetering could become more common in small-bore pipe systems. Figure 2.32 shows an ultrasonic flowmeter intended for low-pressure methane service.

Tracer techniques for gas flowmetering. Some of the most accurate methods, achieving as good as 0.2% in the laboratory, rely on the use of

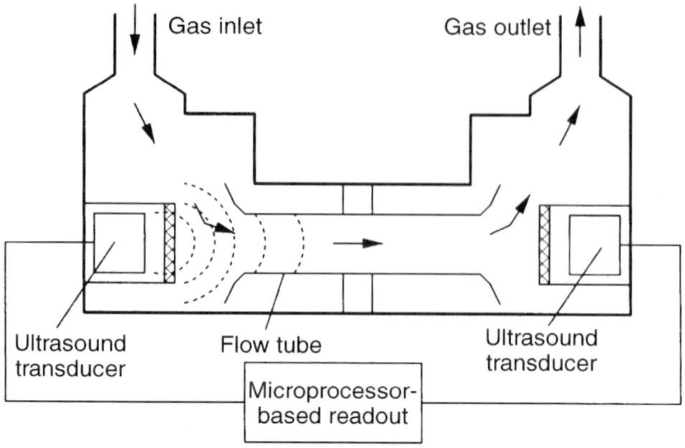

Figure 2.32 Ultrasonic flowmeter.

special tracer techniques. Radioisotopes can be used as the levels needed are low enough to be allowable under health and safety guidelines. Krypton-85 is typical, its β emissions being readily detectable with a plastic scintillator/photomultiplier combination. Other gases can be used as tracers, an unusual noble gas for example, with a sensitive mass spectrometer detector. Even ionisation has occasionally been used in tracer-based flow measurement.

Tracer techniques can employ either the steady-state dilution method or the pulse-injection method. In the first case, the concentration of a known flow rate of tracer is measured well downstream of injection. The known flow rate is then divided by the concentration to obtain the pipeline flow rate. In the latter technique, two detectors, one well downstream of injection, the other even further downsteam, constantly monitor the tracer concentration in the pipe. Small pulses of tracer are added and these give rise to peaks of signal at different times at the two detectors. The distance between detectors, L, the interval between the half-height of the peaks, ΔT, and the cross-sectional area, A, of the pipe yield the volumetric flow rate, V

$$V = LA/\Delta T$$

Both methods rely on achieving thorough mixing across the diameter of the pipe. Steady-state dilution also has to rely on a known mass flow rate of tracer, however, whereas pulse-injection depends only on the physical dimensions of the pipework. Pulse-injection is also more suitable where the tracer has a very short lifetime, as is the ionisation injection technique.

Measuring gases as liquids. Liquid nitrogen and other cryogens can be metered in a number of ways but are often not straightforward to measure because of a tendency to boil and flow in both the liquid and gas phases. Some of the most common techniques for measuring cryogens are:

- buoyancy gauges in small tanks with top access;
- differential pressure (DP) gauge on tanks (this is not very accurate on large flat tanks);
- capacitance level gauge on tanks (this is also not very accurate on flat tanks);
- rotary turbine meters (LN can boil inside them and cause them to rotate at very high speed, albeit briefly);
- calibrated venturi meters;
- load cells indicating tank weight.

Gravimetric measurement of tanks contents is often used as a 'gold standard' to calibrate other means of flow and level measurement. However, load cells are inconvenient and expensive (when accurate ones are required) and are subject to errors during stormy weather because of wind and rain. Where there is a nearby weighbridge, tankerloads of liquid cryogens can usefully be weighed before and after delivery as a means of calibration. It might be thought that the ideal way to measure cryogenic liquids is to use a vibrating loop type of Coriolis mass flowmeter. In principle, these would be ideal, compensating precisely for the presence of bubbles in the two-phase flow. However, these are expensive and apparently do not fulfil their promise in practice because there are little or none in service in the industry.

Differential pressure tank gauges. The differential pressure on a liquid cryogenic tank is in general a very good measure of contents. With an ordinary liquid tank DP is not particularly convenient. Two single-ended gauges must be used and the readings subtracted electronically or the pressure difference between the vapour space and a constantly bubbling dip-tube must be taken. However, with cryogens a single differential gauge can be used, with its two inputs connected to the top and bottom of the tank, because the two lines can be kept clear of any liquid by arranging for them to be uninsulated when the cryogen boils off leaving gas. Most cryogenic tanks have a high aspect ratio, length to width, and have a constant cross-section for over 80% of their length. They are most often placed vertically. Furthermore, tanks are always run with ullage space in the top dished end. These facts all conspire to make a DP gauge a useful device for a cryogenic tank. The variation in density of the liquid in storage does not affect the measured differential pressure

$$\Delta P = \rho g H$$

where ρ is the liquid density, H is the height of liquid in the tank and g is the acceleration due to gravity. If the lower pressure tapping is arranged to be above the lower dished end, then the mass of liquid stored in a tank of area A (lower dished end contents ignored) is

$$M = AH\rho = A(\Delta P/(\rho g))\rho = A\Delta P/g$$

Note that the lines from the tank to the differential pressure gauge must be full of gas to get an accurate reading. The bottom line should be left uninsulated and a restriction such as a needle valve or small orifice can be fitted where the line leaves the tank to guarantee this.

Small cryogenic tanks are often fitted with crude buoyant float gauges. In these, a sprung cylindrical float rod is arranged to project into a small transparent plastic dome with graduations on the top of the tank. When full the float is forced fully up against the spring, whilst when empty its weight falls on the spring and the indicator falls to the bottom of the graduated dome. These small cryogenic tanks are also frequently fitted with more accurate capacitance level gauges. These simply measure the capacitance of a pair of coaxial tubes dipping vertically into the cryogen. The capacitance increases with the amount of cryogen in the annular gap and is measured by a simple electronic circuit.[17] Although less accurate than the DP gauge, these gauges are convenient to fit and inexpensive. They could potentially be offered with an electronic differential output indicating flow directly but this is not an option on devices currently available.

It is unfortunate that one class of tanks for which accurate gauging would be very desirable, the tanks of road tanker trucks, is not so tractable. In Germany, road tanker trucks are fitted with carefully calibrated venturi meters. In the UK and elsewhere turbine meters are used but are often less accurate than a weighbridge.

Gauging high-pressure liquefied compressed gas. Another class of gas storage where contents indication is difficult is inside compressed gas cylinders containing liquefied gas. It would be useful to have a level indication on cylinders containing CO_2 or propane, for example; the pressure gauge on such cylinders only indicates a drop when the cylinder is cool or if it is very nearly empty. Currently, almost all users of these gases have to rely on weighscales.

Special propane tanks, for example those used in vehicles (or even hot air balloons), can be equipped with a mechanical float gauge connected to an indicating needle which is viewed via a small windowed binnacle on the side or occasionally with remote readout. However, CO_2 cylinders are at a much

[17] An even simpler older device immerses a resistive wire into the cryogen. The wire is heated slightly by passing a current. The part immersed in cryogen remains cold and less resistive whilst the part above the level heats and rises in resistance (by a factor of 10 or more if it is heated to near room temperature) so the higher the resistance, the lower the liquid level. The wire is typically mounted inside a sheath tube (open at top and bottom) which protects it and preserves its heat to some extent. It is possible that this type of level gauge could reappear in common use because now the simple wire can be replaced by a length of high T_c (HT_c) superconductor, which would avoid some of the heat dissipation by the wire, that in the cryogen, and be more accurate. One such device, at least, is now available commercially.

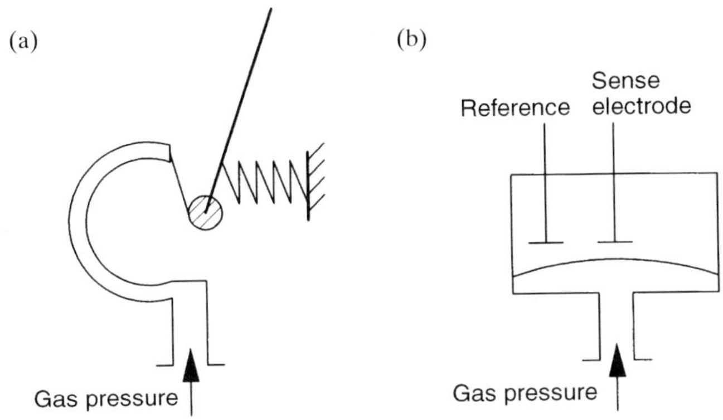

Figure 2.33 (a) Bourdon and (b) capacitance pressure gauges.

higher pressure (60 barg as opposed to 10 barg) than propane and such a device penetrating the cylinder wall is unacceptable. High-pressure versions of the capacitive gauges for use with CO_2 and other high-pressure liquids do not seem to be available currently, perhaps because of difficulties in satisfying the need for ultimate robustness (the gauge head would need to remain on the cylinder during handling, filling, etc.) and necessary safety standards.

2.3.2 Pressure measurements

Bourdon gauges. Pressure measurement in industry still commonly uses the traditional Bourdon gauge where no particular accuracy is needed and where a gauge pressure (relative to atmospheric) is needed.[18] (Figure 2.33) A tube which has been curled into a spiral form when filled with high pressure fluid tends to uncurl. This uncurling movement is linked to a magnifying mechanism which drives the indicator needle across a dial.

Dead volume (the volume of gas which is stored up in the gauge) is a problem with Bourdon gauges. 'Real' dead volume is the actual volume of gas trapped in the gauge and the term 'effective' dead volume is often used to express the fact that a gauge is not a rigid part of the gas pipeline but will increase in volume at high pressure and decrease at low pressure. These changes in volume can, if linear in pressure, be expressed in terms of an equivalent volume of gas.

Effective dead volume leads to inaccuracy in the measurement of pressure

[18] Distinction should always be maintained between 'gauge pressure', measured relative to atmosphere, and 'absolute pressure', measured relative to vacuum. Care should be taken always to express only differential pressures in units like bar or psi, and always to specify single-ended pressures as bara or barg, psia or psig.

in small isolated systems at high pressure. Real dead volume gives this effect but also, more seriously, acts as a reservoir of gas which is not swept by flow. In ultra-pure systems, or in systems handling toxic or explosive gases which must be purged, dead volume is highly undesirable.

In extreme cases, the gauge can be filled with a hydraulic oil, such as silicone oil, reducing the real dead volume, although some effective dead volume will remain. Precautions should be taken if oil is used, however, that contamination of the gas line is avoided. Also, some gases will dissolve in the oil, raising the dead volume again. Some gauges avoid the latter two problems by retaining the oil behind a thin diaphragm or membrane.

The other major defect of the Bourdon tube is the hysteresis intrinsic to the mechanical gearing system applied to the tube end.

For low-pressure work, the Bourdon tube 'motor' for the mechanical pressure gauge is replaced by an aneroid capsule. This achieves greater sensitivity as it comprises one or more thin diaphragms with either a pressure connection or a vacuum inside. The standard domestic barometer is an example of the latter.

Manometers. Manometers using liquid in U-tubes are now used rarely in industry. U-tube manometers are nevertheless still used in laboratories, although the use of mercury as the operating fluid, whose high density makes higher pressures more convenient to measure, has been discouraged by strict health and safety regulations. U-tubes offer simplicity of operation, potentially high accuracy and the minimum of calibration. However, it is usually necessary to use glass sight tubes, with the problems above, and they are quite difficult to read accurately. By sloping the U-tube, and using a low-density liquid such as paraffin, U-tubes can be used for small pressures. These low-reading instruments have now largely been supplanted by digital readout electronic pressure sensors, now available at relatively low cost.

Electronic pressure sensors. 'Electronic' pressure gauges have now become ubiquitous in industry and are gradually replacing many other types of gauge. The basis of a number of related electronic pressure gauges is the electronic sensing of a small movement of a robust stainless steel diaphragm.

Capacitance gauges are the most sensitive, using differential capacitance sensing. They can work anywhere from near vacuum (pressures of 10^{-3} bar) to hundreds of barg, while differential transformers are satisfactory at higher pressures. Capacitance cells are capable of the highest accuracy and are used in equipment where precise pressure is critical. Processes which employ subatmospheric, but not high vacuum, pressures often use capacitance cell sensors. They have supplanted some of the vacuum gauges such as the Pirani gauge where accuracy is needed. At the very highest pressures another form of sensor becomes possible where capacitance change sensed is intrinsic and not due to the movement of a diaphragm: one in which the small change of capacitance of a capacitor with pressure is employed (Andeen, 1971).

Strain gauge detection of the diaphragm deflection leads to a lower cost sensor but one in which drift and noise limit the lower limit of pressure compared to capacitance cells. Electronic pressure capsules can be designed so that the sensing diaphragm is right at the edge of the sensor. With this design electronic pressure gauges effectively achieve much lower dead volumes than any mechanical or manometer gauge can possibly achieve.

There are now tiny silicon-strain-gauge-based pressure sensors. These integrate minute piezoresistive silicon resistors connected in a bridge format with a silicon diaphragm etched out of a piece of silicon a few millimetres across. The silicon is mounted so that gas pressure bells out the diaphragm slightly and differentially changes the piezoresistance values. A simple temperature-compensating and amplifying circuit completes the sensor. These sensors can be made very cheaply; the major cost is often the calibration. (They can even be made disposable; for example, disposable sensors are used to sense dynamic blood pressure pulses in heart surgery.) Again, at higher pressures, it becomes possible to use electrical resistance change in an intrinsic sensor. Alloys such as manganin, originally destined for use as a low-temperature coefficient resistor, can be used in a sensitive bridge circuit to sense pressure via their resistance change with pressure (Anon, 1975).

Temperature measurement. An important element in many measurements on gases is their temperature. This is obviously needed when using volumetric flow measures, for example, but is also needed for corrections in mass flow measurement and is useful in monitoring cryogenic liquids.

At normal temperatures any of the normal or elevated temperature measurements will serve, with convenience, reliability and cost being the main considerations. The small thermal capacity of low-pressure gases occasionally leads to long time constants or errors due to conduction along the thermometer supports, and this is something that needs attention occasionally.

At low temperatures there are more problems. Liquid-in-glass thermometers using mercury are unworkable as they freeze, although coloured thermometers based on coloured ethanol (the lowest freezing of the common alcohols) freeze only at −117°C. Solvent mixtures based on liquid-in-glass thermometers are available down to −120°C (153 K). Liquids with large liquid range, such as ethanol (−117 to 78°C), isopentane (−160 to 28°C) and isopentanol (−117 to 131°C) are the most useful.[19]

[19] Interestingly, ethane (freezing point 90 K) was used at one time, in combination perhaps with other hydrocarbons and a dissolved blue dye, as the active liquid in liquid-in-glass thermometers down to cryogenic temperatures.

Wirewound resistance thermometers are generally bulky but they are simple and function at all temperatures. Cheaper are carbon film resistors, of the type used in electronic equipment. Because of small changes in thermal cycling down to cryogenic temperatures, these are not suitable for accurate work. However, they work right down to liquid helium temperatures from room temperature and are easily calibrated, providing a relatively fast change of resistance with temperature down to the lowest temperatures. Resistivity of metals, such as high purity copper, follow a steady fall in resistance, although below 10 or 20 K or so, depending on the degree of impurities in the metal, the resistivity changes more slowly with temperature.

Thermocouples continue to function to the lowest temperatures but the signal obtained at cryogenic temperatures is small and noisy. A gold/2% cobalt alloy on a copper thermocouple has over twice the thermoelectric power as the standard copper/constantan thermocouple (around 17 mV K^{-1} at 80 K. Compared to thermocouples the signal available from simple diode thermometers is bigger and less noisy. A simple silicon diode basically has forward voltage drop *d* which is proportional to absolute temperature at a constant current, and diodes specially trimmed are now available which give precise temperatures with large, easily used electrical signals. Once these simple and inexpensive devices become as well-known to mechanical and chemical engineers as they are to electronic engineers and physicists, their use will become much more widespread.

2.3.3 *Analysing gases*

The purity of gases is an important part of their specification. For example, in an application where argon is used in a high-temperature furnace to protect against oxidation or nitridation by the air, the presence of O_2, N_2 or H_2O will be very important. Also, the monitoring of gas composition in production plant and gas-using plant is a useful diagnostic for plant problems. For example, the presence of additional helium or neon in nitrogen from an air-separation plant might indicate failure of the column top vent. The presence of hydrogen in the nitrogen from a PSA plus purifier nitrogen supply might indicate failure of the catalyst in the purifier. After a failure incident involving gases, it is important to analyse the gases used for purity and correct majority component. Occasionally inexplicable failures are traced to the wrong gas having been put into a cylinder or to residual gas in a cylinder which was not 'blown-down'. For these reasons, analysing gases is an important part of industrial gases operations and all gases are supplied to an analytical specification with, if required, a certificate of analysis (Table 2.8).

The analysis of gases for their molecular constituents is itself a large subject, with many books and journals devoted specifically to it. As early as

Table 2.8 Argon gas purity specification (typical industrial specification 1995)

Argon	> 99.999%
O_2	< 3 ppm
N_2	< 8 ppm
Hydrocarbons	< 1 ppm
H_2O	< 1 ppm

the late eighteenth century gas analysis had become well-developed, so that Lavoisier could remark in his chemistry book of 1789: 'It is necessary to become familiar with the analysis of gases by long experience.' His book in fact supplanted that need by describing most of what is necessary to carry out bench analyses of gases. Gas analysis is now a very powerful branch of chemical analysis, so much so that many analyses are carried out by converting samples into gas so that the power of sophisticated gas analysers can be turned on to them. This section deals with analysers for industrial gases, however, and with continuous reading instruments, rather than with bench methods and single batch measurements.

Gas analysers can be complex computer-driven instruments capable of simultaneously analysing hundreds of gas components. Multigas analysers, such as gas chromatographs (GC) and mass spectrometers (MS), are used routinely in industry. However, the workhorses of the gases industry are undoubtedly single gas analysers of various types. Most of these have now attained a certain level of sophistication and are supplied by specialist suppliers. These are described in later sections. However, the simplest of these are really exceptionally simple, very low in cost and can often be constructed by the user rather than by a specialist. Some examples of simple analysers actually in use in particular circumstances are as follows.

- *Two electrodes in a distilled water bubbler* (for acid gases such as CO_2 and SO_2, perhaps also for alkalis such as ammonia). With precautions to ensure good dissolution, the conductivity of the water gives a precise measure of the gas which has been bubbled through. This principle, with the addition of hydrogen peroxide to the water to convert SO_2 and NO_x to sulphuric and nitric acids, is used in the statutory measurement of acid gas emissions from exhaust stacks, for example.
- A *hot wire* which, when it is exposed to carbon-containing gas, changes its resistance. A number of common resistive wires show this effect, which is sometimes seen as a nuisance in katharometers used in gas chromatographs. At one time this effect was used in instruments indicating furnace atmosphere in metal heat treatment or carburisation. The wire in effect was being heat-treated and carburised, too, its electrical as well as its mechanical properties being modified.
- A *dual hot wire cell*, the katharometer used for hydrogen detection. The

katharometer simply measures the thermal conductivity of a gas and in its simplest form comprises a bridge circuit with hot filaments in each arm of the bridge. One filament samples the unknown gas, the other a known zero gas.

- *Absorption and titration* is always an option for many gases and has the same advantage as the refractometer of absolute calibration. CO_2 can be estimated with alkalis in this way, oxygen by absorption in pyrogallol and hydrogen by reaction with palladium chloride followed by titration of the hydrochloric acid produced.

The philosophy generally followed in gas installations has been to install a minimum number of on-line single gas instruments of various types. Where these are reliable enough, they can be used both to observe and to control the process on which they are used. The on-line single gas analysers are backed up with occasional off-line multigas analysis, or occasionally by an on-line multigas instrument. The multigas instruments are typically used to check the function of the single gas units and to check for unknown contaminants.

2.3.4 Gas chromatographs

GC is probably the most powerful single technique available to the analyst. Although GCs are most often seen in the laboratory and used for off-line analysis, an automated GC technique is also the basis for many on-line automatic analysers, although this is often not obvious from their external experience. Automated GC analysers, although complex, may run reliably for years with little intervention. GCs for analysing the contents of argon-rich streams in ASUs and GCs measuring trace H_2 and CO in inert gases are examples of on-line use.

The GC concept is straightforward. A steady stream, typically of pure gas, the 'carrier gas' or 'mobile phase', is passed along a long thin tube, the 'column'. The tube is packed with a solid absorbent, the 'stationary phase', such that the gas may still easily pass. The (small) sample for analysis is then injected swiftly as a 'bolus' into the gas stream. The components of the sample bolus are then reversibly absorbed by the first part of the column, then released and carried a little further up the column before being absorbed again (Figure 2.34).

In this way the sample components proceed along the column, progressing more slowly than the carrier gas. Their speed of progress depends on the carrier flow rate and on how much the equilibrium absorption of the component favours the stationary phase. If the component is highly absorbed, then progress is slow; if it is only slightly absorbed, then it is fast.

In normal practice the carrier gas should not be one of the components to be measured. Helium, hydrogen, nitrogen and air are probably the

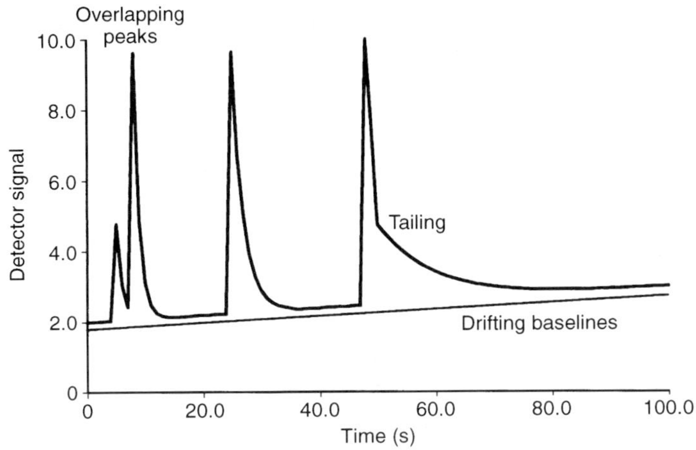

Figure 2.34 Separation of a mixture in a GC column showing typical trace characteristics.

commonest carriers, being chosen in many cases to be the best match to the detector employed, and on economical considerations. The carrier gas should at least be free from troublesome impurities like water (which slowly degrades molecular sieve columns), and should ideally be of the highest purity grade conveniently available. For ordinary work, a standard cylinder gas, perhaps with a molecular sieve trap to remove moisture, is adequate. For ultra-trace work (below 1 ppm) small purifiers (for example, metal getter-types for Ar/He or a palladium diffuser for H_2) should be considered.

The stationary phase may be an inert solid support coated with a thin layer of a low vapour pressure liquid as the active absorbing component. More common, however, are solid microporous stationary phases such as the zeolites or Porapak.

The result of all this, if the carrier gas and stationary phase have been correctly selected, is to separate the sample components so that they come out ('elute') from the column separately. A simple non-selective detector can then be used and the 'area' (integral of concentration with respect to time) of the peaks detected measured to give the relative concentrations of the components.

The standard laboratory gas chromatograph employs a high-quality polymer septum and samples are literally injected with a calibrated syringe through the septum using a needle. However, for on-line GC, and for most industrial routine work, it is simpler and more satisfactory to use a sample loop on a multiport rotary valve for injecting a bolus of sample into the GC column. The loop is filled with flowing sample when switched into the sample position, the valve is rotated to cut off the gas sample and the sample then placed in the carrier gas path (Figure 2.35).

For use as an on-line gas analyser the GC must be automated. Such

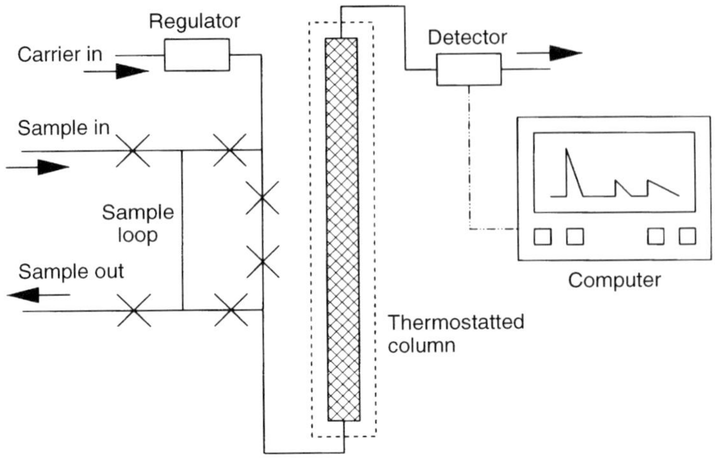

Figure 2.35 Complete GC gas analysis system.

automated or 'process' GCs have been used for many years now and have been substantially perfected. Servo-driven or solenoid valves automatically take samples and carry out any other switching. Resultant peaks are integrated and the results fed either to a digital interface for display on a computer screen or to digital-to-analog converters (DACs) to give a voltage level for each peak, i.e. acting as a pseudo-single gas analyser output. Routine maintenance is confined to occasional replacement of columns and checks on calibration. A typical automated GC application would be the analysis of crude argon streams (containing nitrogen, oxygen, argon and other rare gases) in an air-distillation plant. The process GCs used here are relatively simple systems with a single isothermal room-temperature operation molecular sieve column and a thermal conductivity detector using a helium carrier gas.

GC column choice. By far the most popular stationary phases in GCs for gas analysis are microporous minerals, such as zeolites, or microporous polymers, such as the divinylbenzene copolymer-based Porapak™ materials. These are packed into long tubes of 1/8″ or 1/4″ diameter (2 mm–5 mm or so internal diameter) clean stainless steel, up to 2 or 3 m long, which are then wound into a helical coil if required. It is an advantage to use the smallest column which can be reasonably easily handled and packed. A small column will usually yield the best resolution and is of low thermal mass so that temperature ramping during analysis is possible.

It is possible to use non-packed columns, popular in chemical GC work, in gas analysis. However, these need to be of small internal bore (less than 0.5 mm or so) and considerable length (50 m is typical). Newer configurations include the PLOT (porous layer open tubular) columns with a porous

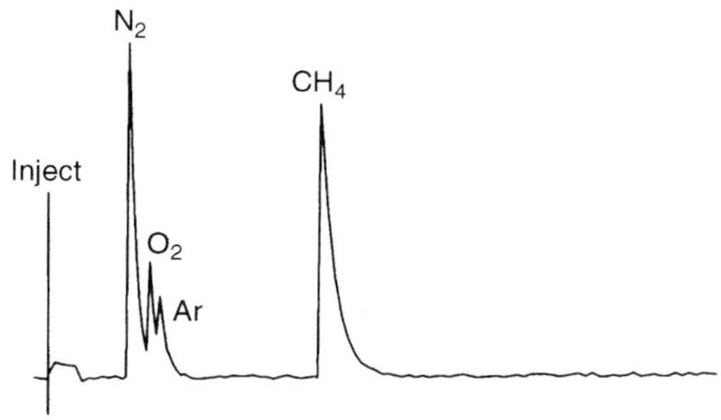

Figure 2.36 Separation of a gas mixture on a Porapak column.

alumina coating on the inside of the glass or quartz capillary tube. The porous alumina can be further coated with another microporous material if necessary. PLOT columns are used commercially in light hydrocarbon separation.

Molecular sieve (zeolite) columns usually employ type 5Å (5 Å pore size aluminosilicate) or 13X (10 Å aluminosilicate) granules in a 1/4″ column. These must either be purchased pre-activated or must be 'activated' before use by driving off water and CO_2, both of which molecules will sit in the active absorption sites of the sieve and prevent correct separation of other gases. The activation is conveniently carried out by heating to above 400°C with dry carrier gas flowing. Molecular sieve columns are used where separation of air gases is required. However, separation of argon from oxygen is difficult, requiring cooling of the column below ambient, and CO_2 tends to absorb and tail badly. Porapak columns use polymer beads, mostly prepared from copolymers with divinylbenzene (Figure 2.36). Chromosorb™ is another trademarked commercial material with similar properties. They provide good separation of most gases, with the exception of the air gases, although air gases are nicely separated at low temperatures on divinylbenzene copolymers. CO_2 can be analysed with these columns, as can water vapour with difficulty (tailing tendency again).

Problems with gas chromatography. Problems with gas chromatography are frequent but it is a measure of the versatility of the technique that they can often be overcome. For example, take the problem of sample size. With most analytical techniques, the smallness of the samples needed is a measure of the efficiency of the technique. However, in GC the sample size analysed is typically only a millilitre or less, which is an advantage that often gives a problem. In many situations the gas is inhomogeneous to some extent. It is therefore necessary to ensure that the tiny sample taken is truly representative of the mixture. An obvious example of this is where a liquefied gas is being sampled: the gas phase may well have a different composition to the liquid phase and samples must be pure liquid or pure gas if the results are to be quantitative.

High boiling point vapours admitted to the chromatograph (for example, oil in compressed gas or solvent in acetylene) will take a long time to exit the column. A zeolite with a retention time of just a minute for nitrogen might have a retention time of hours for a vapour, which would make laboratory GC tedious and process GC impossible. This problem is also present with polar molecules such as water, which in addition tend to cause tailing after their peaks. The longer retention times can be compressed closer in time by ramping the column temperature after sample admission as retention times rapidly decrease at high temperatures. High boiling point vapours can also be dealt with by back-flushing. After 1 or 2 min for analysis of the desired light components, instead of waiting hours for the heavier components to elute forwards down the bulk of the column, the column flow is reversed for 2 or 3 min, eluting the heavy components swiftly backwards from their point of progress near the inlet.

A GC column can be overloaded with the majority component of a mixture in which it is required to measure a minority component or trace gas, which is the sort of problem which is general to any analytical technique. Unlike many other techniques, however, in GC the problem can be dealt with by using two columns. The first column is arranged to feed the second column via a divert valve. During the peak of the majority component, the first column effluent is diverted to waste. The second or main column can then be used for analysis with most of the majority component absent, enabling accurate work even with trace gases.

Some gas mixtures are only partially separated by any single GC column/carrier gas combination. By passing an unseparated peak through a second column of a different material (so-called 'heart-cutting'), the components can be completely separated. In this way, the problem of incomplete analysis of a particular mixture by any one GC column is soluble; provided there are no components which have identical retention times on all column materials, then any mixture can be separated.

Finally, the GC technique is limited by obvious technical factors: it is a very sensitive analytical tool and requires clean systems and consumables. Impurities in the carrier gas, for example, can give rise to a number of problems such as baseline drift, noise and reduced sensitivity. The injection septum or valve needs care to ensure that 'bleed' (continuous baseline due to compounds deposited on the inlet slowly evaporating, outgassing or decomposing) is avoided.

Gas chromatograph detectors. Since the GC column should have separated the sample mixture into peaks separate in time, the only essential requirement for a GC detector is that it distinguishes gas containing any sample component from the carrier gas.

To be a good GC detector, however, a number of other qualities are highly desirable:

- *linearity*: this means that peak area can be used to measure components quantitatively;
- *high speed*: to achieve a reasonable analysis cycle time on mixtures of components with widely varying elution speed, fast-eluting components will tend to elute in just a few seconds;
- *large dynamic range*: with mixtures of widely varying elution speed, high concentrations of components with a fast peak require a good high-level measurement, whilst low concentrations of slow components need high sensitivity;
- *drift free*: slowly eluting components, with peak widths of up to several minutes, will be confused with background drift at lower concentrations, if drift is large.

Variations in carrier gas pressure, flow, temperature and composition will all produce baseline drift. Different detectors are sensitive to different carrier gas changes. For example, flame ionisation detectors are sensitive to H_2 concentration in the flame, whilst flow variations cause drift with katharometers.

The most common GC detectors are as follows.

TCD (thermal conductivity detector or katharometer). A TCD simply passes the column effluent over a hot wire or hot thermistor. The cooling effect of the gas varies a little when it contains some eluted sample, and this varies the temperature and therefore resistance of the filament. This change of resistance is recorded as the signal. In practice, because the resistance change is small, two or four detectors are used, with the resistance of pure carrier gas cells being compared to the resistance of the column effluent cells and additional reference cells out of any gas flow as well if necessary. The dynamic range of the TCD is poorer than the FID, and it is around a thousand times less sensitive. Despite this, it is the most commonly used detector. The possibility of an inert carrier gas such as He, N_2 or Ar and the complete universality of the detection principle means that it is a very attractive detector. Table 2.9 clearly shows the ready detectability of most gases with a helium carrier and shows that for many gases of interest a similar response is seen, minimising calibration problems. (Note that hydrogen gives a peak in the opposite sense to the other gases, which can give problems with amplifiers and using a chart recording efficiently. Also the H_2 peak is nonlinear with H_2 concentration, unlike the other gases.)

FID (flame ionisation detector). A simple FID takes H_2 carrier gas from the column and burns it in pure air, perhaps adding an additional fuel gas stream, measuring the current flowing through the flame between two

Table 2.9 Δ (heat conductivity) for common gases (relative to He)

H_2	95
He	0
CH_4	−287
O_2	−308
N_2	−310
CO	−312
C_2H_6	−318
Ar	−331

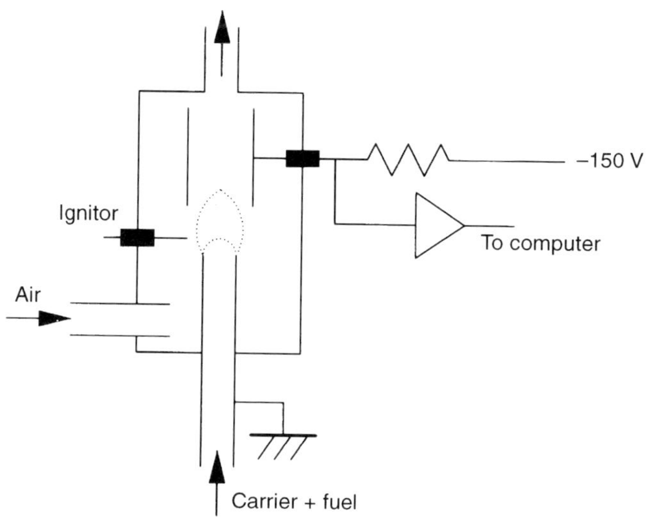

Figure 2.37 Flame ionisation detector for GC.

plates with a differential voltage, up to 100 V or so, across them (Figure 2.37).[20] It is highly sensitive with a high dynamic range and a response which is almost exactly proportional to the concentration of C atoms, which ionise in the flame, in the H_2 carrier gas stream. However, it does not detect non-carbon gases (or indeed CO_x) with usable efficiency, severely restricting its usefulness, and needs a very sensitive amplifier, since the ion current is typically only picoamps or nanoamps.

For detection of low levels of CO_x, it would be useful to be able to use the simple but sensitive FID detector. This can be achieved with the expedient of a 'methaniser' unit. A very simple heated (300–400°C) catalyst tube, containing nickel on an inert microporous mineral support, placed after the

[20] In many ways the FID is the simplest of all detectors to construct and operate; the author made a successful FID device while at school from pieces of wire and laboratory glassware.

hydrogen mixer but before the flame results in almost 100% conversion of CO_x to CH_4, which is then detected in the FID.

The use of additional fuel gas means that the FID flame will not extinguish if a large bolus of non-inflammable gas appears and means that a greater dynamic range can be accommodated by the detector. The fuel often used in FIDs can be pure hydrogen but a mixture of 40 or 50% hydrogen with nitrogen or helium is a useful alternative. Similarly, FID GCs can also be operated with carrier gases other than hydrogen; for example, helium, nitrogen or even air. (If air is used, however, it must be free of hydrocarbons; typically a purifier unit or a cylinder of compressed pure air will be required.)

ECD (electron capture detector). A Nickel-63 β-radiation source ionises the column effluent, N_2 or Ar carrier gas often with methane added, whose electrical conductivity is measured. This detector is sensitive to electrophilic molecules such as halogenated compounds. These compounds reduce the standing current in the conductivity cell, rather than increasing it, and because of this the ECD detector is restricted in its dynamic range.

Ultrasonic detectors. By detecting the change in the speed of sound in the effluent gas, many gases can be non-selectively detected against an H_2 or He carrier as these carriers have a very high sound velocity which drops sharply with heavier gases present; the velocity of sound is inversely proportional to the square root of mean molecular mass. Although ultrasonic detectors are not particularly sensitive, they can be made very robust and can be protected against agressive chemicals and so are often used in the analysis of toxic and corrosive cylinder gases.

In addition to these and other non-selective detectors which are the workhorses of GC detection, most selective gas analysers, such as those described below, can and are used as GC detectors. An example is the flame photometric detector, used for detecting SO_2 and H_2S. It uses an optical filter to look at a hydrogen flame to which the sample gas is added, detecting sulphur atoms via their 394 nm optical emission (phosphorus atoms also emit, at 526 nm) to a photomultiplier.

Conversely, most non-selective detectors normally used with GC can be used as single gas analysers, where the situation confines the gases measured so that no confusing results can be obtained. An example of this is the gas sniffer used to detect gas leaks in the domestic gas distribution network. These sniffers are sometimes mounted on the front of a van. As the vehicle is driven along the road at low speed, gases escaping from the ground near a leak are detected by a flame ionisation detector, alerting the crew to stop and investigate. Another example is oxygen depletion analysis in air using devices based on an ultrasound principle. These devices make the assumption that any increase in sound velocity is due to a decreasing amount of oxygen and sound an alarm at, say, 15% oxygen content on this assumption. The user must ensure that in the situation where they are used there is no possibility of a heavy gas accumulating in the atmosphere to an extent sufficient to mask appreciably the decrease in oxygen content registered.

GC integrators. It is invariably possible to connect a laboratory GC to a chart recorder of some description, plotting detector signal versus time. In many cases, a GC trace on a chart is all that is required to indicate approximate concentrations of different components in a mixture. However, it is difficult to estimate concentrations accurately especially if (as is typical) components have radically different peak widths and if peaks are of radically different areas.

The skilled operator can obviate some of the latter effect by changing the chart recorder gain control and chart speed at the appropriate moment. However, the use of computer-based electronic integrators is now universal. These provide a wide dynamic range, integrate accurately over peaks, no matter their width or if they are asymmetrical, apply calibration factors and give out their answers in convenient form; for example, lists of numbers giving gas concentrations. In the laboratory, it is possible to use a desktop computer, such as a PC compatible, as the integrator, although stand-alone integrators are favoured for convenience and robustness in industrial on-line situations.

For small peaks, and for chromatograms where the baseline signal from the carrier gas drifts, it is important to use an integrator with sophisticated peak detection and baseline correction. A typical algorithm declares a peak to have started when the slope of the signal versus time exceeds a threshold, and stopped when the negative slope value on the return to baseline falls below the threshold. A baseline correction to be subtracted from the peak area can be made by calculating the area of the trapezoid under the peak made by the line from the start value of the peak to the stop value. Corrections can also be applied for tailing and for separating peaks which overlap to some extent (see Figure 2.34).

It is still a wise precaution to run off a chart picture of the chromatogram, at least occasionally, because these sophisticated integrator algorithms can sometimes fail. Most integrators can help in the production of a useable chart recording trace by changing amplifier gain factor and timebase automatically at pre-programmed appropriate moments, or by printing off a record simulating this from memory.

2.3.5 Infrared and related analysers

Simple diatomic and monatomic gases are completely transparent to infrared (IR) radiation. However, most other gases are selective infrared absorbers, with complex bands of absorbtion and transmission. The science

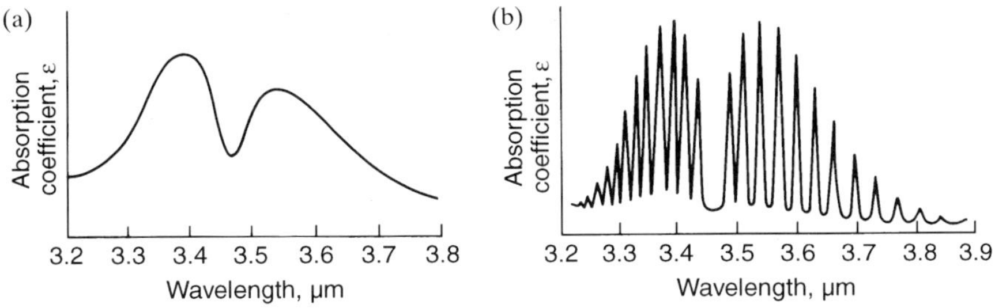

Figure 2.38 Rotational and vibrational structure in infrared absorbtion spectra. (a) Coarse resolution vibration spectrum; (b) fine resolution vibration/rotation spectrum.

of infrared absorbtion by gases is complex but basically the absorbtion is due to the IR radiation exciting vibration (or vibration/rotation) modes in the molecules (Figure 2.38). The frequencies at which vibration occurs depend on the sizes of the masses and springs at the molecular scale, the atomic masses and bond strengths, and are largely characteristic of a particular compounds. The vibration spectrum of each gas has typically a complex comb-like structure, with many sharp peaks of differing heights in the region from 3 to 20 μm. The fine structure reflects rotation excitations, whilst the envelope within which the sharp peaks lie relates to vibrational energy levels.

This fact is employed in two types of infrared analyser: the non-dispersive and the dispersive IR analyser, which measure IR absorbtion at many and at a single frequency, respectively. The non-dispersive types are popular as on-line single-gas analysers, whilst the dispersive types are mainly used in the laboratory.

Non-dispersive IR analysers (NDIRs). NDIR as a principle was developed rapidly as soon as good, reliable electronics were available for the capacitative detectors used in the early 'Luft' analysers. There are many different designs but all are based on a similar set of principles (Figure 2.39).

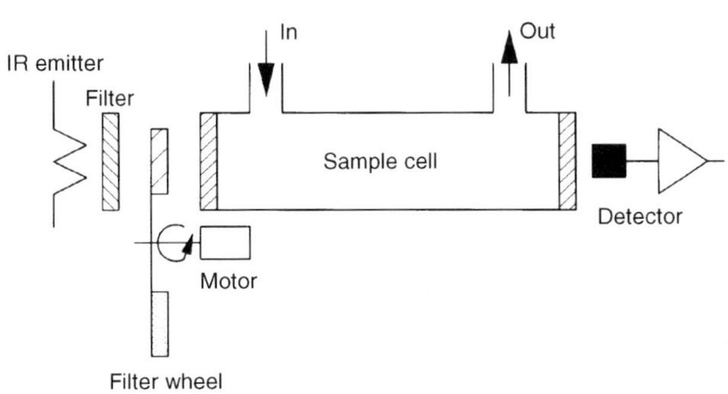

Figure 2.39 NDIR analyser principle.

First an infrared beam is created, typically by filtering out the visible light from a filament lamp or perhaps from a low-temperature 'glower' source. In addition, mid-infrared LEDs and laser emitters which offer improved performance are available today. The beam is then passed through a 'filter wheel' comprising a dummy reference cell and a 'filter cell', which will be a cell full of the gas to be detected. The resultant infrared beam is passed through a coarse infrared filter and then the sample of gas to be measured. Lastly, the beam is absorbed and converted to an electrical signal by means of an infrared photoconductive (lead selenide is typical) or pyroelectric detector.

The filter cell filters out all the radiation lying in the multiple narrow bands to which the gas detected is absorptive. Hence, if any of this gas is present in the sample then the unfiltered detector output will fall while the filtered detector output remains constant. The coarse IR filter makes sure that the band of IR used through the sample cell is identical. The difference signal between the two is proportional to the gas concentration in the sample (at least at small concentrations). The single detector is presented first with the filtered and then with the unfiltered beam in sequence. The a.c. signal produced at the detector is proportional to the gas concentration.

Although convenient, NDIR instruments are nonlinear, sensitive to contamination, have a tendency to drift and are not perfectly selective for the gas to which the filter cell sensitises them. They are normally not used for very high sensivity (below 10 or 100 ppm, say) but can be made very sensitive (ppm level or below) by using long-path, multiple-reflection gas cells. Common instruments analyse for CO, CO_2, NH_3, and N_2O. Car exhaust emissions are now tested in most countries of the world for CO and hydrocarbon content using NDIR analysers. Similarly, most hospitals have in everyday use 'capnometers' using NDIR for measuring a patient's breathing via the metabolic CO_2 coming off the lungs.

IR spectrophotometers. Although normally only used in off-line mode in the laboratory, IR spectrophotometers are important instruments for measurements on unknown mixtures. A wavelength-selected IR beam is first made by means of a diffraction grating. The output from this monochromator is then passed through the unknown gas mixture and the beam detected with a solid-state IR detector. The output as a function of the monochromator wavelength, the spectrum, can then be compared with library spectra of gases.

A qualitative look at what gases are present can often be seen by eye whilst with some computer deconvolution work quantitative results can be obtained.

An alternative form of IR spectrophotometer, Fourier transform IR, uses an interferometer type of set-up, which produces the Fourier transform of the infrared spectrum. This output can then be unravelled into a spectrum and gas concentrations with a computer.

Raman spectroscopy. Until the advent of powerful and inexpensive lasers, Raman spectroscopy for any practical purpose was an impossible dream. Even today, special techniques such as operating inside the cavity of the laser (so-called 'active cavity' working) are often necessary to make Raman useful.

The Raman effect is the small amount of light scattered off atoms that is shifted in frequency by an amount due to excitation of a molecular vibration. Spectral lines with shifts downwards in frequency are known as Stokes lines, whilst lines shifted upwards in frequency are known as anti-Stokes. Using a powerful laser, and rigorously excluding any of the scattered light which is unshifted in frequency as well as the laser beam itself, it is possible to see a unique spectrum for diatomic as well for polyatomic gases. For nitrogen, for example, about 1 part in 10 million of the 488 nm light from a green argon-ion laser is shifted to 550 nm. The spectra for polyatomic gases have strong similarities to the infrared spectra, relating as they do to molecular vibration. Basically, the energy levels in the molecule are the same but the transitions allowed between energy levels differ.

One of the first gas analysis applications tried for Raman spectroscopy was the monitoring of pollution plumes. Very powerful blue or green lasers are arranged to illuminate plumes from power stations or other industrial complexes. Via large aperture coaxial optics, light back-scattered from the plume, several hundred metres up in the air, can be analysed for Raman spectrum.[21]

[21] As recently as 1984, the author was witness to what may well have been one of the first routine practical applications of the technique to gas analysis: the monitoring of oxygen content inside ampoules of a sensitive medical drug sealed under nitrogen. A powerful argon-ion laser was shone through the gas space at the top of the ampoule and the ampoule rejected if more than a pre-set amount of oxygen Raman line was seen using sensitive photomultiplier detectors.

Today the technique is still a difficult one; the laser used in ampoule monitoring was 2 m long and weighed over 200 kg. Its sensitivity to gas concentrations is not high, parts per million monitoring is very difficult, and its commercial value is still not clear. Nevertheless the technique is today more widespread. A Raman-based anaesthetic gases monitor has been developed. Following laboratory work with an argon-ion laser, a commercial instrument is now available using a HeNe laser which will monitor six different gases at the fractional percentage level simultaneously, rivalling mass spectrometric techniques (section 2.3.6) in the percentage range.

There are still problems with such analysers, however. In order to raise their sensitivity to the highest possible level, the gas must be led inside the laser cavity where the optical electromagnetic field is at its maximum. This means that small deposits from the gas being analysed on the optics, or even too large a refractive index difference between the air used to protect the optics and the sampled gas, can lead to the laser cavity gain varying, causing problems. In addition, despite these efforts, signals are still very weak, even on gas concentrations in the fractional percent range. Finally, in order to use HeNe lasers tried-and-tested (and very sensitive) photomultiplier detectors must be replaced by the more difficult technology of avalanche photodiode detectors. The latter are diodes with an intrinsic gain effect, which have a good sensitivity for the Stokes Raman lines, which are in the near IR, being 30 to 150 nm longer in wavelength than the 633 nm HeNe excitation.

Optical interferometric refractometry. One gas detector that is often neglected is the refractometer. The gas refractometer is based on an optical interferometer which is the only ready means to detect the small differences in optical index between gases at ordinary pressures. A refractometer has the advantage over most analysers in that its response can be calculated from first principles and can be used as a standard to calibrate other analysers. The UK National Physical Laboratory and US Bureau of Standards, amongst others, carry out these kind of measurements in standards work, although commercial instruments are now available. The availability of inexpensive laser light sources and electronics may make refractometers more attractive than in the past, although simpler optical components and large-scale series manufacture will be needed to reduce costs to the levels seen in many other analysers.

2.3.6 *Mass spectrometry*

The mass spectrometer (MS) is often a sophisticated and expensive analyser and might not be thought of as useful in industrial gas analysis. However, simple MS devices can cost as little as a few thousand dollars and MS offers some unique advantages, being a technique almost as versatile as GC.

The basis of MS is to take a minute sample gas flow and leak it into a low-vacuum region (the 'source') where an electron beam of a few electron volts energy gently breaks the molecules into molecular and submolecular ions. These ions are then sucked out of the source region by high voltage plates with small holes in them. The accelerated ions then pass through the holes into a region of high vacuum (10^{-6}–10^{-8} mbar or less) where a dipole d.c. magnetic field or a quadrupole a.c./d.c. electric field is used to select them by mass/charge ratio. Only ions of the mass/charge ratio selected by the magnetic or electric fields will penetrate as far as the exit slit. After the exit slit a detector measures the flux of ions. A computer drives the magnetic or electric fields, effectively scanning the mass/charge spectrum of the ions coming from the source region.

The mass spectrometer is, like the gas chromatograph, fundamentally a means of separating molecules rather than a detector *per se*. However, in the case of the mass spectrometer the separation is achieved on charged molecules, molecular ions, which constitute an electrical current flow in themselves. The simplest and most common MS detector, therefore, is a metal plate or cup ('Faraday cup') connected directly to a sensitive electrometer amplifier.

For measurements at very low concentrations, ppm or below, an electron multiplier detector becomes useful. An ion incident on the multiplier causes an avalanche of electrons to be released. The resultant large current is simpler to detect than the tiny molecular ion current. The most common multipliers consist of a highly resistive coating on a curled horn-shaped

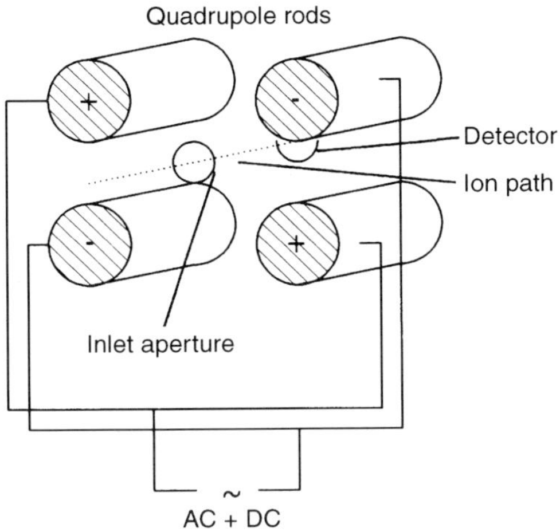

Figure 2.40 Quadrupole mass filter.

insulating tube. For ultra-trace detection, individual electron avalanches are counted; as little as one molecule can therefore be detected.

The highest mass resolutions are obtained by magnetic mass selection instruments. Large magnetic sectors can achieve resolutions of 1:10 000 or more, which means that nominally isomass ions (those with the same number of protons and neutrons, such as N_2 and CO, for example) can be distinguished. Such resolution can greatly simplify the interpretation of mass spectra but is expensive to achieve. In almost all gas analysis work the maximum resolution needed is that to distinguish ions differing by at least 1 atomic mass unit. With a maximum useful molecular ion weight of around 200 or 300, resolutions of 500 or so are entirely adequate for this.

The cost of mass spectrometers has been reduced considerably in recent years with the advent of ~ 500 mass resolution quadrupole instruments and inexpensive electronics and computer control. The quadrupole mass spectrometer principle is that of a molecular mass/charge ratio filter. Four accurately ground parallel cylinders are charged both with d.c. and a.c. (at a few MHz) to create quadrupole electric fields and ions are admitted along the central axis of the quadrupole field. Figure 2.40 shows the quadrupole arrangement. Heavy ions tend to oscillate in the a.c. field with increasingly large amplitude along one plane between the cylinders, while light ions tend to oscillate in the orthogonal plane. Ions of the tuned mass, however, pass through the assembly unaffected. The ratio of d.c. to a.c. electric field determines the mass/charge ratio passed.

The mass spectrometer is completed by the addition of digital-to-analog converters (DACs) and analog-to-digital converters (ADCs) driven by a computer. The computer is programmed to admit gas samples, produce

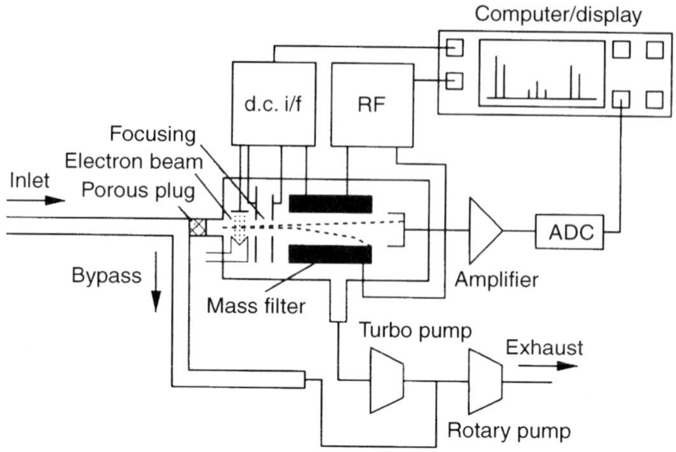

Figure 2.41 Simple quadrupole mass spectrometer.

ramp or staircase voltages to scan mass/charge ratio and record the resultant mass spectrum via an ADC in digital form. The mass spectrum can be displayed directly, output via DACs as a number of pseudo-single analyser voltage signals, or further processed via calibration factors and even mathematical (matrix) processing to yield gas concentrations. The use of matrix inversion procedures often enables the analysis of gases whose mass/charge ratio peaks overlap, provided the mixture components are known. Figure 2.41 shows a complete quadrupole mass spectrometer system.

Despite the low cost of the quadrupole MS, complete MS instruments are still relatively expensive at $15 000 or so. Much of the cost of this kind of mass spectrometer can now be attributed to the requirement to work at high vacuum. Even the smallest instrument needs high vacuum pumping equipment costing many thousands of dollars.

Problems with mass spectrometry. Just as for gas chromatography, there are limitations with mass spectrometry.

- Some gases have virtually identical mass/charge ratio molecular ions. CO and N_2 are the most common examples.
- Ion sources are more or less non-selective. This means that although calibration is simpler, with factors of order unity, it is difficult to measure trace components in the presence of main components. Selective ion sources such as photoionisation have mostly proved to have too many other problems to be useful, although atmospheric pressure silent discharge ionisation is now commercially available and very usefully selective.

- The sample pressure must be reduced from its source pressure to about 10^{-5} mbar or so. This can be done by a simple leak but such an arrangement leads to a very slow response. A solution to this is to reduce pressure in two stages: a leak down to 10^{-1} mbar, typically via a capillary tube, and then via porous ceramic leak from 10^{-1} to 10^{-5} mbar. Most of the sample is thus bypassed to the vacuum pump pumping the intermediate volume.
- Some sample gases will react with the surface of the filament in the ion source, changing the electron beam current and ultimately destroying it.
- Oxygen reacts with residual carbon in tungsten filaments, producing CO gas in minute amounts sufficient to make mass 28 measurements unreliable in the presence of oxygen. The use of rhenium filaments alleviates this problem.

Atmospheric pressure ionisation MS (APIMS). APIMS is a technique which has only in the last five years become commercialised. It is a specialised technique answering the very specialist requirements for measurement of ppb level trace impurities in the bulk inert gases nitrogen and argon, which have come from the ultrapure gases needs of the semiconductor industry.

In its simplest form APIMS involves sucking the sample gas away from a silent discharge needle ionising region and stripping away the bulk of the uncharged gas molecules whilst focusing the small numbers of ions along the axis of a quadrupole mass analyser (Figure 2.42). In practice, this is done by two or more 'skimmer' cones. These are metal cones with a small (100 μm or so) central aperture, insulated so that suitable d.c. electric potentials can be applied. Vacuum pumps reduce the pressure behind each cone until the 10^{-6} mbar or less needed to operate a quadrupole MS is reached. The ions tend not to be pumped away as much as the uncharged ions. At least a small number are retained near the axis and can be analysed by mass-scanning the quadrupole and detected. A multiplier detector is typically needed as the number of ions is small.

The process of ionisation near atmospheric pressure by discharge is very different from electron beam ionisation at low pressure. After initial excitation or ionisation, which is primarily of the bulk majority component, multiple collisions occur between molecules and molecules with low ionisation potentials end up charged, whilst those (such as the inert gases, nitrogen, hydrogen) with high ionisation potentials tend to end up neutral. In this way the ionisation is highly selective for impurities and this is the key to the extreme sensitivity of the method.

There are problems with APIMS, however. Like any MS, it cannot distinguish N_2 from CO by mass analysis and at impurity levels above 100 ppb the ionisation mechanism becomes highly nonlinear in inpurity concentration, making the method unusable except with ultrapure gases. A

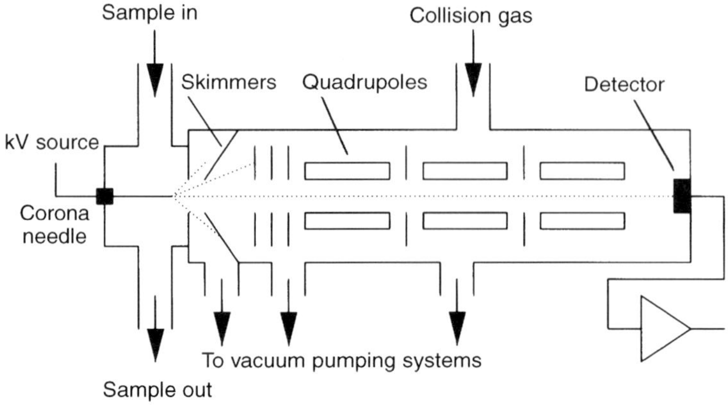

Figure 2.42 Atmospheric pressure ionisation mass spectrometer.

sophistication of the simple APIMS arrangement provided by one company (VG) offers a solution to the first problem. They provide a low-pressure gas cell, which is actually another quadrupole mass filter (Q2) but is set to pass a wide band of mass/charge ratio after the initial quadrupole (Q1), providing a further analyser quadrupole (Q3) after the gas cell to analyse the resultant ion stream. A mixture of N_2^+ and CO^+ ions filtered out by Q1 collide with gas molecules (such as Ar) in Q2 and produce N^+ and C^+ ions which are analysed in Q3. Despite losses in the extra quadrupole stages, performance of about 1 ppb CO in N_2 is claimed.

For calibration purposes APIMS needs a high-quality ultraclean dilution rig. Calibration gases at ppb levels are difficult or impossible so higher level contaminant gas samples are further diluted a factor of 1000 on a rig built into the APIMS to provide suitable low levels for testing, using a built-in purifier and bulk gas supply. The dilution rig can also be used to solve the second problem above, that of the nonlinearity above 100 ppb. To measure contaminants above the 100 ppb level, the sample gas is diluted with ultrapure gas from the purifier by a further factor of 10 or more.

Time-of-flight (TOF) mass spectrometry has been neglected in recent years, reflecting its difficulties compared to quadrupole MS. However, it is being revisited by a number of firms and products offering up to 350 atomic mass unit analysis at low levels are now offered, including an API TOF instrument offering 20 ppt detection limits.

The TOF principle is deceptively simple. If a brief, less than a microsecond, bunch of ions is accelerated out of an ionisation chamber into a vacuum with energy E, say 100 eV, then they will drift in an electric-field-free region towards the collector at a speed v which varies as the square root of their mass to charge ratio q/m

$$qE = 1/2\, mv^2, \text{ so } v = \sqrt{(2qE/m)}$$

In fact, the 'drift' velocity is fast; for an N_2^+ ion, for example, it is 25 000 m s^{-1}, giving a time of flight of only 5 μm or so in a typical flight tube. The measurement must be registered on an oscilloscope for qualitative assessment and then logged electronically with high-speed data handling hardware, the mass spectrum being repeated hundreds of times per second. High speed voltage pulses on the electrodes control the bunch formation and sweeping out of unwanted ions and ion energies of up to a few hundred volts are used.

2.3.7 Other analytical principles

Electrochemical cells. Electrochemical cells are some of the simplest gas analysers and yet can be made to respond selectively to many common gases, often with very high sensitivity. As mentioned in section 2.3.3, the simplest electrochemical gas analyser is just two platinum electrodes dipping into a cell of distilled water. If a basically inert gas containing CO_2 is then passed through the cell, the conductivity measured is proportional to the concentration of CO_2 in the gas. If the concentration of CO_2 subsequently falls, then CO_2 is washed out of the water by the inert gas flow. Although no longer used in practical instruments, this principle shows just how simple electrochemical cells can be. The lead/alkali hydroxide/silver trace oxygen cell is barely more complex than this but it is still very much actively used in industry and is a model for many other similar electrochemical cells. Designed for detecting trace oxygen in inert or reducing gases, the lead/silver cell is capable of detecting sub-ppm levels of oxygen using the following reactions.
At silver electrode, wetted by both gas and hydroxide

$$1/2O_2 + H_2O + 2e^- \Rightarrow 2OH^-$$

At lead electrode, submerged in hydroxide

$$2OH^- + Pb \Rightarrow Pb(OH)_2 + 2e^-$$

Hence every molecule of oxygen which dissolves in the cell near the silver electrode yields one electron, giving a very high sensitivity. Perhaps surprisingly, there are few interfering effects to give rise to false readings. Acidic gases will eventually neutralise the alkali solution and highly reactive oxidant gases such as chlorine might also give trouble but these are relatively uncommon contaminants, and can also be scrubbed out easily. Figure 2.43 shows how a simple O_2 electrochemical cell is constructed.

Slightly more sophisticated versions of this cell, using cadmium for the submerged electrode and a third inert electrode, can achieve sensitivities of a few ppb O_2 in nitrogen. To achieve this kind of performance, however, these trace cells need sophisticated features such as pure nitrogen sparging

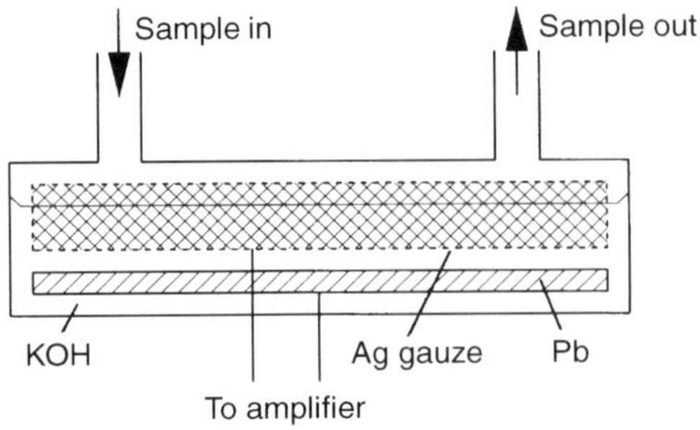

Figure 2.43 O_2 electrochemical cell.

of the reagents, temperature control and very careful sealing. Similar electrode reactions using different metals and reagent solutions can be devised for many other gases (section 2.5.2).

Zirconia oxygen analysers. Although the vast majority of electrochemical cells involve aqueous electrolytes at room temperature, it is possible to run an electrochemical cell with a solid electrolyte. If the solid electrolyte is hot, allowing solid diffusion of ions to take place easily and rendering it reasonably conductive, then workable analysers can result. The most common analyser of this type is based on zirconium dioxide ceramic as the active solid electrolyte. It has the feature that it measures not oxygen but 'effective oxygen content'. Most reducing gases, such as methane or CO, will, if mixed with the gas being analysed, subtract from the oxygen signal as if they were combined stoichiometrically prior to measuring O_2 on the ceramic membrane. This is a useful feature where the analyser is being used to indicated the oxidation potential of a furnace atmosphere, for example, or where it is being used to adjust the fuel-air ratio of a combustion reaction in a burner or engine.

The solid electrolyte is a thin plate of yttrium-doped zirconia ceramic heated to 400°C. Two porous precious metal electrodes are applied to the ceramic membrane, with test gas and reference gas applied to opposite sides of the structure. The arrangement constitutes a concentration cell. A voltage V develops according to the Nernst equation, depending on the logarithm of the ratio (P_a/P_b) of the oxygen concentrations applied

$$V = \ln(P_a / P_b)\, RT/4F$$

where R is the gas constant, T is the absolute temperature and F is the Faraday constant. For example, with atmospheric oxygen as reference and 100% O_2 applied, the cell will give 10 mV. A high impedance voltmeter is connected to the electrodes and typically calibrated as concentration of oxygen, assuming air is applied as the reference gas.

Moisture measurement. The measurement of trace levels of moisture in gases is both important and difficult. Much ingenuity has therefore been expended on moisture measurement and there are dozens of different designs. However, moisture measurements are still slow and inaccurate.

First, it is important to recognise that moisture is difficult to measure in multigas analysers such as MS and GC. Moisture is also difficult to sample; high levels of moisture will condense unless heated sample lines are used. Before a steady reading of moisture is achieved, equilibrium between the gas being measured and the pipeline must be reached but this is slow to achieve. In long pipelines, even with no dead legs and high flow rate, equilibrium is slow to achieve. Measurements of low levels of moisture also ideally use heated lines to avoid errors caused by this invisible adsorbtion along the walls of pipes.

In cryogenic liquids moisture may exist as tiny particles which are stratified in the tanks. Like acetylene or higher hydrocarbons, water is soluble only at low levels in cryogens. Depending on when samples are taken, and where from within the tank, different numbers of these tiny particles will be seen and different moisture levels recorded. Finally, in cylinder gases, the moisture level increases as the pressure falls to low values, i.e. as the saturated vapour pressure of the moisture on the cylinder walls becomes a more appreciable fraction of the cylinder pressure.

The two most popular types of moisture instrument are oxide capacitance and electrolysis. The capacitance instrument is of most general applicability, measuring from a few per cent moisture down to ppm levels. It relies on the fact that a specially prepared microporous oxide, typically aluminium oxide, layer reversibly absorbs a small amount of moisture in a repeatable way, with the amount absorbed being a monotonic function of the ambient gas moisture level. A typical sensor has the Al_2O_3 layer grown on a conducting support substrate, which is the back electrode. The other electrode is then a porous conductor such as a thin gold layer. The electrical capacitance of the assembly is linear in the amount of water absorbed and is easily measured electronically. An alternative construction employs an Al_2O_3 layer on single crystal silicon (Figure 2.44). The silicon can be conveniently equipped with an integral heater, which enables desorption of water by heating prior to measurement. This decreases the 'dry-down' time and makes the silicon sensor a more useful field instrument.

Electrolysis instruments are suitable for low levels (10 ppm and less) of moisture. They consist of a closely spaced bifilar winding of inert metal on an inert rod. The rod is coated with phosphoric acid. The acid dessicates at the operational moisture level to phosphorus oxide, which is an insulator. If a molecule of H_2O approaches, it is absorbed on the P_2O_5 and releases two ions, a hydroxonium and a phosphate ion, which are then discharged on the inert metal to hydrogen and oxygen. The current flowing is thus proportional to the amount of water absorbed by the assembly. The principle

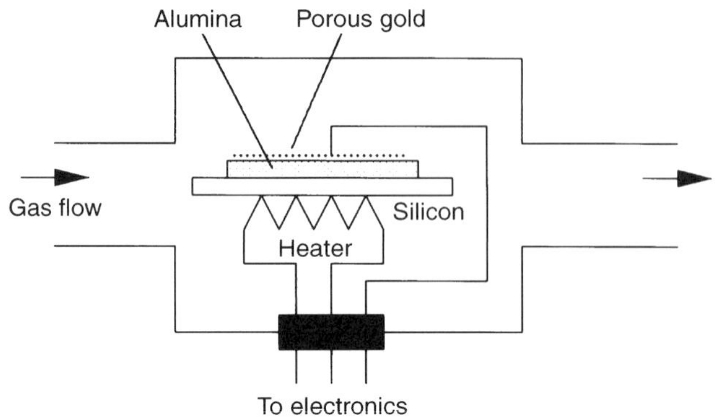

Figure 2.44 Silicon/Al_2O_3 moisture analyser.

was first developed to a usable instrument for measurement of moisture levels in gas in nuclear reactors.

In repetitive processes it is possible to measure moisture changes which are much too rapid to follow with moisture analysers by using specialised sampling techniques. One set of work (Carr, 1994) measured moisture in a rapid thermal processor (RTP), which is a small furnace for a single silicon wafer used in the semiconductor industry. An RTP normally takes in a room temperature silicon wafer and processes it at up to 1000°C in a cycle lasting just a few minutes. Moisture comes in with the wafer, from the air entrained with the wafer and when the door opens, and can cause process problems. To measure it, the IMEC team maintained a slow moisture probe in equilibrium with a calibrated moisture-containing N_2 stream. The level of moisture in the calibration stream was adjusted until a sample switched in for a few seconds from the RTP did not disturb its reading. The technique is analogous to a digital sampling oscilloscope in effect. Presumably another way of obtaining readings from a repetitive process is to arrange for many successive samples from a part of the cycle to be admitted, along with known purity purge gas, measuring the average reading; in effect a 'stroboscopic' technique.

The so-called 'quartz microbalance' can also be used for measuring moisture. In this two quartz crystals are employed. Via their piezoelectric effect, quartz crystals can be set vibrating at a very high constant frequency depending only on the precise size of the crystal. Two crystals vibrating at the same frequency are used such that if a small mass, even a microgramme or so, is added to the surface of one, its frequency will fall slightly relative to the reference crystal. A porous coating on the crystal means that even with parts per million or less of moisture in the gas passing over the crystal some moisture will be reversibly absorbed and give rise to a frequency difference. The frequency shift, Δf, in the crystal resonant frequency, f, seen is given approximately by

$$\Delta f = -f^2 \Delta m / (kA\rho)$$

where Δm is the mass added over area A, ρ is the density of quartz and k is a frequency constant (0.17 MHz cm^{-1} for AT-cut quartz). A 2 MHz crystal, for example, might have a sensitivity of 200 Hz μg^{-1}. By using two identical crystals and flowing reference and sample gas over the crystals (and swapping over these roles every minute or two if necessary), most drift and imbalance effects can be eliminated and the result is a very sensitive moisture analyser which is very inert to chemical attack from reactive components in the gas mixture sampled.

Analysis of scrubber solution. Although not strictly a gas analysis technique, the basis of many laboratory analyses for gases is still wet chemistry. It is often convenient to convert a gas sample into a conveniently handled liquid by bubbling it through a small scrubber or bubbler. Correctly chosen, a scrubber solution may absorb only impurities and not the majority components, providing a convenient concentration effect. No one would measure CO_2 with limewater in industry today. However, a search for, say, trace metals in liquid CO_2 would still most effectively be carried out by absorption in acid followed by wet chemical reactions or ion chromatographic methods.

Measuring silane purity. As well as monitoring scrubbing solutions, some gases can be reacted with solids which can be analysed to give gaseous purity. One example of where this kind of technique yields an effective sensitivity which is higher than any direct analytical technique is in the estimation of silane and chlorosilane purities. These gases are frequently used (section 3.5.6) in the semiconductor industry to lay down epitaxial crystalline layers on top of semiconductor silicon. A measure of the purity of the achieved deposited layer is, in effect, a measure of the gas purity: if the layer has no impurities, its electrical conductivity will be very low, caused only by thermally excited holes and electrons. However, donor and acceptor impurities, such as gallium or phosphorus, donate electrons or holes, which conduct electricity, to the silicon matrix. This turns out to be a very sensitive analytical method. Only 1 ppb of gallium, for example, would give rise to conductivity 3000 times that of intrinsic (theoretically completely pure) silicon. Sometimes, (chloro)silane gases are actually specified by the resistivity achievable, rather than by actual gas purity. With a baseline of something like 400 000 Ωcm for intrinsic silicon, a gas with 1 ppb donor impurities might in the worst case be a gas describable as '130 Ωcm silane'. The purer the grade of silane, the higher the resisitivity. Typical values vary from 1000 Ωcm for an ordinary grade to 10 000 Ωcm and more for an ultrapure grade.

The extremely high sensitivity of this technique raises the question of whether it can be adapted to measure other gases. In a sense, tin oxide semiconductor bead flammable gas detectors are using it because they actually work because of the production of carriers by a redox reaction in the bead. Analysers using a negative effect are also conceivable in which a thin layer of doped semiconductor changes its resistance after the gas reacts with and removes donors or acceptors.

Chemiluminescent (light-emitting reaction) analysers. There are a number of gas analysers which rely on the emission of light by the gas being sensed. White phosphorus suitably prepared glows as it oxidises when oxygen is present. The light emitted at low trace levels is proportional to the amount of O_2 present and can be used down to sub-ppm levels for the detection of trace O_2.

The arrangement for a gas analyser is simply to feed gas into a blacked, tortuous pipe (to eliminate stray light) inside which there is a disc of white phosphorus facing a photomultiplier tube a few millimetres away. The phototube detects the few visible photons emitted by the reaction and boosts the signal to a level of reasonable current.

Many NO_x analysers also rely on chemiluminscence, in this case the light emitted when NO_x reacts with ozone, also achieving a high level of sensitivity.

Mercury reduction analysers. The modus operandi of these is to pass the sample gas mixture through a tube containing mercuric oxide. At a few hundred °C, any reducing gas present will cause the release of minute amounts of mercury vapour. Mercury vapour can be detected very sensitively (in ppb or even ppt) by its very high optical absorption, which is especially high for optical emissions from a mercury lamp. Although rarely used on their own, they are found in specialised gas chromatographs where they can monitor for H_2 and CO while ignoring inert major components such as N_2 or Ar. The simpler FID detector could not detect H_2, would require a methaniser to pick up CO and might be swamped by CO_2 with a methaniser.

Magnetic oxygen analysers. These rely on the extraordinarily high magnetic forces which oxygen gas shows in a magnetic field, the paramagnetic susceptibility of oxygen, which is far higher than any other gas. Oxygen molecules behave as small dipoles, i.e. in a magnetic field they have a force exerted on them which is proportional to the field gradient and in the direction of increasing magnetic field.

Some instruments simply place a tiny lightweight hollow dumbell (filled with a gas other than oxygen, or under vacuum) in the gas stream, with one half of the dumbell in a strong magnetic field. In the event that the sample stream contains oxygen, a force is exerted on the dumbell, since the sample stream is then attracted into the magnetic flux, whilst the dumbell, being

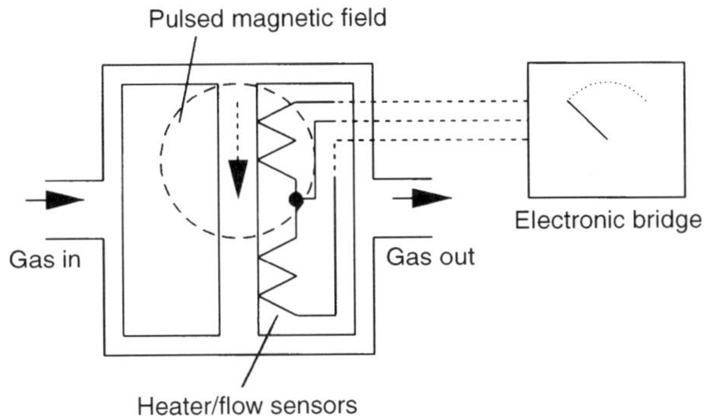

Figure 2.45 Magnetic wind O_2 analyser.

magnetically inert, is not and is therefore repelled. The force of repulsion is measured by feeding back from an optical position sensor a twisting force via moving coils as used in electrical meter movements such that the dumbell does not move. The force, F, exerted is approximately given by

$$F \simeq XAB^2/2$$

where X is the paramagnetic susceptibility, A is the area of the dumbell in the non-uniform fringe of the magnetic field and B is the magnetic field.

Some feeling of how difficult these instruments are to construct can be gained by noting that in a 0.1 Tesla field (the most readily achievable with a modest permanent magnet), a 20% O_2 sample will only exert a net force of order 10^{-7} N (10^{-5} g) on a dumbell of 5 mm diameter.

A simpler type of instrument (known picturescuely as 'magnetic wind' oxygen analysers) detects flow through a heated pipe in a solenoid field, and depends on the fact that the paramagnetism of oxygen is temperature dependent (Figure 2.45). Again the 'wind' force measured is very small, corresponding to a differential pressure of the order of microbars. The results from this form of instrument are affected by temperature, gas pressure and carrier gas type as well as oxygen concentration, so careful calibration and correction algorithms need to be applied for accurate work.

Other optical absorbtion analysers. In the section on IR analysis above it was noted that some gases at least absorb electromagnetic radiation of other frequencies. The greenish hue given to the air by Cl_2 that is due to absorbtion in the optical wavebands, for example, can be detected electronically. There are microwave rotation absorption bands, well-documented by research on high resolution military radar. For example, oxygen has absorption lines at 5000 μm. However, special techniques would be needed to effect detection: RF circuits at 60 GHz are relatively difficult, secondly, the maximum absorption is only of the order of 10 dB per kilometre, or 0.01 dB (0.23%

absorption) in a practical size, say 1 m, resonator. Experimental oxygen analysers have been used using the ultraviolet (UV) absorption of oxygen. While N_2 is a non-absorber, O_2 has a high absorption starting at 190 nm. Claimed advantages are simplicity, high speed of response, useful in medical applications, and high selectivity to oxygen, although no commercial analysers on this principle are available currently.

2.3.8 Monitoring air for gases

In addition to the commercial requirements for analysis, the gas industry and gas users have a responsibility to monitor the air for emissions hazardous both to the health and safety of their staff, and also for emissions ultimately harmful to the wider environment.[22]

Oxygen depletion monitoring. Oxygen depletion monitoring is employed to prevent exposure of personnel to air with too little oxygen. Depleted air is relatively common (section 2.5.2) and very dangerous. Just a few breaths of depleted air, a matter of a fraction of a minute, is dangerous so monitors should ideally be mounted near the mouth of persons to be protected and should have a response time of a few seconds.

These requirements are difficult to satisfy. Only electrochemical cells normally meet the portability requirements of small size and power requirements, whilst only instruments such as the paramagnetic oxygen detector have really quick response times.[23] However, the latest generation of electrochemical sensor is faster than ever and meets the requirements of safety in many cases.

Flammable monitoring. Safety requirements frequently lead to a demand for rapid response instruments which respond to all inflammable gases and vapours. Ideally such universal flammable gas monitors should show a response proportional to the lower explosive limit.

The heated metal oxide semiconductor bead detectors, such as those pioneered by Taguchi in Japan, are now ubiquitous. They function by registering a change in electrical conductivity. An oxide, often tin oxide based, bead is heated to 300–500°C. On exposure to a flammable gas in air the electrical conductivity increases. The mechanism would appear to be a catalytic oxidation resulting in charge carriers being injected into the oxide.

Although inexpensive and useful in alarm circuits, their quantitative

[22] Science and industry used small animals to monitor air quality until quite recently. British mines had canaries, for example, while Priestley (Gibbs, 1965) describes how the number of mice used in experiments could be minimised by keeping hold of their tails and withdrawing them when they appear stressed. The practice is not, however, really acceptable today.

[23] As noted above, an ultrasound device, which can be made portable and rapid in response, is a possible option for some situations.

accuracy is, despite many years of development, dubious. First, the output from such detectors is not particularly stable. Second, water vapour often gives some spurious response, the response to non-hydrocarbons such as CO is often poor and poisoning reactions are common. Particular poisoning reactions are frequently seen with reactive silicon-bearing materials such as silicone aerosols (commonly used as polishes) and with chlorofluorocarbons (still used in some specialist aerosol cans and in cleaning).

Toxic monitoring. Much toxic gas monitoring is still performed by chemical tape detectors. For nearly all gases a reaction can be devised whereby the contaminant produces a colour change on a piece of porous paper impregnated with a reactive solid. A very simple example is the monitoring of the highly poisonous hydrogen sulphide which, despite apparently being easily smelt at low concentrations, can be dangerous

$$\underset{\text{white}}{PbAc_2 + H_2S} \qquad \underset{\text{black}}{PbS + 2HAc}$$

A roll of tape is unwound slowly from a reel across an active aperture where air either diffuses or is pumped. A photodiode and lamp then checks the colour of the exposed tape and rings an alarm if a critical level of coloration is exceeded. The exposed tape is then wound on to a take-up spool where it constitutes a reviewable record of the toxic gas concentration.

Simple electrochemical cells are also excellent in this job. Rather than pumping the air sample across the electrolyte as in analyser practice, however, typical cells allow the sample to diffuse into the cell. The cell is also typically sealed from moisture loss by means of a semi-permeable polymer membrane. Polythene and polytetrafluoroethane, for example, are excellent moisture barriers but allow many gases to pass readily. Many different cells are available, covering many of the common requirements for toxic monitoring. For example, one manufacturer lists cells available for the following gases in air: O_2, NH_3, CO, Cl_2, F_2, Br_2, HCN, H_2S, NO, NO_2, SO_2, HCl, HF, SiH_4, AsH_3. These cells are all claimed to operate at sensitivities of 1 ppm or less. Most are based on a simple two-electrode electrochemical cell, choosing the electrode and electrolyte to selectively respond to the gas of choice. A typical example is the CO cell. This has the reaction

$$CO + H_2O \Rightarrow CO_2 + 2H^+ + 2e^-$$

at the sensing anode, a porous conductive type. The electrolyte, a strong sulphuric acid solution, conducts the hydrogen ions away and ensures that the CO_2 is not absorbed but can diffuse away.

Although many of these cells cannot make accurate measurements and are subject to many interfering effects, they excel in the role of air-contamination alarms. Exact quantitation is not needed. Also, provided that interfering gases do not interfere in a negative way, subtracting from the true reading, and provided that false alarms are not produced by normal concentrations of innocuous gases, interference is not important in this application.

The principle of photoionisation by UV mercury lamp illumination has also been used for toxic gas monitoring. Photoionisation detectors respond to any contaminant in the air which has an ionisation potential below that of the UV photon energy. With a photon energy of 10 eV, air gases and some pollutants are not ionised but many important pollutants are, such as H_2S and NO_x. Photoionisation is perhaps best used as a GC detector for detection of hazardous gases at trace levels in air, however, because the addition of GC separation enables a more confident measurement. The fact that the detector does not respond to air gases (N_2, O_2, CO_2 Ar, and H_2O) means that it can achieve a high sensitivity to these and can use filtered/purified air as a carrier gas if necessary.

The remote sensing of pollutants in air by laser Raman spectroscopy was mentioned in section 2.3.5 but this sort of instrumentation is not common in everyday industrial use because of its cost and complexity.

Mass spectrometer multipoint systems. These rely on a central mass spectrometer analyser connected by a multiway valve to a series of sample pipes. Typically a central display unit plots the samples versus time for each sample point. In addition, an electronic signal representing each measurement made is typically returned to a display next to the sample point.

Great care must be taken to make sure that the toxic gases monitored are not absorbed by the pipework system and to make sure that the pipework does not induce undue delays. Because of these considerations, the pipes are continuously pumped to maintain flow and improve the response time of the instrument.

Heated sample lines are also popular because they avoid problems with moisture condensation, for example by continuous heating of the sample tubing. A convenient way of achieving this is to pass a current down a stainless steel tube such that it acts as a resistor. With some thermal insulation around it, this can be a very convenient system. Even more convenient is a coaxial assembly with a narrow-bore stainless steel sampling tube down the middle, which is insulated and a copper braid and outer covering used. The heating current from the low-voltage transformer travels up the braid with little loss and the stainless steel is heated in the current return path, the sampling end of the tubing being soldered to the braid. In this way, the heated sample line can be used just like a simple tube, can be flexible to some extent and can be installed more easily.

Occasionally, the central mass spectrometer will use a membrane separator. A dimethyl silicone rubber membrane will effectively remove much of the air from the sample stream, whilst absorbing and transmitting many toxics to the MS source region (see remarks on the PETRA instrument in section 2.1.7). Although quantitative measurement is then more difficult, the concentration factor makes the instrument much more sensitive and this may well be the decisive factor, rather than accuracy, in a toxic monitoring application.

The human sense of smell. The sense of smell as a multigas analyser is often overlooked. The nose is not a quantitative analyser, tires easily (Moncrieff, 1967), does not detect asphyxiant and many more dangerous gas mixtures, and is thus mainly useful for qualitative monitoring and toxic gas warning. It is, however, enormously important. Were analyses of a large enough number of industrial accidents involving gases carried out, they would almost certainly show that more lives are saved each year by people using their noses than are saved by the most sophisticated monitoring systems; which is not to say that monitoring systems are not also essential. There are also many industrial processes where smell is an important process or quality control input. Efforts should be made both in the design and operation of plant to use the sense of the smell as a safety feature. For example the design of systems should avoid, where other considerations allow, the routine presence of masking odours that might prevent a leak or other malfunction being detected via smell. Similarly, operators of plant should never ignore unexplained smells. (A brief introduction to the sense of smell is given in Appendix B.)

2.4 Gas handling and storage

The generation of large amounts of electrical power was already possible by the mid-nineteenth century. However, the widespread use of electricity was much delayed because the means to deliver that power were not sufficiently developed. Even the comparatively simple requirement for a well-insulated cable was not easily met at that date, and the many familiar ancillary devices needed (transformers, fuses, switches) were not available in suitable form. Transport of gases from the point of production to the point of use is no less important. Gases must be delivered to the point of use at the correct temperature, pressure, flow rate and chemical composition, and a host of analogous ancillary devices are needed, e.g. pipes, regulators, bursting discs and valves.

Fortunately, gases can be efficiently stored, contrasting with the difficulties that attach to electricity storage. Storage has meant that pipelines from the point of production to the point of use, except for coal-gas and, today, for methane fuel, are not generally needed. Originally gases were simply stored at low pressure in 'gasometers', which are simply water-sealed piston and cylinder arrangements. Today, efficient storage relies on storing cryogenic temperature liquid gases at low pressures or storing gases in high pressure cylinders. Industrial gas for a typical application is therefore produced, delivered either in cryogenic form or in high pressure cylinders by road, then piped a short distance, via pressure and flow controls and perhaps via filtration and purification, to the point of use. Gas pipelines, for all that they are usually short, however, still require very careful design.

The materials used with industrial gases vary from the unsophisticated 'black iron' pipe to high technology oils such as the perfluoropolyethers. The choice of materials for pipes, seals, lubricants and other gas-wetted components depends on the gas and pressure to be contained. Gas suppliers generally give tables of materials recommended for use with different gases and these should be consulted during the design of a system. These tables can be supplemented with information from suppliers of special materials, such as elastomer seals, who will often have carried out extensive testing for suitability.

2.4.1 Pipelines and pipeline components for gases

Pipelines. The extensive use of steel 'black iron' pipe for long-distance pipelines for (relatively) inert gases in bulk depends on its strength and very low cost. For dry gases inside, and with paint and perhaps electrical protection on the outside, it is a highly satisfactory material.

Choice of materials is always a combination of technical suitability, cost and convenience in application. The extensive use of copper tubing for small bore installations of all description for many different gases is largely due to the latter factor. Copper tube is by far the easiest material to handle and install for lower pressure work, has a fairly low cost and is chemically inert to nearly all gases. Stainless steel tubing has been slowly growing in popularity for many years. Soft-annealed stainless steel tubing can be handled nearly as easily as copper, while it will retain an attractive external appearance, keep very clean inside over time, and can be used with many gases. Its use is routine in semiconductor gas installations, being valued for its low particulate and corrosion resisting properties, although even semiconductor installations historically used copper pipe without problems for inert gases.

Gas normally contacts ('wets') only the inside of a pipe or vessel. It is therefore not necessary to construct pipes or vessels out of highly inert bulk materials; a coating or lining, if sufficiently thick and contiguous, will often suffice. Although the relatively small sizes of pipes used in industrial gases usually preclude any kind of liquid-based coating process, gaseous 'passivation' processes are used, for example fluoride passivation with a weak fluorine in argon mixture.

Metal pipes are made in many different ways: extrusion, rolling followed by welding and hot or cold drawing down of a short fat pipe are those seen most commonly. For the more easily melted softer metals, such as aluminium or copper, extrusion is popular, whilst welding is the cheapest method for thin-walled steel tube. The very highest quality pipes are today made by the drawing down of a short fat pipe called a 'bloom'. This is produced by forcing a red-hot billet against a mandrel, which pierces the billet. Cold drawing can yield very high-quality, strong pipe.

The methane fuel gas industry has recently taken to installing small and large diameter (from 20 mm up to 500 mm diameter) simple unreinforced polymer pipes for underground transmission of gas. These pipes are made of polyethylene (PE) and are tough, inexpensive, and completely rustproof. Underground, they are estimated to have the potential to last for a minimum of 50 years. Part of their attraction is that they can be heat welded *in situ*. The heat welds in PE are easily as strong as the original material and are much less likely to leak than, for example, jointed iron pipes. Welds are achieved by cutting the pipe ends off square, heating them by contact with an electrical hotplate and then ramming the two softened pipe ends together with a hydraulic rig to form a butt joint.

The PE material has a usable strength of about 10 MN m^{-2}, giving them potential pressure ratings of the order of 20 bar in pipes with walls which are not too thick. In fact, however, methane distribution is low pressure (100 mbarg) to domestic users and only 2 barg or so to normal industrial users, allowing a generous safety margin.

So far, few industrial gas users have taken advantage of the possibilities of pipes made from these materials. There are concerns about moisture, hydrocarbons and other contaminants, for example, getting into pure gas streams, whilst the use of plastic pipes with oxygen has obvious safety problems and the pressure ratings developed so far have been lower than are generally used in industrial gases. However, they are now a tried and tested system, offering considerable advantages for installations with very long runs of pipe.

Flexible pipes. Standard rubber hoses, consisting of fabric-reinforced synthetic or natural rubber, are normally made with lower pressure ratings, e.g. 20 bar or less. Although rubber hose with some higher pressure ratings is available, these materials are not particularly popular with industrial gases because they are heavy if they are strong enough and subject to degradation from UV or oxidation.

For flexible connections in systems, bellows hoses, typically made out of stainless steel, are often employed. Care should be taken with bellows that they are not cracked. If they are subjected to a sharp bend and then bent back again they can fail; in applications where repeated flexure is needed, the amount of flexure needs to be strictly controlled within the limits which the supplier gives. In addition, for lower pressures, straightforward unreinforced tube in nylon or another strong polymer can be used. Simple polymer tubing is certainly very convenient, even if it is semi-rigid, for example for pneumatic control connections, and there are several inexpensive 'instant connect' systems which allow very simple connections for pneumatics. However, the rapid loss of mechanical qualities with temperature in polymers should be borne in mind; nylon becomes surprisingly soft even in hot water. Ratings of up to 10 bar are readily available, however, in unreinforced tubing.

For high-pressure flexible connections, a bellows can be armoured with a multilayer braded stainless steel overcoat, giving pressure ratings up to 200 bar or so. More cheaply, a 'pigtail' or coil of a smallish bore pipe is commonly used and, if a reasonably long length of pipe is coiled, this is surprisingly flexible. However, for many applications genuine flexibility is needed and this is where armoured polymer hose is mostly used. This is supplied with special crimped end-fittings by the manufacturers, and normally comprises two or three layers. The inner tube, containing the gas, is made from a polymer such as nylon, which is both strong and also a good gas barrier. Teflon is often used because of its chemical inertness and therefore wide applicability with nearly all gases. The middle layer is a strength layer, typically a fibre-reinforced rubbery polymer, or it may be braded stainless steel. The outer jacket, if used, protects from abrasion, UV degradation and oil or solvent attack. A neoprene or polyurethane material is typical. Polymer hoses can offer pressure ratings up to 300 or 400 bar.

Patented 'quick-connect' systems are available for use with any pipework but obviously are particularly aimed at use with hoses. A quick-connect aims to provide a gas connection within a second or two, typically by simply pushing the pipe fittings together, perhaps with a twist or disengagement ring to pull to lock them in place, and also has provision for automatically sealing off the ends of either or both of the pipes when the joint is disassembled again.

Pipework joints. The most satisfactory pipework joints for many purposes are those which braze or weld pipe sections together, provided the minimum possibility of a leak developing subsequently and maintaining the pressure rating of the pipe itself. The lead/tin soft soldering of copper pipes used in low-pressure domestic gas installations is unacceptable in the much higher pressure system pipelines used in industrial gases in the latter regard. Instead, copper tube may be brazed, using similar joints to the more familiar soldered types, but using the much stronger and higher melting brass-based brazing alloys instead of solder for the filler metal.

Although brazing can be used on stainless steel pipes, these are generally joined by tungsten-inert gas (TIG) orbital welding, using automatic machines from companies such as ARC. Orbital welders, once correctly set up, can perform butt welds on prepared pipes in a matter of seconds, with little in the way of fumes or fuss, and the joint is bright and almost indistinguishable from the pipe itself from a distance. They clamp the two pipe ends tightly together, apply argon inerting then, under computer control, a TIG tip arc is struck and the tip is then 'orbited' around the pipe at a precise speed, stopped as the arc returns to the starting point and the joint is completed. A shield renders the wearing of goggles unnecessary and the amount of heat used is surprisingly low; the welded joint can be handled in a matter of seconds. Pipework can also be welded by using socket welding. Fittings are arranged to have a socket to match the pipe used, the socket supporting the joint while welding takes place. Automatic TIG welders can also usually handle socket welds. Some fittings are arranged to be a compromise between butt weld and socket weld, using a very short (1 or 2 mm) socket, which aids in providing a little extra material as a filler in the completed weld but does not have the disadvantage of containing a crevice (which could harbour contaminants) in the finished joint as there is in a deep socket weld.

For joints in larger pipelines, it is clearly appropriate to use welded-on flanges, which may then be bolted together with a flat composition sealing washer or perhaps a more complex seal. Flanged joints are highly inconvenient, however, except in the largest sizes, taking much time and labour to assemble for the first time and for subsequent assembly/disassembly, and must be tightened carefully and evenly to be effective. For slightly smaller pipes which can be handled suitably, at modest pressures, it is of course possible to cut tapered threads on the end of the pipe and joint them using bushing or union nut couplings. For long pipelines made of inexpensive black iron pipes, this method is often used. However, cutting threads on the outside of a pipe reduces the pressure rating of the pipe considerably.

Conventional compression fittings of the Swagelok™ type are very widely used (there are many other brands with a variety of pressure ratings and prices) (Figure 2.46). These are convenient and can be made highly leak-tight. Their convenience lies in the fact that the pipes need no special preparation other than cutting to length and no special equipment is needed. Also, once the joints have been made, they can be easily demounted and then reattached. The tubing must have a good outside surface finish, must be kept free of scratches, and the sawn-off end must be carefully deburred, however. The tightening of the fitting itself, when first tightened, attaches the 'olive' (the sealing and mechanical restraint ring) firmly on, giving a joint which will meet very high pressure ratings to match comparatively thick walled pipes. Standard 6 mm pipes in stainless steel with such joints can be used up to 500 bar or so. Their *modus operandi* is to crimp the olive down onto the pipe by an annular wedging action. Because the pipe material is selected to be softer than the fitting material, the olive is crimped inwards onto the pipe, which is slightly deformed inwards, retaining the olive. Care should be taken not to overtighten, however; the joint will not be more leaktight, and the female cone of the fitting can be splayed out, making it difficult to demount and reuse on subsequent occasions.

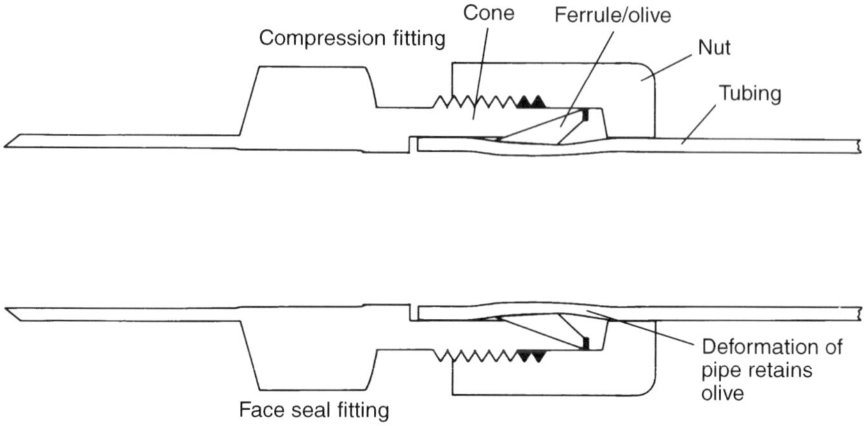

Figure 2.46 Compression fitting used for joining pipes in the gas industry.

Fittings of this type are available from 1/16″ (2 mm) through to 2″ pipe sizes. For the larger sizes pressure ratings are lower, although with rather thick-walled pipe 200 bar ratings can be achieved. A strong arm and large spanner is needed to swage the olives onto the pipes in the larger sizes and electric or hydraulic machines are available to render this task easier. These have the additional advantage that no strain need be placed on the fitting to which the pipe will be attached, other than the more modest torque required to effect a gas seal.

With all these proprietary joints, the manufacturers' tables of fittings and fitting/pipework combinations should be consulted as to the usable pressure rating and piping to match the fitting manufacturers' requirements should be obtained.

High pressure pipeline and storage vessel design and materials

Pressure vessels. A pressure vessel can be taken to mean a container for gases of above a certain size (say 0.5 l or so) and above a certain pressure (perhaps as low as 1 barg or less). After many spectacular and usually fatal accidents in the nineteenth century (e.g. Anon, 1993) with exploding boilers, pressure vessels are now subject to a large number of regulations. The regulatory codes are listed by government authorities across the world, with many countries adopting the ASME codes of the USA (ASME VIII part 1), the DIN codes of Germany or the BS (BS 5500 for fixed vessels, BS 5045 part 1 for seamless and part 2 for welded movable vessels) codes of the UK. The codes do differ in some respects, but are broadly similar.

ASME and the other codes specify:

- materials: a limited number of metals, mostly grades of steel, are acceptable;

- safety factors;
- structural design, calculation methods and record-keeping;
- design details, such as welds and fittings;
- inspection and testing;
- corrosion and fatigue life;
- conditions of use (pressure cycling, temperature, pressure, intended contents and trace contaminants);
- ancillary equipment, such as pressure relief valves.

The codes themselves should always be consulted before constructing a vessel, although pressure vessel suppliers always work to one of the codes and generally will work to any code on request. However, it is probably worthwhile noting at least some of the details contained in the codes:

- dished ends are produced by rolling of flat plate with a high power hydraulic ball forming system: this does produce some thinning of the plate used and codes specify what allowance must be made for this;
- longitudinal joints in cylinders are under higher stress than circumferential joints and this must be accounted for;
- vessels must show a safety factor of 4 versus maximum design pressure before failure and must show a safety factor of 1.5 versus maximum design before plastic flow (yield) (ASME);
- where connections to pipework, manholes or other devices pierce the vessel wall, formulas are given for putting in enough strengthening material to compensate for the loss of wall material;
- impact testing such as Charpy V notch at a low temperature may be specified for materials that can become brittle.

Cracks and testing. It is very rare indeed for a pipeline or gas cylinder to fail under ordinary conditions. Materials selection is one reason for this but an additional reason is that materials, components and complete installations are all tested.

All cylinders, many other components and some pipelines are tested by filling with water and pressurising well beyond the design pressure. As noted above, hydraulic testing is comparatively safe because if failure happens to occur, the energy release is low compared to the explosive release that follows failure with gas. Hydraulic testing, however, has its own problems. Water or other fluid must be drained out and the vessel dried after testing. This presents problems with some designs, which must be drainable with water-compatible components. Furthermore the temperature of testing must be carefully chosen. If it is low, then the design may fail due to brittle failure but this may constitute a better proof test. If a higher temperature is chosen, then failure is unlikely but some amount of distortion may occur if the material yield stress is exceeded.

Another important means of testing is that of crack detection. Two simple

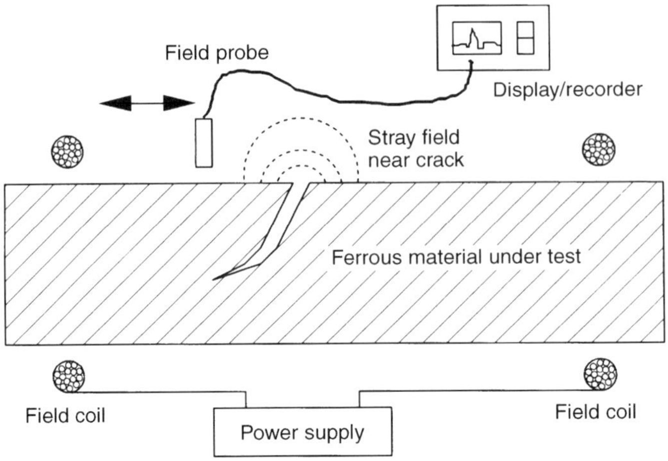

Figure 2.47 Crack detection with a magnetic sensor.

techniques for finding cracks which extend to the surface are sprinkling with iron filings and the use of a fluorescent dye penetrant. The dye is sprayed on to the surface to be tested and any excess wiped off. A crack retains some dye which then bleeds out on to the surface and becomes visible.

Iron filings sprinkled on to a magnetic vessel or pipe which has some magnetic flux imposed on it from an external magnet or coil will tend to accumulate where there is a rapid change in flux. Where there is a crack, even quite a small one hidden by subsequent painting, iron filings will tend to stick since the crack causes a big change in magnetic flux.

More sophisticated than these, but still using essentially simple equipment, are electric or magnetic methods (Figure 2.47). The latter amounts to little more than an instrumental method to replace iron filings. A 'transmitter' which induces a d.c. or a.c. magnetic flux in the test piece is attached by a clamp to the latter and a search coil shows where d.c. or a.c. magnetic flux at the surface changes rapidly. An eddy current detector (similar to the treasure-hunter's 'metal detector') can be used as an a.c. magnetic instrument. The eddy current detector uses a search coil to induce an a.c. field in the material being examined and this (typically) changes the self-inductance of the search coil and hence the a.c. generator frequency. Any change in frequency signifies a problem. Cracks only 0.5 mm deep can easily be detected but further sensitivity is typically inhibited by the surface finish on the vessel examined. Electrical resistance methods are rarely used but these can detect many flaws in pipelines or vessels in a very simple fashion, if care is taken with contact resistance problems, for example by using multiple finger contacts and four-point (or more) resistance measurement.

X-ray inspection. X-ray radiographs (sometimes actually made with a radioactive source such as cobalt-60) are now a standard routine for welds, at least for a sample of welds carried out. X-ray equipment is fast and special X-ray tubes are available for different geometries of weld; for example sources emitting a disc-shape fan of rays for irradiating circumferential welds in pipes and vessels with one shot.

Gamma-ray sources such as cobalt-60 require longer exposure times and more sensitive film than X-ray radiographs. However, they can provide images of thicker pipes and can be very small and lightweight, enabling them to be pushed up inside even quite small pipes (with precautions to avoid radiation exposure to personnel). They can also be mounted on a small motorised crawler or 'pig' which can be manoeuvred hundreds of metres along a pipeline.

Weld radiography is required in many codes for pipes and vessels carrying toxic materials, and is good practice where any doubt whatever exists on the quality achieved in welding. However, it is worth perhaps pointing out some of its limitations:

- the need to clear personnel away from a radiography site is awkward in situations such as a construction site;
- off-line processing of film is time-consuming;
- image quality is often limited by fogging (due to scatter of X-rays), grain (due to low achieved dose) and source size;
- thicker samples need higher voltage X-ray tubes or cobalt-60.

Ultrasonic crack detection. Ultrasonic systems operate on the time-honoured principles of military radar or sonar, sending out a short burst of MHz ultrasound via a piezoelectric transducer, then listening for reflections from it. With the speed of compression waves (the main transmission mode) in steel 6000 $m\ s^{-1}$ or so, an 'A scan' of a 5 mm thick pipe wall will be only 1 μm long.[24] Commercial systems are as little as $10 000 or less to buy, although a computer-based system with two-dimensional plotting and signal processing might be $50k–100 000. They are particularly suitable for material without welds and, of course, provide an accurate readout of metal thickness.

A particularly useful kind of ultrasonic transducer is that which can launch a wave at an oblique angle to the surface. This can then be placed on the smooth walls outside of any metal-working zone and used, for example, to

[24] This sounds difficult but in fact the electronics is fairly easy with today's electronics; the author made a special purpose system of this kind to work with a laboratory oscilloscope in a few days.

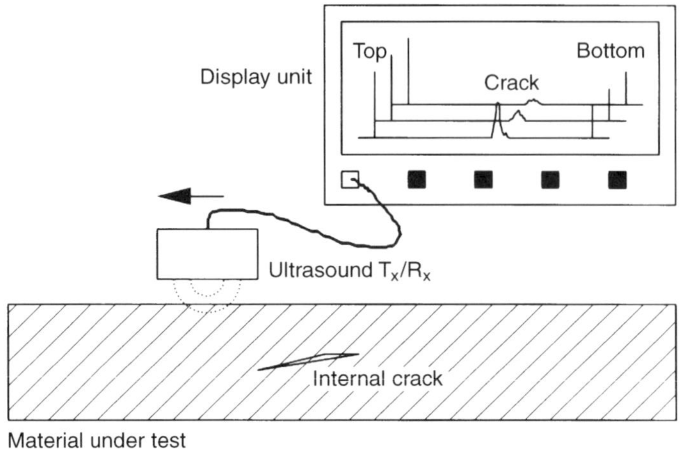

Figure 2.48 Cracks in A-scan detected by acoustic reflection.

examine a neighbouring weld zone. This kind of transducer is used on gas cylinders.

The principle difficulties with ultrasonics often revolve around the skill of the operator in interpreting the images and oscillograms obtained, and in maintaining good sound-conducting contact with the test piece. Unfortunately, the thinnest imaginable air film between the transducer and the test piece completely reflects the transmitted signal and prevents reception of the ultrasound signal; this is why the system detects small cracks so easily. A thin film of oil or gel is usually necessary to ensure an adequate contact with the substrate being examined. Also, there are often a surprising number of spurious reflections echoing around a pipe or cylinder to cause confusion to the unwary, so interpretation is quite skilful, although multiple tests on similar objects like cylinders do allow a 'fingerprinting' approach to be taken. Computer-based testing of gas cylinders can now be carried out by relatively unskilled operators, using a computer to interprete the waveforms, knowing the types of reflections obtained. Computer systems also allow the display of a two-dimensional map of reflections, giving a rather better idea of where defects are than simply chalking on the cylinder where the 'A-scan' showed a result.[25] Figure 2.48 shows the principle of A-scan acoustic reflection for crack detection.

Acoustic emission (AE) is one of a number of more unusual methods which are also occasionally used for tests of cylinders. A crack does not

[25] Those who have seen ultrasound in action for antenatal care in hospitals will no doubt have been duly impressed by the pattern-recognition skills of hospital ultrasound technicians. They can confidently comprehend the apparently meaningless blobby image and assess the development of a human foetus with surprising accuracy.

lengthen smoothly but progresses by small increments related to the grain structure of the material. The progress of a crack as pressure is increased on a cylinder can actually be detected by picking up acoustic impulses which occur as the crack elongates. Furthermore, by timing the arrival of the impulse at each end of the cylinder, the position of the expanding crack can be detected. It is only very recently, however, that acoustic emission has been sufficiently sensitive to be useful in the gases industry in small equipment. AE equipment is slowly becoming accepted and may eventually be accepted as a standard test method for cylinders and pressure vessels, along with hydraulic testing.

The size of crack that is permissible depends on the alloy used, as well as on the pressure of the service required. The alloys used for gas cylinders are chosen not just for their strength but also for a high 'work of fracture per unit crack area', W. With simple assumptions, the maximum size of crack, L_{max}, can be estimated approximately from

$$L_{max} \simeq 2WE/s^2$$

where E is Young's modulus and s is the average tensile strength. Beyond this critical crack size, according to the simple theory, the energy generated by further extension of the crack exceeds that needed to create more crack area so the material fails catastrophically. In a typical example, the maximum size of crack in typical mild steel sheet is an appreciable fraction of a metre by this formula – clearly well beyond the size at which a leak would be clear – but the formula is still useful to remind us that the smaller the work of fracture, the smaller the critical crack size. Thus, because the work of fracture for brittle metals such as the high strength, high carbon steels is just a few thousand J m^{-2}, they are unsuitable for cylinder manufacture.

Another potential problem is fatigue cracking. This is typically a problem with aluminium and has been extensively researched because of the extensive use of aluminium in aircraft. Given the fairly uncontrolled conditions of use of the hundreds of millions of gas cylinders in use worldwide, extreme caution has been exercised in the use of Al alloy cylinders. In fact, this conservatism has led to Al cylinders having very thick walls, and being about the same weight as a steel vessel of the same volume except at the smallest sizes (section 2.4.7). Nevertheless, some difficulties have been seen in aluminium cylinders, typically due to problems not related to the basic cylinder containment problem. Corrosion due to alkali exposure, fairly harmless to steel cylinders, and neck-cracking caused by excessive stress induced by the valve fitment are two of these.

The embrittlement of metals caused by hydrogen is another potential problem. Steel cylinders for hydrogen are specially heat-treated to minimise this effect and are inspected especially frequently. Hydrogen pipelines, at least in smaller sizes, frequently avoid this problem because they are made in thick-walled cupro-nickel alloy ('tungam') tubing.

Pipework design. Industrial gases such as nitrogen are frequently piped around sites to multiple user points as much as other commodities such as hot water or electricity. Indeed, some of the same principles apply to N_2 and other bulk gas installations, for example:

- pipe sizing should be such that undue pressure drops are not seen with all equipment operating (also, pipe sizing should allow for additional requirements such as the gas velocity speed limit for oxygen, 8 m s^{-1});
- when multiple drop points are needed, for example, a 'ring main' offers the possibility of smaller bore pipe than a similar single-ended pipe;
- on-line (pressurised) connection/disconnection facilities, as provided by outlets with isolating valves or, better, with self-sealing quick-connects, are an advantage, enabling a new equipment to be connected without interruption of other equipment.

Pipework must take into account often conflicting demands:

- convenience in use: pipework should ideally be easily accessible for maintenance purposes, for example;
- regulations on pipe locations (fuel gases should not be run near to oxidants, for example) and labelling;
- ease of fabrication: anyone who has tried to weld or solder a pipe in an underground or underfloor run will attest to the difficulties of some pipework layouts;
- pressure losses: an excessive number of sharp elbow bends, for example, or the use of too many high loss seat valves, for example, can give rise to high pressure drop over a complex system;
- in liquefied gas installations, in addition perhaps to cryogenic problems, the ordinary features of liquid pipework, such as 'water hammer'[26] or cavitation effects, must be avoided.

Where multiple similar units such as filters or heat exchangers are connected together in parallel with pipework, it is possible for an undesirable maldistribution of flow to occur and the pipework designer should be aware of this. For example, in heat exchangers it is good practice to arrange parallel units in Z-manifolds rather than U-manifolds in order to avoid maldistribution.

[26] Water hammer occurs typically when a valve on the end of a long pipeline running at speed is closed too quickly. A potentially damaging high pressure shock wave runs up the pipeline, as the fact of the closure of the valve is communicated to the pipeline liquid, in order to stop it. The effect can repeat cyclically at certain values of valve opening in long pipelines, a highly undesirable situation. Water hammer is most easily prevented by ensuring the valve opening time is long compared to the characteristic time constant τ of the pipeline, given by

$$\tau = 2L/v$$

where L is the pipe length and v is the shock wave velocity, which is dependent on both pipe and liquefied gas properties but is quite typically around 1000 m s^{-1}.

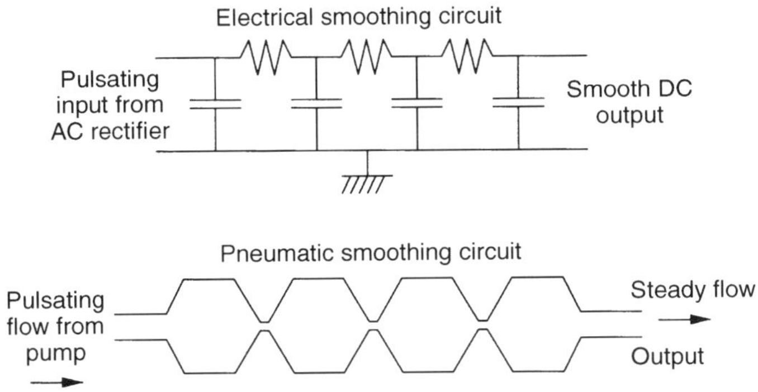

Figure 2.49 Gas pulsation filter and its electrical analogue.

The connection of pumps in parallel has other problems, as when one of the parallel units suffers a breakdown and may be driven in reverse by the others. Best practice here is to install a check-valve in each of the pump outlets. Pulsations caused by pumps are common and may be dealt with using fluid circuits analogous to the d.c. power supply filtering circuits used in electrical engineering. The classic d.c. power filter supply circuit has one or more series resistors with capacitors in parallel. An analogous circuit in gas plumbing comprises restrictions such as orifices or small diameter pipe sections interspersed with larger sections of pipe or small pressure vessels in series (Figure 2.49).

2.4.2 Valves and other pipeline components for gases

Transistors are often compared to valves controlling gas flow in their ability to control the flow of electric current. However, gas valves are intrinsically much more imperfect in their abilities with respect to gases than transistors are with respect to electricity. Valves often suffer from some or all of the following problems:

- flow restriction
- leakage across seat
- leakage to outside
- high actuation force
- nonlinear flow versus actuator position
- wear and short life
- trapping of particles leading to leaks, wear and failure
- slow actuation time.

The results of emphasis on different aspects of these problems is that many different forms of valve are available, which vary enormously in cost and complexity (Figure 2.50).

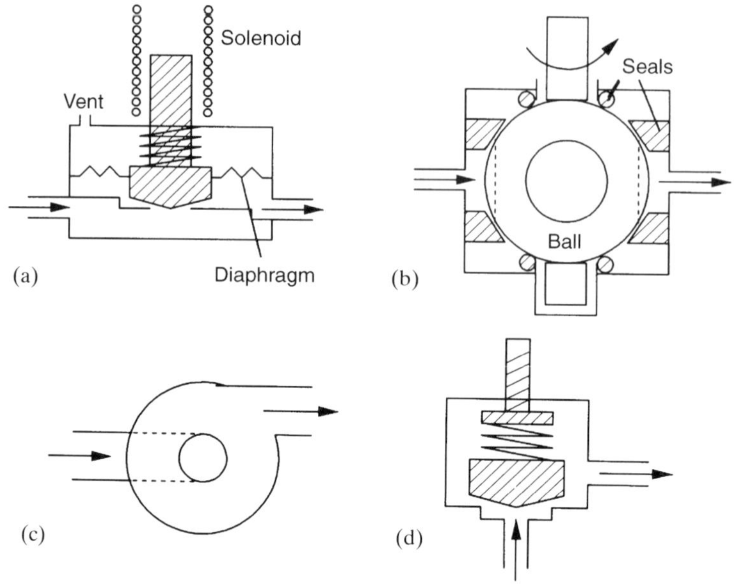

Figure 2.50 Some common types of valves: (a) poppet; (b) ball; (c) fluidic diode; (d) standard spring check/relief.

Poppet valves should be considered for many plant designs where reliability and frequent valve actuation is required. They can be made very reliable and offer a high resistance to leaks under high differential pressure. Only a very small movement is required for actuation, bellows or diaphragms can be used to prevent leakage to the actuating stem, and sliding contact can be designed out almost completely, giving low wear rates. Poppet valves can also be designed for very rapid actuation in safety applications.[27] However, they are normally placed so that increasing differential pressure leads to increasing actuation force. They also offer a relatively high resistance to flow and do not normally make particularly good metering valves.

Valves such as needle valves or gate valves can be designed so that their

[27] The actuation speed of a valve is dependent on the power driving it and its size in general. One project the author was involved with required ultrarapid pulsation of an oxygen supply. The most effective controllable high-speed valve was an electronically driven solenoid poppet valve. This could switch from open to closed in a few milliseconds by driving it with more electrical power than normal and by driving it down as well as up, rather than relying on the return spring. However, for a system requiring a continuous stream of pulses, a rotary valve driven by a high-speed electric motor proved to be a better system. This system was capable of over 15 000 pulses per minute. A special design of valve with many ports (rather like the mechanism of a siren sounder) could have raised this high frequency even further. However, with regular pulses of gas there are severe problems with resonating effects and also a loss of pulse amplitude because of smoothing effects in the pipework.

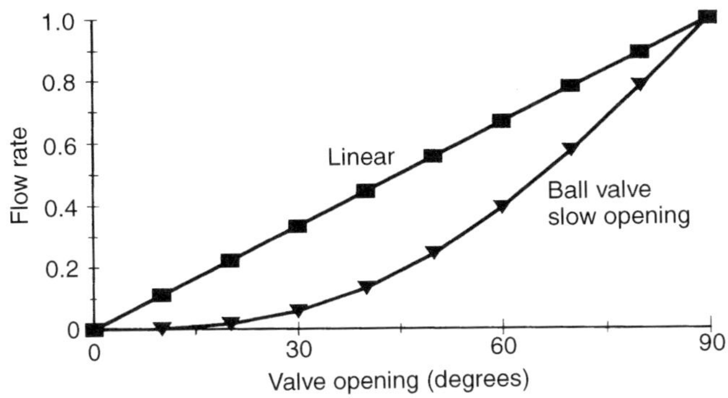

Figure 2.51 Examples of valve opening characteristics.

effective orifice area is a linear function of actuation. However, the rate of flow on opening a valve is rarely in practice a linear function of the opening of the valve because of the effects of the pipeline in which they are placed. Other valves, such as ball-valves, have a nonlinear slow-opening characteristic, while still others, such as butterfly valves, have a nonlinear fast-opening characteristic (Figure 2.51).

An approximate idea of the nonlinearity expected in the flow out of a valve with upstream and downstream pressures P_o and P can be obtained by noting that for small opening values the valve will typically behave as a critical orifice, like a small orifice plate in a large pipe, at least for supply pressures much greater than downstream pressure. In this case, the flow rate expected will be proportional to the valve area of opening (A) and the absolute supply pressure

$$\text{mass flow} \simeq kAP_o$$

As the valve approaches the bore of the pipe in opening area, a low pressure-drop orifice formula is appropriate

$$\text{mass flow} \simeq k'AP_o(P_o - P)$$

where P is the downstream pressure. However, for most valves, the flow at larger opening angles is distorted by drop in the supply pressure and by flow resistance in other components.

The effect of the surrounding pipework can be understood if a combination of orifices and linear flow elements in series with the valve is imagined (Figure 2.52). The effect is that most valves tend to have a fast-opening characteristic in practice when fitted in a pipeline. As a result valves having a slow-opening characteristic are favoured if it is required to provide better control of low flow rates.

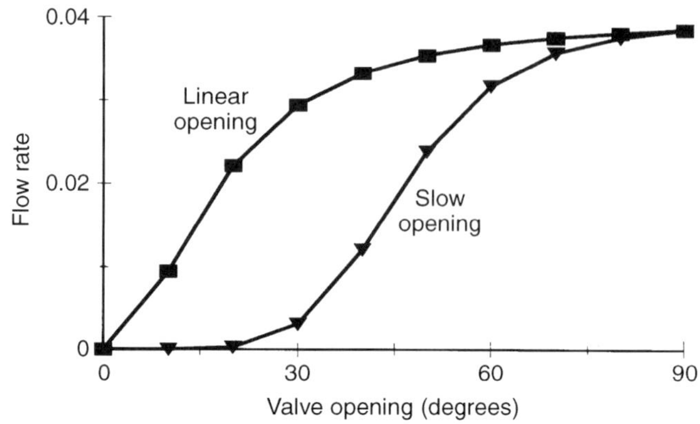

Figure 2.52 Flow versus pressure for a valve in a pipeline.

Non-return/check valves. Non-return (NR) or check valves are an essential feature of many pipeline systems. They are the fluid equivalent of the electrical diode, conducting gas in one flow direction but not the other. They serve to prevent or restrict reverse flow in many operational situations; for example pump outlets and PSA outlets. NR valves are frequently required for safety reasons, too, as in gas bubbling systems, where the NR valve prevents the process liquid entering the gas line.

It is possible to make devices with no moving parts, sometimes dubbed 'fluidic diodes', which have some of the function of an NR valve. These rely on the inertia of the fluid (most often liquid but gas diodes are possible) in a swirl chamber. In the forward direction, the fluid flows smoothly into the centre of the chamber and out of the periphery. In the reverse direction, the fluid flows into the periphery and sets up centrifugal forces which greatly increase the reverse-flow resistance.

Seals. Many gas components cannot be entirely constructed out of the most ideal material – solid metal. Although all-metal valves and pumps are possible, and some valves used for high purity semiconductor gases are all-metal, most practical devices require rubber or plastic compliant and sometimes sliding surfaces to provide a gas-tight seal.

Distinction should be made between seals which are static and subject only to the small, slow movements associated with differential thermal expansion, and those which are required to undergo more rapid compression or stretching or are, worse, subject to sliding. Static seals may be made of materials which would be useless for dynamic seals, materials which would show little protection against abrasion, for example, or which swell.

Distinction should also be drawn between seals which are at ordinary pressures and those at such a high pressure that they can be destroyed by being extruded out of the volume in which they are confined. At pressures above 100 bar it is often still possible for ordinary 'O'-ring seals of common composition to be used. However, at such high pressures it is necessary to use them with techniques such as 'backing rings', which are metal rings placed so as to prevent 'O'-ring extrusion. The backing rings are on cylindrical seals and are often wedge-shaped in cross-section ('mitre rings'). They function by being rammed by the operational pressure against another ring or a shoulder on the cylinder to form a very close fit with the bore of the cylinder, allowing no gap for extrusion of the polymer seal ring. Alternatively, specially shaped soft metal seals, such as the 'wave ring' or 'delta ring', can be used, or a direct metal–metal sealing assembly used, such as a bull-nose fitting. With careful attention to detail, good sealing up to 10 kbar or more is possible (Anon, 1975).

Although not difficult compared to high-pressure devices, even modest pressure seals require some design work. For packed or stuffed seals, the basic principle is that the pressure exerted by the seal on the surfaces must exceed the fluid differential pressure applied. If this is not the case, then the packing will be pushed back by the fluid and leaking will occur. The necessary pressure is often provided for simply by allowing an adjustable pressure on the stuffing via a gland nut, which is pretightened to a known torque, or simply tightened until leaking stops to an adequate degree. Often a lubricant is included in the composition of stuffing seals used in gas apparatus, as lubrication from a minute leak, used when containing pressurised oil, is not possible.

There is both general published and manufacturers' data on the required size and form of grooves for 'O'-rings. These published guides achieve a precompression of the seal which is sufficient to guarantee no leaks under usual operating conditions. Seals for pistons or rotating shafts, such as sprung lip seals, which must both seal and not impose a significant torque or heat up are more difficult to design but specialist suppliers can advise on suitability and fitting (section 2.1.3).

Elastomers (rubbers) are high molecular weight polymers (Table 2.10). Their molecules are held together by a relatively small number of primary bonds (cross-links) between polymer chains, and a large number of secondary bonds, which are much weaker links. It is the latter links which are stretched or broken during stretching of rubbers and reformed when the applied force is relaxed. The primary bonds ensure that the rubber article returns to its former shape when force is removed and they are what distinguish an elastomer from an ordinary plastic.

Swelling is the natural reaction of any elastomer when its secondary bonding breaks down under the influence of a solvent. The solvent, which would in the absence of the elastomer's cross-links dissolve it completely, is absorbed into the material and destroys the secondary bonds to an extent depending on potency. The result is a swollen elastomer. Swelling may be

Table 2.10 Elastomer seal materials and their properties

Elastomer	Maximum temperature (°C)	Strength	Bad solvents
PTFE	260		
PVC + plasticiser	80–105		All, except alcohol
CTFE	200		Slight swelling in halosolvents
Silicone	288	$<10\ MN\ m^{-2}$ (weak)	Ketones, but resistant to most organic solvents
Nitrile		$10\ MN\ m^{-2}$	Ketones cause bad swelling
Butyl			Resistant to most
Isoprene (natural rubber)		$40\ MN\ m^{-2}$	Oils, many solvents
Neoprene			Ketones
Polyurethane		$40\ MN\ m^{-2}$	Slightly attacked by ketones
EPDM			Resistant to most

PTFE: polytetrafluoroethylene; PVC: polyvinylchloride; CTFE: polychlorotrifluoroethylene; EPDM: ethylene propylene diene elastomer.

slight, reversible and associated with only slightly degraded elastomeric properties. Frequently, however, swelling is gross, 100% or more increase in dimensions is possible, and associated with undesirable degradation such as stickiness or crumbling.

Seals are a compromise between the following issues:

- tight gas seal versus high frictional forces;
- elasticity versus slipperiness;
- elasticity versus solvent resistance;
- cost versus lifetime

There is no perfect seal material. Swelling on exposure to the wrong solvent, as noted above, leads to problems. Swollen seals often suffer further knock-on problems caused by swelling such as frictional heating or abrasion leading to failure. Many elastomers will lose elasticity and fail as a result of oxidation in air or oxidants. Many plastic materials suffer 'creep', i.e. permanent deformation, which frequently leads to loss of gas-sealing capability if not outright failure. Cryogenic service has even more problems, with most polymers becoming brittle long before liquid nitrogen temperatures. A graphic demonstration of this is easily achieved by dipping a rubber item in liquid nitrogen and hitting it cautiously with a hammer; even the toughest rubbers generally shatter with little provocation.

Grades of elastomers vary enormously in their actual performance and it is nearly always worth contacting the manufacturer for specific information. Also, many practical elastomers are blends of several different compounds and may vary in their degree of cross-linking, varying their performance over wide limits. Highly cross-linked polyurethanes, for example, are more resistant to chemical attack than low cross-linked polymers with a similar chemical make-up. Fillers vary the property of the basic polymer in practice, too. Properties which may be important in practice, such as wear resistance and coefficient of friction, are affected by the addition of fillers, such as carbon black, and additives, such as rosin, respectively. Again the polymer manufacturer (or the firm which carried out the compounding and processing) should be able to help.

One method of sealing joints in gas systems is also a means of lubrication. Threaded joints in gas systems, which rely on a tapered thread to seal them, although not in general the first choice for a new system, are completely satisfactory if correctly assembled. The threads, particularly of stainless steel components, will not in general tighten satisfactorily without a lubricant grease; they have a tendency to 'galling', in which small areas of the steel faces cold-weld to each other and are then torn apart leaving a damaged finish.

The choice of lubricant is very restricted (Figure 2.52) and, in fact, the leak tightness of the joint can be improved by the use of a thread sealant. These both provide lubrication and fill in any minute gaps left in the assembled taper joint. For many purposes PTFE tape is the best choice for thread sealant. It is compatible with a large range of gases and is satisfactory at surprisingly high pressures.

Gas compatibility. Gas companies all issue gas compatibility tables, giving metals and seal/hose materials suitability for each gas (Table 2.11). Note that fluorine gas can only be used in systems that have been passivated for fluorine service (section 3.5.3). As described in section 2.5.9, hydrocarbon grease and oil can form an explosive when oxygen is dissolved into them at high pressure. Even apparently non-inflammable materials, including, for example, chloro and fluoro plastics and rubbers, are, in high pressure pure oxygen, highly inflammable or worse.

Oxygen compatibility is tested by placing a sample in a 'bomb' (a small high pressure vessel), applying high pressure oxygen and then attempting ignition electrically. The temperature can be steadily increased until a point is reached at which the test compound sustains combustion and ignites. In pure oxygen, many ordinarily stable materials are unstable and burn easily with very low autoignition temperatures. Some examples are given in Table 2.12.

Lubrication. Whilst for many gases ordinary mineral oils and greases give satisfactory service in gas equipment, for some gases, and especially at high

Table 2.11 Gas compatability

	Aluminium	Copper	Brass	Steel	Stainless	Zinc	PTFE	PTFCE	PVDF	Polyamide	Polypropene	Butyl rubb	Nitrile rubb	Chloropren	Viton	Silicone	Ethylene propylene
Acetylene	A	NR		A	A		A	A		A	A	A	NR	NR	NR	NR	A
Ammonia	A	NR	NR	A	A		A	A	NR	A	A	A	A	A	NR	NR	A
Argon	A	A	A	A	A	A	A	A	A	A	A	A	A	A	A	NR	A
Arsine	A		A	A	A		A	A	A			A		A	A	A	A
Boron trichloride	NR	A		A	A		A	A	A	NR	A	NR	NR	NR	A		
Carbon dioxide	A	A	A	A	A	A	A	A	A	A	A	NR	NR	NR	NR	A	A
Chlorine	NR	A	A	A	A		A	A	A	NR		NR	NR	NR	A		NR
Dimethyl ether	A	A	A	A	A	A	A	A	A				A	NR			A
Ethane	A	A	A	A	A	A	A		A				A	NR	A	NR	NR
Ethylene oxide		A	A	A	A		A	A				NR	NR		NR		NR
Fluorine	A	A	A	A	A	N	NR	NR	NR	NR	NR	NR	NR	NR	NR	NR	NR
Fluorocarbon 14	A	A	A	A	A	NR	A	A	A	A	A	A	A	A	A	A	A
Helium	A	A	A	A	A	A	A	A	A	A	A	A	A	A	A	A	A
Hydrogen chloride		N	N	A	A	N	A	A	A	NR		NR	NR	NR	A	NR	A
Hydrogen fluoride	N	A	NR	A	NR	A	A	A	A	NR		NR	NR	NR	A	NR	A
Methane	A	A	A	A	A	A	A	A	A	A	A		A	A	A	NR	NR
Nitrogen	A	A	A	A	A	A	A	A	A	A	A	A	A	A	A	A	A
Nitrogen dioxide		N	N		A		A	A	NR	NR	NR	NR	NR	NR	NR	NR	NR
Nitrous oxide	A	A	A	A	A	A	A	A	A	A	A	NR	NR	NR	NR	A	NR
Oxygen	*	A	A	*	*		F	F	F	F	F	F	F	F	F	F	F
Propane	A	A	A	A	A	A	A	A			A						
Silane	A	A	A	A	A	A	A	A	A			A		A	A		A
Sulphur dioxide	A	A	A	A	A	N	A	A	A	NR	A		NR	NR	NR	A	A
Sulphur hexafluoride	A	A	A	A		A	A	A	A	A	A	A	A	A	A	A	A
SUVA HFC 134a	A	A	A	A	A	NR	A	A					N				
Tungsten hexafluoride	A	A	A	A	A		A	A	A			NR	NR	NR	NR		

Key: A, acceptable; N, not compatible; NR, not recommended; * normally compatible but may ignite under extreme conditions; F, risk of fire/explosion.
Information courtesy of BOC.

Table 2.12 Autoignition temperatures of polymers in oxygen

Polymer	Autoignition at 1 bara (°C)
Nylon 6.6	195
Neoprene	185
Fluoroelastomer	330
PTFE	>425

pressures, special compounds are necessary. Oxygen service, and service with oxidant gases, are generally the most stringent. Lubricants should be certified or tested for ignition temperature before use.

Corrosive gases, such as hydrogen chloride, are not in general a problem with lubricants and seals, unless, like, for example, chlorine or fluorine, they are also oxidising agents. Fluorine is exceptionally difficult to work with, requiring seals and lubricants with a high state of fluorination such as the perfluoropolyethers. For many purposes, the perfluoropolyethers, such as Fomblin™, can be used. These are made in a range of molecular weights from light lubricating oils to greases. They are fully fluorinated and exceptionally stable.

A typical polyether might have the formula

$$CF_3O\text{–}(C_3F_6O)_m\text{–}(CF_2O)_n\text{–}CF_3$$

These compounds have a lower viscosity than their molecular weight would suggest. A compound of molecular weight of 2600 a.m.u. has a pour point of 42°C, for example. (The pour point is 5°C above that at which the oil begins to 'set' into a jelly or wax.) They are water repellent and have the curious property of being able to dissolve surprisingly large amounts of oxygen; 50 or 100 ml ml^{-1} liquid is not untypical, a fact which may have applications in medicine (section 3.9.1).

The perfluoropolyethers are, however, rather poor in their lubricating properties. They have a low surface tension and a very high viscosity index. This means that their viscosity decreases more steeply with increase in temperature than ordinary oils. Machinery using perfluoropolyethers is thus more susceptible to seizing because an increase in temperature can lead to a decrease in lubrication, leading to further increase in temperature and so on until failure. They are thus suitable only where conditions (gas pressure, bearing pressure, relative speed of bearing surfaces) are not too arduous.

Unlubricated bearings can be used in many gas applications at low speeds. Bearings such as Glacier DQ, which combine PTFE self-lubricating plastic with graphite and soft metals, are particularly suitable as they are substantially non-reactive and can also be used at cryogenic temperatures satisfactorily, when normal lubricants such as oils are frozen and ineffective.

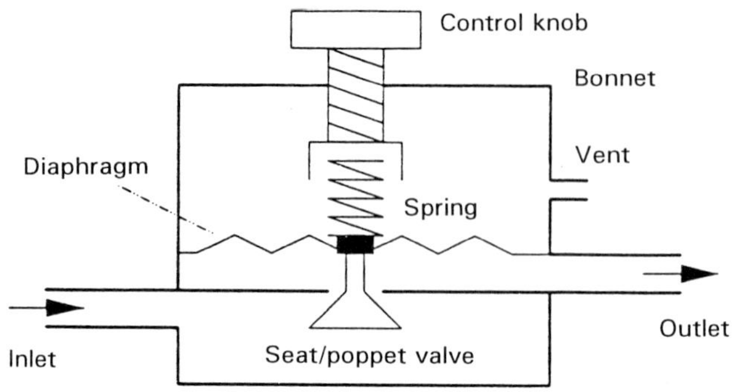

(a)

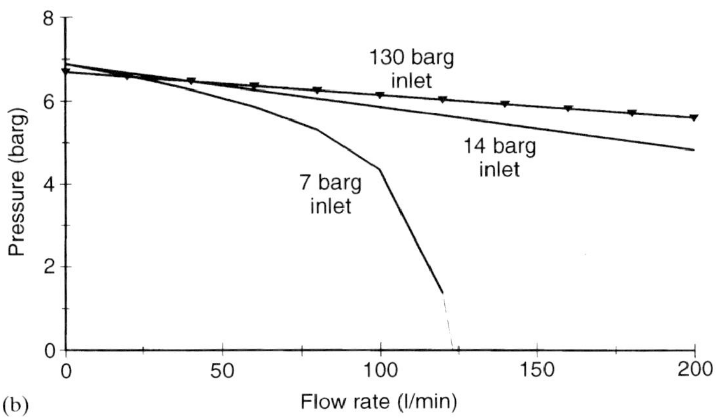

(b)

Figure 2.53 (a) Pressure regulator. (b) Output from a single-stage pressure regulator versus flow and inlet pressure.

Regulators. There are many different types of gas pressure and flow regulators. However, the term 'regulator' is normally reserved for poppet valves connected to a control diaphragm, as in Figure 2.53a (a practical embodiment is shown in Figure 3.22). The spring can be adjusted in compression with the screw to adjust the outlet pressure or changed for a different one for different pressure ranges. This simple mechanical combination has been the basis of gas pressure regulation almost since the conception of high pressure gas. A standard regulator is simple, inexpensive and reliable. However, it has all the problems of a poppet valve, together with the problem of the diaphragm, which can leak. In addition, the pressure regulator performs its function only imperfectly, as is shown in Figure 2.53b.

Gas regulators, finally, can also lead to pressure instabilities in systems, i.e. oscillations and pulsations, which can be very detrimental to a system's function and lifetime.

Gas regulator theory. The ideal gas regulator would employ a very large diaphragm to operate a very small poppet valve and would be opened by a very long spring. It would open very wide until the downstream pressure approached the set point given by the spring pressure, when it would rapidly close down to the correct orifice for the flow rate needed at that pressure.

The dynamic theory of gas regulators is complex, with the inertia of the moving parts interacting with acoustic resonation of the gas in the pipeline. However, the static theory of pressure regulators is simple and still useful. A standard regulator, with diaphragm area A and poppet valve area B tends to have an outlet pressure which rises with fall in inlet pressure and falls with flow rate as follows

$$P_o = [F_o + AP_a - BP_i - RQ/(k\sqrt{P_i})]/A$$

where the outlet, inlet and atmospheric pressures are P_o, P_i and P_a, the spring exerts force $F_o - Rx$, where x is its displacement, and the constant in the orifice flow versus pressure and displacement curve is k, and the flow rate is Q.

A two-stage regulator is much better than a single-stage because the inlet pressure is stabilised, reducing to immeasurably small (0.1 psi or so in typical designs) the variation in P_o due to P_i changes.

Other devices which can be useful in pipeline design are:

- The FLOSTAT: a device which tries to maintain constant outlet flow, rather than outlet pressure;
- the excess flow shut-off: used for safety purposes, this shuts off when flow exceeds a preset limit, and can only be reset by manual intervention;
- flashback arrestors, which prevent flames travelling back along pipelines carrying inflammable or other reactive gases. These are valves which shut off if an excessive pressure rise or backflow occurs. They also contain a metal grid 'snuffer' assembly, which cools a propagating flame and extinguishes it. Some are designed to autoreset, others require a manual reset. They are mandatory with some inflammable gases such as acetylene.

2.4.3 *Removing particulates: filters and cyclones*

In most gases, particles above a few microns are eliminated during production or purification. For many processes, this is sufficient; a few micron-sized particles, for example, are unlikely to affect the functioning of a steelmaking process. However, there are many more processes where particulates even at the submicron level are unacceptable.

The vital parameters of a filter from a user's point of view are:

- its efficiency in removing particles versus particle size;
- its pressure drop;
- its compatibility with gas;
- its re-entrainment;
- that any possible contamination, gaseous or particulate, caused by the filter itself must be acceptable;
- susceptibility to clogging or other lifetime limitations;
- its cost.

Many of these parameters interact. Filter drop and efficiency are typically complementary, i.e. very efficient filters will have a greater pressure drop than less efficient ones. Filters amount to a finely divided form of material which can be expected to deteriorate much more rapidly than normal bulk material if there is any degree of degradation reaction between filter and gas. Compatibility of filter material is thus very important. For example, any filters used with oxygen gas must be certified as suitable for that service; many organic filter materials form a serious fire hazard when filled with high-pressure oxygen gas.

A filter is superficially a very simple component. It is not, however, simply a sieve on a microscopic scale. It might be thought that if a particle is smaller than the sieve size, it will penetrate, while if it is bigger, it will not. However, this is not the case. As the size of a sieve is shrunk to a few microns or less, its functioning changes. It is possible to produce a submicron filter which functions as a sieve but this is most unusual practice.[28] Most submicron filters work as sieves for the very largest particles but rely on other mechanisms for medium-sized and very small particles.

Macroscopic diffusion is one process, whereby small particles, so small that they are subject to Brownian motion, wander into small interstices of the filter and are then permanently captured by short-range intermolecular forces such as van der Waals'. The other important mechanism is that of inertial impaction. Larger particles, which are still smaller than the larger holes in the filter and might therefore penetrate it, are retained by impacting into the filter medium as they pass through the filter. Once impacted, the particles remain trapped by electrostatic and van der Waals' forces.

Submicron filter action: most penetrating size. A useful concept in assessing inertial impaction on filters is the Bremstrecke or stopping distance. This is the distance which a moving particle takes to slow down to essentially zero velocity in a stationary medium. For a particle radius r, density ρ travelling initially at velocity U_o in a fluid of viscosity μ

$$\text{Bremstrecke} = 2\rho r^2 U_o /(9\mu)$$

which stems directly from the integration of Stoke's law.

Next consider that a particle must be swept around a filter fibre of diameter R at about the stream velocity U_o. If the stopping distance of the particle is large compared with R, then it will be impacted and the particle will carry on in a straight line rather than following the stream

[28] Nuclepore™ filters are sieving filters at the submicron level. They actually consist of a plastic membrane with accurate diameter non-overlapping holes randomly etched into them. They have a very high pressure drop relative to filters of the same rating, are expensive, easily overloaded and relatively fragile. They are used mostly to collect samples of particulate matter for examination under an optical or electron microscope. How they are produced is quite interesting. Blank plastic membranes are heavily irradiated in a nuclear reactor or other facility. The ionising particles leave the polymer of the membrane with some chemical bond damage but only in the region where the particles passed through. These damaged regions are attacked much more swiftly by a strong chemical etch bath, which thus 'drills' a pattern of holes through the membrane, all the holes being the same and of a size determined by the etching time.

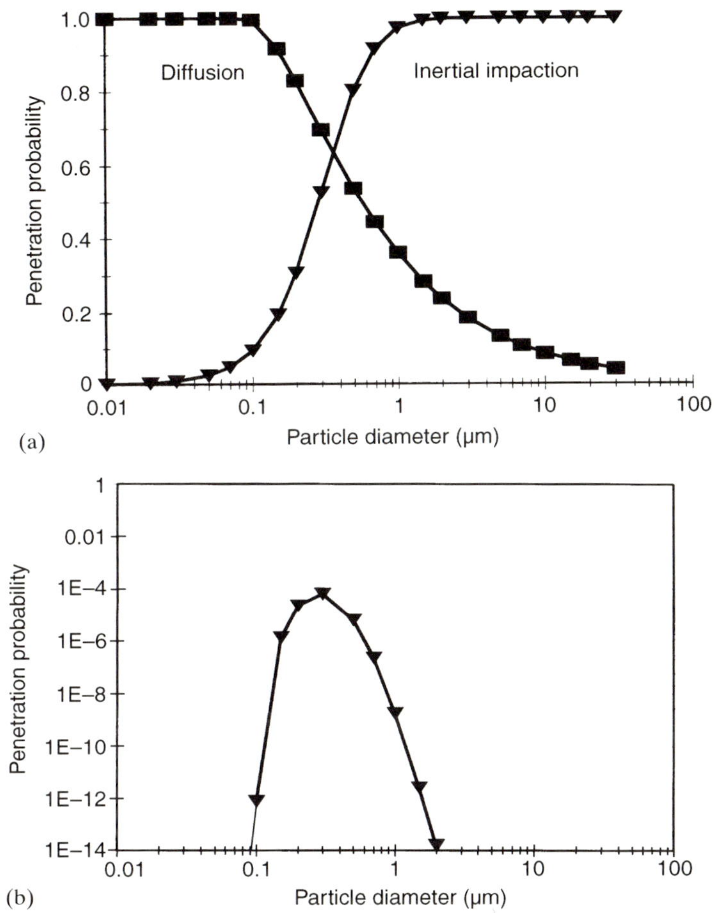

Figure 2.54 Filter efficiency versus particle size. (a) Filtration mechanisms; (b) penetration in practical filter.

$$R \simeq 2\rho p r^2 U_o / (9\mu)$$

i.e.

$$d_{crit} = 2 \times \text{critical radius} = \sqrt{[18\mu R/(\rho U_o)]}$$

Thus d_{crit} will be obtained. This is the critical size below which the particles will not impact but will be carried along streamlines through the filter, in the absence of other effects. Typical filters have critical sizes in the hundreds of microns to 0.5 μm or so for impaction.

At very small sizes, the diffusion capture process operates. Here the Brownian motion scale must exceed the filter interstices typical size, in order that in the time allowed during passage through the filter, the particle can diffuse to a solid surface and stick. The distance x gone by a particle under Brownian motion is proportional to the square root of the time allowed, as in other diffusion problems. For particles bigger than 60 nm or so (the mean free path for nitrogen at NTP), the Einstein formula gives (Einstein, 1924)

$$x \simeq \sqrt{[kTt/(3\pi\mu d)]}$$

Setting $t = R/U_o$ and x as R gives

$$d_{crit} = kT/[3\pi\mu U_o R]$$

Typical filters give in the order of a fraction of a micron for d_{crit}, below which particles are captured.

Between the almost perfect retention of very small particles by diffusion and good action in the sieving and impaction size ranges, there is for most filters a size for which the filter is least effective (Figure 2.54). The most penetrating size will vary, as will the efficiency or penetration coefficient at that size. However, for example, a certain electronic gas filter has a most penetrating size of about 0.3 μm, at which penetration is 0.0001%.

Filter materials. The majority of filters are made of some kind of fibre. Filters can be made of thick beds of material or can be 'membrane' filters, where the bed can be as little as a few tens or hundreds of microns thick. Fibrous filter cloths are woven or felted from materials as diverse as glass fibre and Teflon, whilst other filters rely on porous ceramic or partially sintered metal powder compacts. The latter types, typically made from alumina, stainless steel or nickel are discussed in section 3.5.5. They are expensive and used almost exclusively in wafer fabrication plants. A filter cloth or membrane is often given a face velocity rating. This is the maximum gas velocity through the medium that will not result in entrainment on the outlet or, worse still, break-up of the filter material.

In filters where some clogging might be anticipated, two or more different filters, of varying degrees of fineness, are laminated together or a continuous variation in fineness throughout the thickness of the filter is produced. The coarse filter layers are on the inlet side and the fine filter layers are on the outlet side so that clogging and re-entrainment are minimised.

Gases are generally very clean when delivered from production plant or from bulk or cylinder sources; many orders of magnitude cleaner than the several mg m^{-3} loading commonly seen in air. Gas line filters are therefore by no means always necessary although a coarse-sintered filter can stop occasional large particles, for example swarf from machined components, from getting to the point of use. Gas filters can in any case be designed for minimal loading, while often a pressure drop of a fraction of a bar is not significant in a system. As a result gas-line filters are often very small units. By contrast air filtration is almost inevitably required on the air inlet of gas production and in gas applications needing air, and furthermore both loading and pressure drop are considerations. Fibreglass filters are often used, with fibres of around 2 or 4 μm supporting much finer fibres in a paper-like matrix held together with a small proportion of adhesive. A very

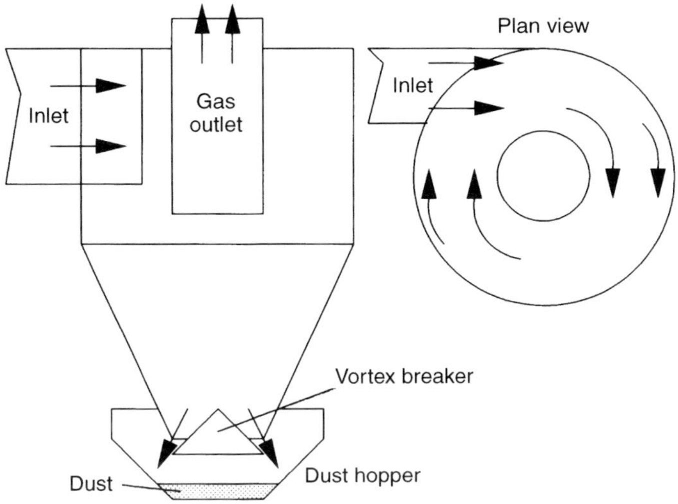

Figure 2.55 Cyclone dust separators.

large area of this filter mat is then folded up into a compact module which offers a very low pressure drop (often measured in inches in a water gauge) and long intervals between replacement. Similar HEPA (high efficiency particulate air) filters are mounted in the roof and used for cleanroom air in wafer fabrication plants.

Occasionally, in dusty parts of the world, it may be necessary to employ coated or 'viscous' air filters. In these the fibres of the filter are coated with a heavy oil to which particulates adhere strongly and will not re-entrain. In these types a relatively coarse filter can still be effective and the dust loading they can withstand is higher than that of uncoated 'dry' filters.

Cyclones. Cyclones are devices that can clean gases via a centrifugal separation action (Figure 2.55). They are rarely used in industrial gas applications because these mainly use clean sources and cyclones are mainly applicable to gases with a high burden of relatively course particulate contaminant. However, cyclones do find application in exhaust treatment in many gas applications, as well as in some gas production processes. They are capable of separating out dust as fine as 1 μm or so. The most common type is the conventional reverse-flow type, whilst straight-through cyclones with fixed or moving impellers are rather unusual.

The balance between Stoke's law drag on a particle and the centrifugal force to which it is subjected in the cyclone is the essential nub of cyclone theory. A conventional cyclone analysis leads to an inverse square-root dependence of critical particle size on gas velocity and cyclone size, i.e.

$$d_{\text{crit}} = K/\surd[(\text{flow rate}) \times (\text{cyclone radius})]$$

Cyclones normally have a much lower pressure drop than any workable filter design. The pressure drop is approximately proportional to gas flow squared in the operating range of many designs. However, cyclones have little capability for very fine dusts; the minimum particle collected is typically in the 2–10 μm diameter region. Another problem is that the particulate receiver must be sealed, otherwise large flows around the receiver may cause re-entrainment and loss of efficiency. It is interesting to note that cyclones have a 'fluidic diode' effect (section 2.4.2), i.e. they have a much higher pressure drop in the forward flow direction than in the reverse direction.

A typical cyclone is used for processing waste gases; perhaps ensuring compliance with dust emission regulations or recovering by-products in the collected dust. The dimensions of cyclones are critical design parameters and must be accurately calculated in practice. Typical proportions might be 1.4 m height, 300 mm throat diameter, 200 mm × 70 mm inlet, 70 mm diameter gas outlet. In such a typical cyclone, with an inlet gas velocity of $15\,m\,s^{-1}$, the cut-off size expected would be about 3 μm diameter at a dust density of around $2\,g\,cm^{-3}$.

It is now possible to model cyclone action with simple computer models, which are now used to predict performance and enhance design. These models have made cyclone design more scientific and, incidentally, have shown clearly how small, apparently innocent, design changes in the cyclone shape can have significant effects. Even as small a change as a dished end on the top, useful for strength in a pressurised system, can change cyclone performance markedly.

Electrostatic precipitation. Electrostatic precipitation involves electrically charging particulates and then using electric forces to attract them to a plate where they are discharged and collected. Electrostatic precipitators are more useful in cleaning up dirty gases (section 2.5.14). They can have very much lower pressure drops than filters or cyclones, whilst still removing large amounts of particulate (Brauer and Varma, 1981).

The simplest precipitators apply a negative voltage of 10 kV or more to a 'wire' electrode, which may simply be a wire of small diameter or a more substantial tape or rod with sharp points at intervals along its length. The corona discharge produces electrons and ionises the gas, and some of the mainly negative gas ions produced transfer electrons to particulates. The negatively charged particulates are then attracted to the positive collection electrode. Vibration, either continuous or regular impact 'rapping', must be applied to ensure that the precipitator does not get too thickly coated in dust and to collect the dust.

2.4.4 Gas flow control systems

Simple gas panels. The standard paradigm for control systems must today be electrically actuated valves controlled by microprocessor-based digital electronic logic circuits known as programmable logic controllers (PLCs). However, for a modest gas control system (perhaps only two gases need to be supplied, one for purging, one for running, at constant pressures) a simple set of manually operated valves and regulators, together with some ancillary components, will often suffice. Where the component sizes are conveniently small (say below 30 mm pipe diameter), the components are often mounted on a metal board as a separate subsystem built in a factory rather than assembled on-site. Such subsystems are often called 'gas panels'. Cylinder gases will typically use pipe of 6 mm external diameter so gas panels are popular for compressed gas service.

Gas panels are often required to govern the ratio of gas flows so that the mixture of gases delivered is of constant composition, despite changes in the total gas flow. This can be achieved by regulators with critical orifices or needle valves, or, more accurately, by mechanical flow ratio regulators using balanced diaphragms. Alternatively, and increasingly today, two or more thermal mass flow controllers are used. These can be used to govern flow rates separately or, with an electronic link between them, to maintain constant ratio. Filters are often mounted on gas panels, together with non-return valves to prevent backflow and possible contamination of the bulk gas store. In the case of cylinder gases, the input to the panel is via a pigtail or the more expensive Teflon-lined hose, with the cylinder connection compatible with the chosen gas type on the end.

Gas cabinets. It is preferred on safety grounds to place toxic and corrosive gases outside in a protected area but one in which ventilation to the outside air is free and plentiful. However, this is not always practicable. Cleanliness requirements work against this simple solution, as do production site layouts where equipment using gases is often placed far away from outside walls. It is for these reasons that toxic or corrosive gases in cylinders are most often handled by a gas panel inside a 'gas cabinet'. This is a metal cupboard with space for two cylinders of gas (one active, one purge gas) and the gas control panel.

Gas cabinets are strong, of coated 1 or 2 mm steel construction, fitted with an extractor fan which sucks air (and any process gas which may have leaked) out and off to a vent stack or, where quantities may be significantly large, to a scrubber system which will remove the dangerous gas. Purging of the pigtail is particularly important for these gases, which otherwise give out a puff of a few hundred cm^3 of gas each time the cylinder is changed, which is unpleasant and potentially unhealthy for the operator. Good purging is achieved by evacuating the pigtail using a blast of purge gas through a venturi (this can give 100 mbar or so vacuum, enough to improve the purge cycle radically over simple blow-down) then repressurising with inert purge gas and repeating the process several times. The purge gas can also be used to purge out the downstream pipeline ready for maintenance operations, for example.

There are generally two doors on a gas cabinet, one to allow limited access for adjusting one of the gas cabinet pressure regulators or valves, and the other, larger, door for the addition or removal of cylinders. Figure 3.23 shows a typical compressed gas handling cabinet, showing two process gas cylinders and one purge gas cylinder for high purity gas service in the semiconductor industry.

Electronic flow controls. For larger systems electronic controls are justified. Some will have a complete PC or similar small computer, with display and keyboard, as their master controller. A plethora of software and hardware is aimed at satisfying this market and prices have tumbled in recent years. In many markets inertia forces are great but electronic controls will eventually take over from pneumatics almost completely.

For variable flow rates the valves are controlled by sensor feedback loops to proportional/integral/differential (PID) controllers whose set points are in turn set by programmable logic controllers (PLCs). For on/off valve control electrical signals direct from PLCs are used. Sensors are typically flow sensors or pressure sensors with less often (cryogenic) level detectors, thermocouples or more exotic probes being used. Mass flow controllers (MFCs) are the mainstay of applications such as semiconductor manufacture where many fairly small flow rates must be accurately controlled. The MFCs in semiconductor practice are usually directly connected to the computer that controls the rest of the equipment, rather than having their own PLC.

Despite the predominance of electronic control, however, there are occasions where an alternative approach may be justified even on new equipment.

Pneumatic controls. As stated, conventional control systems today often use electronic logic to decide which valve should be actuated and by how much. However, it is often inefficient to try to use electronic valves, such as solenoid valves, directly on process gas flows; large solenoids drawing significant amounts of power would often be required. 'Pilot-operated' pneumatic valves are therefore used for higher flow applications. Small electric solenoid valves are the interface between the electronic world and the gaseous world. These 'pilot' valves switch a compressed air supply to the actual process valves, which are powered by pneumatic piston or bellows working off a gas or compressed air supply rather than electricity. There may even be two stages of pneumatic actuation involved in large process plant, the second stage being the solenoid switching pneumatic valves which in turn switch compressed air onto the process valve actuator.

A pure pneumatic control system is simpler in concept than these more typical systems because it uses compressed air not just for the actuation of the process valves but also for the logic and amplification normally done electronically. Given the low cost and versatility of electronic controls, it may be doubted that pneumatic controls have any application whatever and it is certainly true that they are now rather unusual. Pneumatic controls, however, have a number of advantages over conventional general purpose electronic controls, at least for small systems. For example, they:

- are intrinsically safe, no sparks being generated in normal pneumatic systems;
- can sometimes function without an external power source (if the process gas itself can be used instead of air in the control system);
- are not subject to electromagnetic interference;
- do not radiate electromagnetic signals which can interfere with other controls;
- are capable of functioning in dirty, hot and environments with suitable filters;
- are often less expensive than similarly reliable electrically actuated valves;
- allow relatively high actuation forces to be easily obtained without compromising on actuator speed.

Against these advantages are considerable disadvantages where complex functions are concerned. It is not easy to generate a complex sequence of actions with pneumatic controls; these sorts of sequences are much more easily carried out by the electronic programmable logic controller units described above. One might add a further disadvantage: that the younger engineering designer today may well be rather unfamiliar with pneumatics.

Fluidic controls. In principle the terms 'fluidic' and 'pneumatic' should be synonymous, since fluidic components invariably employ air or gases as the working fluid, just as pneumatic components do. The term 'fluidic' is, however, now reserved for systems which employ a set of non-moving part pneumatic components; the fluidic diode mentioned in section 2.4.2 is an example. A simple example of a fluidic active component is the commonly used wall-attachment (or Coanda) effect bistable (Figure 2.56). This device has a compressed air input, two signal inputs and two outputs. When the input tube at the active output is puffed momentarily on, the input stream is switched to the opposite output. These devices can easily be made an inch or so in size and a fraction of an inch thick. Their size is usually small compared to the interconnections and pipework in a typical system. The fact that fluidic devices have no moving parts is ingenious and makes for high reliability. Fluidic devices are efficient devices for switching flows, switching at high speed (ms) and with high amplification factor (power controlled/power controlling).

Analogues of most electronic components can be produced in fluidic form. Amplifiers, OR and AND gates, timers and oscillators are all possible. Essentially, fluidic controls are applicable where pneumatic controls are appropriate, especially in situations which value the advantages

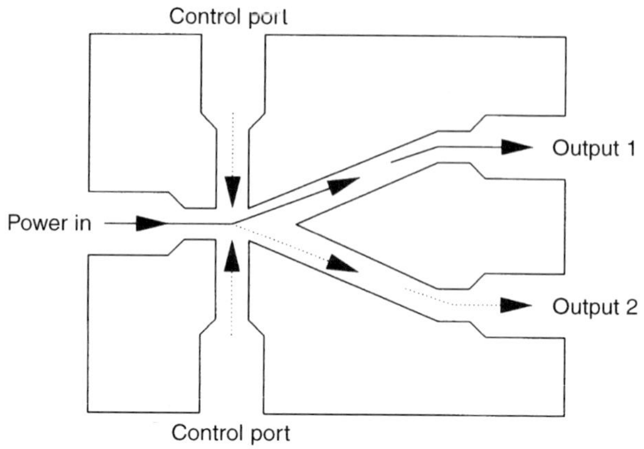

Figure 2.56 Wall-attachment fluidic amplifier.

which stem from the absence of moving parts, i.e. resistance to vibration, long lifetime, and an absence of a requirement for lubrication. The major disadvantages of fluidics today, however, are their poor availability and lack of support from suppliers.

Although fluidic systems are unusual, there are certain useful fluidic subsystems which are more commonly seen; for example the fluidic oscillator molecular weight sensor. These are made using fluidic switching technology and achieve good precision. In effect they are the same in principle as an organ pipe running off a carefully regulated gas supply whose pitch varies with the square root of gas molecular weight. The frequency of oscillation can be easily converted into a readout of average molecular weight.

Back diffusion. Back diffusion refers to the following situation. Suppose that the exhaust end of a gas system is open to the air and air is an undesirable impurity in the system. If the exhaust flow is low, or the exhaust pipe too short, then it is possible for small amounts of air to diffuse back up the exhaust, against the general gas flow, and spoil the process.

Why this arises is as follows. Each molecule of air and gas is moving at around 400 m s^{-1} per second but, of course, it soon hits another molecule and this changes its direction. In this way, it progresses according to the diffusion equation described in section 1.3.10, rather than linearly forward at 400 m s^{-1}. In fact, for the average molecule to go any reasonable distance (section 1.3.10), the time allowed must be very long. However, a tiny fraction of the fastest molecules can travel quite a long way.

If back-diffusion occurs in a pipe with plug flow, then the concentration of back-diffusant rapidly falls exponentially ($\exp(-x)$) to zero. However, if back-diffusion occurs in a pipe with viscous flow, then 'wall creep' occurs and the back-diffusant falls only hyperbolically ($1/x$). (It is well worth remembering, too, as noted in section 1.3.10, that diffusion as a phenomena is pressure-sensitive; back-diffusion will be more serious in low-pressure systems.) Back-diffusion effects seen in practice are between the exponential and hyperbolic in effect.

Leaks in pipework. Finding leaks in completed pipework is an essential activity both during original installation and after any modifications. If the pipeline can be pressurised with an inert gas such as nitrogen, then simple tests, variants of the time-honoured soapy water, can be employed. The use of synthetic surfactants or non-water-based bubbling solutions can occasionally be useful. Bubbles as small as a few millimetres across forming in a minute can easily be seen, corresponding to leak rates of the order of 10^{-4} mbar s^{-1}, which is a fairly small leak rate; a standard 10 m^3 (200 barg, 50 l water capacity) cylinder would be exhausted only after hundreds of years.

Helium, being a smaller and faster-moving molecule than nitrogen, will find smaller leaks with soapy water. Furthermore, helium can be used with a helium leak detector.

Helium leak detectors are basically simple mass spectrometers. They are tuned to helium, which is a rare gas (< 1 ppm) in the atmosphere, giving a small background signal. They can be used in several ways but the most effective way is as follows.

1. The pipework is evacuated and connected to the He leak detector.
2. A small cylinder of He is used to play a stream of He on the outside of the pipework via a hypodermic needle.
3. When He is played onto the pipe at a leak, He atoms are almost instantaneously sucked down the pipe towards the leak detector vacuum pump and detected.
4. The needle source of He is used to pinpoint the leak precisely.

In this way, leaks of the order 10^{-9} mbar l s^{-1} can be detected, and often located to within millimetres.

Many pipework systems cannot be evacuated; they may have seals which only seat correctly if under positive pressure or they may include vessels. Many large vessels, even those designed for a several bar overpressure on the inside, may fail by buckling under atmospheric pressure on the outside if subjected to vacuum. If the pipework cannot be evacuated, then a different type of leak detection strategy is called for. Pressurising the system with an easily detectable gas is the way normally employed; for larger leaks pressure drop can be observed.

Helium sniffing of the air around the vessel, using a mass spectrometric detector as used in the vacuum leak detector but adapted to take an atmospheric pressure sample, is possible. However, a cheaper instrument, relying on the detection of a halogenous gas, typically a CFC or SF_6 or one of these diluted with nitrogen, using a sensitive electron capture detector is an alternative which will suffice for many applications. Even less sensitive than this are simple thermal conductivity-based leak detectors, which, however, have the advantage that they will detect any escape of gas with a thermal conductivity differing from that of air; no sniffing of gas is necessary (although sniffing gas which is very high or low in thermal conductivity will improve the sensitivity of the method).

Probably of similar sensitivity is the technique of listening for leaks. In this a microphone tuned to receive ultrasound (typically around 40 kHz, at which frequency inexpensive transducers are available) is used as the sniffer. When the microphone is placed near a leak from a gas source at high pressure (over 1 barg or so), it receives the small ultrasonic noise signal caused by the turbulence of the escaping gas. This is amplified and operates an audible signal and visible analog meter for the operator. Low-pressure systems can be checked without applying any differential pressure by placing a 40 kHz transmitter inside the vessel or pipework and probing the outside with the microphone.

Repairing gas leaks. Although the pressures used in cylinder gas systems may seem high, up to 300 bar, and the size of bulk gas equipment large, the principal sorts of leaks seen, and the ways to repair them, are not dissimilar to that of any pipeline system. Before attempting any repair, it is always best to shut off the gas supply and, if a dangerous gas is in the line, purge with nitrogen or other inert gas.

High-pressure fittings of the Swagelok™ type can often be sealed by simply fastening them up tighter, if possible after checking for dirt on the seal faces. However, overtightening the fittings will bell out the face opposite the pipe olive and make it difficult to dismantle the equipment. High-pressure fittings with face seals of the VCR™ type sometimes require a new seal washer to be fitted. This type of fitting should always be fitted with a new seal washer after dismantling; they may leak if a new washer is not fitted.

Valve stems, after time or extensive use, will tend to leak because their seal gland packing is no longer sufficiently tight. A modest tightening of the gland nut will normally effect a seal. The ease with which the valve may be operated should be checked after this, as an overtightened gland can make the stem difficult to turn. Cylinder valve outlets are often subject to leaks. It is best to turn off at the cylinder valve and look for a cause rather than tightening the joint past its intended torque. Typically, the cause will be dirt on the mating faces or a worn sealing gasket, if one is fitted. If a cylinder valve will not seal shut, it may be possible to fit a gastight metal plug or cap in the outlet. (The plastic dust plugs and caps provided with some cylinders to keep dirt out of the cylinder outlet are, needless to say, not sufficient.)

2.4.5 *Liquid gas storage and handling*

Liquefiers. In former times, the demand for oxygen in gaseous form was so large that the small amounts of liquid nitrogen needed could easily be drawn off a large air-distillation plant without affecting its performance. However, the relative demands for oxygen and nitrogen have reversed and much of the nitrogen is required in liquid form. The most efficient way of satisfying this new demand is to add a liquefier to an existing gaseous nitrogen stream or to design from scratch a plant with a liquefier. Liquefiers are designed purely and simply to liquefy a pure gaseous nitrogen stream on a large scale (up to hundreds of tonnes per day) and can be very efficient. Hydrogen liquefier plants are usually smaller (10 tonnes per day) but are similarly specialised and can also be relatively efficient although, as noted in section 2.2.5, hydrogen requires a large energy input for liquefaction.

Apart from the large liquefiers seen on gas producers' sites, however, small nitrogen and helium liquefiers are often found on-site with their application. He liquefiers are usually designed as part of a sealed cryogenic refrigeration package and are usually maintained by the manufacturers. Small nitrogen liquefiers are often seen on remote sites, however, and are usually operated to provide liquid nitrogen externally. They are certainly a convenient[29] source of small quantities of LN, although the economic case for them is often weak.

Cryogenic insulation. Liquid nitrogen kept in any ordinary vessel quickly evaporates. Furthermore, the vessel containing it will be swiftly cooled to near 77 K, at which temperature many vessels crack or are dangerously brittle. Even worse, liquid oxygen can accumulate on the cold surface, especially in dry air, and the liquid oxygen can then run off and form a serious hazard by being absorbed into organic matter. It is not enough, at least for smaller equipment, to make the insulation thicker. It is not difficult to show, for example for a cylindrical geometry, that the heat flow Q from a long tank of radius R with insulation of conductivity S with thickness T is

$$Q = 2\pi S/[\ln(1 + T/R)]$$

Hence large increases in T, far beyond R, do not yield very much in the way of decreased Q and large thicknesses of insulation will not save small cryogenic tanks from having an undesirably large degree of boil-off. Instead, the quality of insulation must be improved.

As storage tanks or plant increase in volume their surface area increases

[29] They do have inconvenient features, too. Normally a Dewar vessel is wheeled up to these units for filling. With large vessels precautions should be taken to stop when filling is complete, as filling is a slow process with these units, unless they have a good size accumulator vessel, and the operator may be tempted to leave the filling unattended, giving rise to overflows.

only as two thirds of the power of the volume, so that large tanks and plant can use the crudest insulation.[30] Liquid methane, for example, is often stored in 10 000 ton tanks insulated with simple cement-based insulation blocks. These would be completely useless for a 25 l vessel for liquid helium. For some applications, such as insulating large cryogenic plant, the use of good quality mineral insulation, such as perlite, is acceptable. (Even here, however, there are problems: a nitrogen or dry air purge is necessary to stop atmospheric air depositing water in the Perlite, for example.)

Ordinary insulating materials are not acceptable, however, in more stringent applications in cryogenics such as storage or transportation tanks, or pipes. Most, such as insulation-grade concrete blocks or cork tiles, are simply not good enough, while others are too expensive or degraded at 77 K.

Very early in the history of cryogenics the fundamentally simple observation was made that what is not there (a vacuum) cannot conduct heat and the concept of the Dewar vacuum flask was invented. These are nowadays familiar as a domestic item. However, they were originally glass double-walled hand-made glass flasks, blown and then evacuated, made for laboratory purposes.[31]

Vacuum insulation has undergone several improvements since Dewar. The reduction of radiative heat losses via silvering of the glass was followed by the discovery that the quality of vacuum needed was considerably reduced if the vacuum space was filled with a highly porous powder. This is because of mean free path effects similar to those discussed in section 1.3.13. Basically, the low-pressure gas in the spaces between the porous powder will have a low conductivity if the mean free path exceeds the dimensions of the interstices. The reason for this is easily seen by inspection of the two formulae for heat conduction per unit area between two plates

At high pressures

$$Q \simeq \Delta T/L$$

While at low pressures

$$Q \simeq P\Delta T$$

where

P is the absolute pressure, ΔT is the differential temperature and L is the plate spacing.

[30] Haldane's famous essay about scale effects in biology (Haldane, 1927) shows the same effect in reverse: warm animals that live in the cold Arctic without a shelter have to be large in size if they are to have reasonably low metabolic rates.

[31] A standard domestic vacuum flask is still a perfectly serviceable device for storing cryogens. However, the lid should not be screwed on after filling because after a little while the cryogen will reach a surprisingly high pressure and crack the glass vessel inside, followed by cracking and leaking of the outer plastic casing. In addition, these flasks will occasionally break on filling as a result of thermal shock.

Note that in the high-pressure case the thermal losses will be independent of pressure. There is no point in achieving a poor vacuum between the vessel walls; this low-pressure gas will conduct heat away from the cryogen just as effectively as atmospheric pressure gas.

At low pressures, when the mean free path becomes large compared to the insulation cell size, the thermal losses are reduced with pressure, making the achievement of a better vacuum useful. It is also worth noting that there is no dependence on spacing L between plates in this case; many plates can be placed closer together in series and thermal insulation improved without making the spacing larger. The use of microporous material is thus indicated.

The cheapest and most popular microporous material is perlite. This is a natural whitish rock which contains internally trapped water. On firing it quickly at 800°C or so the water boils, creating small pores inside the rock. The final material is like very lightweight coarse sand. Similar, but coarser (5–10 mm pieces), is a brownish material called vermiculite. Glass fibre and similar, but cheaper, rockwool fibrous materials are also useful insulation materials which can be evacuated to achieve high insulation quality.

The same principles are also exploited in the use of Mylar™ multilayer wrapping insulation. This uses multiple wrapping layers of a plastic film, such as the polyester Mylar™ or Melinex™ products, which is then evacuated. The film has a thin-film deposit of shiny aluminium on its surface which achieves almost zero emissivity and perfect reflectivity, reducing radiant heat losses. Evacuated Mylar™ multilayer insulation is often dubbed 'superinsulation' and is invariably used for helium vessels and for the small portable/wheelable vessels often used for cryogens in the laboratory, which generally lie in the range 20–200 l.

Aerogels. The principle of microporous insulation may eventually be carried to its ultimate extent in the use of 'aerogels'. These unique materials are very curious to handle, in appearance something like 'solidified smoke', they have densities of the order of 0.1 g cm^{-3} a very low refractive index (1.05) to match, and are transparent with a pale grey/light blue appearance (Hrubesh, 1987; Fricke, 1988). Their remarkably low refractive index was the reason for their development. They were first made for their optical properties, being used in Cerenkov radiation counters in nuclear physics research. An aerogel produces Cerenkov radiation only with particles travelling very close to the speed of light.

Aerogels, however, are also extraordinary insulation materials. Instead of pore sizes in the 0.1 to 1 mm range, these are generally silica-based optically transparent materials with pore sizes of the order of a few nanometres (there are also organic aerogels). Aerogels are made by polymerisation in solution followed by the supercritical extraction of the solvent. The supercritical extraction is critical, as otherwise the pore structure would simply collapse from surface tension forces or decompose during conventional dessication, while the high viscosity of the liquid would mean that it could not be extracted in any reasonable time.

The currently favoured chemistry for aerogels is to react a silicon compound, such as tetramethoxysiloxane, with water. The methanol formed in the reaction is driven off in the supercritical extraction, leaving a silica 'nanosponge'. An older technology uses silicate precipitated from sodium silicate (water glass) with hydrochloric acid. The dimensions of the pores in aerogels are comparable with the mean free path of atmospheric air and they are being developed to have insulation qualities in the range of 10 times better than conventional plastic foams, without the use of vacuum.

Radiation losses in cryogenic insulation. The losses in cryogenic insulation due to infrared radiation are considerable. The T^4 dependence in the Stefan–Boltzmann radiation law means that these losses can be considerably reduced if multiple layers of low emissivity surfaces can be used; the rather curious fact emerges that a cryogenic tank can be insulated by means of layers of metallic foil.

How this arises can be seen if a set of equations describing the heat transfer q across a set of foils 1 and 2 from T_c (cryogen) to T_h (room temperature) is considered

$$q\ (\text{without foils}) \simeq (T_h^4 - T_c^4).$$

When foils are used, there are three linked equations for q

$$q \simeq T_1^4 - T_c^4$$
$$q \simeq T_2^4 - T_1^4$$
$$q - T_h^4 - T_2^4$$

hence, adding the equations,

$$q\ (\text{two foils}) \simeq (T_h^4 - T_c^4)/3$$

In general then, adding N foils reduces the IR heat losses by a factor of $N + 1$. However, to suppress IR radiation without adding to conductive heat losses a submicron reflective metal layer on a plastic substrate, typically vacuum-deposited aluminium, is the most effective technique. Hence multilayer Mylar superinsulation is always aluminised and flakes of metal can be added to perlite evacuated powder insulation to minimise IR losses. Perlite as normally made transmits a certain amount of visible and infrared light, i.e. it is semi-translucent. Perlite with a little aluminium powder or flakes added to decrease IR loss is called 'opacified'. Metal flakes can add to the solid's thermal conductivity, however. Perhaps today users of perlite should consider adding flakes of aluminised glass or Mylar (something like the 'glitter' used on Christmas cards).

Table 2.13 Thermal activities of constructional materials

Material	Thermal conductivity ($W\,m^{-1}\,K^{-1}$)	Comment
Aerogel	$\sim< 10^{-2}$	
Foam plastic	10^{-2}–10^{-1}	
Vermiculite	$\sim< 10^{-1}$	
Cork	$\sim< 10^{-1}$	
Glass reinforced plastic	10^{-1}–1	
Fibreglass insulation	10^{-2}–10^{-1}	
Stainless steel	10–25	
Aluminium	50–250	
Perlite	10^{-2}–10^{-1}	
Perlite	$\sim 2 \times 10^{-3}$	vacuum 10^{-5} bara
Opacified perlite	$\sim 5 \times 10^{-4}$	vacuum 10^{-5} bara
Mylar superinsulation	10^{-5}–10^{-4}	high vacuum 10^{-8} bara

(Note that these figures are very uncertain, depending on measurement techniques as well as showing very large batch to batch variations.)

Insulation materials are rarely good mechanically and it is often necessary to provide better support to the inner wall of tanks. 'Fibreglass' (i.e. glass reinforced plastic) supports are often used because fibreglass has a surprisingly low thermal conductivity compared with metals (although the stainless steel favoured for cryogenic work has one of the lowest thermal conductivity values amongst metals) (Table 2.13).

Vacuum insulation is often maintained over long periods of time at low pressure by including a getter material in the annular space of the vessel. Getter materials are highly reactive alloys which often include finely divided titanium or calcium (section 2.2.11) and work by chemically combining with common gas inleaks and outgassing, e.g. nitrogen, oxygen, moisture and carbon dioxide.

Liquid helium handling. Liquid helium is much more difficult to maintain in storage than LN or LO. The reason for this lies in its very low specific heat and low density. For each kilowatt hour of heat absorbed, 22 l of LN will be boiled off or 1400 l of liquid helium. Mylar superinsulation and multiple wall vessels, with liquid helium insulated from an LN outer vessel, are usually employed for long-term storage of liquid helium or for delivery of liquid helium over considerable distances. Very large helium tanks will lose considerable quantities of gas, even in superinsulated vessels, and these vessels are often fitted with helium recovery. The boil-off gas can be purified and reliquefied or, alternatively, compressed into cylinders for use as balloon gas.

Cryogenic storage tanks. A typical storage vessel will have a stainless steel inner vessel and a mild steel outer, with the interspace filled with one of the

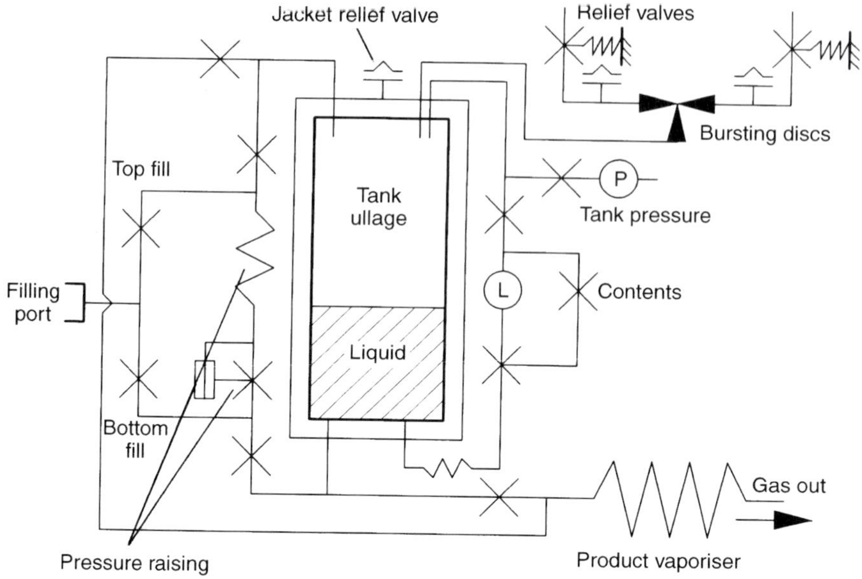

Figure 2.57 Schematic of 25 tonne LN storage tank.

cheaper cryogenic vacuum insulations. The interspace is evacuated to well under 1 mbar, or whatever is appropriate to the insulation, and sealed off. As well as simply storing the gas, the cryogenic storage tank has to:

- be safe
- display level or contents
- provide an on-line filling facility
- supply constant pressure liquid cryogen.

To meet these requirements a simple double-walled vessel must have a number of other features (see Figure 2.57).

The requirements of safety dictate a number of pressure relief devices. The interspace must have a non-return valve, which is often simply a plate held down by the vacuum within on to an 'O'-ring seal. The inner vessel must be vented by two parallel safety relief devices; typically a bursting disc set at a few bar less than the safe vessel pressure and a relief valve set at a few bar less than that. This bursting disc/relief valve set is usually connected via a two-way selector valve with another similar valve set on the other arm so that a burst disc or faulty valve can be replaced without blowing 25 tonnes of product into the atmosphere. These vent valves are connected to the vapour space above the maximum liquid level. Finally, all sections of pipe connected to the tank (including the process draw off and vaporiser) which can be valved off at two ends, must have separate additional safety relief valves because quite small quantities of trapped LN can rapidly vaporise, exerting up to several hundred bar and violently bursting pipes and hoses.

The tank contents will usually be read out by a differential pressure gauge (section 2.3.1) and its operating pressure via another gauge attached to the same line. It is an advantage to be able to remove faulty gauges without loss of product, so isolating valves are needed.

It is an advantage to be able to fill a vessel that is actually in operation supplying gas without disturbing the gas-using process by interrupting supply or changing pressure. This can be achieved by pumping a new cryogen into the tank via two valves: a top-fill valve spraying liquid into the vapour space and a bottom valve discharging into the liquid. If the delivered cryogen is cooler than the tank, then top-filling will lower pressure in the tank and bottom-filling will increase it. In this way disturbance to the tank pressure can be minimised. Although control via manual valves may seem crude, it must be remembered that the pressure is governed by the pressure-raising circuit and by downstream gas regulators in any case.

To fulfil its function properly, liquid cryogen must be supplied to the process or vaporiser at a constant elevated pressure, typically up to about 15 bar or so. This is achieved by governing the tank pressure via the pressure-raising circuit and drawing off liquid for the gas-using process via a dip-tube connection. The pressure-raising circuit takes liquid from the bottom of the tank, puts it through a pressure regulator and then to a vaporiser connected to the tank vapour space. If tank pressure falls too low, liquid cryogen is admitted to the vaporiser and expanded to warm gas. This sort of pressure-raising circuit has a relatively low pressure feed due only to the cryogen hydrostatic head (only a few psi if the tank level is low) so good designs have fairly generous valve and pipe sizing to allow the tank to be utilised as fully as possible (see Figure 2.58).

Pumps for cryogens. Pumps for cryogenic liquids have some unique features which are worth noting. A reciprocating pump is simple enough to design in principle for liquid nitrogen for gas cylinder filling, for example. It is simply a reciprocating piston with enough drive to give 250 bar or so on the outlet and check valves to prevent backflow. However, the check valves must function in liquid nitrogen, the cylinder must be vacuum-jacketted to avoid boiling, and the piston cannot be lubricated with oil and yet must seal reasonably well; leaking past the piston will entail gas loss and cause heating, leading to boiling. Materials for the cold parts must be strong at cryogenic temperatures with particular caution over steel and polymer components. Finally, thermal design must include blocking cooling of the room temperature parts (the electric motor, gearbox, conrod and their lubricant grease and oil) and allow for rapid contraction of parts as they cool down.

Commonly, centrifugal liquid pumps are used for low-pressure liquid transfer around cryogenic production sites and for filling transport containers. These may rely on thermal blocking from a conventional drive. However, a neat solution to the problem of a high-speed rotating seal between motor and cryogen is to enclose the motor in the cryogen; the so-called 'canned motor' pumps. With many less aggressive cryogens such as liquid nitrogen and liquid argon (but not liquid helium, where even a small power dissipation is thermodynamically unacceptable), this is quite useful. The copper coils in the motor lose much of their resistance at cryogenic temperatures and run much more efficiently, whilst the heat inleak for the whole assembly can be lower than the heat input due to resistive Joule heating of the motor coils (copper resistivity at 77 K is roughly one seventh of its room temperature value).

Operating cryogen pumps is not as straightforward as operation of ordinary pumps. During start-up, huge volumes of gas will have to be blown through the system to cool it down and this must be safely vented. During cool-down, differential thermal contraction may occur and this may delay the cool-down. Start-up of the pump itself must be done with caution to avoid damage whilst the system is cooling down and pumping only gas.

Using liquid cryogen pumps involves some care with design. Most fundamentally, the suction side of the pump needs to be considered. The suction side of a pump lowers the pressure on the liquid and, at least during start-up, the inlet pipework will raise the cryogen temperature. Boiling is therefore likely, followed, once the pump begins to cool, by cavitation.[32] The increase in speed is accompanied by a decrease in pressure according to the Bernoulli equation. 'Cavitation' in this context means bubbles formed in or before the pump which collapse at the outlet of the pump. The bubbles typically collapse milliseconds after formation. The collapse of a bubble is accompanied by a sharp pulse of ultrasound which is powerful enough to erode, over time, even hard metal pump parts and casing. Although some cavitation is probably inevitable, it should be avoided as much as possible as it is detrimental to the pump life. The tendency to boil is restrained by the amount by which the liquid can be superheated but this should not be relied on. Placing the pump at the lowest point in the system, and using a straight run of pipe without siphon sections from the supply sump or vessel, is good practice. Vacuum-insulated pipework is often an advantage in avoiding too much two-phase flow and heat losses.

The net positive suction head (NPSH) should be calculated for all pump installations. This is the system pressure (H), plus the tank pressure (P_t), minus the cryogen vapour pressure (P_v) at operational temperature and

[32] Cavitation occurs at any point in a system where liquid is subjected to a pressure lower than its vapour pressure. Cavitation is by no means confined to pump-suction problems. It occurs around sharp edges in pipework, sudden changes in pipe diameter, elbows, in fact anywhere a liquid with a high vapour pressure accelerates rapidly.

allowances for pressure fluctuations at the inlet of a reciprocating (P_f) or other unsteady pump and flow frictional losses (ΔP), i.e.

$$\text{NPSH} = H + P_t - P_v - P_f - \Delta P$$

NPSH for cryogen pumping is, of course, problematical because of the large value of P_v. With P_v at least equal to P_t in value, H must be large enough to overwhelm the remaining negative terms. Many pump manufacturers will quote minimum NPSH values (often in the region of 1 to 5 m water gauge) and these can be used for estimating whether or not a particular installation will function correctly.

Vaporisers. Vaporisers are a form of heat exchanger but one which is not designed with the usual minimum ΔT considerations. In fact vaporisers operate with a very large ΔT but are designed to be as simple as possible and to have minimum capital cost.

Vaporisers are often simple coils of tubing left in the open for the air to warm the liquid gas to near ambient. This is fine for systems where the exact gas temperature does not matter and where sustained high flow rates are not required. It is a common sight in cold but humid weather, typical west European conditions, to see massive blocks of icy snow deposits submerging vaporisers. Ice blocks are often a sign that gas-flow requirements were underestimated during the installation planning or sometimes a result of successful company growth leading to gas consumption rates in excess of the vaporiser capacity. Exceptionally large balls of ice need to be removed because they can block access to valves and thermally insulate the vaporiser pipework to some extent, allowing very cold gas through to the point of use. The effect of the ice ball on thermal efficiency is not perhaps as big as might be expected, however, because of the effect noted in section 2.4.5; increasing insulation thicknesses are not very effective.

For higher flow rates it is necessary to add fins to the tubes of these ambient vaporisers and to operate two banks of vaporisers so that one unit can be de-icing while the other is in use. A fan and ducting can be added to enhance the air convection, as flow rates increase, and a hot water de-icing ('deluge') system used to remove the ice build up. If flows exceed the rate at which even these deluge vaporisers operate, then it is necessary to use heated water bath vaporisers. These simple but effective devices are capable of enormous gas outputs. They immerse vaporiser tubing in a bath of water supplied with site process hot water or in which are also immersed electrical or steam heating elements.

Siting of cryogenic tanks. Most gas companies have their own rules about the siting of larger cryogenic storage tanks. For example, for 20–200 tonne liquid oxygen tanks a typical company recommendation would be for the following safety distances:

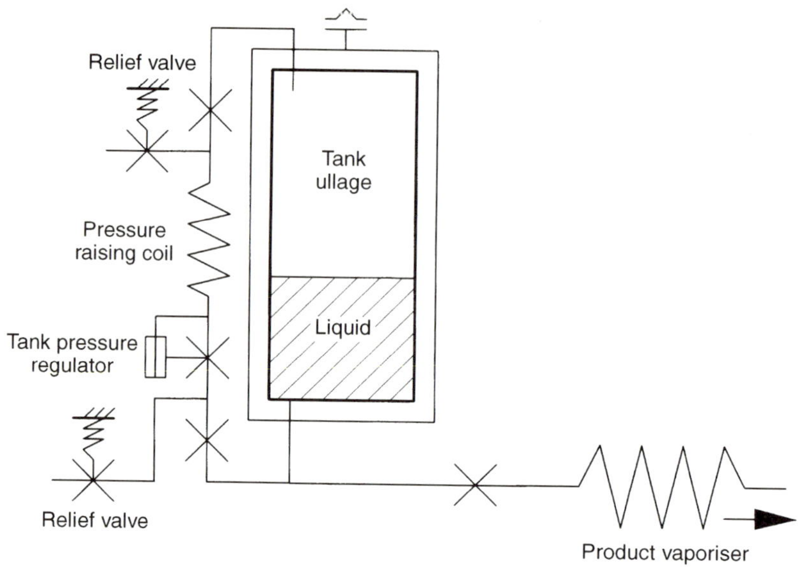

Figure 2.58 Pressure raising in cryogenic tank.

> 3 m: continuous lengths of flammable gas pipelines;
> 8 m: pits, drains, flammable gas pipelines including fittings, roads, car parks, smoking-permitted areas, electrical substations;
> 15 m: large wooden buildings, areas where crowds of people congregate, railways, compressor intakes;
> 30 m or more: bulk flammable gas storage must be placed at even greater distances, depending on the quantity stored.

Large cryogenic tanks. Unlike the standardised cylindrical shape used for vacuum-insulated vessels up to a hundred tonnes or so, very large tanks, up to 10 000 tonnes or so, assume very different shapes and designs. As noted above, non-vacuum insulation becomes more common. Spherical shapes are also common as the very much higher cost of fabrication can be offset against savings in material and lower losses.

'Cluster' tanks are now used where it is necessary to provide a set of tanks for different cryogenic liquids as, for example, on air-distillation sites where different grades of liquid nitrogen, oxygen, argon, some at high pressure, some at low pressure, often require storage. The cluster tank simply groups all the required storage vessels inside one large cylindrical envelope, the large interstices between the tanks and the envelope being filled with non-vacuum perlite or vermiculite.

Cluster tanks also avoid the worst effects of 'roll-over'. Roll-over manifests itself as an unexpected huge increase in pressure in a large tank. It occurs when the contents of a tank are stratified, with dense cold liquid lying above warmer, less dense liquid; a liquid version of the unstable 'temperature inversion' often seen in the atmosphere. This stratification can be surprisingly stable until a small disturbance causes some rotation motion to start. Quickly, the motion builds up momentum, with the cold upper layer pushing down into the hotter lower layers and vice versa. Suddenly, the upper layers are hotter, but without the additional hydraulic pressure which formerly contained them, and they flash off and boil, raising the pressure suddenly. Roll-overs cause a large amount of gas to be vented and can even overwhelm safety venting devices and damage an installation.

Large tanks cannot economically be made to withstand much more than the hydraulic pressures of the liquid sitting in them. They must therefore be vented to the atmosphere, leading to losses. Losses arise from several causes:

- heating of the cryogen, resulting in vapour boil-off;
- 'breathing' of tanks, caused by expansion and contraction of vapour, air and liquid during temperature cycling.[33]

Venting gas can sometimes be employed in some process near to the storage tank. However, where this is not possible, and where the gas is valuable, such as helium, then venting losses are dealt with by extraction from the vent stream and reliquefaction.

2.4.6 Storing gas in gasometers

All over the industrial world large thin-walled steel cylinders with slightly domed tops, some large enough for a football pitch on the top, used to be a common sight. These were water-sealed gasometers or gasholders. They are still a simple and effective means of storing low-pressure gas, and have the convenience that they accept gas and give it out at a more or less constant low pressure. They can also be built inexpensively with rather relaxed standards because of that low pressure.

Mechanically they are simply large versions of the bell-jar in a water trough system used in chemistry classes at school, save that the bell-jar in this case actually floats on the water. The pressure from a gasometer is low (atmospheric pressure plus the pressure necessary to support the bell top and sidewalls) and rises only slightly with the state of fill as the submerged walls emerge from the water seal and lose their water buoyancy. To avoid moisture in the stored gas, other sealing methods can be used: in small

[33] The losses due to 'breathing' are not restricted to cryogenic liquids. The typical losses due to breathing of a solvent tank during filling and emptying, and due to thermal effects, are discussed in section 3.7.3 with respect to solvent vapour recovery. Solvents in bottles in the laboratory have surprisingly high losses, ether and acetone are quickly lost, and in hot weather there are large losses of petrol/gasoline from vehicle fuel tanks.

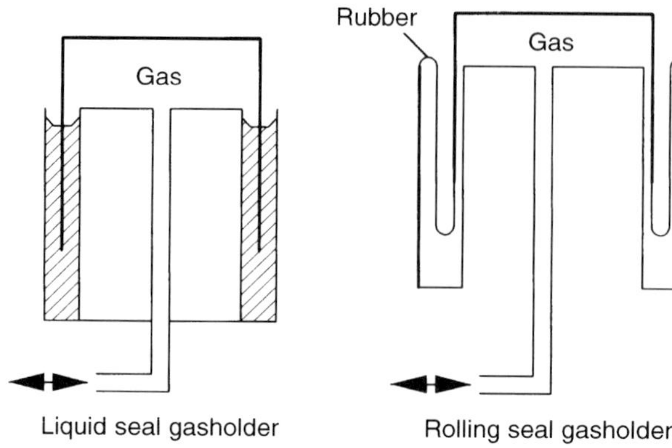

Figure 2.59 Gas holders using water and rubber seals.

systems, a flexible rubber tube which everts as the gasholder increases in volume can be used (Figure 2.59)

Today, the storage of methane fuel gas in large bulk is done by liquefaction of the gas and storage of the liquid methane in a cryogenic tank. For diurnal storage gasometers are still used, although storage underground in a wide portion of transmission pipeline (the 'line-pack') is now the most common method. During the day, when usage is heavy, the pipeline pressure is run down, whilst at night it is ramped up. With pressures of up to 75 bar and pipelines of over 1 m in diameter, huge amounts of gas can be stored diurnally, reducing the pipeline flow losses and decreasing the required pipe diameters.

2.4.7 Making and using compressed gas cylinders

The earliest commercial supplies of gas in transportable containers were probably delivered in gas-bags. Gas-bags, despite their bulkiness, are still in use for small amounts of gas, for example for calibrating analysers or for transporting samples for analysis. Today gas-bags are made from tough polymer film laminated with metal, which prevents escape of gas by diffusion or contamination by air.

Gas can be supplied in disposable containers such as 'aerosol' spray cans. These can withstand only modest pressures, however (10 bar or so) and are available only up to about half a litre water capacity, giving contents in permanent gases of only 5 l, a somewhat inefficient package. Oxygen for 'recreational' use is supplied in spray-type cans in Japan. For readily liquefiable gases such as butane, aerosol cans are more efficient, although the pressure rise with temperature limits the vapour pressure of the gases which can be delivered in this way.[34] However, the mainstay of compressed gas usage is, and will remain, the high-pressure refillable gas cylinder.

Compressed gas cylinders made from strong forged or welded steel or aluminium, capable of containing 50 bar or more pressure, with reasonable wall thicknesses have been around since the late 1800s, when the Taunton and Erhardt processes for deep-drawing thick steel cylinders became available.

Gas-to-packaging ratio. An early cylinder might have achieved $0.05\,m^3\,kg^{-1}$, in other words a cylinder weighing 100 kg would contain only $5\,m^3$ of gas at atmospheric pressure. Although the materials used now are stronger, pressures higher and wall thickness somewhat less, it is still the case that the gas-to-packaging ratio is not very high for compressed gases; the cylinder far outweighs the product it contains.

This is worst for hydrogen, where a 44 tonne articulated truck can deliver only about 100 kg of H_2. Even for the more favourable example of the oxygen cylinder, the ratio is only about 1:10; the standard O_2 cylinder itself contains less than 20 kg O_2 in a 100 kg package. The liquefiable gases such as carbon dioxide are more efficient, with the efficiency decreasing as the vapour pressure at 70°C (the normal upper temperature limit for safety purposes) goes up. Gases such as butane in thin aluminium refillable or throw-away spray-can containers are the most efficient in their gas-to-packaging ratio.

Unfortunately, there is little that can be done about the problem of efficiency of gas cylinders for gases which will not liquefy at room temperature, except the development of stronger container materials. If a long gas cylinder is considered, for example, it is readily shown that the weight of the cylinder is proportional to its volume for a given pressure, i.e. there is little advantage in larger cylinders. In practice, cylinders of below 20 or 30 l water capacity are less efficient. Above this size, however, cylinders are made up to 1000 l or more but are no more efficient than the standard 40–50 l size.

This can be understood by studying the tension in the walls of a gas container. The tension T per unit length (hoop stress) in a sphere is given by

$$T = PR/2$$

[34] For larger capacities, plastic containers might be used in the future. With metal or other coatings to slow loss of gas by diffusion, biaxially-stretched polyethylene terephthalate (PET) bottles of a few litres capacity could be used. Ratings of 3 barg are readily available (bursting pressure 12 barg) and probably at least 5 barg is safely possible. These containers would offer an inexpensive and lightweight throw-away package for up to 15 l of permanent gas and much larger quantities of liquefiable gas. Safety problems with using plastic gas containers may be a limitation, however.

where P is the pressure applied and R is the radius of the hoop. This can be seen most easily if a sphere is imagined to be cut in half along a diameter. The total tension force across the cut must be $2\pi RT$ and must be equal to the force due to the pressure inside across the cut area, i.e. πR^2. The minimum cross-section of the walls must be proportional to the hoop stress, i.e.

$t = kT$, where k is a constant.

therefore

$$t = kPR/2$$

Hence the total mass M of the sphere must be

$$M = 4\pi R^2 t = 2\pi R^2 kPR = (2kP)\pi R^3$$

i.e.

$$M = 3/2(kP)V$$

Hence it is seen that the mass of the container is proportional to the volume enclosed and there is no advantage to using larger containers. Larger containers, although needing less wall area, must be made with thicker walls and this latter fact exactly compensates for the lower area needed. It is also interesting to note that for approximately ideal gases, there is also no weight advantage in using higher pressures because they simply need thicker and heavier cylinders. Higher pressures, however, do allow a smaller volume package.

Gas compressibility (Z-factor) affects these conclusions to some extent, giving an advantage to higher or lower pressures depending on Z behaviour with pressure, and liquefiable gases are obviously usually stored in cylinders strong enough to render them liquid.

The above discussion is centred on spherical vessels but similar considerations apply to cylindrical vessels. However, in cylinders the longitudinal stress is that of a sphere of radius equal to that of the end caps, but the circumferential (hoop) stress is doubled, so that cylinders must be made proportionally thicker. The greater hoop stress also means that cylinders tend to fail via cracks along their length (like a sausage splitting longitudinally in the frying pan).

Fabrication of cylinders. Standard gas cylinders are made by a multistage deep-draw forging process involving massive rotating machinery and steel heated until it is bright yellow (Figure 2.60). First, a thick-walled pot is formed from the heated steel billet in a 'back-extrusion' or piercing process, then the pot is drawn over a mandrel with powerful concave rollers, which give the precise cylinder shape and draw out the walls to the correct thickness. The base of the pot is thinned and dished so that the finished cylinder will have a concave base which will allow the cylinder to be placed

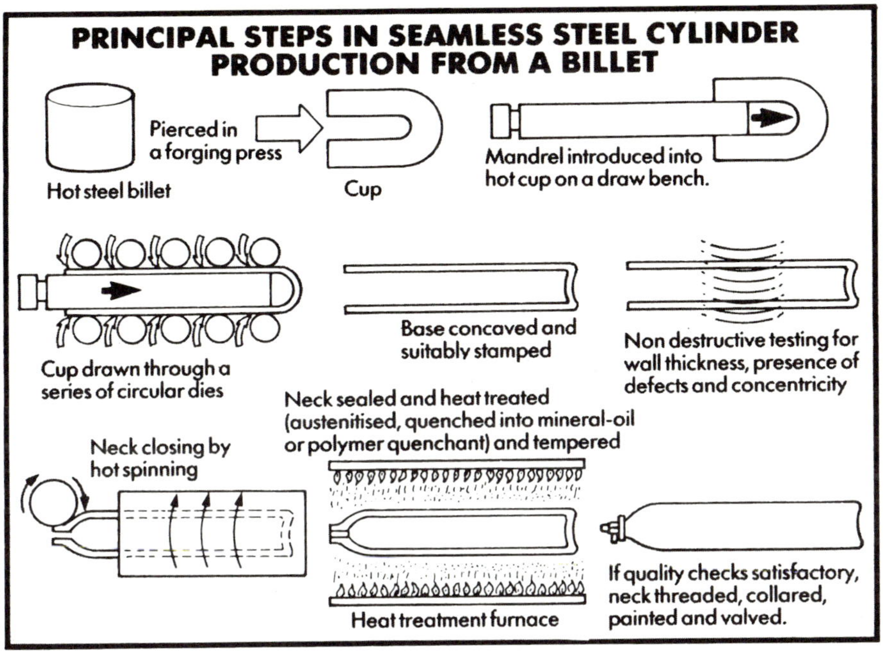

Figure 2.60 Making a seamless gas cylinder.

free-standing upright. Previous to the latter process, single cylinder handling in depots and during fillings was rendered considerably more difficult because cylinders could not be stored upright except in frames or trollies. Round-bottom cylinders are still favoured for some purposes; when used, in manifolded cylinder sets, for example, the round bottom is quite acceptable. Stress concentration occurs in free-standing cylinders which have to be carefully modelled to ensure that the material is not overstressed, and the base must be made much thicker than a round-bottom cylinder.

The crucial neck end of the cylinder is then fashioned. Following cutting to length, the thin-walled pot is then reheated if necessary (by an electrical resistance heating method in some works) at the neck end and spun on a lathe. Heavy shaping wheels then squeeze the walls at the end together into a thickly reinforced neck area. A neck ring can be added at this stage. The whole cylinder is then heat-treated by furnace heating and oil (or polymer) quench, cooled to ambient temperature, and ultrasonically tested and visually examined with a borescope, before being stamped with a unique number and then shipped. Polymer quench heat-treatment may become a standard in the future because a polymer/water quenchant mix can be tailored in the cooling rate it applies, and is intermediate between oil and water quenching. The cylinder serial number, together with other information such as pressure rating and empty weight, are then stamped on the wider 'shoulder' part of the cylinder neck. A record of each gas cylinder serial number is kept, so that in principle they are as traceable as a motor vehicle, with test records also recording the cylinder number, just as for motor vehicle testing.

Cylinders are made and tested to comply with codes of practice and regulations which, although similar in effect, differ in numerical details and procedures around the world. Cylinders in the US must meet ASME or DOT standards, in the UK they must meet BS standards, and in Europe TUV, DIN and EC codes. It is essential to refer to these requirements for details.

Ultrasound testing may be carried out, perhaps magnetic testing, too, either during or after finishing hot-working. Occasional sample cylinders are tested to destruction. Hydraulic burst testing, or, occasionally, pneumatic or hydropneumatic burst testing is carried out, giving the ultimate maximum pressure of the cylinder. Hydraulic failure tests are conducted behind a stout wooden screen as failure does not engender much release of energy; a simple split occurs in the cylinder resulting in a harmless gush of water. Pneumatic tests may emit fragments and involve a violent explosion, and must be conducted in a properly constructed bunker with, for example, rows of 15 cm thick railway sleepers to protect the bunker itself from damage.

The ultimate failure pressure of a cylinder will always be considerably beyond the test pressure, itself normally from 150% to 175%, dependent on local regulations, of working pressure. Cylinders are designed to fail usually on a longitudinal line along the body. Cylinders which fail in the neck or base area are regarded as undesirable. The cylinder's cylindrical wall is around 5 to 6 mm thick in the case of standard large portable cylinders, whereas bottom and neck regions are up 12 mm or so thick. These destructive tests (especially the pneumatic tests, which model more accurately behaviour after failure) also reveal how dangerous a cylinder burst may be. Cylinders which allow fragments to be emitted like a military bomb are regarded as unsatisfactory, whilst cylinders which simply split in a ductile manner are regarded as failing in the best possible way. In this way, in an accident, although personnel very close to the ductile bursting cylinder might be injured or killed by the blast or by the gas itself, personnel a somewhat longer distance away will escape unscathed (see Figure 1.2).

Hydraulic tests may involve 'plethysmography' – observing volume changes of the cylinder while it is pressurised. This is achieved by jacketing the cylinder in a rigid jacket filled with water, and noting changes in the water level in a manometer tube attached to the jacket. The plethysmograph readings can be used to set maximum permissible strains for cylinders with a simple and consistent apparatus.

Standard gas cylinders have a thread machined inside the neck with a suitable thread-cutting machine. The valve, which has a tapered thread, is then screwed, with a suitable thread sealant, into the cylinder to a specified (very high) torque with another motorised machine. Because of the risk of neck-cracking if overtightened, or leaks if undertightened, gas cylinder valves should not be regarded as removable by the user and should always be fitted by the gas company or cylinder supplier. The cylinder guard, a tough plastic or metal shield, is fitted over the valve. This provides protection against the valve being broken off in the event of the cylinder being dropped, while also providing a grip for more convenient handling and allowing easy access to the cylinder valve and connection. It is worth avoiding a cylinder valve breakage; if a cylinder valve is broken under pressure, the broken parts will fly off dangerously, whilst the cylinder itself, if untethered, may be propelled (briefly) along the ground, depending on the size of the orifice formed, causing further damage in addition to that which the gas leak may cause.

An ordinary steel cylinder is often left with its natural finish on the inside: fairly clean bare steel with a modest amount of mill-scale (black iron oxides which are formed due to oxidation while the cylinder is hot during the forming process). The scaling on the outside of the cylinder is typically rather worse than on the inside and often storage periods before sale of a cylinder mean that it is often already beginning to develop a thin coat of rust before delivery. This and mill-scale is removed, usually using shot-blasting, and one or more coats of paint applied to the outside. Gas companies often pass their new cylinders through the same shot-blast and paint line that old cylinders pass through. Cylinders are regularly retested hydraulically and then repainted; a coloured neck-band indicating last date of retest helps facilitate this retest procedure. The paint colours follow a code related o the cylinder contents, which varies from country to country, and is important in allowing speedy identification of potential hazards in an emergency (see Appendix D).

Special gases often require an improved finish on the inside. Despite having to carry out all operations through the neck hole, it is possible to carry out practically all the normal metal-finishing processes on the inside of a gas cylinder. Shot-blasting to remove mill-scale, followed by some sort of plating, is quite common. For stainless steel cylinders, internal electropolishing is often carried out, while aluminium cylinders may be anodised.

There is a large variation in the sizes and shapes of gas cylinders used. Forged steel cylinders have steadily advanced in pressure over the years, from 50 to 137, 175, 200 and 230 barg in most applications. The latest generation of steel gas cylinders to be introduced in the UK are now certified for filling to 300 barg; other countries around the world are following suit. The lowest pressures normally possible now are around 125 bar, whilst the highest pressure single-wall cylinders are 400 bar. Cylinders vary in size from the 25 mm diameter and 200 mm length of an ethylene 'tube' cylinder, used for a single dose of ethylene for ripening fruit such as bananas, to the

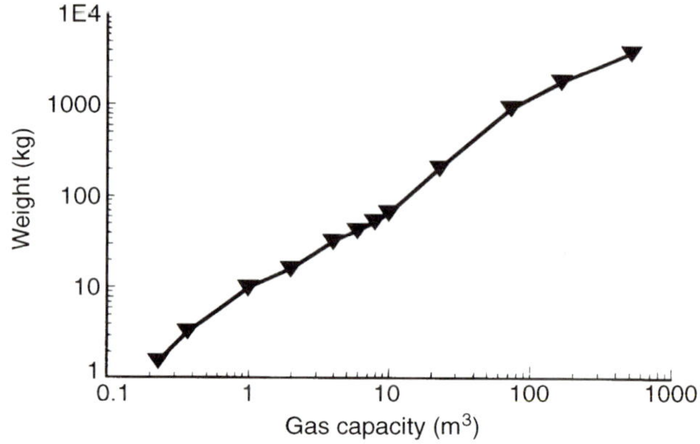

Figure 2.61 Efficiency of gas cylinders.

immense ‘torpedo’ cylinders of 700 mm diameter and 10 m length used for bulk storage of compressed hydrogen. Capacities thus vary from 3 l in the ethylene tube to 500 m^3 or so in large hydrogen torpedo cylinders.

Depending on the design and material required, seamless gas cylinders can be made by methods other than the standard deep-drawing of a pot as described above. Seamless tubing can be used, by necking down both ends, for the manufacture of cylinders with valves at both ends. Simple circles of plate can be cold-drawn into lightweight cylinders.

Stainless steels can be used for containing more corrosive and more valuable gases, as can Monel and other nickel alloys. Gases such as tungsten hexafluoride, which are both valuable and corrosive, are invariably packaged in stainless steel or Monel cylinders. However, these are not useful for more ordinary gases as the materials are dramatically more expensive than steel (or aluminium) and have lower strength, and so have to be made very thick, giving a very poor gas-to-packaging ratio at exceptionally high cost. The most common material for cylinders is still steel, sometimes carbon steel, but most often with alloying elements such as nickel or, especially, chromium/molybdenum or manganese. The highest pressure cylinders available are now in chromium manganese steel and these achieve the highest capacity weight ratios, around 0.15/0.18 (or a little higher) m^3 kg^{-1}. The new 300 bar cylinders achieve a result which is nearly 30% better still, at least in the largest (15 m^3) size. Manganese steel cylinders are still at around 0.15 or so m^3 kg^{-1}. (Figure 2.61)

For lower pressures seamless cylinders are not essential. Steel cylinders can be prepared by welding from heavy gauge plate for service up to 100 bar or more. Similarly, aluminium alloys are used in welded form for low-pressure liquefied hydrocarbons, chlorine and similar gases. Welded cylinders are subject to many of the stringent rules to which drawn or forged cylinders are subject, with additional requirements for X-rays of welds for higher pressure variants. Welded cylinders have been moving upward in pressure for some years now, a trend which may be expected to continue.

Aluminium versus steel for gas cylinders. Aluminium has advantages in small sizes but is less obviously effective in large sizes, for example, because thick walls are required and fatigue cracking may occur. In larger sizes an Al cylinder is typically only a little lighter than a steel cylinder with the same mass of gas. New alloys are needed but these will need to be very thoroughly investigated before they are accepted. Neck-cracking on some types of current aluminium cylinders has had the effect of slowing what might perhaps a few years ago have been perceived as a drift towards Al cylinders for more general use. (In a threaded gas connection, the greatest stress falls on the threads closest to the high-pressure side (this is generally where failure will occur around the stress concentration near the first thread) if no precautions are used. High-pressure reactor vessels have been designed using a spring coil insert or a tapered thread to avoid this problem.)

Aluminium is used in lightweight forged cylinders for scuba diving because their slightly lighter weight is a cost advantage in an application where even small savings are valuable and it also offers some advantages in corrosion resistance. There are also occasions when aluminium cylinders are preferred on chemical grounds as they are less reactive towards most products (except alkalis) than steel cylinders. In addition, aluminium cylinders can be made to deliver much lower moisture content than steel cylinders, even stainless steel. Moisture levels of less than 0.1 ppm can easily be reached in well-prepared Al cylinders.

Ultracylinders. Some military requirements are for cylinders able to withstand much greater pressures than are normally supplied in cylinders; up to 400 or 500 bar, for example, in the gas bottles involved in missile gyroscopes. There are a number of ways of improving the performance of the standard gas cylinder and, although none is commercially viable for general use at the moment, some are used in military and specialist roles.

Cylinders can be wrapped with fibre-reinforced plastic (FRP). If the FRP chosen is, like carbon fibre, both strong and stiff, then it can take much of the stress imposed on the cylinder walls. The wrapping can be a simple body wrapping, which reflects the greatest stress direction of the cylinder and enables thinner metal walls to be used. Alternatively, the wrapping can be all-over wrapping, allowing a thinner base and neck, too. All FRP gas containers can be made, but are unusual. They normally require a gas-impermeable lining, as almost all common structural plastics are permeable or reactive with gases to some extent, and there is even a danger that gas could penetrate the plastic matrix and weaken the matrix/fibre interface or the fibre itself.

Wrapped cylinders were developed most obviously in the 1960s by NASA to serve its needs for ultralight pressurised containers for fuel and gases in space vehicles. The stress in a cylindrical pressure vessel's sides in the tangential direction is double that in the longitudinal direction. This is easily visualised as follows. A cylinder of length L, radius r and thickness t has to withstand a longitudinal tension of $P\pi r^2$ from its end caps with a cross-section of metal of $2\pi rt$, giving a longitudinal stress of $Pr/(2t)$. The cylinder has to withstand a tangential tension of $P2rL$ in a cross-section of metal of $2Lt$ (imagine the cylinder is sliced longitudinally up the middle), giving a tangential stress of Pr/t, twice the longitudinal stress. The wrapping of fibre is therefore arranged in a tangential direction to prevent longitudinal splitting of the cylinder. The pressure rating can thus be increased by a factor of up to 2.

In practice, this doubling of pressure rating can be achieved with a 10% increase in mass, owing to the much greater strength/weight ratio of the glass fibres over the cylinder-grade aluminium. To achieve this wrapped cylinders are set up to be under a certain amount of prestress tension in the fibreglass when unfilled. This will ensure that the fibreglass and the aluminium each take their fair share of the stress when the cylinder is at its maximum pressure. The prestress tension is achieved by a technique called 'autofrettage', i.e. hydraulically inflating the cylinder to above the yield point of the aluminium so that the Al flows slightly. On releasing the hydraulic pressure, the cylinder relaxes back into a state of compression in the Al and tension in the fibreglass.

Fibreglass-wrapped cylinders are made by first specifying cylinders with extra-thick bases, then wrapping them evenly with a 'tow' of continuous glass filaments with a winding machine which applies a constant tension. The tow is passed through a coating and roller squeezing mechanism prior to being wound, to ensure that the matrix resin thoroughly impregnates the fibres. Computer control ensures that the wrapping is done consistently. The 'green' epoxy/glass laminate is then hardened by a gentle bake at 100°C or so for 24 hours. Autofrettage is carried out to prestress the fibreglass. The normal inspection and hydraulic test procedures are then carried out. In order to avoid difficulties with cylinders bursting in fires because of the glass-fibre failing, the cylinder valve may require a thermal safety vent plug, perhaps made from a fusible solder material (fusible plugs are also common on many welded gas containers).

A typical military application for a wrapped cylinder would be for a small lightweight compressed nitrogen bottle. The nitrogen would be used in the Joule–Thomson effect cooler to cool the infrared detector of a heat-seeking missile, or used to power a microturbine to spin a stabilising gyroscope up to speed in a second or so. Premium breathing cylinders for divers (and also for land-borne breathing sets) are now also supplied with glass-fibre composite wrapping, and there is just a hint that cylinders in some portable applications may move this way, too, with one or two products being advertised in lightweight wrapped cylinders. The L'Air Liquide company offers a lightweight oxy-acetylene welding set (Rollerflam) using an FRP-wrapped cylinder for the oxygen supply. However, wrapped cylinders remain, for the moment, exceptional.

Spherical compressed gas containers. Something that may surprise the newcomer to the gas industry is the shape of cylinders. Gas cylinders would have a better performance (around 30% less weight) if they were they spherical,[35] so why is it that in the gas industry spherical shapes are rarely employed?

As noted above, some very large liquid storage facilities are indeed spherical, as are some helium Dewars internally, but these are most exceptional. Some gas cylinders were spherical (one example was the wire-wrapped compressed air container in the V1 cruise missile) and examples of these are still preserved by the science museums in London and Munich.

There are a variety of practical reasons why spherical gas cylinders are almost unheard of, however. The gas cylinder valve needs to be installed in a specially strengthened and thickened part of the cylinder wall, which spoils the perfection of the spherical symmetry anyway. Spherical containers are difficult to fabricate, whether by forging and rolling or by welded seam construction, and, once fabricated, are difficult to store and handle efficiently. The fabrication of spherical cylinders is clearly difficult to do with conventional metal-forging and drawing plant, but it is surely to be anticipated in the future that, using another fabrication technique, spherical or more nearly spherical cylinders might become attractive.

Cylinder valves and connections. Cylinder valves are required to be substantially tough, foolproof, long-lasting, corrosion proof, resistant to heat and extremely leak-tight. A cylinder with a damaged valve, either leaking or jammed on or off is a potential danger, so standards for valve quality have to be very high. For service in semiconductors an additional requirement is demanded that no particulate matter, even a submicron particle, should be generated in the valve.

Most gas companies now possess (and supply) 'gas cylinder coffins' for dealing with the occasional problem cylinder of a dangerous gas, typically caused by a valve failure. These are large, strong, corrosion-proof pressure vessels with a large access door. A leaking cylinder can be loaded inside and the access door sealed shut. Even if the cylinder completely discharges

[35] The performance would be improved as follows. The hoop stress on the walls of a cylinder of radius r is, say, T. Suppose a cylinder is $9r$ long, it will have volume $31/3\ \pi r^3$ and weigh about $22c\pi r^2$. A sphere of twice that radius would have a similar hoop stress T and volume of $32/3\ \pi r^3$ but weigh only about $16c\pi r^2$. The latter would also have no heavy base, although a similarly heavy neck would probably be needed.

inside the coffin, the pressure build-up is limited due to the larger volume of the coffin. It can then be safely transported away for disposal at leisure. A wet scrubber system can be connected to the 'coffin', loaded with the appropriate water, acid, oxidising or alkali wash solution (e.g. sulphuric acid, sodium hypochlorite or sodium carbonate) and then the 'coffin' depressurised safely. These are, however, very rarely used. Some gas companies also supply scrubber systems mounted on wheeled trollies for this kind of emergency.

Most cylinder valves are of a simple, conservative design, proven over many years to be reasonably effective. A typical design has a simple steel spindle, its shaft sealed with a packing gland, sealing on a soft or hard seat in the brass body. The outlet connections for different types of gases are of different diameter, thread pitch and direction. Some outlets rely on a bull-nose-to-cone seating arrangement and these require tightening with a substantial spanner to achieve a leak-tight fit. Other cylinder connections employ an polymer ring seal, such as an 'O'-ring or a metal-reinforced seal, in various ways and this can often be tightened to leak-tightness without a spanner, although the seal ring must be inspected and replaced from time to time.

The use of cylinder valves involves a certain amount of common sense. 'Snifting' (briefly turning the valve on and then off again to allow a brief blast of gas out) can still be done for the more benign gases, especially where cylinder valve outlet ports may have collected road dust, rain, etc. The 'snift' removes the debris, which might otherwise interfere with the seating of the outlet connection or the pressure regulator downstream. However, easily flammable gases such as hydrogen, or dangerously reactive or toxic ones such as silane or chlorine, should *not* be snifted. If outlet connections are dirty on these gases, less common now, as many cylinders are supplied with plastic or metal covers or plugs, then a blast of dry clean compressed air and a dry lint-free cleaning cloth are the appropriate solution. As mentioned above, cylinder valves are normally on a tapered thread in the end of the cylinder but, because they are screwed in with a machine, they are exceptionally difficult to unscrew. In any case, as remarked above, removal of cylinder valves should only be carried out by the gas company or cylinder supplier.

With liquefied gases, dip-tubes may be attached to the valve internally. Long dip-tubes allow liquid to be withdrawn from the cylinder in an upright position. Small curved dip-tubes ('goose-neck') from the valve to the cylinder side just below the shoulder allow liquid withdrawal in the horizontal position. With low rates of withdrawal, cylinders of liquefied gases such as carbon dioxide simply absorb a small amount of heat from their surroundings and vapour pressure will be maintained. However, if large withdrawal rates are needed, then it is better to withdraw liquid via a vaporiser pipe so that cylinder temperature (and therefore pressure) is maintained. Liquefied gases normally also require protection against overheating in case of fire. Either a pressure-operated bursting disc (for higher vapour pressures, as with liquid carbon dioxide) or a fusible metal plug (for lower vapour pressures such as propane cylinders) is required and is usually fitted to the cylinder valve.

For some gases 'multiport' valves are required, typically two ports, one connected to a dip-tube which extracts liquefied gas, and the other connected to the vapour space. For the more dangerous gases there is some advantage in the use of a pneumatically operated valve. This cannot be turned on by an incautious 'snifter' while the cylinder is in transit but can be turned on from a safe distance once at its destination. Cylinder valves incorporating a check-valve function have occasionally been suggested and would have the advantage that cylinder contamination, by rainwater, oil or other liquids sucked back after a cylinder is emptied, would be less likely. However, common cylinder valves do not protect against suck-back, and so it is useful to include check-valves in cylinder gas manifolds and other systems.

Finally, it seems likely that the slow adoption of 300 bar cylinders may take place. If they are found acceptable in every way, they could replace 200 and 230 bar cylinders, making a pressure-reducing cylinder valve useful. So much equipment in the marketplace is compatible with up to 230 bar but is not sufficiently rated to be safe at 300 bar, that gas companies are reluctant to introduce the higher pressure cylinders in case users inadvertently connect them to old low-pressure equipment and this leads to an accident. The idea of a small, relatively crude regulator that can be installed within the cylinder valve to reduce the output down from 300 bar is therefore very attractive. The new cylinders can then be sold, with all their advantages of higher capacity per cylinder and better gas/weight ratio, as fully compatible with existing equipment. Designs from the Ceodeux and Broeder Mueller companies are now being used in small numbers. They regulate to 60 bar, well below the maximum rating of all equipment, and give an added bonus that they will provide a steadier output pressure in applications where a highly constant pressure is needed during cylinder emptying. They also have a check-valve function. The only significant disadvantage would appear to be the loss of a convenient small gauge giving cylinder contents, although weighscales can obviously be used.

Manifolded cylinders. Multiple cylinders connected by a manifold of pipework is a convenient way of delivering larger amounts of gas than a single cylinder can handle. A bundle of K-size cylinders is not inferior in weight for the same gas volume to a single large cylinder but can be made into a convenient cuboid shape. Such a bundle is dubbed a manifolded cylinder pallet (MCP) and may have, say, 12 or 15 K-size cylinders, containing 150 m^3 of gas and weighing 1.5 tons, firmly bolted to a frame on a

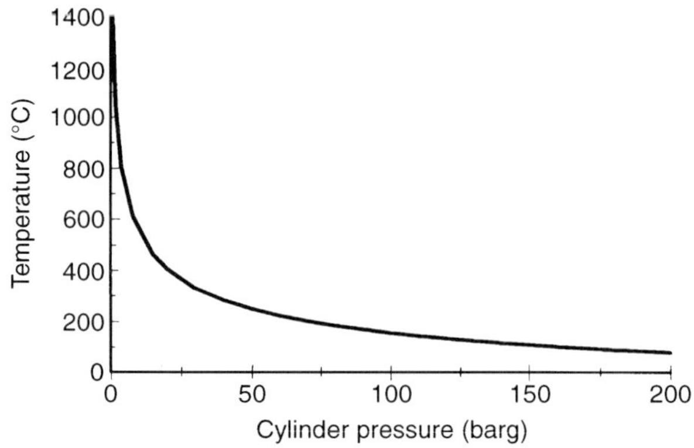

Figure 2.62 Theoretical adiabatic temperature rise while cylinder filling.

metal pallet for fork-lift truck handling. The cylinders used in MCPs can be older round-bottomed types now regarded as unsuitable for individual service. For hydrogen service large numbers of cylinders are often manifolded together and bolted to a 30 tonne vehicle trailer. Such a hydrogen trailer, containing up to 1000 m^3 gas, can then be driven to the user and left there.

Also very common is a two-cylinder manifold that can be used to provide continuity of gas supply. Whilst one cylinder (or MCP) is being changed, the other cylinder is used to maintain the supply pipeline pressure. Many such systems are manually operated. However, it is increasingly popular to fit pneumatically operated valves to the manifolds and to connect these to a controller box that accepts cylinder pressure transducer or electronic weigh-scale inputs. The controller can be arranged either to simply change over from one cylinder to another after pressure falls from the first on-line, or it can be arranged to switch to a 'standby' cylinder only while the 'primary' supply pressure is low and to revert to the primary when pressure there recovers. The latter mode is used, for example, when a single cylinder backs up supply from a large manifolded cylinder pallet.

Cylinder filling. Cylinder filling cannot in general be carried out simply by connecting a bulk gas store to a cylinder with a pipe and opening the valve. Temperature rise during filling is just one of the complicating factors. Temperatures rise during most cylinder-filling operations (Figure 2.62). This can be understood in gas filling by considering the first 'slug' of gas that has been filled to be under a 'piston' composed of the next slug; the first slug will be compressed and will heat up. If the cylinder were perfectly insulated and of low thermal mass, the heating would be adiabatic.

The ideal adiabatic temperature rise during filling of an ideal gas can be roughly calculated as follows. First the rise in temperature T of gas compressed from P_{in} to P_{out} in an insulated cylinder is noted

$$T_{out} = T_{in}(P_{out} / P_{in})^{(g-1)/g}$$

The initial 1 bar fill of the cylinder will be compressed adiabatically to 200 bar, whilst the next 1 bar slug will be compressed from 2 to 200, and so on, with the temperature rises then being averaged as the gas in the cylinder mixes. If perfect mixing is assumed to be taking place during the fill, then the calculation is fairly simple. A simple calculation assuming the above law and considering the volumes added from 1 to 200 results in the above temperature curve, assuming no loss of heat to the cylinder and no mixing between added 'slugs' of gas; in effect 'plug flow'. Although the peak temperature and the exact shape of the curve will differ in practice, the calculation brings out some very pertinent points:

- the maximum possible peak temperatures are surprisingly high: this will not surprise anyone who knows how a diesel engine ignites its fuel by simple adiabatic compression of air (adiabatic compression can also act as a source of unintended ignition in oxygen systems; section 2.5.9);
- the final temperature with fast filling, room temperature gas can easily be such as to make the cylinder distinctly hot to touch;
- the temperature rise is most marked just after first starting the fill and declines thereafter.

Of course, when the gas is supplied via a narrow valve restriction, there may be a Joule–Thomson cooling at that valve but this small amount of cooling will be mostly negated by heat exchange down a long delivery pipe. The cryogenic method of permanent gas cylinder filling has the advantage that it can be arranged to vaporise but not to fully heat the gas, so it can supply the cylinder at a significantly low temperature, compensating for the adiabatic heating to some extent.

In filling of liquefiable gases such as propane and carbon dioxide, which exist in liquid form under cylinder pressure, there is little heating. However, some precautions are still necessary. Cylinders should be filled from the bulk liquid supply leaving an ullage space above the liquid. The ullage space should be sufficiently big that the cylinders can be heated to 70°C before exceeding their pressure rating. However, they can easily be overfilled, unlike cylinders filled with permanent gases, and then they may burst even if only heated moderately.

Acetylene cylinders warm up during filling, because of the heat of absorbtion of the acetylene in the acetone. This is a problem because at higher temperatures insufficient acetylene will dissolve. Acetylene cylinders are therefore cooled, typically by water spray, during filling.

The temperature rise during fill of a gas into a liquid, such as acetylene into acetone, can be roughly estimated from the vapour pressure curve for the mixture as follows. Assume the vapour pressure P follows the law $P = P_o\exp(E_a/NkT)$, where k is Boltzmann's constant, N is Avogadro's number, T is the absolute temperature and E_a is a constant related to the activation energy for releasing a gram-mole of gas from dissolution. Hence the thermal energy released on solution is simply $E_a(V/22.4)$, where the volume of gas absorbed (V) and the temperature rise (T) are calculated depending on the thermal capacity of the solvent and solid absorbent. For example, for dissolving carbon dioxide in acetone (acetylene is similar), a maximum temperature rise of around 30°C is anticipated (ignoring the solid absorber in this example).

Cylinder filling with cryogenic liquid. In this technique, liquid gas is drawn from a low-pressure storage tank, its hydraulic pressure raised with a reciprocating liquid pump to well above the maximum 230 bar filling pressure, and then a metered amount of liquid is admitted to a batch of cylinders connected on a 'milking machine' style set of flexible connections. Using liquid filling, 15 cylinders may be filled simultaneously in a few minutes. Pumping cryogenic liquids up to high pressures is also an efficient way of filling cylinders because the pump only has to pump a small volume of liquid; the energy to produce the much larger volume of high-pressure gas for the cylinder is produced by the vaporisation of the liquid. Attention to safety is paramount, however, because a cylinder overfilled with liquid gas will soon rise in pressure beyond safety limits and fail disastrously; the fracture will be brittle if the steel cylinder has been cooled down by the cryogen sufficiently before failure.

Gas decant filling of hydrogen. Hydrogen is inefficiently stored in cylinders but is difficult and expensive to liquefy. Considerable ingenuity has therefore gone into bulk hydrogen delivery. As an alternative to delivering cylinders, some gas companies drive a vehicle trailer of manifolded cylinders, comprising either standard or $20' \times 1'$ diameter 'tubes' full of hydrogen, around to a customer. Instead of mechanically off-loading the 20 tonnes or so of cylinders, only the 100 kg or so of gas is off-loaded by a pipe to cylinders fixed on the user's site. However, all the hydrogen cannot be unloaded. Rather, hydrogen flows from the trailer to the user's cylinders until the pressures in each are the same. This 'gas decant' process does have some disadvantages, however, in that it is either inefficient in its use of truck capacity or does not fill the user's cylinders to full pressure. It also requires more cylinders to operate than straightforward cylinder or trailer exchange. For example, consider the following situations:

Trailer capacity	1000 Nm^3, initial pressure 200 bar
User storage	200 Nm^3

Settled (delivered) pressure	~167 bar
167 Nm^3 delivered	17% of trailer load
Trailer capacity	1000 Nm^3, initial pressure 200 bar
User storage	500 Nm^3
Settled (delivered) pressure	~133 bar
333 Nm^3 delivered	33% of trailer load

For large users the most common system is trailer exchange; this is efficient provided a whole 20 tonne trailer is a reasonable match to demand.

Residual gas: pre-evacuation of cylinders. For lower quality gases it is sufficient to simply blow down cylinders to close to atmospheric pressure and then refill them. However, high-quality gases must not be filled on top of the previous contents of the cylinder. Instead, the gas already in the cylinder must be blown down to 1 atm, through a scrubber exhaust system if necessary, and then evacuated with an oil-immersed high-vacuum pump to a low pressure before refilling.

Carbon dioxide. CO_2 is one of the few common substances which sublime. Solid carbon dioxide, 'dry ice', at −95°C evolves gaseous CO_2 directly without an intermediate liquid being formed, which is very useful in various dry ice applications (sections 3.4.2 and 3.7.4).

Standard high-pressure gas cylinders and equipment can be used for handling liquid and gaseous CO_2 at room temperature. At cryogenic temperatures CO_2 can, with some precautions, be handled as a liquid, just as liquid nitrogen, argon or oxygen can be. Alternatively, cryogenic carbon dioxide can be handled as a solid, just like ice, which it resembles to some extent. There are, however, some variations on the standard equipment especially for CO_2 and some special rules for handling the gas. One of the latter is that in handling cryogenic liquid CO_2 a certain minimum pressure, 4 bar gauge at least, is always maintained above the liquid. It is quite possible to vent the gas above refrigerated liquid CO_2 in such a way that the remaining CO_2 boils and then solidifies. This solid CO_2 can then only be removed by allowing it to sublime. This limits the possible rate of gas withdrawal to a very low value and entirely prevents liquid CO_2 withdrawal. It is therefore highly undesirable. A possible rescue procedure is to repressurise the tank containing solid CO_2 with a fresh supply of CO_2 gas. The gas will slowly warm up the frozen solid until it reliquefies but this procedure could take days or weeks.

The most common storage tanks for bulk CO_2 are insulated pressure vessels which store the gas at 5 to 15 bar and, if they are to provide hold times of more than a week or so, are provided with a small refrigerator system which can provide enough cooling power to counterbalance any heat inleaks.

2.4.8 Transport of gases

Delivery by pipe. The most efficient way of moving gas depends on how far it has to go and how much is needed. For large quantities and short distances a pipeline makes sense. Industrial gases are rarely used in such large quantities as methane fuel, and can often be made on site unlike methane fuel, so pipelines for industrial gases are unusual. However, there are extensive pipelines for other industrial gases in some regions of the world. Belgium and northern France, for example, have a network of pipes for nitrogen and oxygen. 'Silicon Valley' (the area around San Francisco and San Jose in California renowned for the large number of semiconductor manufacturers) has a pipeline for nitrogen running to most of the wafer fab plants there. However, pipelines are limited in their capacity by pressure and maximum flow rate. Maximum pressures are typically limited to less than 100 bar; often only 10 or 15 bar. There are further limitations on pipelines because of the necessity of obtaining 'wayleaves', i.e. the rights to dig a hole underneath somebody else's property. The cost of erecting a pipeline in a desert is completely different from costs in a typical industrial or city site, which can be measured in hundreds or thousands of dollars per metre.

Road transport of industrial gases. For the smallest users of gases the most economical mode of delivery is almost always road transport of cylinders. Despite the high weight of the gas cylinders, this is most often the best way of obtaining a dry, pure, gas in quantities of up to a 100 m^3 or so at a reasonable cost. The high cost of road transport means that much of the cost of the product boils down to the cost of delivery. A complete cryogenic road tanker unit (tractor plus specialised tank semi-trailer) will typically today (1996) cost around a quarter of a million dollars. This must be expensively manned, fueled and maintained, and these costs can easily exceed production.

From the air-distillation column oxygen may be costing only ten dollars or so per tonne. The same tonne of oxygen in cylinders is a thousand dollars or so, a hundred times more expensive. Gas manufacturers trade off the cost of local production with transport and usually arrive at a compromise where gas from a handful of ASUs is delivered in liquid form to widely scattered 'compressed sites' where the liquid gas is packaged in cylinders and transported only the last 50 miles or less in that form.

For larger volume users of gases, the best form of gas delivery is mostly delivery and storage of liquefied gas in cryogenic tanks although on-site generation of gas is often an alternative. After its liquefaction at a large ASU or liquefier unit, gas is loaded onto trucks which typically are large articulated vehicles with the maximum permitted road weight and thus maximum payload. The scaling laws for the cost of building and operating a road vehicle are such that only a small number of smaller vehicles can be justified, for delivery to cramped city sites, and most of the fleet must be the largest types permitted.

Despite the utmost attention to transport costs, it is rarely economic to transport gases such as nitrogen and oxygen more than 100 miles or so by road. As a result, most gas manufacturers try to achieve a sprinkling of production plants this kind of distance apart in their key markets.

The limitations of road tanker capacity can be obviated by resorting to rail transport, although with some disadvantages. A rail tanker train can be a valid alternative to a back-up plant because it can deliver gas over hundreds of miles in quantities of hundreds of tonnes per day.

Liquid hydrogen and helium. Liquid hydrogen and liquid helium are exceptional with respect to the other gases for a number of reasons. They both require very large amounts of energy for liquefaction but are also very inefficiently stored in high-pressure cylinders.

Both gases are delivered in liquid form in some circumstances, despite the huge costs of liquefaction. Road tankers and ISO containers are used, which are superinsulated and may also have a liquid nitrogen cooled jacket built into the insulation. Helium is only commercially extracted in large quantities in a few places, and therefore is most frequently shipped long distances, while hydrogen is only rarely required more than a hundred miles or so from a production site. It is rarely moved in liquid form outside the USA, although the advent of wafer fab plants requiring the high level of purity found in liquid hydrogen has changed this picture somewhat. Helium is frequently required as a cryogen when it clearly must be delivered in liquid form. Small superinsulated vessels are available with a reasonable hold time for smaller users.

Gas storage in absorber cylinders. The low density of hydrogen storage in cylinders is a particularly serious problem for developers of vehicles using hydrogen engines or fuel cells. Metal alloys which store hydrogen have been extensively investigated for their potential use in hydride-based rechargeable batteries. There are a number of formulations based on nickel, iron or magnesium metals which store hydrogen at a sufficiently high density (comparable to that in compressed cylinders) that they can be considered as hydrogen gas storage alloys. Relatively lightweight cylinders are used as the hydrogen equilibrium pressure above the hydride is low. There are a number of R&D programmes, for example that of Mercedes-Benz and the hydride specialist firm HWT in Germany, researching the application of hydride storage systems to fuel-cell and other hydrogen fueled vehicles.

However, although these absorbers offer a conveniently lower pressure than compressed hydrogen, there are problems. Firstly, the alloys must be prepared in finely divided form, followed by a sintering process so that both gas flow and surface area are maximised. Considerable heat is evolved during absorption, slowing the filling process, whilst on desorption cooling occurs and heat must be put in to maintain gas flows. Also, on cycling, many of the hydrides swell and contract with the addition and subtraction of the gas, and this causes fracturing of the sintered granules, creating problems with dusting. So far, no commercial hydrogen absorber cylinders are available.

An absorber cylinder system which is commercially available is the Nippon Sanso/Matheson SDS (safe delivery source). In this, phosphine or arsine are absorbed onto zeolite in a gas cylinder at atmospheric pressure. Zeolites and activated carbon absorbers become 'saturated', i.e. little more gas is absorbed for further pressure rises after 10 bar or so, at a rather low gas/solid ratio (a few per cent is typical). As a result of this, these absorbers are more useful at low pressures and for modest amounts of gas. The zeolite in the SDS cylinders increases the amount of gas available from the cylinder at 1 bara by a factor of six. A filter in the cylinder neck prevents dust from the zeolite leaving the cylinder in use.

Arsine and phosphine are both normally used at reduced pressure, in the vacuum of an ion implanter for semiconductor manufacture, so the low source pressure is not inconvenient, whilst the safety advantages are considerable since cylinder leaks will not result in these poisonous and pyrophoric gases escaping. Such escapes, quite apart from their danger to staff, have an economic implication. If a high-pressure cylinder leaks even a small amount of arsine, for example, the whole of a wafer fab plant, with hundreds of workers, would have to be evacuated while further gas escape was stopped and checks were made before sounding the 'all clear'. One potential problem with storing such reactive gases in this way is the potential exothermic reaction of air with the stored gas. However, the fairly low concentration of the gas in the absorber means that the temperature rise expected is only 35°C, according to the manufacturers.

2.4.9 On-site generation of gas

An alternative to moving gas is to make it where it is needed. If the demand for nitrogen or oxygen is very large, then an on-site air-distillation plant can be erected for the gas supply. However, few demands outside large steel and chemical works are this large. As a result, many decisions about on-site generation for oxygen or nitrogen revolve around the acceptability or otherwise of the much smaller and less expensive PSA plant discussed in section 2.1.6.

The decision to employ a PSA unit will depend on the whole economic equation of the site where gas is needed and on the purity of gas required. Economic factors would include:

- the cost of electrical power needed
- the cost of maintenance
- the number of hours per annum gas is required
- the peak flow needed
- whether any subsequent purifier stage is needed.

In addition, some economic factors will also occur, such as the noise pollution potential of the plant. Not only will there be a continuous buzz from the compressor, but also many designs produce a loud rush of escaping gas on a bed changeover; a cryogenic tank is much quieter.

Basically, a PSA unit will match well to a gas demand that is substantial (say, greater than $10\,m^3\,h^{-1}$), continuous, low in purity needs, with no peak flows and, on a site well away from residences or offices. There is a trade-off in PSA designs between purity and production rate which favours low purity (section 2.1.6).

Large-scale methods for the chemical generation of gases are not often used where physical methods are available (section 2.1.5). However, many chemical generator methods can be made to work in small sizes, too, and can in these sizes be competitive with 'merchant' gas. However, many chemical methods require the delivery and storage of chemicals, which may have disadvantages similar to those of merchant gas, and all have some purity problems, for example:

- H_2 from electrolysers: gas produced contains oxygen and is saturated with water;
- H_2 from methanol crackers: methanol needed and a PSA type of plant is often needed to eliminate CO_x from output;
- acetylene from carbide: purity is unacceptable for welding, for example, without further processing;
- N_2 from inert gas combustion generators: purity is very poor.

2.4.10 Gas losses

It is generally accepted that a few per cent of most gases, especially bulk cryogenic gases, are lost in transit from the production plant to the final end use. Occasionally, buyers of gas become concerned about this and sometimes some of the loss can be economically reduced. Some loss mechanisms have already been discussed; those from large static storage tanks, for example. There are, however, more possibilities for losses:[36]

- leaks: underground leaks are difficult to both detect and fix (housekeeping figures, generated by summing equipment gas consumption and relating it to overall flow rates are often not useful due partly to the inaccuracy of flow meters and partly because only a fraction of equipment will typically be fitted with metering);
- filling warm cryogenic tanks: release of the vapour is often needed if the pressure rise is such as to prevent pumping in new liquid;
- purging gas: essential to maintain purities after connections or disconnections have been carried out;
- safety relief valves occasionally dump fairly large quantities of gas;
- continuing gas supply to equipment which is out of service: this may be necessary for some equipment to keep it purged ready for action, for example;
- decanting of pressurised hydrogen (as discussed above): the gas delivery truck always drives away with a sizeable fraction of the hydrogen delivery, although this gas will typically not be lost but simply recycled as an undeliverable inventory of product.

2.4.11 Vacuum pump technology

Vacuum pumps are an important part of industrial gases, both in production and application. For example PSAs can be made advantageously with a vacuum cycle, the vacuum swing absorption (VSA) process, while purging of gases is more effective with vacuum. In industrial applications the use of vacuum pumps is legion; many processes can be carried out in low-pressure gas, often with considerable gas savings. Low-pressure applications are often restrained, however, by the relatively high cost (compared to mass throughput) of vacuum pumps and the additional cost of building containment which will withstand air pressure at 1 bara.

To appreciate the cost of vacuum, consider that a large vacuum pump, costing £20 000 or more, might typically be pumping a 10 mbar load at 80 m^3 per hour. At this pumping rate, this large machine is only pumping a kilogram or so of air each hour. To appreciate the cost of vacuum containment vessels, imagine a 1 kg bag of sugar balanced on every square centimetre of a large piece of equipment. It is not difficult to understand why

[36] The author has on several occasions gone to review gas consumption on complex sites where there are many different uses for the gas because there has been a suspicion that gas usage was higher than necessary. In most of these reviews some unnecessary losses were quickly found but nevertheless the overwhelming majority of the losses found could not be reduced economically. There were even one or two pieces of equipment where gas flow had been turned down too low for safe operation or was reducing equipment lifetime, and an increase in consumption was recommended.

much stronger and more expensive construction is needed. A warning might be added here that ‘improving’ a process by adding a vacuum cycle needs careful calculation on construction strength; simply applying a modest vacuum to equipment not designed for it may cause buckling.

The relatively low cost of inert gases such as nitrogen, or even argon, means that they compete strongly with vacuum as a means of avoiding oxidation of sensitive or hot oxidisable materials. The use of inert gases at atmospheric pressure allows vessels of the thinnest metal sheeting to be used and also allows relatively imperfect sealing to be adequate, with careful attention to the leaks created. Inert gas would therefore be the correct choice for a belt-furnace. The vacuum seals needed for a vacuum-belt-furnace would be almost impossible to engineer for practical cases.

Distinction should be made in vacuum pumps between those capable of evacuating to atmosphere and those, such as diffusion pumps and turbo-molecular pumps, which only function with another ‘backing’ pump to boost their output pressure up to 1 bara.

Venturi, ejector and diffusion pumps. One of the simplest means of generating a low pressure is to use a high-pressure jet. A jet of gas or water can entrain another gas stream or the low pressure caused by a venturi can be used to suck another gas stream. One of the simplest embodiments of the entrainment principle is the water-jet pump commonly used in laboratories for vacuum-aided filtration funnels.

A water-jet pump connected to a rapidly flowing stream of cool tap water can generate less than 50 mbar at flow rates of a few litres per minute; a litre of gas is pumped for every three or four litres of water passed. Water-jet pumps rely on entraining and then lowering the pressure of the pumped gas. Ejector or venturi pumps which use gas instead of water, are very similar and exploit the entrainment and venturi principles in a straightforward way.

Diffusion pumps (Figure 2.63) carry the entrainment principle to ultrahigh vacuum pressures. Although less popular nowadays, these are very simple, reliable pumps with no moving parts. Once pressure is below a few millibars or so (where diffusion pumps work best) gas molecules do not undergo collisions very often, at least over a short distance. It is therefore possible to consider the diffusion pump as a jet of heavy molecules simply knocking the light gas molecules out of the pipe being exhausted. A diffusion pump has a cascade of such jets fed with a copious supply of heavy molecules arranged in concentric fashion, the higher jets pumping gas to the next set of jets and so on. The jets further down the cascade become smaller, more numerous and closer to the pump wall, so that they remain on a scale comparable with the mean free path of the gas as the gas pressure is built up. The assembly is capable of pumping down to 10^{-8} mbar or so from 1 mbar; an impressive compression ratio. The jets are fed from the boiler unit and filled with oils that are resistant to decomposition at their boiling point. The

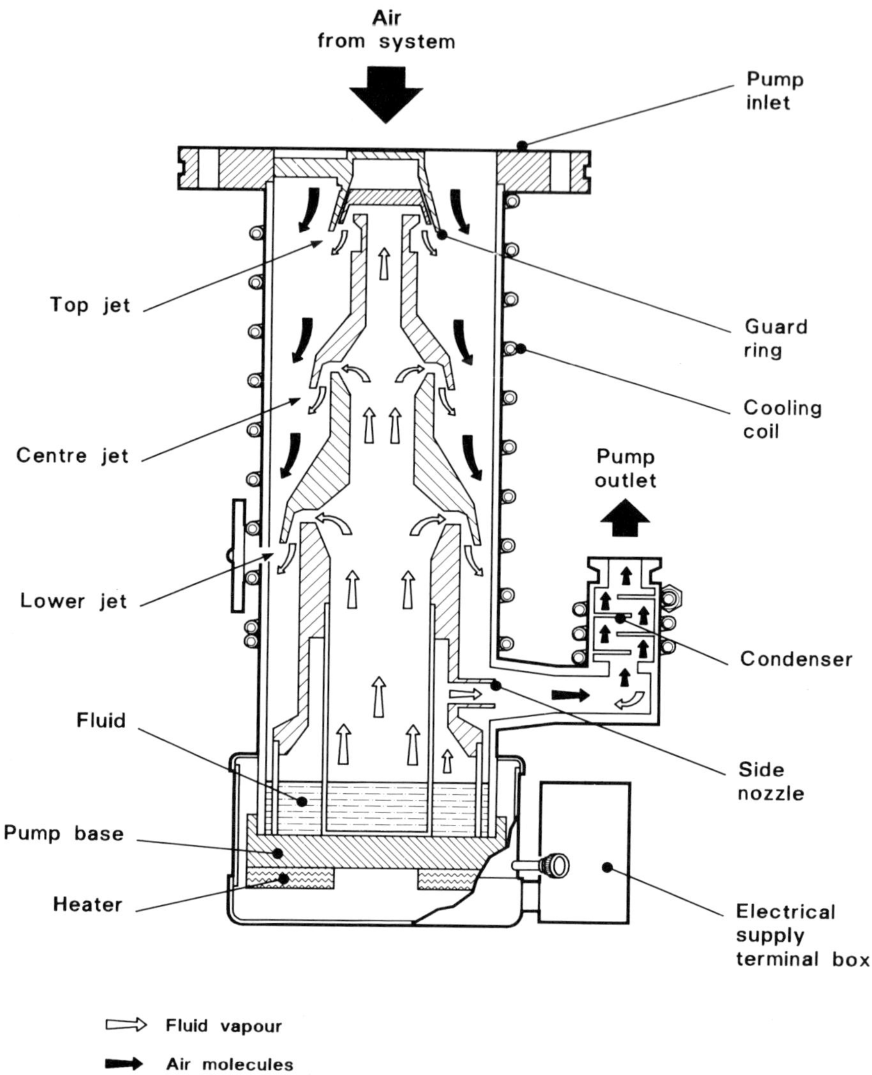

Figure 2.63 Diffusion pump (courtesy of Edwards High Vacuum, Crawley, UK).

oil boils and the vapour is fed to the jets. Mercury was formerly used but has a higher vapour pressure than vacuum oils and is now regarded as a significant health risk.

Mechanical vacuum pumps. The standard mechanical vacuum pump is still, after a century or so, the oil-immersed rotary vane pump (Figure 2.64a). This is robust and, thanks to being bathed in oil, fairly long lived. A central rotor rotates in an eccentric housing, whilst spring-loaded vanes or flaps seal the annular gap into compression chambers. An inlet check valve is usually needed to prevent oil being sucked back along the inlet line when the pump stops with its inlet line under vacuum. Two stages driven on the same shaft are normal, achieving spectacular compression ratios of 10 000 000 or so.

One company (Edwards High Vacuum) is now marketing two innovations in the primary pump market. One of these is a dual-stage oil-immersed pump which has two switchable modes to optimise either its mass-flow throughput or the ultimate vacuum achieved. The other innovation is a revisitation of the 'scroll pump' principle. This avoids some of the oil-immersed vane pump problems while being usable in a much smaller size than the Northey claw pump described below.

For large pumps, the oil-immersed pump is challenged by the Northey interlocking claw concept, the so-called 'dry pump' (Figure 2.64b). Particularly in the semiconductor industry, where pumps are required to keep the vacuum process chamber absolutely clean, and where some of the gases pumped can react with the inside of the pump and its oil, these have important advantages. Three stages, with the first stage often a high-throughput Rootes blower, all driven from the same shaft, is a typical arrangement. Because the rotor tips and ends are not sealed with oil, these pumps are made with very fine clearances and must be carefully temperature controlled to avoid problems with clashing rotors. They are the most expensive type of vacuum pump.

The type that is rapidly becoming the standard high vacuum and ultrahigh vacuum pump is the turbomolecular pump (Figure 2.64c). This is a straightforward turbine pump normally arranged with multiple rotor fan wheels on a single shaft interleaved with stator fans. At the input (high-vacuum) side, the fans have large, wide, high incidence angle blades. The wheel and stator fan blades then progressively become narrower and of lower incidence moving down the shaft towards the outlet, reflecting the fact that they must pump denser gas.

Turbomolecular pumps necessarily rotate at high speed with small clearances between rotor blades and the body, and this is an arduous service for conventional ball and roller bearings. This has led to the adoption of magnetic bearings in many units (section 2.1.3). Turbomolecular pumps of standard design cannot pump gas up to pressures beyond a few millibar and are therefore only operated with a rotary or Northey pump to exhaust their throughput at atmospheric pressure.

There are two other kinds of vacuum pump, ion pumps and cryopumps, which are mostly seen in very low-pressure work and therefore of less interest to us here. The former function by trapping gases into a sputtered layer of metal in a plasma discharge electrode unit. The latter simply use liquid helium temperatures to freeze out gases onto simple baffles or into a liquid nitrogen or liquid helium cooled carbon absorber (section 2.5.3). Operational pressure ranges for vacuum pumps are given in Table 2.14.

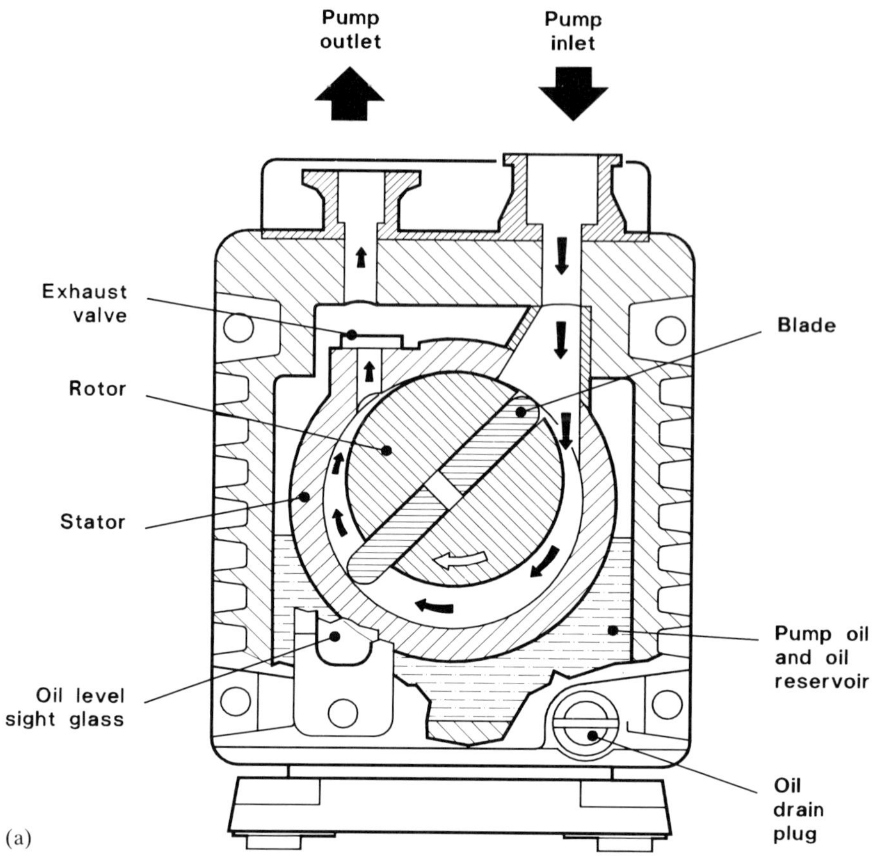
Pump
outlet
Pump
inlet
Exhaust
valve
Blade
Rotor
Stator
Pump oil
and oil
reservoir
Oil level
sight glass
Oil
drain
plug

(a)

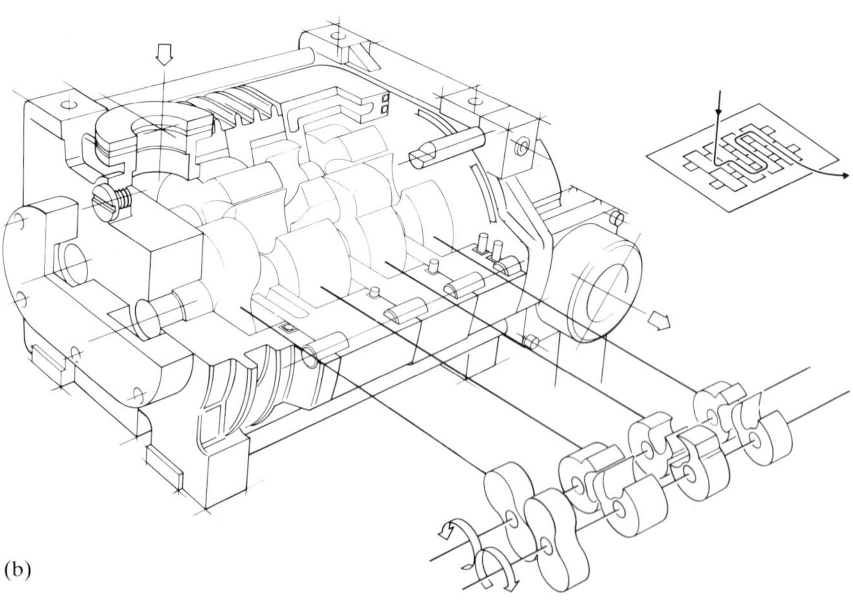

(b)

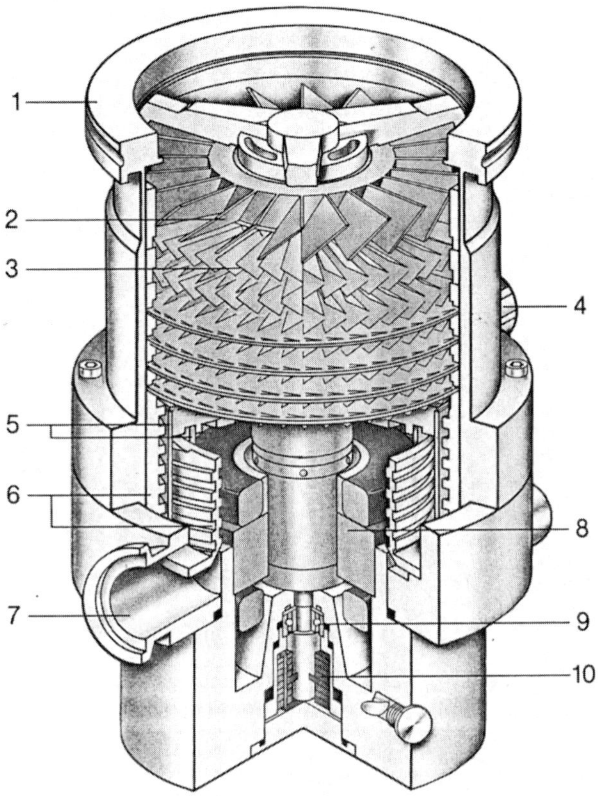

(c)

Figure 2.64 Mechanical vacuum pumps: (a) rotary vane pump (courtesy of Edwards High Vacuum); (b) multistage claw (Northey) vacuum pump (courtesy of Edwards High Vacuum); (c) turbomolecular vacuum pump: 1, high vacuum connection; 2, rotor disk; 3, stator disk; 4, venting connection; 5, molecular drag rotors; 6, molecular drag stators; 7, fore vacuum connection; 8, motor; 9, high-precision ball-bearing with ceramic balls; 10, operating medium reservoir (courtesy of Balzers Pfeiffer).

Gas ballast. Gases at low pressure can consist partly or entirely of vapours which will condense at room temperature and pressure to liquids. If this happens inside a mechanical vacuum pump it can be a problem because the liquid formed will mix with the pump oil, reducing its effectiveness and carrying vapour back to the inlet. This problem can be reduced by running the pump warm or hot; up to around 100°C is easily achieved. However, the use of a gas ballast is often more effective. The gas ballast, typically a small flow of nitrogen, can be added either to the second stage of a multistage pump or to an auxiliary inlet port in a single-stage pump. The effect of the added gas flow is to increase the temperature of the pump, which helps indirectly, and to increase the proportion of non-condensable gas in the pump outlet when it exhausts to atmosphere.

Table 2.14 Operational pressure ranges for vacuum pumps

Pump	Inlet/outlet (mbar)	Capacity ($m^3 s^{-1}$)
Water-jet ejector	20/1000	0.001–1
Gas or steam ejector	100/1000	0.01–100
Diffusion pump	100/1	0.01–50
Oil-immersed vane	10^{-4}/1000	1–500
Northey claw	10^{-4}/1000	50–500
Turbomolecular	10^{-10}/1	0.03–10

To illustrate how gas ballast works, consider a pumping workload consisting entirely of water vapour. This cannot be exhausted to atmosphere unless the pump is run at over 100°C. At high throughputs, the pump oil would soon be reduced to an opaque emulsion with a high vapour pressure, reducing its ultimate pressure. Contrast the situation with nitrogen ballast added at 100 times the inlet water vapour rate. The pump can now run at a safer 30°C and exhaust to atmosphere a nitrogen gas with 10 mbar water vapour, since the saturated vapour pressure of water is only 42 mbar.

2.5 Safety and environmental issues

There is a justifiably greater emphasis on safety today than existed even 10 or 20 years ago.[37] The larger gas suppliers all have technical safety departments who will be pleased to help any customer with safety issues, often free of charge. If there is in any doubt whatsoever about safety in a gas installation then professional advice of this sort should be sought.

In many cases, before the gas suppliers will connect up large supplies of gas, or dangerous or unusual gases, they will frequently insist on qualifying the user and his equipment as suitable. Occasionally they will perhaps carry this as far as a 'safety audit' or HAZOP: two species of precise analysis of the failure modes, their probabilities and consequences for a plant or piece of equipment.

2.5.1 Where does gas go in an accident?

The most frequent cause of a gas accident is gas escaping from its containment, either into another part of a process or to the atmosphere.

[37] Who today would inhale a lungful of heating gas and proceed to blow it down a Bunsen burner with the burner lit up? Yet this sort of lunatic party trick, which could easily kill the perpetrator, was regularly tried by schoolboys in England earlier in the century. Even official work was not so conscious of safety issues in the past. A common lecture demonstration was the filling of a heavy glass gas-jar with a stoichiometric mix of acetylene and oxygen. The audience would be rendered temporarily deaf by an ear-splitting detonation, whilst the gas-jar, inside a length of iron pipe, was reduced to a heap of powdered glass. A remote concrete bunker equipped with a robot to ignite the gas would probably be *de rigueur* today.

Where the gas goes in the event of a leak or equipment failure is often the most fundamental question in a hazard analysis. Is the gas lighter or heavier than air? Is there a pit in the floor or a trap in the ceiling where accumulation could take place? Is there a source of ignition in the place of accumulation? If an operator were rendered unconscious by the gas, would he fall into a zone of more danger or a zone of less danger?

That gas does sometimes get into very peculiar places is not in doubt. Much has now been written about the percolation of radon gas from uranium in some granitic areas such as Cornwall in England. Huge increases in lung cancer were related to radon inhaled by the occupants of houses with stone floors. Less well known (Hendrick and Sizor, 1992) are incidents due to old mine workings, some hundreds of years old. Mine workings undergo 'breathing', i.e. exhaling the carbon dioxide- and oxygen-depleted air contained in them, during excursions in atmospheric pressure. In some areas of northern England this gas, picturesquely known as 'stythe', is seen in houses above the workings even years after closures, and can be dangerous.

Nitrogen gas from cryogenic leaks flows downward; although N_2 is lighter than air at the same temperature, the low temperature of N_2 means that it will be denser than air. This has led to fatal asphyxia accidents when attempts have been made to fix cryogenic leaks occurring in pits or in enclosed buildings. Strong ventilation, i.e. opening doors and windows and if possible a powerful fan, is needed to clear such gas accumulations by blowing in fresh air if an enclosed or pit area must be entered.

Finally, along with elaborate precautions to ensure safety with gases should go a reasonable degree of prudence with respect to the more ordinary accidents it is possible to have with gas equipment. Probably the most common accident in the gas industry needs no more than an empty cylinder and a little ordinary carelessness: the toppled gas cylinder. A gas cylinder is sufficiently heavy (up to 100 kg or more, depending on the exact size and contents) to be capable of spraining ankles, breaking toes and causing unpleasant flesh wounds, irrespective of its contents. Gas cylinders should be secured to prevent toppling when not actually being moved.

2.5.2 Toxic and asphyxiant gases

When asked about gases, heating gases usually come to mind first. Second thoughts might revolve around the poisonous nature of some gases.[38] It is an unfortunate fact that very few gas mixtures are safe to breathe. Even gases which are not actually toxic may have other undesirable consequences and

[38] My grandfather saw chlorine gas used on the Western Front in Belgium in the First World War and talk of gases would quickly remind him of this, whilst millions of not-so-old people remember the gas masks that were carried, but fortunately never used, in the Second World War.

Table 2.15 Toxicity of gases

Gas/danger level	Comment
Nitrogen	Nontoxic, asphyxiant
Nitrous oxide TLV 50 ppm	Nontoxic, asphyxiant, anaesthetic (at 80% in O_2)
Butane TLV 800 ppm	Nontoxic, asphyxiant, mildly anaesthetic (at 25% in air)
Carbon Dioxide IDLH 5%	Slightly toxic, asphyxiant
Ozone IDLH 10 ppm TLV 0.1 ppm	Toxic
Chlorine IDLH 30 ppm	Highly toxic
Phosgene ($COCl_2$) OES 0.1ppm 8 h TWA	Very toxic

all gas mixtures containing insufficient oxygen will cause asphyxia. Some examples of typical classifications of gases and their toxic effects, giving the USA National Institute for Occupational Safety and Health TLV (threshold limit value) or IDLH (immediately dangerous to life and health value) are given in Table 2.14. The TLV is the maximum concentration permitted during the working day and is similar to the OES (occupational exposure limit) for 8 h TWA (time weighted average). As can be seen, descriptions do not follow a totally consistent pattern. Butane is probably a more powerful anaesthetic than N_2O, for example, whilst 'highly toxic' chlorine has a higher IDLH than merely 'toxic' ozone. Some of these discrepancies reflect general practice. For example, ozone is rarely used in concentrations higher than a few per cent, so leaks in practice are less serious than leaks of 100% Cl_2.

There are large numbers of gases that are so dangerously reactive or toxic that most countries insist on special licenses for businesses that use them. The military poison 'gases' (actually, most of them are liquids with an appreciable vapour pressure at room temperature) are some of these. After a initial crude attack in 1914 with cylinders of liquid chlorine, the deadly business of war moved on to expressly designed chemical poison molecules such as phosgene ($COCl_2$), Lewisite and mustard gas. These fairly simple compounds do have some peaceful uses. Phosgene is a relatively common chemical intermediate and other materials of this type are occasionally required in small quantities for scientific research purposes. Mustard gas, for example, can be used to induce mutations in cell genetic material.

The organophosphorus poison 'nerve gas' compounds Sarin and Tabun are much more complex molecules that work by interrupting normal biological cell function. They inhibit the enzyme cholineresterase used in biological nerve systems and are extremely dangerous. They are related to the organophosphorus insecticides developed in the 1930s and almost unused in war, save for some incidents in the Iran/Iraq war and a recent terrorist attack in Tokyo. Their ability to interrupt particular biological reaction sequences means that there may be occasions in which milligram amounts are needed in biochemical research.

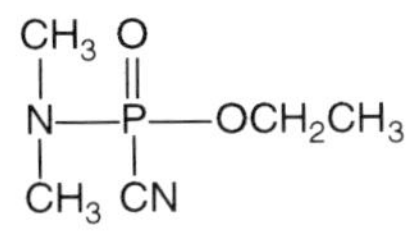

Tabun

There is general agreement worldwide about toxicity of gases. However, in each country there are separate and different regulations about TLVs and IDLHs, and levels vary somewhat, as do the descriptions applied to gases.

Double-pipe containment. Double-pipe containment is occasionally used for highly toxic gases. Longer runs of pipe, usually those in between the source cylinders and the equipments using the gas, are fabricated in two concentric pipes. The normal gas path is down the inner pipe. The outer pipe is purged continously with nitrogen and vented to an exhaust stack or to a gas scrubber. Ideally it will be monitored for toxic content.

2.5.3 Cryogenics

Safety with cryogenics is mostly a matter of common sense. If cryogenic surfaces are treated with the respect accorded to hot surfaces, then few problems will in general be encountered. Cool surfaces should not be touched except with gloved hands and contact with cryogenic liquids should be avoided.

Some additional phenomena need to be allowed for, however, relative to hot liquid handling; one is the welding of human flesh to cold metal and another is the brittleness of many materials at low temperatures. Most people have touched the walls of a domestic freezer and are none the worst for their experience. Similarly, most of the time, the metal components of a cryogenic liquid system, even those at very low temperature, can be touched briefly without apparent harm. This is because cold surfaces in a typical ambient atmosphere quickly condense moisture, accumulating a protective film of ice. However, just occasionally, the metal touched will be relatively clear of the condensed ice that quickly forms on a cold surface, and this clear surface is very dangerous. The welding of flesh to the cold metal is instant and often the cause of unpleasant wounds.

There is a temptation, when a moving component is rendered immobile, 'frozen', by low temperature, to hit it with a hammer. This must, however, be strenuously resisted because normally robust steel or plastic parts may shatter and the small problem of sticking caused by the low temperature may become a much bigger problem of broken components.

Leidenfrost phenomenon. There is an effect, the Leidenfrost phenomenon, which can give a degree of safety in handling cryogenic liquids. The occurrence of the Leidenfrost phenomenon is not confined to cryogenics and in fact occurs in many places in physics. Leidenfrosting is seen on scales from the largest to the smallest; from the collision of galaxies made of antimatter and matter, to the hovering of droplets of water on a hot saucepan. Basically, a cryogen such as liquid nitrogen (LN) dropped onto a much warmer surface forms and maintains a thin film of boiling vapour which isolates the LN from the surface. In this way, personnel are to some extent protected from small accidental spills and droplets landing on them from boiling LN because the droplets do not wet skin easily because of Leidenfrosting and tend to run off harmlessly, even more effectively than water from a waxy surface.

This can lead to a false sense of security, however. If the LN pools in the surface that it lands on or otherwise remains there long enough, it will cool that surface until the rate of vapour production slows. At some point the protective film of vapour disappears, allowing a suddenly much increased rate of cooling and resulting in painful and sometimes dangerous burns to personnel. Finally, it is worth mentioning that surfaces cooled in dry air by liquid nitrogen can condense liquid oxygen, which is itself a relatively dangerous substance (section 2.5.9).

2.5.4 *Flammability*

Many gases are inflammable and all flammable gases have the potential to cause explosions. Some gases ignite or explode spontaneously, such as silane. Gases specifically designed to ignite on contact with air or oxygen are sometimes dubbed 'hypergolic'. Compounds such as pentaborane or triethyl aluminium reliably ignite and have been used in small rocket motors (silane ignition is not reliable, particularly for pure grades).

Flammability is a very complex subject in general but has been intensively studied over the years. For combustion, certain conditions must be met which are characterised by more or less quantitative parameters:

- the fuel/oxidant ratio must be correct within limits: these are the upper and lower explosive limits (UEL and LEL);
- the oxidant stream must contain a minimum amount of oxygen: this percentage of oxygen is called the oxygen index;
- a minimum amount of energy must be provided to the combustion mixture: this is the ignition energy;

- a minimum temperature of ignition source must be applied: this is the ignition temperature.

Flammability of mixtures in air at atmospheric pressure, at least in safety matters, is the most common situation so this is mostly what is considered here. Flammability can be considerably enhanced by oxygen enrichment (section 2.5.9).[39]

Flames. It is possible to react together fuels and oxidants using a heterogeneous catalyst; 1% platinum on a ceramic fibre plate will allow the combustion of common fuels from around 300 or 400°C upwards. However, a catalytic system is not necessary because of the phenomenon of stable combustion flames.

A flame is the reaction zone between a fuel and an oxidant gas (Figure 2.65). The zone typically appears more or less stationary but is in fact a dynamic phenomenon formed out of the fast flowing gas mixtures. A conventional flame burns at a rate which equals the flow rate of the gases so it is in dynamic equilibrium. A flame is a self-sustaining chain reaction: once started, the flame generates heat, which raises the temperature of more fuel and oxidant so that it, too, combusts, generating more heat, which raises the temperature of more fuel and oxidant, and so on.

The most useful distinction that can be made is between 'diffusion' flames and premixed flames. In the former case, pure fuel flows into the atmosphere, and the flame is formed where fuel and air mix – a large, typically luminous, flame. In the latter case, the gases are premixed and allowed to flow from their containing tube into the atmosphere, where they are ignited. The fuel and oxidant are raised in temperature above the ignition temperature for the mixture as they approach the flame and the reaction zone begins. A complex molecular dance then occurs, with molecules losing and then gaining electrons and atoms, and the temperature rises rapidly, limited only by reaction kinetics to reach typically 1800°C or more. Within the space of a millimetre or so the luminous part of the flame has begun. Finally, reactions proceed to reach equilibrium and the final temperature (the 'adiabatic temperature' if the flame is not cooled) is reached in the bulk of the flame.

A majority of relatively stable species, such as carbon dioxide and water vapour, are then formed as ambient gas entrainment and radiation cools the gases beyond the tip of the flame to below 1000°C. Mixed in with stable

[39] Similarly, flammability is affected by pressure; the Victorian 'fire piston' was a demonstration showing just that. In the fire piston a piece of highly flammable solid material at the bottom of a cylinder, which can be made of glass, is ignited simply by compressing the air above it suddenly with a piston. This both raises the temperature adiabatically (see section 2.5.9 adiabatic ignition of oxygen mixtures) and raises the oxygen partial pressure, which itself renders the mixture more flammable. Fire pistons were one kind of portable convenient source of ignition before gaseous or petrol fuelled lighters lit by misch-metal flint or piezoelectric spark became available.

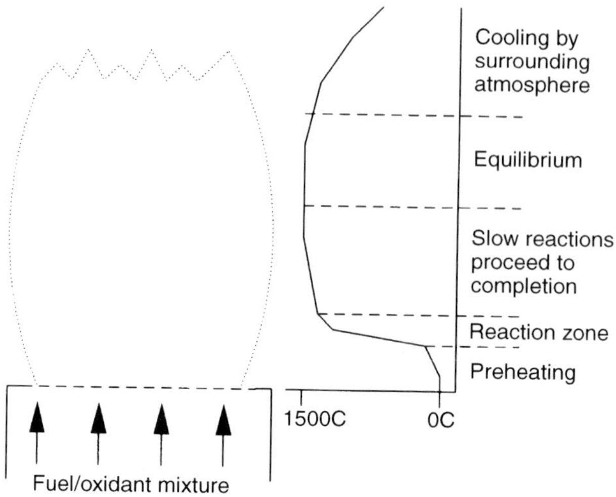

Figure 2.65 Simplified flame structure.

species will be low levels of undesirable gases, NO and NO_2 (collectively called NO_x) if air has been the oxidant, along with partially combusted fuel, sulphur dioxide from any sulphur in the fuel and, with hydrocarbon fuels, microscopic carbon particles. The heat released sustains the chain reaction and tends to make it propagate back down the supplying tube at a fairly well-defined velocity known as the 'flame speed'. If the flow speed is lower than the flame speed, the flame will travel back up the tube; this is called 'blow-back'. On the other hand, if the flow velocity is too large, the chain reaction cannot propagate back against the flow fast enough and 'blow-off' or 'lift-off' of the flame occurs. Stabilisation of the flame is achieved in small, simple burners by the cooling and flow slowing effects of the edge of the supplying tube. In larger burners specific measures such as a 'quarl' (a small ante-chamber into which a furnace burner fires), rapid rotation of the gas stream or a bluff object in the gas stream, are used. These cause flame recirculation, giving stabilisation.

Diffusion flames are only rarely used. Diffusion flames are typically large, luminous, unstable and, if using a carbon-containing gas, deposit carbon onto cool solid surfaces placed in them. Premixed flames are those normally employed in most applications. They are typically much less luminous than diffusion flames, are stable and are much smaller than diffusion flames with the same thermal output. Partially premixed flames are very common and they have two often distinct flame zones: a small inner zone of reaction of the premixture and a large, less luminous outer zone where oxidant diffusing into the still combustible reaction products of the inner flame allows further combustion.

There is a further distinction amongst flame types depending on their flow characteristic. While small burners (such as cooking burners or laboratory Bunsen burners) tend to have laminar flow, larger industrial burners show turbulent flow. In turbulent flames, the reaction zones described occur in small transient combustion cells formed in the intensively mixed gas. Turbulent flames are described as premixed or 'jet' flames. In the latter, although the fuel supply is separate from the oxidant, the turbulent mixing means that, in effect, they have characteristics very like premix flames.

Turbulent flames can achieve greater combustion intensity (throughput of reactants per unit volume of flame) than laminar flames and many industrial designs rely on this to achieve high rates of heating and effective heating at high temperature. High intensity flames are unstable and have a strong tendency to lift-off. One stabilisation system for intense flames is an 'oxygen pilot'. A small side stream of oxygen is used and the oxygen locally increases the flame speed, allowing a more intense flame before lift-off occurs.

Ultrahigh intensity burners sometimes use a combination of laminar or low-velocity flames with high-velocity highly turbulent flames. The lower velocity flames stabilise the unstable high-velocity flames, relighting them continuously.

To achieve a premix flame from a liquefied gas source involves vaporising the liquid stream effectively. Simple burners rely on the burner parts heating up to do this. Other burners use an ambient air-heated vaporiser. Hot-air balloons have special vaporiser coils near the lower part of the flame to ensure that the liquid propane fuel used is thoroughly evaporated into the gas prior to combustion (see Figure 3.45).

Rocket motors have a similar arrangement whereby liquid oxygen, and liquid hydrogen if used, is fed into pipes welded onto the outside of the thin metal propulsion nozzle, ensuring complete conversion into gas using the formidable heat of the rocket flame. The latent heat of vaporisation of the cryogens cools the rocket motor walls, allowing them to survive the intense conditions for longer.

2.5.5 Flammability limits

Most substances that are not fully oxidised and therefore inert will display a positive heat of combustion, according to standard chemical enthalpy tables. However, compounds such as perfluorinated alkanes have very low heats of combustion (less than 8000 $J^{-1}g$) and will not normally ignite. Other compounds may be so diluted with inert materials, as with the asbestos /rubber composition gaskets popular in gas systems, that their net heat of combustion is low. With these exceptions, all compounds are flammable and solids such as polymers PE or nylon (combustion heats 42 000 and 34 000 $J^{-1}g$ approximately, respectively) will burn vigorously. Solids burn by reaction at their surface with the air or oxygen in the same way as a

Table 2.16 Flammability limits of gases in air

Gas	LEL (%)	UEL (%)
Methane	5.0	15.0%
Propane	2.1	9.4%
Butane	1.9	8.4%
Acetylene	2.5	80.0%
Hydrogen	4.0	74.2%

candle, the tendency to ignite being formalised in the oxygen index measurement. Liquids and gases can also be ignited in this way and burn at the surface, giving a diffusion flame. However, in the case of vaporisable liquids or gases they can mix with air or oxygen and combustion can take place within the bulk of the mixture, often explosively. The tendency to burn in bulk in this way can be given a quantitative shape in the form of 'flammability limits'.

If a gas is mixed with air in stoichiometric proportions, i.e. the oxygen in the mix is exactly that needed to oxidise the fuel, then, if the gas is at all flammable, it will ignite. For instance, a mixture of 29% hydrogen to 71% air ignites explosively. However, if the gas is mixed in proportions far from this ideal mix, it may also ignite explosively. The limits of percentage in air which ignite are an important safety parameter. The lower explosive limit (weakest fuel mixture) is important in situations where fuel gas leaks from a pipeline or vessel into the air, where there may be a source of ignition. It is used in setting flammable gas detectors; these are frequently set to alarm at 20 or 40% of the lower limit. The upper explosive limit is important in situations where there may be an ignition source within a fuel stream or vessel. Furnaces for steel heat treatment, for example, involve temperatures well in excess of ignition point with a fuel gas atmosphere. In the event of incomplete purging of air from the system, for example during start-up, the proportion of fuel in the mix may sink below the upper explosive limit and become dangerous.

Some gas flammability limits are listed in Table 2.16. Outstanding amongst them are the extraordinarily wide limits of hydrogen and acetylene. The flammable limits of mixtures are not in general predictable and should be measured in cases where they are important. However, early in the investigation of combustion processes in France, le Chatelier proposed an approximate formula on the following lines. A mixture is flammable if

$$(N_1/\mathrm{LEL}_1 + N_2/\mathrm{LEL}_2 + \ldots) > 1$$

where N_1, N_2, . . . are the mole fractions in the atmosphere of flammable fuels 1, 2, . . . and so on. This common sense and simple formula can, however, be quite inaccurate in particular cases.

2.5.6 Ignition energy

If a gas mixture is subjected to a high-temperature spark, it may not ignite. Anyone who has snuffed out a lighted match in a saucer of paraffin (an experiment to be tried only outside with small quantities) will have an idea of what is going on here. That same saucer of paraffin will be easily set alight by a roaring gas burner, whose flame may well be cooler than the lighted match but whose energy is larger.

For most gas mixtures flames only occur if a certain minimum amount of energy is supplied: the ignition energy. For the hypergolic fuels, the ignition energy is effectively zero; a passing cosmic ray or a random conjunction of molecules is sufficient to start the chain reaction. Examples of hypergolic fuel mixtures are impure silane and air or unsymmetrical dimethyl hydrazine and dinitrogen tetroxide. The latter mixture is frequently used in aerospace applications where cryogenic tanks are not wanted.

For some gases, such as hydrogen, the ignition energy is exceptionally low, just a few microjoules. This range of ignition energy can easily be generated in static electrical discharges which occur in everyday activity.[40] For other fuel gases, ignition energies are in the millijoule range but still smaller than the static voltages which can occur. Proper precautions to avoid all high voltages are necessary to avoid accidental ignition of gases where it is necessary to use apparatus in an inflammable atmosphere (section 2.5.8).

The minimum ignition energy is usually estimated by gradually increasing the size of an electrical spark in a stoichiometric mixture of fuel and oxidant. The spark energy is related to the energy stored in a capacitor discharged across the spark gap.

2.5.7 Oxygen index

The oxygen index (OI) of a substance is the lowest percentage of oxygen in the air which still allows self-sustaining combustion of that substance. It is determined by igniting a sample strip in a steady stream of O_2/N_2 mixture and seeing whether the sample stays lit like a candle or extinguishes. Like the LEL and UEL, the OI is a fairly quantitative concept, although OI is dependent on pressure and is affected by a number of other considerations as well as ignition energy and temperature. It is interesting to note that many polymers, even those such as PTFE which are generally reckoned to be inert and nonflammable, have an oxygen index, although only at elevated pressure in the case of PTFE (Table 2.17).

[40] A plastic biro pen rubbed on a shirt sleeve will typically generate enough electric charge to deflect a stream of water from a tap or pick up small pieces of paper electrostatically. Taking out a biro to note where a flammable gas is leaking out of a pipe could thus be disastrous.

Table 2.17 Oxygen index of polymers

Polymer	1 bara (%)	20 bara (%)
Nylon 6,6	30	20
Neoprene	35	26
Fluoroelastomer	61	31
PTFE	(>100)	43

Fluoroelastomers and PTFE are regarded as satisfactory materials for oxygen systems because of their high oxygen index. Copper, nickel and their alloys are regarded as the most satisfactory materials for arduous duty in oxygen systems, such as pumps, valves and other moving parts, and where a finely divided material is needed, such as in filters.

Another means of measuring the sensitivity of materials to ignition in oxygen systems is the autoignition temperature. Finely divided samples are slowly (200°C per minute) heated in a heavy-walled reaction vessel at a standard pressure, often 100 bar, until they ignite spontaneously. Most polymers ignite under these conditions at between about 150 and 350°C, although PTFE does not. This sort of test gives a similar ranking to the oxygen index in most cases.

2.5.8 Safe working in flammable atmospheres

It is possible, and often essential, to be able to work in areas where there may be a flammable atmosphere continuously or intermittently present. If it is necessary to work in an atmosphere containing a flammable mixture of hydrogen, precautions against accidental ignition must be stringent. The statutory requirements for flammable atmosphere working (i.e. Zone I requirements) were therefore drawn up with this need in mind.

In a Zone I area all high-power electrical apparatus must be completely isolated from the gaseous atmosphere. It should be either hermetically sealed or well-sealed and then pressurised continuously from within by an inert gas such as nitrogen. Low-power electrical apparatus may be either isolated or 'intrinsically safe'. The latter is achieved by ensuring that the maximum voltage in the circuit is only a few volts and that the maximum instantaneous current possible is in the milliamp region or below. Small electronic devices, for example, may be used if they contain no large capacitors and have batteries of a high internal resistance. (Caution must be sounded over electric torches and flashlights, however; on these the bulb must be isolated, as the 2500°C+ tungsten filament in it would be a splendid source of ignition.) Barrier circuits are available, using Zener diodes and resistors, that limit the voltage and current that can be transmitted into the zoned area along wiring.

All metal surfaces should be electrically bonded together so that ignition effects from static discharges or electrochemical currents can be avoided. Plastic materials should be of the slightly electrically conductive 'anti-static' type. Finally, the use of ferrous tools may need to be controlled. Striking of ferrous alloys on other metals, stones and other hard materials can lead to pyrophoric particles. These are microscopic particles of iron which ignite spontaneously in air or oxygen and can easily lead to ignition of a flammable mixture.

2.5.9 Gas release accidents

Burning gas leaks. Gas suppliers should always be consulted as to fire precautions, along with local fire service officials if appropriate. Different gases will require different approaches, as will different sizes of installation. However, some general principles are perhaps useful to know.

It is not infrequently the case in large and extensive pipeline systems that small leaks of flammable gas occur. A small burning leak, especially if it is outside or in an accessible place inside a building, is often not particularly dangerous and if prompt action is taken it can be easily extinguished before real problems occur. First, the gas supply must be shut-off, and, if possible, the pipeline depressurised via a suitable vent stack. The fire can then be extinguished in the way appropriate to the circumstances of the leak. It should be pointed out, however, that extinguishing a gas leak fire, if the gas supply is still on, is dangerous. If the fire is successfully extinguished, gas will carry on accumulating and will mix with air, presenting the risk of an explosion much more dangerous than the original gas leak fire.

Hydrogen installations are particularly subject to leaks, because H_2 leaks very readily, and installations such as thin membrane purifiers and arrays of many hydrogen cylinders offer many opportunities for leaks. Hydrogen installations are often placed outside where the impact of a small leak is normally very small. One of the difficulties of H_2 leaks is that they are exceptionally difficult to see. Outside in bright sunshine the flames are almost invisible. Old fashioned manuals advise the use of a piece of paper or a rag on a stick as a flame detector. With glowing carbon particles and atoms added in this way, a hydrogen flame is indeed very much more visible.

Detonation, deflagration and BLEVEs. The effect of an ignited gas cloud depends on the speed at which the flame front travels. Gas/air mixtures can burn rapidly emitting copious heat but causing little immediate mechanical damage. BLEVEs (burning liquid expanding vapour explosions) are a special case of this and are typically seen where a container of flammable liquefied gas such as propane is ruptured and ignited. The mechanical damage seen is relatively slight, extending only a little beyond the zone of the visible fireball. In these cases the flame front in the mixture of gases travels slower than the speed of sound so heat released by the ignition is dissipated smoothly in the expansion of the mixture. This situation is referred to as a deflagration explosion. As the burning mixture approaches a stoichiometric composition, the speed of the flame front will increase and a deflagration explosion becomes more dangerous. If a sealed vessel full of such a mixture is ignited then the vessel may well be ruptured by the adiabatic expansion.[41]

Much rarer are those accidents where a large volume of near stoichiometric mixture is accumulated before ignition occurs. In these cases, the flame front travels faster than the speed of sound and, with or without a containment vessel, there will be detonation explosion. Large volumes of gas can be burnt and high temperatures and pressures accumulated before this energy can be dissipated by expansion of the surrounding air. As a result, the energy of the burning gas is efficiently converted into an expanding shock wave of immense destructive power. Military high explosives are designed to produce this kind of detonation explosion. Military devices (fuel-air bombs) have been devised to utilise the power of gas-air explosions. Gas burner and boiler manufacturers, and fuel gas suppliers are often equipped with special explosion test chambers. These are used to investigate the effects of small detonation and deflagration explosions as part of their development efforts towards safer burner and boiler design and use.

Oxidant accidents. Two of the common oxidants in industry are relatively harmless gases: oxygen and nitrous oxide. Unlike reactive oxidants, such as hydrogen peroxide or fuming nitric acid, both the gases are stable when mixed with many oxidisable materials. This makes them in some ways more hazardous than the obviously perilous nitric acid because they can leak out in large quantities before ignition occurs.

Liquid oxygen (LO) has more potential than other oxidants for big accidents because it is used in such large quantities on a routine basis. LO is easily soaked up by organic material, forming a 'low explosive'.[42] A typical LO/organic mixture will release more energy per unit mass than an explosive, e.g. combustion of

nitroglycerine $= 6.8\ \mathrm{kJ^{-1}g}$
octane + O_2 mixture $= 10.6\ \mathrm{kJ^{-1}g}$

[41] If the vessel is vented a great rushing sound, like a jet engine, will be heard as a very rapid expansion takes place, followed by the expulsion of a sudden spout of hot gases.
[42] For example, one of the party tricks performed by schoolmasters used to be the soaking of a cigarette in liquid oxygen. An ordinary Benson and Hedges King Size, normally smouldering for 10 minutes or so, is gone in two seconds, showering sparks everywhere like a firework. Given this behaviour from a few grams of material, it is clear that the contents of a 50 tonne liquid oxygen tank soaked in an organic substance is a major hazard.

The flammability limits of inflammable substances with oxygen are wide, ignition energy and temperature is low, flame temperature is high, and flame speeds are high. These effects all conspire to make a fire or explosion in the presence of oxygen both likely and serious.

Lubricating grease or oil should not be applied to compressed oxygen gas systems. The density of oxygen at 230 bar is only about a quarter of the density of liquid oxygen and high-pressure gaseous oxygen tends to dissolve and form an intimate mixture with organic materials. A typical compressed oxygen failure occurs when an oxygen system becomes contaminated with an organic compound such as lubricating grease. The grease may have been applied with good intentions, for example, to a sticking valve. However, high-pressure oxygen dissolved in the organic consitutes an explosive, possibly even a detonating explosive. A small but very dangerous explosion then occurs, which will rupture the pipework system and perhaps fire out small parts, such as valve components, like bullets.

It is essential not only not to use lubricating greases and oils in oxygen systems but also to eliminate any adventitious organic compounds by a rigorous cleaning procedure for all components used in oxygen service. There are standard degreasing/cleaning procedures for oxygen components defined in the national and international codes, and these should be followed. It is worth noting that cleaning of a complete system, after assembly, may well be impossible. There may well be dead-legs, parts of cylinders and loops of pipe which cannot be degreased *in situ*. In the event that a system is contaminated with oil or grease, it may well be necessary to dismantle and degrease each component, followed by reassembly.

As noted in section 2.4.2, the composition of seals in oxygen systems is also problematical, with perfluorinated polymers most suitable, at least for less extreme service. Asbestos and related composition seals are often specified for oxygen service as these are essentially non-combustible, at least if they are not contaminated.

A much rarer failure occurs when the gas velocity of oxygen down a pipe is very high, perhaps because of a serious leak somewhere else. In this case, iron or steel pipes can be ignited, and once ignited will burn and melt in the stream of oxygen. The pipe then behaves exactly like a thermic lance and can cause great damage. To avoid this possibility, oxygen pipes should be made from brazed copper or high-alloy stainless steel, such as 316L, or restricted in pressure and flow rate. Even stainless steel is not recommended above 70 bar. An accepted rule of thumb is that for steel pipes at lower pressures, below 40 barg, flow velocity should be kept below 25 m s^{-1}, whilst at higher pressures flow velocity should be kept even lower, less than 8 m s^{-1}. All oxygen valves and fittings for service at higher pressures should ideally be made from brass, copper or 316L. However, note that local construction codes should in all cases be obtained and followed in the design, construction and operation of oxygen systems.

Sources of ignition in oxidant accidents. It is usually not possible to exclude all ignitable material from an oxygen system. Polymer seals and flexible hoses, for example, will always have some degree of flammability in high-pressure oxygen; even Teflon will ignite at high enough pressures with sufficient ignition effort.

It is surprising, when a survey of investigated oxidant accidents is made, how many ways there are of generating sufficient heat to give ignition. The most unlikely mechanisms have been implicated; for instance, heat generated by particles of rust or dust in the pipe hitting an obstruction has been shown to be sufficient under adverse circumstances in a high-pressure oxygen pipe. The frictional heat generated by turning a valve in a tight stem seal has also been shown to yield enough heat; these sorts of accidents are particularly unpleasant because the operator turning the valve may well be injured. It should be remembered that many powdered materials are 'pyrophoric' in pure oxygen. Just as tiny particles of misch-metal from a lighter 'flint' ignite in air as they fall through it, so minute steel particles, for example generated by abrasion, can ignite in oxygen. Electrical or electrostatic sources of ignition are commonly blamed for igniting volatile organic vapours in air and although such sources should have been eliminated from contact with oxygen systems by design they can occasionally cause accidents.

Adiabatic compression ignition in oxygen systems. The most common source of ignition in oxygen systems at high pressure, however, is probably adiabatic compression heating following rapid opening of a valve onto a restricted low-pressure system. As discussed in section 2.4.7, this effect occurs because a small volume of low-pressure gas is sufficiently confined (perhaps it is open to the atmosphere but only through a relatively small orifice) that when high-pressure gas is suddenly admitted, the low-pressure gas is compressed, with a flow pattern that in narrow pipes does not allow much mixing, unlike the case in cylinder filling. Whereas in cylinder filling transient temperature rises of a few hundred degrees or less are typical, in pipes the full 1000°C or more calculated above might be seen.

In any oxygen system with a high-pressure source, valves opening flow from that source should ideally be slow-opening screw-jack operated seat valves; cylinder valves are obviously suitable in this regard. Valves with particularly slow-opening characteristics (typically by an extended plug, somewhat like a needle-valve type) are available which will guarantee particularly slow opening.

The mode of operating of a manual system (or, equivalently, the software in an automated valve system) can also provide protection against adiabatic compression ignition problems. For instance, procedures can ensure that all valves which may have high pressure behind them are opened slowly or parts of the system which include flexible hoses are always maintained at low

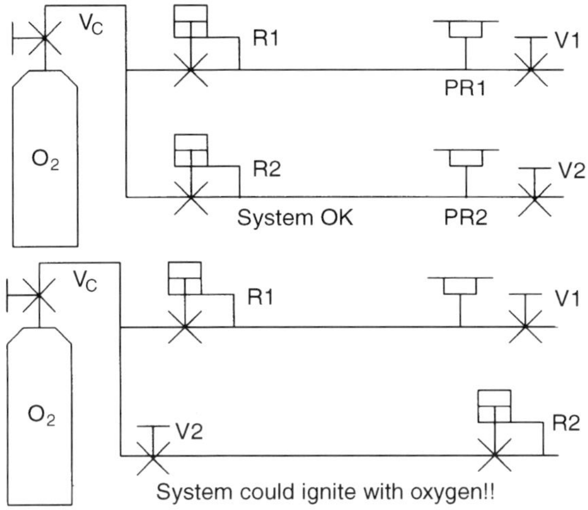

Figure 2.66 Design considerations in a small oxygen installation.

pressure. However, it is most unwise to rely on procedural protection; it is far better to design a system which cannot easily be put into danger. An example will show the sort of design considerations that apply (gas suppliers as well as national guidelines and regulations such as ASTM or CGA should be consulted before designing or modifying oxygen systems which involve any high-pressure parts).

In Figure 2.66 a single oxygen cylinder or bank of cylinders is used to provide relatively high pressure oxygen at two pressures to two user stations. High flow ball-valves are specified for V1 and V2 so that large flow rates can be catered for. System 1 will be safe in most circumstances; opening Vc will not cause adiabatic heating in the diaphragms of the regulator because it is a slow-opening type. Opening V1 and V2 does not cause a problem because the pressure (< 30 bar) behind them is unlikely to be high enough to cause a problem at the user station. In the event of either of regulators R1 or R2 leaking into the downstream (low-pressure) side when V1 and/or V2 are off, the pressure relief valves PR1 or PR2 will lift, thus assuring low pressure behind V1 and V2. V1 and V2 could be paralleled by screw-operated plug valves which have to be opened (slowly by their nature) before opening V1 and V2 for high flow rate, providing protection against adiabatic heating in the user device.

System 2, although using the same components, has the potential for an oxygen ignition accident. If Vc has been opened and V2 is then opened suddenly, high-pressure oxygen from the cylinders and the pipeline to V2 will compress the oxygen just before R2, and R2 may ignite or even explode.

Other rules have evolved from the gas companies' long experience with making and using oxygen systems based on the science explained above, for instance:

- to limit oxygen gas velocities pipes should be generously sized and places where velocity is artificially increased over the mean, such as at throttling valves, must be made of inert materials;
- high-speed rotating equipment should be designed with positive clearance under all service conditions to avoid frictional heating;
- pressures used at all stages in a system should be minimised.

Nitrous oxide and other oxidants. In general, nitrous oxide is a very safe gas to handle, with fewer problems than the more ubiquitous oxygen. It does show all the oxidant accident potential of oxygen, however, with a lesser degree of reactivity giving a greater safety margin. (However, it should be noted that some rubber materials are decomposed by nitrous oxide, so caution must be exercised with seals.)

In addition to its oxidant properties, nitrous oxide has been known to be the cause of accidents due to thermal decomposition. Occasional failures of this type happen as follows. A section of pipe carrying nitrous oxide is seen to be hot, perhaps for no apparent external reason. This section of pipe quickly becomes red hot and then fails, releasing a large volume of nitrous oxide at very high flow rate, together with a shower of molten or burning metal. This sort of incident can occur because nitrous oxide, like acetylene, is an endothermic gas, i.e. it has a large positive heat of formation ΔG (see Kaye and Laby, 1995). When it is decomposed at a few hundred °C into its constituent nitrogen and oxygen, it releases heat, which causes further decomposition. The released oxygen may result in combustion of pipe material, further exacerbating the situation.

Finally, it should not be forgotten that other reactive gases have the potential to cause oxidant accidents. Fluorine is such a powerful oxidising agent that most ordinary substances spontaneously ignite or explode. Only specially passivated stainless steel systems are normally suitable and gas suppliers' specific recommendations for handling this exceptionally hazardous gas should be adhered to precisely. Nitric oxide and nitrogen dioxide are both relatively unstable compounds, easily yielding their oxygen atoms in combustion reactions. The other commonly found oxidant gas, chlorine, also has the power to support vigorous combustion and, furthermore, combustion in chlorine generally leads to the formation of copious quantities of hazardous hydrogen chloride gas.

Reactive gas accidents. Purging of all reactive gas pipelines is essential, with effective venting during purging. Ideally, purging is by a pressure or vacuum cycle (section 2.5.10), with care taken in venting of the purged gas.

Failing this, purging for a much longer time by passing an inert gas flow will usually be acceptable from a safety point of view.[43]

Silane plants and other dangerously pyrophoric or explosive gas plants are commonly constructed in 'friable buildings'. Operators observe from behind solid walls with heavy armoured glass windows. The outside walls and roof of the building are light, however, to allow any pressure waves to escape outwards and reduce damage to the operators' room. Ultrapure silane, after its distillation step, is particularly dangerous because it ignites spontaneously only occasionally. It can therefore build up to a large volume before ignition, often explosive, takes place. Frequently, the area around potential leak sources is exhausted into a duct where there is a continuous source of ignition, e.g. an electric spark or flame. This 'burn box' then guarantees that ignition will occur early during the development of a leak, so that any explosion which may occur will be very small. If a large flame is used, the burn box also acts as a scrubber, oxidising any suitable reactive gas present to less harmful oxides.

Helpfully, leaks of many reactive gases can be easily located because of the reactivity of the gas. Many reactive gases react with moisture in the air to produce a mist of particles or droplets (the boron trichloride mentioned earlier is an example). Small leaks of acid gases, such as hydrogen chloride or sulphur dioxide, can be easily seen if a sponge soaked in ammonia solution is brought near to the leak; a tell-tale white smoke will appear.

2.5.10 Safe equipment design

Materials selection for gas pipelines and vessels was dealt with in sections 2.4.1 and 2.4.7 but it may be worth saying something in addition about materials for cryogenic service with respect to safety. Many properties of materials are better at low temperatures but some are not. The problems of materials selection for cryogenic use boils down to two prime considerations:

- will the material become undesirably brittle at the operational temperature?
- will thermal contraction during cool-down from room temperature match other components or, if it will not, has the component been designed to allow for this?

Thermal expansion is more serious in cryogenics because of the simultaneously increased stiffness and brittleness of materials; greater strains are induced and these strains are more likely to lead to failure. Whereas even

[43] For example, boron trichloride hydrolyses with moist air to form a dense white cloud of boric oxide and hydrochloric acid droplets. On purging with nitrogen a very large cloud of unpleasantly acrid white smoke will appear before the nitrogen purge finally exits the pipe.

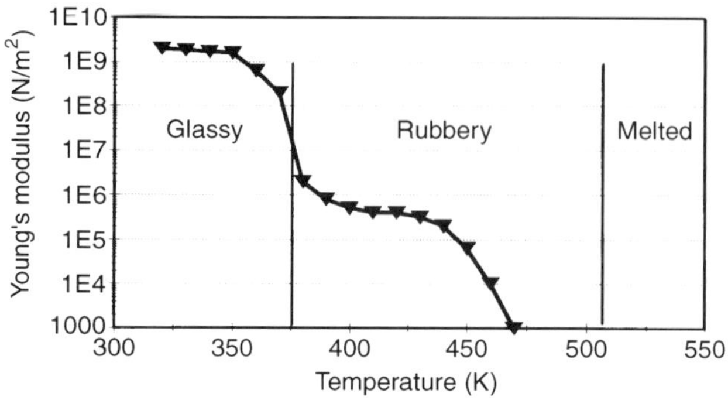

Figure 2.67 Elasticity of polystyrene showing glass transition from glass to tough polymer to rubber and liquid states.

serious mismatches would be unimportant in most applications where temperatures are cycled 200°C upwards, this is not necessarily the case for 200°C downward cycles.

Metals with body-centred cubic (BCC) structures are in general unsuitable for cryogenic service, as they become as brittle as glass long before LN temperatures. BCC metals include, unfortunately, most grades of steel. Carbon and many alloy steels are serviceable in cryogenics only if very lightly loaded because of these problems and are only used if economic factors are overwhelming. Consequential failure due to cryogenic embrittlement of steel or other parts because of a cryogenic spill should always be borne in mind in confined layouts. A small LN spill or leak may lead to excessive cooling of the outside of nearby parts under tensile stress (pipes under pressure, steel beams or cables, motor or compressor castings, for example) leading to their failure in service.

Similarly, most plastic polymers become very brittle at a little below room temperature. The movement of one polymer chain past another, which accounts for the flexibility and toughness of polymers, becomes increasingly difficult as temperature is lowered, until the polymer becomes a rigid solid with a random arrangement of atoms, i.e. a glass. The so-called 'glass transition' of most useful plastics is below room temperature but only a little below for most. PTFE is useful to the lowest temperatures (Figure 2.67).

Polymers also have the problem that they have large coefficients of thermal expansion and poor thermal conductivity. This means that rapid cooling, for example dunking them in liquid nitrogen, can result in them undergoing differential thermal expansion leading to distortion and cracking. Polymer adhesives frequently peel off adherent surfaces after cycling between room temperature and LN.

Safety in pipework. A set of guidelines for safe pipework is a standard part of the stock-in-trade of industrial gas firms. These guidelines will cover:

- labelling: labels should be attached every few metres of pipeline and given a colour code if possible;
- end-fittings: ideally these should be selected to match unique gas cylinder or tank fittings so that the wrong gas cannot be connected;
- location: location of pipework in the ceiling of a factory, for example, generally leads to warming of the gas, which may be useful when the gas has come from a cryogenic source; accessibility for fixing leaks should also be a consideration for joints in the pipework;
- material suitability for gas;
- material suitability for pressure;
- suitability for contaminants in gas: e.g. particulates in silane leads to the use of tied diaphragm regulators so that blockage and leakage cannot take place with particles underneath control valve so easily;
- bending, minimum radius, type of bending machine to be used, anneal state of material needed;
- joining: welding, compression joints, flanges, soldering, grades of solder;
- flow rates: e.g. flow rates of high-pressure oxygen through steel pipes allowed are limited.

It should not be thought that all of these guidelines are completely hard and fast but any deviations from these mostly well-thought out and time-honoured rules should be carefully justified and agreed with everyone concerned.

Purging. Purging of a gas system means the replacement of the initial gas mixture (often air) in a container or pipeline network with the correct process gas at the highest possible purity. The purging of flammable or toxic gas systems is necessary for safety reasons. There are three basic ways of purging:

- flow purging: simply passing the process gas through the network;
- pressure purging: the system is repeatedly pressurised and then let down to a lower pressure (typically atmospheric);
- vacuum purging: the system is evacuated then backfilled with gas, sometimes repeatedly.

Flow purging is the least efficient because gases do not flow as perfect plugs; there is always a slow-moving boundary layer. Also, any dead-legs in the system will act as long-term reservoirs of impurity. Pressure purging and vacuum purging deal with these problems, with vacuum purging being by far the most efficient (Figure 2.68).

Purging theory follows a simple harmonic progression. If the maximum pressure is P_+, and the lowest pressure is P_-, then the concentration of impurity C will reduce from an initial C_o in the following way, for n purge cycles

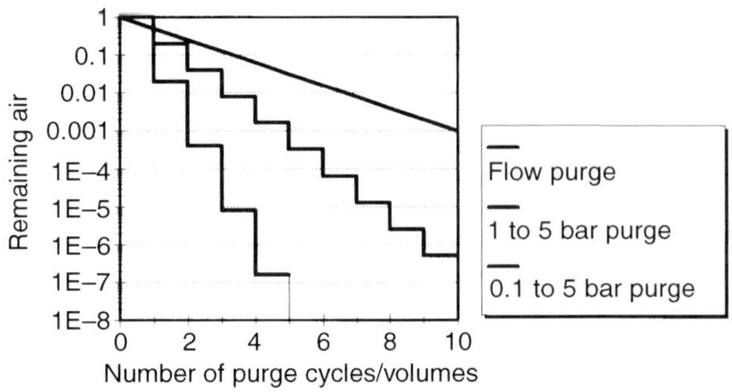

Figure 2.68 Purging a gas line of air by vacuum and pressure.

$$C = C_o\,(P_-/P_+)^n$$

Thus a vacuum purge of 1/1000 bar is equivalent in efficiency to pressure purging to an impracticable 1000 bar. With 1/1000 bar vacuum and 10 bar process gas pressure, the concentration of C_o will decrease by a factor of 10 000 for each cycle in this simple theory. With a simple 1 bar vent each time instead of vacuum, C_o will go down only by a factor of 10 for each cycle.

The harmonic progression of purging depends on 'perfect mixing' of the added gas to the lower pressure gas in the pipeline. Clearly this will not be the case if pressure cycling is carried out from one end of a long pipeline; the 'dead-leg' end of the pipeline will behave partially as a plug flow and simply expand and contract as the single-ended purge proceeds. If a line can be pressurised from one end, and vented from the other, purging will be much more effective and better than a perfect mixing assumption.

Naturally, practice is more complex than this theory indicates. For example, if the impurity is water, this tends, at ordinary temperatures, to lie in substantial amounts in a thin film over the inside of a pipeline or cylinder and will only slowly evolve under the influence of heat and vacuum. Heat and vacuum must therefore be used on cylinders for ultrapure semiconductor gases where ultralow moisture content is needed.

Bursting discs and pressure relief valves. No gas system can withstand infinite pressure and in fact most systems cannot withstand the approximately 1000 bar pressures which liquid nitrogen can exert if trapped in a 'dead-leg' and evaporated, so relief valves or bursting discs are essential.

For high-pressure systems, 'bursting discs' (thin diaphragms of a corrosion-resistant metal) can be fitted into special mounting jigs. These act as a weak link in the system, bursting before any major component can be damaged and releasing the gas pressure. Only a small amount of debris is ejected when a disc blows; an important safety feature, although the event is spectacularly noisy and not a little frightening for nearby spectators.

Bursting discs are carefully stamped out of metal sheets, with a testing regime to ensure that they will burst at the correct pressure. Small variations in alloy composition and in heat treatment or rolling can lead to big variations in burst strength, so very careful quality control and testing is essential. They are available only in a limited number of pressure ratings matching common requirements, as they are economically made only in large numbers. They should always be mounted in a supplied or recommended holder, as even small changes in the holder design may affect the effective pressure rating. Related to bursting discs are the fusible plugs often fitted to cylinders. These are made of a solder alloy which melts and releases gas at modest pressures in the event of a gas cylinder being overheated in a fire.

At very high pressures (100 to 10 000 bar), bursting caps or tubes may be seen. These are more common on hydraulic than gaseous systems because of the problem of debris ejection. They can be machined out of solid rod material and made highly predictable. A bursting cap is simply a machined dead-end with an accurately machined groove around the pipe circumference below the end, giving a weak point of known width and thickness. The end cap detaches on bursting and must be retained, however. A machined piece of thick-walled tubing with an accurately made thinner walled section can also be employed, which splits axially on overpressure. In addition, at the highest pressures, bursting discs may be seen which are thinner around the edges than in the middle. Provided this is done carefully, it offers the advantage that the disc will reliably burst leaving a completely clear flow path and maximum venting flow (e.g. Anon, 1975).

Lower pressure systems, and higher pressure systems which need a resettable pressure relief system, employ spring-loaded check valves. The standard type of pressure relief valve is related in its mechanical design to the check valve described in section 2.4.2. In it a spring holds down a seat-type valve; in the event of excessive pressure (more than the 'cracking pressure') the seat lifts and the gas is vented. Simple relief valves reset when the internal gas pressure falls to a little below the cracking pressure, allowing the spring to close the seat valve. Other relief valves require a reset pin or button to be pushed to reseat the valve; this is only possible when the internal gas pressure has fallen.

Pressure relieving devices of some type should be employed:

- where cryogenic liquids might become trapped;
- where chemical reaction or increase in temperature might cause a large increase in pressure in a trapped volume;
- where the failure of a pressure regulator might result in overpressurisation of the system downstream of the regulator;
- wherever there is a large vessel or pipe whose failure would release a large amount of pneumatic energy;
- on systems used to contain a high-pressure liquefied gas such as CO_2.

Relief valves must not only be set to relieve at the correct pressure, they must be sized to allow a reasonable relief flow rate. In most cases the manufacturers give flow curves but approximate calculations are simple and reasonably accurate on a standard orifice plate basis (see, for example, section 1.3.8) for many designs of bursting disc or valve.

For systems which may rise rapidly in pressure, reactor vessels with rapid exothermic reactions, for example, the size chosen must be very generous, as reactions generally increase in rate with rise in pressure and temperature. A rapid build up of P and T thus leads to a 'thermal runaway' and bursting of the reactor vessel if relief is not rapid. For other systems, a cryogenic tank is an example, where the maximum possible required relief flow rates are more easily calculated and in any case more modest, much smaller relief valves or bursting discs are appropriate. An overlarge bursting disc could itself lead to problems, especially as the bursting disc, unlike the relief valve, will not reset after a while but will allow all the gas in a system to escape. The huge cloud of freezing fog produced when tons of LN is released suddenly, the possibility of an oxidant accident with a sudden massive release of oxygen, and the toxicity of a large release of carbon dioxide are all examples of where a bursting disc actuation could itself cause a problem.

Sizing of pipes leading to pressure relief valves must always be generous. If possible, relief devices are located immediately adjacent to the largest vessel or pipe in a system. Consideration must also be given to where the relief valve discharges to. Ideally a tall, large-diameter vent stack discharges the relieved system into the open atmosphere several metres above where any human might possibly be.

Gas storage location. Gas cylinder location and installation should never be arbitrary. Toxic and flammable gases can often conveniently be located outside along with a primary pressure regulator, a rain shelter above and low-pressure gas piped inside to the point of use. Any leaks during cylinder changeover are then easily diluted in the open atmosphere. Cylinders should in any case be secured so that they cannot topple and precautions such as a guard or cap should be taken to ensure that the cylinder valve cannot be sheared off in the event of it being dropped.

Gas cylinders on vehicles are another area where safety must be considered. Gas cylinders should not be carried in the same compartment as the driver and passengers. A modest leak of gas could lead to incapacity of the driver and a vehicular accident. Alternatively, a vehicle accident could result in valve rupture or leak and present an additional hazard to the occupants. The elevator or lift in a building is equally a vehicle from these points of view, and cylinders and cryogenic containers should be sent up in the service lift without a human attendant if possible.

'Empty' cylinders can still contain a significant amount of gas under considerable pressure. One reason is increasing moisture level at low pressure which prevents the use of the last 20% or more in high-purity applications. Another problem is the minimum pressure required to run a particular system. In any case, a declared empty cylinder may well turn out to be full; there is little difference in weight, so unless a pressure gauge is applied and the cylinder valve opened, it is difficult to be sure that a cylinder is empty. Empty cylinders should therefore be accorded the respect paid to full cylinders.

2.5.11 Safety audit and HAZOP/HAZAN analysis

Every system in regular use for whatever purpose has a probability of failure. Sometimes this probability is exceedingly small, or the consequences relatively light, and this risk is simply accepted. More often the failure probability is high enough, or the consequences sufficiently serious, that the system is designed to include back-up components. Alternatively, the system is designed to 'fail-safe', in other words to fail in such a way as to avoid serious consequences; reliability is impaired but not safety.

Complex systems can end up being surprisingly unreliable and unsafe if insufficient thought is given to the probability and effect of failure of components and combinations of components. For example, suppose that a liquid oxygen burner system had 50 components that were all individually critical to its safe operation; components like perhaps the tank pressure regulator valve, the vaporiser coil, the burner cooling water, the flame failure sensor, and so on. Further suppose that each of these fails randomly and has an MTBF (mean time between failures) of 20 years, which might seem reasonable if a system lifetime of 10 years is needed. The MTBF of such a system would in fact be only a few weeks. If even a small number of the failure modes were unsafe, this would be unacceptably bad performance. (Real systems will often, but not invariably, be better than this because many failures are wear related and have a more or less definite lifetime; put another way, their MTBF in their early years is much lower than their average MTBF.)

The reason for the bad system MTBF in the above case is the result of multiplying together the probability of no failures. This magnifies even small deviations from perfection because any one defect anywhere in the system causes complete failure.

$$P_{\text{tot(sys fail)}} = 1 - (1 - P_i(_{\text{subsys fail}})) \times (1 - P_{ii(\text{subsys fail})}) \times \ldots$$

For small component failure rates, the system failure rate ($1/\text{MTBF}_{\text{tot}}$) is simply the sum of the component failure rates ($1/\text{MTBF}_{\text{i}}$). This kind of design is, of course, avoided in practice, and systems are designed where possible to provide, for example, a back-up subsystem. Provided the back-up subsystem is truly independent of the primary, and this must be proved beyond reasonable doubt, then the probabilities of failure are multiplied and the system failure rate reduced

$$P_{tot(sys\ fail)} = P_{p(primary\ fail)} \times P_{b(secondary\ fail)}$$

For small failure rates, the MTBFs for the subsystems are added.

Hazard/operability study: HAZOP. The basis of HAZOP is a systematic account of all the possible failures of individual components or subsystems. This essentially qualitative method can then be followed by quantitative calculation, calculating the probability for consequent failure modes from estimates of individual component failures; this is then HAZAN (hazard analysis).

HAZOP has proved to be a valuable discipline for bringing together and formalising the thoughts of the designing engineers and scientists, and the operators of gas systems with respect to safety and efficient operation.

Particularly when a new design of plant is being designed, the right format must be arranged.

Correct group of people. It is important to collect together the right group for HAZOP. It must include the designers, the users (including the 'shop-floor' users), the scientists who developed the new process, and an independent chairman, perhaps a delegate from the safety department or a consultant. The best teams probably include creative thinkers with little detailed knowledge of the plant, as well as those closer to the project. In the case of applications of industrial gases, those who work for the gas user should be involved, as well as gas company personnel.

Correct level of authority. It is a waste of time if the HAZOP group does not have the necessary authority to implement at least small changes in plant and its operation.

Correct orientation and guide words. The starting point for HAZOP is to think up all the possible ways the plant can go wrong. To do this, it is helpful to use a set of guide words such as are listed below. These words are then applied to the parameters of the gas system in question. The question 'What could go wrong if . . . ?' should be asked for each parameter:

None	e.g. pressure/flow/reverse flow/lubricating oil
Part of	e.g. omission of part of gas mix
Less of	e.g. lower temperature, flow, pressure
More of	e.g. lower temperature, flow, pressure
More than	e.g. impurity or air/water/oil inclusion
Other than	e.g. other than normal mode, such as shutdown, startup

In each case, the interaction of any abnormalities with operator action must be considered. Some abnormalities will be obvious to the operators, others will be measured and relayed to them by specifically provided instruments, whilst others may occur *incognito*.

The application of these principles must then be applied at selected points throughout the plant under study. It is of course a matter of judgement how far apart those points lie. However, points separated by reliable intervening components, valves, regulators, etc., are probably a reasonable guideline.

Hazard analysis: HAZAN. There are essentially three elements in any quantitative study of hazards: failure probability, damage consequent on failure and a judgement on the acceptability of the 'average consequences', i.e. the product (failure rate) × (consequent damage). Each of the failure mode probabilities in a HAZOP may have been measured. However, failure data is subject to statistical error, with only relatively small numbers of failures typically having been observed. Also, rare failures may not have been observed at all but must be inferred and probabilities estimated. For these reasons, failure rates are always subject to considerable error.

Even bigger errors creep into HAZAN, however, when the consequences of failure must be considered. How big will the leak be when it happens? How far will the fire spread? It is almost impossible to answer these sorts of questions quantitatively from experience but estimates must be made to complete the analysis.

Take the example of a pressure vessel safety relief valve. Suppose a relief valve is tested every year. Further, suppose that on average, 1 in 100 relief valves tested has failed stuck down, i.e. failed dangerous, as opposed to failed safe, and that the pressure vessel it relieves goes overpressure every three years. The rate of hazard occurring on the pressure vessel as a result of this is therefore $0.3 \times 0.5 \times 1\%$ every year, i.e. 0.0015 hazards per annum or about one accident every 700 years. (The valve is likely to fail half-way between tests on average, i.e. it will be failed for 0.5 years before discovery on average. The vessel has a 1% chance of having a failed valve on it each year and it tries the valve out 0.3 times per year. A more complex formula is appropriate where the test interval exceeds the interval between tries.)

It is not sufficient to stop at this point and give the relief valve a clean bill of health, however. Further calculation is indicated. Suppose the vessel concerned contains a harmless compressed gas but the rupture would almost certainly kill or badly injure the operator working with a machine incorporating the pressure vessel. The fatal accident rate (FAR) resulting could be as high as 15, i.e. probably higher than would be acceptable in the gas industry, which is aiming at FARs in planning of 0.5 or less. The relief valve would therefore need to be paralleled with another similarly reliable device, preferably of a different technology, for example a bursting disc.

Finally, at the end of a HAZAN, the acceptability or otherwise of the average consequences must be decided. This area is curiously subjective. It has been estimated, for example, that the nuclear industry spends £10m or more for each life saved on safety equipment, whilst the healthcare industry finds it difficult to justify spending more than £100k or so to save a life. The FAR is defined as the number of fatalities resulting from each 10^8 working hours. The FAR of many industrial activities is below the FAR involved in ordinary life (in which many fatal accidents are due to motor vehicle use), which is effectively around 20. Only a few workers, such as fishermen, offshore oil workers and building erectors, have a FAR significantly above 20.

In the past there were many more dangerous industries. Today dangerous activities are not tolerated, at least while it is reasonably easy to make them safer, which appears to be possible for all but exceptional work such as offshore oil production. Fortunately, the gases industry in most countries is reasonably safe, with FAR in the 1–10 region. It is also worth remembering that many accidents in the gases industry are in fact road accidents while, as remarked above, the most common accidents in the gases workplace are unremarkable incidents during handling such as dropping a gas cylinder on the toe or cuts on the hands. Some simple safety equipment such as gloves, goggles, safety spectacles and safety shoes with steel toecaps are very worthwhile to minimise risks.

2.5.12 Environmental issues

The stratosphere is defined as beginning at 10 000 m. While the atmosphere below, the troposphere, contains most meteorological phenomena and contains most of the mass of the atmosphere, the stratosphere is important, too. Although the stratospheric air is very thin (only 1/4 bar at 10 000 m and 1/100th bar at 30 000 m) the sunlight which reaches us is filtered by substances in it such as ozone. It is only comparatively recently, however, that the importance of the stratosphere has been realised.

Ozone in the stratosphere. Ozone exists in the stratospheric layers and absorbs most of the ultraviolet (UV) radiation produced by the sun. Without this reduction in UV light, widespread skin cancer would occur. There had been worries for some time amongst atmospheric scientists that the stratospheric ozone might be affected by man-made gases released into the atmosphere, especially chlorofluorocarbons (CFCs). The publication in 1985 of results of the British Antarctic Survey confirmed the worst fears, showing that for five successive springs Antarctic ozone had been plummeting to unprecedented low levels at two locations; the so-called 'ozone hole' (Figure 2.69).

There is now comprehensive evidence that ozone holes are forming above the polar regions. Some of the Nimbus (see Houghton *et al.*, 1984) series

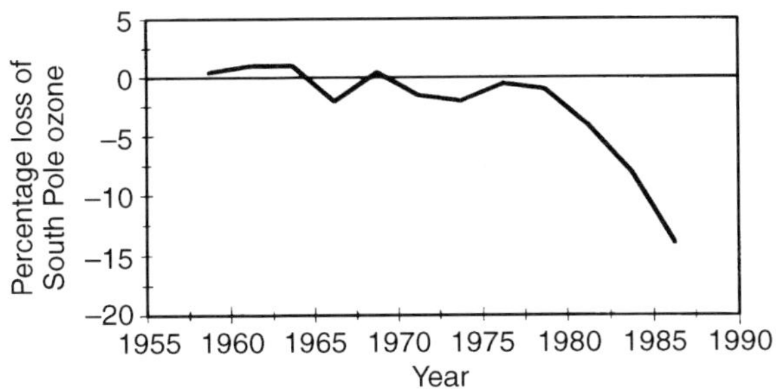

Figure 2.69 Development of ozone hole over Antarctica (Bridgeman, 1990).

satellites have mapped ozone distribution thoroughly and more monitoring stations on the ground are filling in details in the pattern of change. The ozone holes have been predicted to spread to lower latitudes, according to models of ozone depletion which have been proposed, and there is already evidence that this is happening.

Ozone depletion is primarily caused by halogen atoms causing chain termination of the ozone formation reactions. Photodissociation of a CFC is followed by reaction of the chlorine atom with ozone. Reactions such as

$$Cl + O_3 \Rightarrow ClO + O_2$$
$$ClO + O \Rightarrow Cl + O_2$$

occur. The latter reaction recycles the Cl atom to cause more ozone destruction and destroys an oxygen atom that might otherwise go on to form ozone

$$O + O_2 \Rightarrow O_3$$

The CFCs lead any list of ozone-depleting substances. Their estimated relative effects (ozone depletion potential, ODP) on stratospheric ozone are given in Table 2.18. Newer compounds, regarded as replacements for traditional CFCs, are considerably better than the bench mark CFC-11 and CFC-12 (Table 2.19).

CCl_4 probably has a much lower impact because it is more reactive than the other substances listed in Tables 2.18 and 2.19. In sunlight it reacts with oxygen in the air to form the highly reactive (and toxic) phosgene. It is the chemical stability of CFC compounds that paradoxically causes them to be the potent source of ozone destruction. Their low solubility and low reactivity with water or oxygen in the troposphere means that they are not washed out there, unlike more active halogen compounds such as HCl or HF. Their lifetimes in the troposphere are thought to vary between 65 and 400 years. These very properties, resistance to hydrolysis and low flammability, are those for which they are valued.

Many industrial spokesmen were vehement in their opposition to the

Table 2.18 Ozone depletion potential of CFC gases

Compound	Boiling point (°C)	ODP
CFC-11, CCl_3F	+23.8	1.0
CFC-12, CCl_2F_2	−29.8	1.0
CFC-13, $CClF_3$	−81.9	?
CFC-113, CCl_2F-$CClF_2$	+47.6	0.8
CFC-114, $CClF_2$-$CClF_2$	+ 3.5	1.0
CFC-115, CF_3-CF_2Cl	−39.1	0.6
HCFC-22, $CHClF_2$ (formerly called CFC-22)	−40.9	0.05
CCl_4	+76.7	?low

Table 2.19 Ozone depletion potential of CFC replacements

Compound	Boiling point (°C)	ODP
HFC-134a, CH_2F-CF_3	−26.5	0
HFC-23, CHF_3	−82.2	0
HFC-123, $CHCl_2$-CF_3	+27.9	0.02
HFC-124, $CHCl$-CF_4	−11	0.02
HFC-125, C_2F_5H	−48.1	0.02

banning of CFCs and vehement that these molecules, with their remarkable properties of zero flammability and toxicity, and wide range of boiling points, were irreplaceable. The availability of the new HCFC range of molecules, also non-flammable and of at least as low toxicity, but with zero or almost zero ODP and of acceptably low reactivity, shows how groundless this opposition was. Most applications formerly using CFC compounds can now use HFCs. Azeotropic or near azeotropic mixtures of the HFCs are now available, which has further extended the range of available boiling points.

NO_x and VOCs. NO_x (nitrogen dioxide and its dimer dinitrogen tetroxide, nitrogen monoxide and nitrous oxide) is formed during high-temperature combustion processes. VOCs (volatile organic compounds) are anything from natural tree resins seen near conifers in the summer to petrol vapour evaporating from vehicle fuel or isobutane from aerosol spray cans.

Ozone is as undesirable in the troposphere as it is desirable in the stratosphere. It is toxic and causes chronic respiratory ailments even in very low concentrations. Ozone occurs near ground level in the troposphere. It is often found in the ugly brown-stained air above industrial areas which are situated in a natural bowl, especially in clear, sunny, windless weather. Normally the atmosphere at ground level contains little ozone. However, the presence of large quantities of NO_x and VOCs, in the several parts per million region, changes this. NO_x and VOCs react with oxygen in bright sunshine to produce ozone at low concentrations. High concentrations of

NO_x are themselves toxic, so effort has always been directed against them. However, VOCs have only recently been recognised as the cause of ozone formation at low levels. Investigators looking at serious smog problems in the Los Angeles area first proposed VOCs as the main cause and subsequent studies now confirm this. VOCs are now rated on the POCP (photochemical ozone creation potential) scale, with the simple saturated hydrocarbons rated intermediate between oxygenated compounds, such as alcohols, which have low POCP and complex aromatics, which have high POCP.

Other atmospheric pollutants. Carbon monoxide is chief amongst other pollutants. CO is produced during incomplete combustion of organic material. However, external combustion devices, burners, do not produce much CO unless they are badly adjusted. Most CO comes from internal combustion engines. Particularly when it is idling, the car petrol engine produces CO copiously; depending on the adjustment of the carburation system, from 1% to 8% CO is common. Although aimed at NO_x, the catalytic exhaust treatment units now common on vehicles also reduce CO. Despite this recent improvement, many countries are now concerned about CO production from vehicles and are insisting that minimum standards are met.

These new government regulations mean that cars must be tested for CO and then adjusted to minimise it. This has led to a worldwide demand for calibration gases for the analysers, often the non-dispersive infrared type, employed in the CO testing. Special gas divisions of the industrial gas companies supply CO mixtures for simpler analysers and mixtures with accurately known contents of moisture, CO_2 and hydrocarbon vapours for the more sophisticated multigas (hydrocarbon + CO) analysers. In the UK currently, for example, the Ministry of Transport vehicle testing stations are supplied with a mixture containing 2400 ppm propane, 6% CO and 15% CO_2 in nitrogen for the calibration of gas analysers.

The greenhouse effect. A greenhouse keeps plants warm on a sunny day by allowing visible light through its transparent windows to warm up its contents but does not allow infrared radiation and convection to cool the plants. The earth can be regarded as a kind of giant greenhouse. The atmosphere of the earth, although its principal constituents are transparent to visible sunlight, contains infrared absorbing gas molecules such as H_2O, and especially CO_2, which block the radiation of heat from the surface to space. Emissions of many man-made gases, especially CO_2, into the atmosphere are increasing the greenhouse effect and it is thought to be possible that these emissions are therefore causing significant global warming.

The emission of carbon dioxide is an intrinsic part of fossil fuel (oil, gas, coal) burning. The emission of CO_2 is so copious that no practicable scrubbing system can be devised. (Suggestions that power station exhaust stacks could be cleaned up and the resultant CO_2/nitrogen stream bubbled

Table 2.20 Global warming potential of selected gases

Gas	Global warming potential (GWP) (20 years integration)	Halon-11 related GWP (HGWP)	Estimated lifetime in the atmosphere (years)
Carbon dioxide	1.0		120
Ammonia	? low	0	? short (rain soluble)
Nitrous oxide	270		150
Methane	63		10
Halon 11 (CCl_3F)	4500	1.0	60
Halon 12 (CCl_2F_2)	7100	3.1	130
Halon 22 ($CHClF_2$) (HCFC-22)	4100	0.33	15
HFC-134a (CHF-CF_3)	3200	0.28	16
HFC-125 (C_2HF_5)	4700	0.84	28

(The figures are 1994/5 figures derived from Friends of the Earth, London, and BOC Group Plc. The exact figures are subject to frequent updates and these revisions even occasionally change the ranking of gases in their potential for global warming. All the relevant factors, in lifetime, for example, are still inexactly known.)

through a few square miles of made-to-measure mangrove swamp have to be regarded as still at the concept stage.)

The effect of increased efficiency in any energy-intensive processes is to require less CO_2 to be emitted. Efficiency increases are therefore one of the few currently effective weapons against excessive CO_2 emissions. Combustion using O_2 for high-temperature industrial processes and furnaces also has, in general, a positive beneficial effect on CO_2 emissions, *inter alia*. This arises in the following way. Suppose a process needs, say, a temperature of 1500°C to operate. This can be provided by fuel/air at 2000°C, but in this case 75% of the heat of combustion is wasted in the exhaust stack gases, which exit at 1500°C unless complex regenerators/recuperators or other energy recovery systems are in use. However, an oxygen-enriched air stream at 2500°C would find only 60% of its heat of combustion wasted.

Although most global warming fears are centred around CO_2, many other gases are much more potent infrared (IR) absorbers, and could give rise to problems at very low levels; much lower than the few 100 ppm of CO_2 that are necessary to influence global warming. Long-lived IR absorbers that could in principle be a problem include the CFCs, VOCs and nitrous oxide. A glance at the list in Table 2.20 of the global warming potential of a selection of gases, together with the approximate concentrations of the gases, reinforces the view that CO_2 is only the principal problem because of the large volume of CO_2 emissions.

The correspondence between carbon dioxide and the halocarbons depends on assumptions about the lifetime of gases in the atmosphere as the global warming potential is an integral of the warming effect produced over the lifetime of the gas. It is assumed here that CO_2 and the halocarbons are injected at time zero and their effect measured over 20 years. Short lifetime gases have a disproportionately large effect on these figures. If a longer time such as 100 years is taken, then the long lifetime gases will have a larger GWP relative to CO_2, which is reckoned to have roughly a 120 year lifetime. The longer integration time is not appropriate because technology and usage of gases changes significantly over 20 years and will most probably have changed dramatically over 100 years. However, some of these lifetimes are so long that current technology may leave an undesirable legacy, as has happened for CFC-11 and CFC-12 in ozone depletion.

Local pollution issues. Although unimportant on a wider view, there are potent local pollution problems which still need resolution in many industries. River and lake pollution is an example of this. Oxygen (section 3.6) and other gases used to control waterborne pollution problems can have a significant positive impact on local pollution problems.

2.5.13 Industrial gases and the environment

The gases industry itself has been, and remains, in general one of the most benign to the environment. With few exceptions, industrial gases have themselves been made in clean processes and have enabled cleaner processes to be used in the rest of industry.

Examples of where damage to the environment has been reduced by the use of industrial gases include the following:

- solder inerting with nitrogen where CFC solvent use is no longer required for cleaning off flux residues (section 3.5.14);
- oxygen processes to make steel have reduced the formidable capacity of steelworks to pollute the environment with smoke and dust and reduced the use of coke ovens.

Less impressive examples, where the gas industry has led to increased pollution, include the production of acetylene. Calcium carbide production via electric arc furnaces produces copious smoke, whilst the reaction of the carbide to make acetylene results in large quantities of lime sludge.

Environmental audit and life-cycle analysis. Environmental audit aims to determine the ultimate benefit, so-called life-cycle analysis, to the environment of alternative products and processes. The environmental issues can be summarised in an analogous way to a financial statement comparing the monetary pros and cons of projects. For example, the analysis of an aerosol can versus a pump-spray unit is shown in Table 2.21. From this rough example, it is not clear, contrary to claims often made, that a pump spray package has any real overall environmental benefit over an aerosol package. The results of a careful environmental audit are often surprising in this way.

Table 2.21 Environmental comparison of aerosol with pump spray

Feature	Energy (kJ)	Degrades in waste rapidly?	Troposphere pollution?	Stratosphere pollution?
Aerosol can				None
Steel can	700	Yes		
Isobutane	2400	No	Yes, POCP	
Valve	100	No		
Pump spray				None
Plastic body	2400	No		
Valve	100	No		
Additional solvent	2400	No	Yes, POCP	

Ecopoints. With life-cycle analysis it is clearly now possible to compare the overall environmental impact of options. There then arises the question of whether any measure can be devised which enables the comparison to be made quantitative. Is it better, for example, to emit NO_x from a furnace and save the CO_2 that would have been used if electricity had been consumed to generate oxygen for running that furnace with low NO_x emission? The Ecopoint scheme, being developed by Dutch and German groups (e.g. Pré Consultants, Amersfoort, Netherlands) is one of a number of recent attempts to do just that. It weights energy used and wastes emitted in order to drive decision-making in a consistent way. Initially, the assignments are as shown in Table 2.22.

2.5.14 Scrubbing

Gas scrubbers are vital in modern industry. Although more efficient and intrinsically cleaner processes are making great strides, and gross polluters like the LeBlanc process have disappeared, many plants have unavoidable effluent gas streams. The term scrubber in the gases industry has a slightly wider usage than its meaning of a gas/liquid contacting tower in chemical engineering and encompasses basically any device, whether using liquid or solid, which removes a pollutant.

Table 2.22 Ecopoint values for waste materials

Unit	Material	Ecopoints
kWh	Energy	0.13 (electric energy)
kg	CO_2	3
kg	Acetone	10 000
kg	Waste oil	100 000
kg	PCBs	1 000 000 (polychlorinated biphenyls)

Filters are, of course, employed in many effluent streams; their functioning in the role of clean gas supply was noted in section 2.4.3. In the effluent role, different filters come into play to address the different problems posed by dirty gas. Although efficiency is still important, low pressure drop and ease of removal of filtrate become equally important in dealing with exhaust gases. Bag filters can be used to solve the latter problem, together with arrangements to shut off flow and vibrate the bags to release the precipitate from time to time. Instead of low-capacity membrane filters, 'depth' filters, with a thick bed of relatively larger fibres, are more attractive. Electrostatic precipitators or cyclones are likewise more useful on dirty effluent streams than low-capacity filters.

However, filters and electrostatic precipitators are sometimes a messy solution to the problem of particulates in effluent gas streams and no solution at all for the problem of unacceptable gaseous contaminants. Scrubbers proper are then required. Some 'scrubbers' actually achieve a completely benign catalytic conversion. The precious metal catalyst in the motor car is an example of this; once up to operating temperature, it converts NO_x into nitrogen by catalysing the reduction reaction. In industry there are examples, too, such as the SCR (selective catalytic reduction) process for the reduction of NO_x in furnace exhaust stacks and the Claus process for removing hydrogen sulphide.

Wet scrubbers. Scrubbers were used to absorb evolved HCl from the LeBlanc process in the nineteenth century. Since then, designs have advanced in execution, though not in basic principle. Wet scrubbers remove gas impurities by absorbent liquid sprayed down a tower or trickled over packing material in a column through which the gas stream is passed in an upward direction. With many impurities, water washing can be used. Many particulates, acid, alkaline and some organics, can be effectively absorbed and removed. With other impurities, special solutions, such as ethanolamine for carbon dioxide or hydrogen sulphide contamination, or sodium hydroxide for arsine and phosphine, can be employed.

The largest scrubbers in use are those for removing sulphur dioxide from the exhaust of electric power stations burning coal. These scrubbers typically use a slurry or powdered limestone reactant, which is converted by the exhaust into a rather unattractive calcium sulphate/sulphite by-product (although the latest processes are capable of producing a better sulphate, a usable grade of gypsum for plasterboard manufacture).

With liquid scrubbers there remains, of course, a liquid effluent disposal problem. In many cases, the highly dilute solution produced is completely acceptable in water courses and no further treatment need be applied; this is 'open cycle' scrubbing. In others, the contaminant gas may be desorbed and usefully employed (as in CO_2 or H_2S stripping from ethanolamine, e.g. Kohl and Riesenfeld, 1960) and the liquid recycled; this is 'closed cycle' scrubbing.

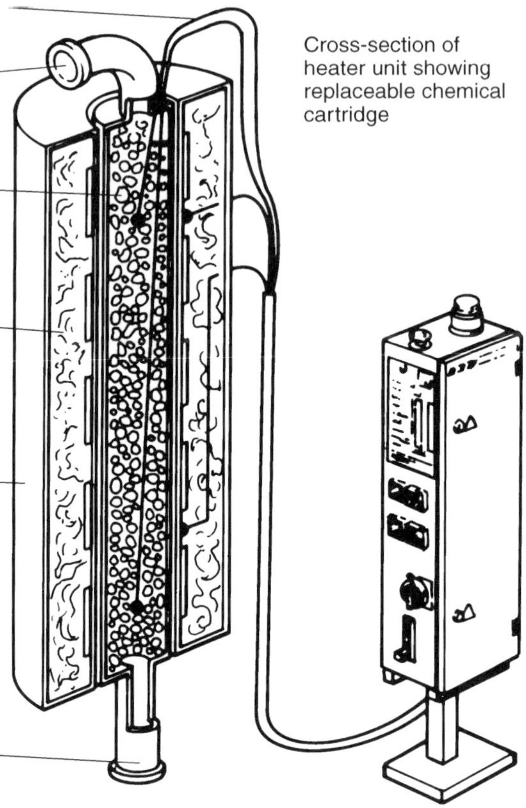

Figure 2.70 Absorber bed system with exchangeable absorber cartridges (courtesy of Edwards High Vacuum Ltd, Crawley, UK).

Wet scrubbers are often combined with another technique. For example, a wet scrubbing system can be combined with an incinerator technique whereby the incineration is carried out at a 'submerged' burner. Dry scrubbers, e.g. incinerators and plasma-based scrubbers are mostly used in the semiconductor industry.

Dry scrubbers. Dry scrubbers are simple adsorbers for gaseous contaminants, typically based on porous carbon. They are popular in the semiconductor industry where small percentages of occasional contaminant are removed from large ventilation air or nitrogen purge flows, typically with a specially activated carbon bed. Similarly, in the coating industry, nitrogen contaminated with solvent can be processed through a carbon bed adsorber where large amounts of organic solvent can be adsorbed, particularly at low temperature (Figure 3.36). The bed can be regenerated by heating in a pure gas stream for a while and the solvent recovered. In some systems the nitrogen or air scrubbed is also recycled.

Other scrubbing/recovery systems are based on using liquid nitrogen or other means to cool the exhaust gas, causing condensation or enhanced adsorption of the solvent vapour or gas. The nitrogen added is used as the carrier gas and recycled. Ingenious use of 'cold recovery' from the gas stream minimises the coolant usage (see section 3.7).

The capacity of carbon absorbers for low concentrations of impurity is reduced, as examination of a simple equilibrium equation shows. For low levels of impurities such as the ppb levels of mercury in amalgam brine electrolysis cells, it is useful to add reactive chemicals to the porous carbon. In the case of mercury, for example, the addition of iodine to the absorber bed blocks mercury for much longer than a simple carbon bed.

Heated scrubbers are also available and these also rely on chemical reaction with the effluent gas. These can be very simple; heated calcium oxide (lime), for example, reacts with acid gases to form stable harmless salts. The low cost of the lime means that it does not need to be recycled. One system sold for removing semiconductor vacuum pump exhaust gas claims to remove more than 20 common (mainly acid) gases (see Figure 2.70).

3 Applications of gases in industry

3.1 Gases in metal smelting and extraction

3.1.1 Oxygen in ferrous metals

The biggest air oxygen plants in the world supply oxygen for steel making. A single liquid air distillation column can produce up to 2000 tonnes or so of oxygen per day for a steelmaking site. This thirst for oxygen of the steel making process is prodigious: it is equivalent to the oxygen breathed by over half a million people. The fact that so much oxygen is used may seem puzzling at first glance because the basic process of producing steel is the chemical reduction or deoxidation of iron ore, essentially by adding a fuel such as carbon

$$Fe_2O_3 + C \rightarrow Fe + CO_2$$

Some oxygen is used for cutting steel (steelworks are the biggest users of oxygen cutting equipment). However, applications for oxygen such as these use in total only small amounts of oxygen. Much more oxygen is used in the oxidation of impurities in iron and much more than that again in blast furnaces.

3.1.2 Blast furnaces and the smelting of iron

A typical blast furnace is basically a very large continuously fed smelting furnace (Figure 3.1). Solid raw materials, iron ore, coke, limestone and various minor additions, are all carefully prepared by sophisticated grinding and pelletising processes. These solids are added to the top of the furnace via a complex arrangements of large bell-valves. Also at the top is the gas exhaust, now not allowed straight into the atmosphere as it once was.[44] The exhaust is now diverted via various filters and scrubbers to clean up both particulates and gases, some of the latter being recycled. In some processes the top-gas from the blast furnace is an important by-product itself. At the high temperatures (1500°C or so) in the furnace, the iron ore is reduced to

[44] Only a few years ago this exhaust provided steel towns such as those in southern Wales in the UK, with spectacular smoke plumes that could be seen from 50 miles away on a clear day and cast a continuous, stratified, often reddish-hued dark cloud over their surroundings.

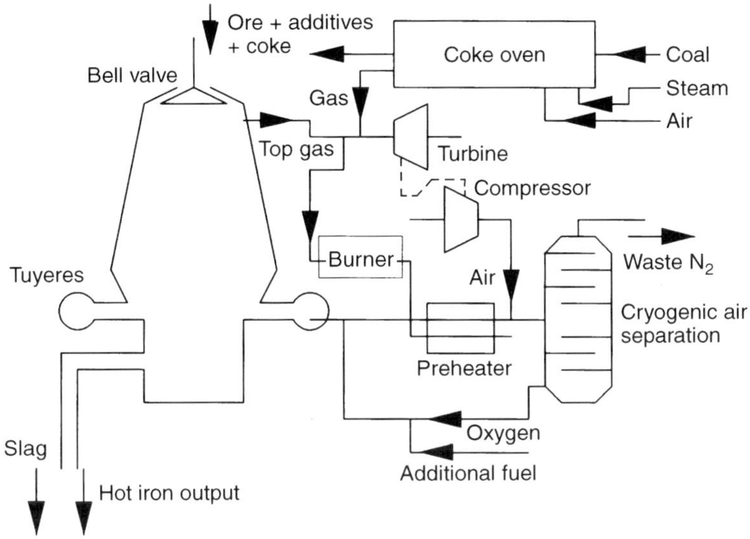

Figure 3.1 Conventional blast furnace ironmaking.

molten iron, which then trickles down through the porous coke to the bottom. The limestone and other materials react with phosphorus, sulphur and so forth in the ore and also melt and trickle down. The coke, having a much higher melting point (3700°C) provides the solid substrate for these reactions.

Down near the bottom, molten iron, alloyed with up to several per cent carbon from the coke, is tapped off, either directly into steel making or, less frequently in modern works, into temporary storage by being poured into pig-iron ingot moulds and allowed to cool. A little above the pig-iron tapping at the base are a set of water-cooled inlet pipes, the 'tuyeres', for blasting the mixture with oxygen, hot air or other blast gas mixtures. Other tappings allow for sampling of the charge during operation and removal of molten slag (the limestone and its products of reaction and impurities in the iron ore).

Standard blast furnace design and operation has to follow a number of rules, at least approximately, as follows:

- The furnace must vary in cross-section, first increasing in diameter from the filling bells gently and then waisting towards the bottom more sharply. If this kind of variation in cross-section is not employed, then there will be problems of the charge collapsing on itself and blocking gas flow or, conversely, the charge will not descend quickly enough during reaction, reducing the furnace throughput.
- The gas flow must be as high as possible to maintain throughput, but not so

high as to blow too much of the smaller pieces of charge out of the exhaust or so high that the burners heating it cannot maintain temperature.

- The temperatures chosen have to be high enough that the oxygen in the blast can oxidise the coke to CO. The CO then performs most of the actual reduction of the iron oxide ore to metallic iron. Temperatures must also be hot enough everywhere that the high-carbon (5% or more) alloy produced is melted, at least in the lower part of the furnace.

The injection of oxygen provides a valuable additional variable in the process. Oxygen enrichment of the blast air heating the furnaces enables higher temperatures at relatively low capital cost, and some increases in capacity, and has been a 'retrofit feature' in many steelworks upgrades. Today, however, oxygen enrichment is fitted into new blast furnaces from day one, a design change which can be regarded as mostly due to the increasing cost of coke.

Coke is made by the pyrolysis of coal. The special grades of 'coking coal' needed to prepare coke have become increasingly expensive and have had their own effect on coke cost. However, in addition to coke, the process also produces prodigious amounts of noxious sulphurous gases, and new coke ovens can only be installed with extensive exhaust gas treatment to limit environmental impact. The result of this is that coke ovens can now cost more than 30% of the total cost of a steelworks, more than the blast furnaces they serve.

There is therefore an increasing emphasis today on reducing the need for coke in blast furnace operation. One way of reducing the amount of coke needed is the injection of fuel through the tuyeres, providing an alternative source of heat and CO gas. Both finely powdered coals or a spray of hydrocarbon oil can be used, with the former being very attractive for its low cost and high carbon content. However, this cannot be carried very far before the heat absorbed by the endothermic pyrolysis of the fuel means that the flame temperature achieved by the tuyeres is reduced, which is where the requirement for oxygen comes in.

The use of both fuel and oxygen additions to the blast air is crucial in allowing large amounts of fuel to be added. The oxygen enrichment raises the tuyere flame temperature. The overall effect is to allow a much lower coke/iron ore ratio, reducing coke costs more than the added fuel/oxygen costs. Fuels may be heavy petroleum oils, natural gas, powdered coal or coal/oil slurry. Enrichment factors of 5% or so are typical and 60 m^3 of O_2 and 100 kg of fuel per tonne of iron is needed to save 150 kg of coke. With a 10 000 tonne per day blast furnace, therefore, an oxygen demand of 600 000 m^3 or 500 tonnes per day is indicated. In fact, oxygen plants of much greater capacity, up to about 2000 tonnes and often several plants, are used. This reflects the fact that other uses have been found for oxygen in steelworks. In addition, the year-on-year increase in the size and number of

Figure 3.2 Bottom of blast furnace showing the ring of tuyeres which inject fuel and oxygen-enriched air into the burden.

blast furnaces using oxygen and their increasing oxygen usage has led to plant sites with a more complex supply system than would have been the case had an entire new gas plant been designed to satisfy current usage.

3.1.3 Increasing prospects for oxygen in blast furnaces

There is ongoing work at conventional blast furnaces to increase further the amount of injected fuel, decreasing further the need for coke. Coal injection rates of 250 kg per tonne of iron now seem achievable (see Corbett and Smith, 1993) and this will need oxygen enrichment of 10% with blast temperatures of 1100°C or so. This kind of continuous evolution will undoubtedly increase the demand for oxygen in steelworks in the future.

More revolutionary developments, such as ore injection or the carbide route, may also affect oxygen usage in the blast furnace. The standard blast furnace cannot directly utilise iron ore. The fines in the ore would block the furnace when it was loaded at the top or be blown out by the gas flow. Instead, the ore is ground finely and then sintered, with additives, into lumps ('sinter') which are then broken down into roughly equal size. However, in a blast furnace steelworks much of the energy usage, and the capital investment (up to 30% or so), will be needed for the ore grinding, sintering and pelletising plant. It makes sense, therefore, to think of adding fines directly to the furnace, thus saving the sintering and pelletising processes. Fines can be blown into the tuyeres, where they will be reduced and melted before they cause blockage or loss problems.

Beyond the use of blast furnaces, it is quite possible that the new processes

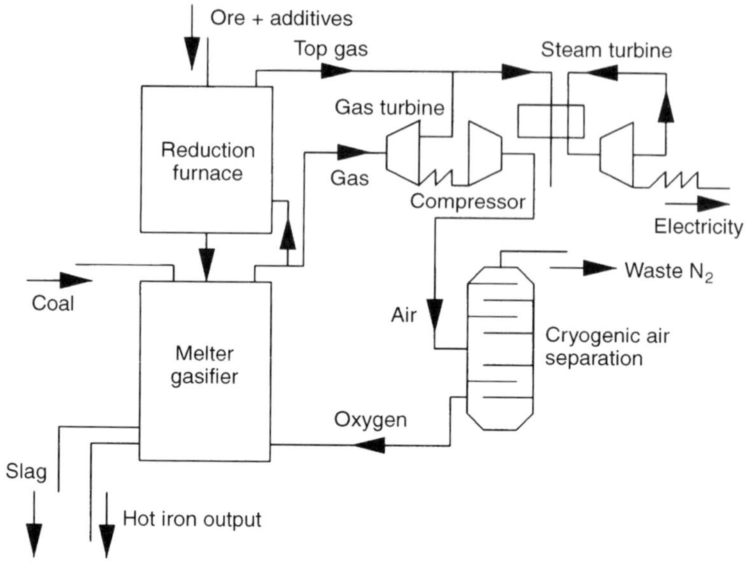

Figure 3.3 Simple schematic of the COREX process.

under development, which are moving away from the blast furnace, may become significant in the future. Some of these possibilities are:

- DIOS: this is a Japanese project, funded by the central government agency MITI, to use ore fines, coal and pure oxygen all directly; a 500 tonne per day pilot plant is now being completed by the Japanese company NKK at Keihin.
- CCF: the Cyclone Converter Furnace is being developed by a British Steel/Hoogovens/Ilva/CSM consortium, using a two-reactor approach; an upper 'cyclone' reactor using ore fines for preheat, reduction and production of an intermediate reduced product, which feeds a lower melting reactor heated with oxygen and powdered coal that produces gas for the cyclone and the final liquid iron product.
- NUCOR/US Steel: a carbide reduction route is another prospect for increased oxygen usage in smelting; this proposal from US Steel is to use smelting of iron carbide pellets made near the ore production site (section 4.3.5).
- COREX: a full-scale plant is operational and this system is examined in more detail in the next section.

COREX: pure oxygen blast and direct coal fueling. The key feature of the COREX process, now in full-scale trials, is an integrated plant tying together both the iron-smelting units and the power-generation and oxygen-generation plant. The first major COREX plant was started in

Pretoria, South Africa, in 1990 and now produces over 300 000 tonnes per annum of molten iron. There are around ten more plants in detailed planning and it seems likely that this process will have some impact on annual production by the end of the millenia.

The heart of a COREX plant is the melter-gasifier (Figure 3.3). Here a pure oxygen blast melts reduced iron pellets and gasifies powdered coal. The iron is tapped off, along with slag at the bottom, whilst the copious gas produced at the 1000°C operating conditions is taken off to a dust separator and thence, still hot, to the iron-pellet reduction furnace. Here pellets of ore (together with slag additives) are reduced from Fe(II) or Fe(III) minerals to pellets of iron in the hot gas stream. The gas mixture from the melter-gasifier is designed to be mainly CO and H_2, with 30% CO_2, and provide complete pyrolysis of all the coal components to these simple gases. Slag-forming additives such as dolomite and limestone in the reduced iron burden have a double duty. As well as forming slag with unwanted elements in the ore, they absorb sulphur from the gasified coal. The resultant sulphuric slag is no more offensive than standard slag.

The rest of a COREX plant has nothing to do with iron-smelting but is essential if the process is to be competitive. First, the amount of gas produced exceeds that necessary for reduction and melting because of the large amount of coal it is necessary to inject This gas is burnt in a combined cycle gas turbine/steam turbine power plant, producing 100–300 MW spare electric energy for a typical 300–1000 ktonne p.a. plant. The electrical output is reduced because part of the power produced must be used for the oxygen plant compressor. Alternative uses may be found for the COREX gas, depending on neighbouring plants, but a good match must be found to its output.

3.1.4 Conversion of iron to steel

The conversion of the weak and brittle raw pig-iron into the hard but tough steels is basically a process of adjusting the alloying elements in the iron. Carbon must be moved downwards, along with other undesired elements such as sulphur or particulate inclusions such as unreduced ore, and some additions must be made.

The basic conversion process is the elimination of most of the excessive amounts of carbon present in blast furnace iron. This is essentially achieved by oxidation, today invariably using oxygen blast. The relationship between the oxygen and carbon content of steels generally is an inverse one, as expected from consideration of a simple chemical equilibrium

$$C + O \Rightarrow CO$$

At equilibrium, $[C][O] = k[CO]$, where k is a constant. Hence, increasing oxygen levels in the steel by blowing more in will decrease carbon content.

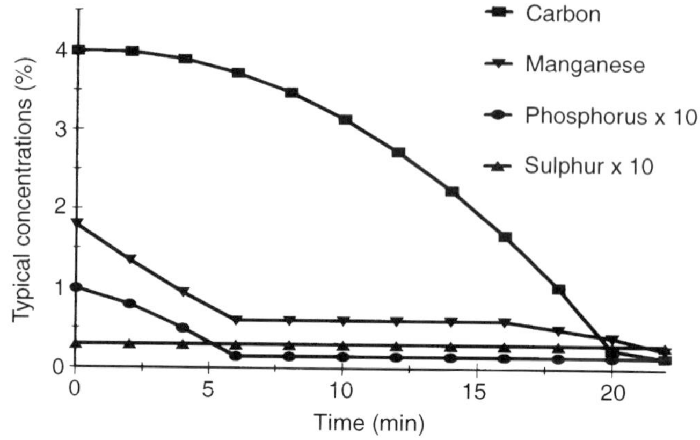

Figure 3.4 Reduction in tramp/alloy elements in steel in BOS process.

The older methods of steel conversion, such as Bessemer converters, involve blowing air through the bottom of the molten steel vessel (bottom-blowing). Air now cannot be used as the iron produced has too high a nitride content. The modern versions of the Bessemer technique generally involve an oxygen blow along with a coaxial shield of hydrocarbon such as methane. Recent converters have been dubbed Q-BOP (quick basic oxygen process). The endothermic dissociation of the hydrocarbon near the injection pipes causes a degree of cooling which enables the injectors to survive (with oxygen in the blow they can be burnt away very quickly) a matter of a few minutes.

Other undesirable 'tramp' elements, such as manganese, phosphorus and sulphurs are also removed from molten iron in steel conversion (Figure 3.4).

As well as gases, some additional solid chemical refinements may be added in the purification process. For example, cerium removes residual sulphur and oxygen. In addition, there are reactions between the refractory container and the molten steel; for example, phosphorus is removed by reaction with basic lining refractories, hence the term BOS (Basic Oxygen Steelmaking).

3.1.5 Basic oxygen and Lind–Donawitz/oxygen-lancing steelmaking processes

As noted above, oxygen blown through an iron melt will tend to oxidise sulphur and carbon to gaseous SO_2 and CO. The Lind–Donawitz (LD) process avoids the need for technically difficult holes in the bottom of the vessel by using an effect of the oxidising process. LD, or oxygen lancing, is so-called because it is done by lowering a water-cooled lance, a rod fed with high pressure oxygen, into the steelmaking vessel. The CO produced in the

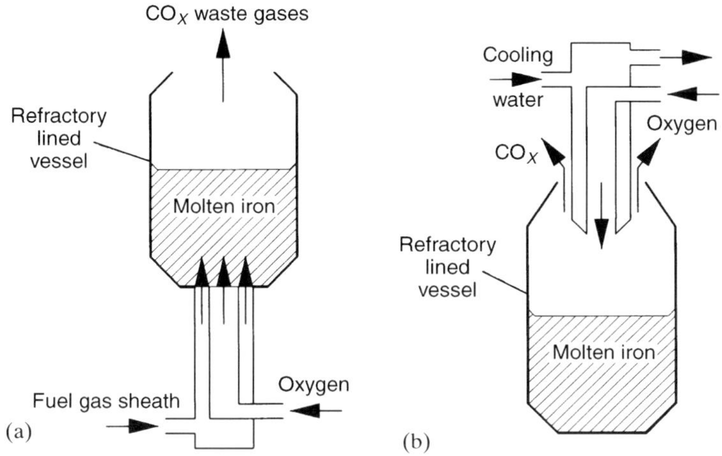

Figure 3.5 (a) Bottom-blown converter. (b) Lind–Donawitz converter.

high-carbon melt results in a stirring action, distributing the added oxygen. As a result the lance does not need to be immersed in the melt, neither is it necessary for the jet of O_2 to penetrate deeply into the melt. In fact either of these circumstances would lead to destruction either of the lance or of the vessel lining.

In practice, the LD process is rapid and efficient; a converter vessel of 200 tonnes or so can be reacted in 20 min or so. The process is also highly exothermic and the heat output can be used to enhance the process output by melting added scrap metal up to around 10% of the process. The produced steel is poured out into a ladle for alloying elements to be added and then taken off for casting. Variants of the LD process, which is known generically as basic oxygen steel manufacture after patent battles, are now the most popular processes in the industry. Variations include additions to the oxygen blast, for example powdered calcium oxide (lime). This reacts with excessive phosphorus in the steel melt.

3.1.6 Mini-steelworks and arc furnace steelmaking

There are enormous advantages of scale in blast furnace iron-smelting. For this reason, small steelworks use scrap steel or, more rarely, prereduced pellets of iron. Starting from cold metal of fairly low carbon content, there is no possibility of using the carbon content to raise the vessel temperature, as in the oxygen injection of the LD process. Melting is therefore carried out with an arc furnace. Subsequent processing may include vacuum, deoxidisers and oxygen/argon blowing. The latter constitutes the argon–oxygen decarburisation (AOD) process (section 3.1.7).

The high cost of electrical heat, and the very large carbon electrodes needed, make the arc furnace a relatively expensive source of liquid iron. A number of steelworks now operate arc furnaces with additional fuel burners. These are switched on early in the steelmaking cycle to supplement the electrical heat. They use oxygen and coal or oxygen and methane, and have advantages in terms of spreading heat over the burden more evenly during the early part of the cycle. As the burden melts, the burners are extinguished and the arc relied upon to complete the melting process.

Direct reduction. Pellets of sintered iron, produced by 'direct reduction' of iron ore by hydrogen or methane without melting of the iron, are already an important feedstock of iron for mini-steelmills.

They have many advantages: they can contain more than 95% iron, with very low sulphur content. Sulphur is usually removed to a large extent during the reduction process as volatile SO_2 or H_2S. Direct reduction can also avoid much of the high cost (section 3.1.3) of pelletising ores as it can operate directly on ores with fines.

3.1.7 Argon for stirring

Molten steel is highly reactive, so much so that the only gases which can be guaranteed not to react with it are the inert gases such as argon. Argon can therefore be used for agitating steel melts without the danger of unwanted reaction. 'Sparging' with argon can thus be used for lowering absorbed levels of gases such as CO and N_2 in melts. For stainless and alloy steels, chromium, nickel and the other principal alloying elements are also reactive in melts and argon is useful here, too.[45] The use of argon as an inert blowing gas is not confined to steels. Non-ferrous metals and alloys which react with nitrogen, such as aluminium and magnesium, can also be stirred and degassed by using a blow of argon.

The popularity of low-carbon stainless steels (such as 316L) has meant that processes for the economical reduction of carbon in bulk stainless steel melts have had to be developed. The argon–oxygen decarburisation process is now common for the lowering of carbon levels in stainless steels. Blowing the highly alloyed stainless steel melt with pure oxygen sufficient to reduce the dissolved carbon down to low levels is likely to oxidise too much of the alloying chromium and nickel. An obvious solution is to blow with an inert gas containing only a proportion of oxygen. However, nitrogen would lead to nitride formation and so is often ruled out. AOD involves bubbling a

[45] The reactivity of molten stainless steels is such that for high-grade stainless steels it is now fashionable to carry out vacuum remelting and sometimes even repeating this process. 316L stainless steel treated in this way has fewer particulate occlusions, lower dissolved gases and can be polished/electropolished to mirror quality finish more easily than regular grades.

mixture of argon and oxygen through the steel alloy melt where the oxygen combines with the carbon and is carried off as CO in the argon.

The AOD process is the largest scale use so far of argon gas; furnaces used have output measured in hundreds of tonnes per day. Argon usage of up to 20 m^3 per tonne hot metal processed is reported. Fortunately, argon is available in large amounts from the liquid air distillation separation plant used to supply the oxygen to other parts of a steelworks.

3.1.8 Oxygen non-ferrous metal extraction

Many non-ferrous ores are present in nature as sulphides in the reducing conditions of deep rocks. Many of these, including nickel, cobalt, copper, zinc and molybdenum, can be successfully extracted by breaking up the rock and then applying a hot solution of chemicals. Typically ammonia, sodium hydroxide or, most likely, sulphuric acid, along with air or oxygen, are used in this hydrometallurgical 'winning' of metal compounds.

For example, O_2 is commonly used in copper extraction. Liquors used for extraction of copper can often be regenerated continuously by bubbling oxygen through them.[46] Typical reactions are

$$2Cu_2S + 12NH_3 + 2(NH_4)_2SO_4 \Rightarrow 4Cu(NH_3)_4SO_4 + 2H_2O$$

$$4CuFeS_2 + 5O_2 + 4H_2SO_4 \Rightarrow 4CuSO_4 + 2Fe_2O_3 + 8S + 4H_2O$$

Oxygen allows a useful degree of process intensification to be attained and this is the basic justification for its use over straightforward compressed air. The gas must be pumped into the suspension of mineral and chemicals, and it is often cheaper to use a small vessel with concentrated ore/chemicals than a dilute solution. The gas bubbles through, replacing the dissolved oxygen that has reacted. For a concentrated mix the replacement of oxygen must be performed quickly. At least five times more air must be pumped than oxygen, because of the O_2 content of air. In practice, much more air than this is normally required because the bubbles of air become depleted in oxygen and are then less effective at transferring oxygen to the mix. If a pressure vessel is employed for the reaction, then, other things being equal, a five times higher pressure must be used, raising the cost of the vessel and pipework. Finally, oxygen systems are designed to have minimum gas

[46] Another commonly used copper extraction process is that of etching printed circuit boards. Here the standard process involves the removal of bare copper from between the tracks on a copper-plated plastic circuit board. The tracks are protected by photosensitive film or lacquer. The etch solution is usually ferric chloride, which is reduced to ferrous chloride during the etch

$$Cu + 2FeCl_3 \rightarrow CuCl_2 + 2FeCl_2$$

However, the activity of the etch bath can be extended by bubbling oxygen through the solution, which reoxidises the Fe(II) to Fe(III). However, build up of copper and various reactions and precipitates prevent continuous regeneration.

exiting the reactor, whilst air systems must have most of their flow rate passing right through, carrying off heat, water and chemical spray. These side-effects all have a cost and, in practice, oxygen, despite its high cost, can frequently be justified. This was the conclusion of a number of studies by Bechtel, Anaconda, Inco, Cominco and others (Denny, 1977).

Oxygen in gold leaching. An unusual hydrometallurgical reaction which uses oxygen is that of gold leaching. The GOLDOX process, developed in South Africa by AFROX, uses oxygen to assist in the extraction of gold in South Africa and elsewhere. The process operates basically as for the sulphide ores above, with the exception that the starting material is very poor (only approximately 5 parts per million of gold) and contains tiny particles of solid gold. The chemicals added are lime and cyanide, and oxygen is supplied to complete the reaction

$$4Au + 4CN^- + O_2 + 2H_2O + 2Ca(CN)_2 \Rightarrow 4Au(CN)_2^- + 2Ca(OH)_2$$

The aurocyanide compound is soluble whereas the hydroxide and many by-products (due to reaction with gangue minerals) are largely precipated. Oxygen may be added before cyanide addition to oxidise gangue minerals such as iron suphides, which waste cyanide. The aurocyanide is recovered in carbon absorber units. The process can use oxygen from the air. However, with pure oxygen there are advantages in smaller leaching vessels, increased gold recovery, savings in cyanide and improved carbon absorber lifetime.

Ore roasting. There are problems in dealing with large volumes of mineral-laden water in the processes described above, however, and hot smelting of sulphide ores is an old-established and popular process. Just as oxygen can help in the iron-smelting process it is also used in the smelting of other metals. For example, using oxygen or O_2-enriched air to burn it, chalcopyrite ($CuFeS_2$) can be converted largely to impure copper (ready for electrolytic refining) and sulphur dioxide (useful for sulphuric acid production).

Oxygen in non-ferrous smelting again seems a curious anomaly. The use of oxygen in non-ferrous sulphide smelting revolves around issues such as the reduction in fuel used to reach high temperatures and the reduction in waste exhaust gases. The use of oxygen reduces gas clean-up, gives a reasonably clean source of SO_2 and reduces all capital equipment size.

3.1.9 Oxygen flame enhancement

Flame emissitivy. The addition of oxygen to a flame increases its 'adiabatic temperature', i.e. the theoretical maximum to which it can rise. The high adiabatic temperature for 100% oxygen addition is very important for pure

Table 3.1 Effect of oxygen enrichment on radiant heat from flames

Enrichment (%)	Flame temperature (°C)	Radiation factor
0	1800	1.0
20	1965	1.4
40	2163	1.9
60	2405	2.8
80	2709	4.3
100	3100	7.0

These are approximate extrapolated figures for hydrocarbon/oxygen/air mixtures.

oxy-fuel flames in metal cutting and welding (section 3.2.3), while the addition of oxygen at less than 100% is useful elsewhere in processes such as iron-smelting (section 3.1.2).

Another example of the use of oxygen addition is in heating tanks of molten glass. Here addition of oxygen raises the flame temperature. It might be wondered why this is necessary, given that the melting point of commercial soda-glass is 900°C, well below the adiabatic flame temperature for air/methane. The answer lies in the inefficiency of heat transfer, much of which stems from the low emissivity of the flame. A flame transfers heat from its core to the object being heated by both convection and conduction. In many situations, where the hot gases from a turbulent flame flow around the object until they are appreciably cooled, heat convection is important. However, in other cases heat transfer is mainly via radiation. This is the case with a glass tank; conduction is limited because the tank is heated by a flame above the surface of the molten glass and the primary heat transfer[47] is radiative. The increased flame temperature due to oxygen enrichment leads to increased radiation, which is large compared to the rise in temperature (Table 3.1).

The effectiveness of oxygen-enhanced flames can be increased further in many instances where increased heat transfer is needed by the use of oxygen lances, separate from the main burner (Monnot, 1985). For example, where a tank of molten liquid is being heated from above by a flame, an oxygen lance directing oxygen into the underneath part of the flame just above the liquid ('under-flame lancing') will enhance heat transfer to the liquid whilst not increasing the general temperature of the furnace top. It also economises on

[47] Noting that flame heat transfer is dominated by radiation, some researchers have attempted to improve the radiation from a flame by means other than increasing its temperature, i.e. increasing its emmissivity. Noting that it is the infrared-active gases such as water vapour and carbon dioxide that dominate infrared emission in many flames, some attempts have been made to add even more infrared-active gases to the flame. Obvious candidates would be additional fuel gases or vapours such as acetylene, or polyatomics, such as the freons. Although in principle they would possibly be effective at small concentrations, such methods have not been shown to be effective.

the use of fuel and oxygen. Local heating enhancements of at least 150°C are achieved in practice.

Oxygen versus regenerative/recuperative burners. Regenerative burners save fuel by more fully utilising exhaust stack heat from a process and achieve higher temperatures than the simple adiabatic temperatures (around 2000°C) achieved in the combustion of fuel in air. They do this by preheating the air, most often with exhaust gases. This can be achieved by heating the heat exchanger with an auxiliary flame or exhaust (recuperative) or, more commonly, by heating in a bed of a refractory mineral (regenerative). There should be two or more beds of refractory so that one bed can be heated whilst another is being used to preheat the combustion air in the furnace.

This sort of system is most common in large installations such as blast furnaces or glass works, where the heat 'regenerators' are vast cylinders, 'stoves' full of honeycomb brickwork. Blast furnace regenerators can operate up to around 1300°C. Regenerative burners are complex, particularly in smaller sizes, and prone to problems but are potentially cheaper to run, in general, than oxy-fuel burners. Oxy-fuel burners are cheaper to buy and maintain but need costly oxygen supplies.

Oxy-coal burners. The use of oxygen with coal outside of the steel industry is still unusual. However, the potential advantages are high: coal, even micronised coal, is cheap and coal flames have a very high radiant emissivity and a high adiabatic temperature potential. Oxy-coal is now being tried in some parts of the world in large steam-raising furnaces.

3.2 Gases in metal fabrication

3.2.1 Gases in metal casting

A surprisingly large proportion of steel and non-ferrous castings are still produced by pouring molten metal into a mound made out of sand with the aid of a pattern. The major exceptions to this are relatively small non-ferrous castings produced in enormous numbers, for example many car parts. Sand moulds are not simply made like a child's sandcastle from beach sand bound with sea water, however. It is very difficult to mound sand precisely without a binding agent.

In the foundry, the sand used is superior in its aggregate make-up to beach sand. It contains a wider variety of particle sizes, including fine clay-like material which helps it to bind together and form smoother surfaces, and is carefully adjusted for moisture content. For many purposes this is sufficient. For more predictable properties, more robust moulds and better casting finish, however, much sand-casting uses binders either in the bulk sand or just on the mould surface. These binders are typically organic materials. Starch is one simple possibility for stiffening up the sand, although higher performance is obtained from resins, which can be induced to set hard, at least on the surface, by heat (such as phenolic resins), a catalyst sprayed onto the surface (epoxies can be sprayed with organic peroxides to achieve this) or by a reaction with a gas. In the latter category are silicate binders which set with the application of carbon dioxide. The surface-hardening binders can withstand the impact of molten metal, which can simply wash away weak sand surfaces.

Sand used for moulding is largely reclaimed and used again. Here organic binding resins present problems as a proportion of the sand used must always be dumped and this disposal is a problem. Another problem is the fumes emitted by the sand during the casting process; organic resins evolve unpleasant fumes. Finally, the bound sand must be removed from the casting and this can be a problem for sand stuck in the recesses of a casting if there is an appreciable amount of binder present. One process (EFFSET™) has been developed which neatly sidesteps all these problems and provides at once higher quality and environmental friendliness: ice binding of sand moulds. This involves preparing the usual sort of sand/clay/moisture mixture, shaping the sand around the master pattern by the usual ramming process, then removing the pattern and freezing the sand with a spray of liquid nitrogen. Once the surface is frozen the mound is very robust, being an aggregate/ice mixture. It is mainly aggregate and at the opposite end of the scale from the impressively strong pykrete (see section 3.7.2), which is mainly ice.

Surface oxidation and dross formation in casting can be reduced if an inerted path can be arranged from the melting ladle to the mould. Gas inerting of moulds is possible, as is the use of inerted enclosed ladling facilities. Even when only the mould is inerted, quality improvements are seen. Argon gas is a good choice for this because is completely inert and because, being heavier than air, it can be used to fill up mould spaces open at the top (left there to allow liquid metal in and for overflow), rather like topping up with water. Argon inerting is commonly used on both sand-casting and on die-casting installations.

3.2.2 Furnace atmospheres for metal processing and heat treatment

Heat treatment of metal parts has in the past often been carried out by heating parts coated with a substance which will react at furnace temperatures in a suitable way but the coating process can be laborious and the parts may be disfigured by oxidation in places due to air in the furnace. Even more laboriously, carburising used to be commonly achieved by packing parts into carbon inside a furnace. Today, a suitable furnace atmosphere can achieve heat treatment without oxidation or other undesired surface reaction, and the gaseous nature of the treatment allows all part surfaces to be processed. A mixture of inert and reactive gases, either oxidising or reducing, is employed. Mixtures based on a preponderance of nitrogen are inexpensive and often effective. Nitrogen is nowadays most often supplied in pure form by gas companies, although inert gas generators (section 2.2.1) may be employed. Argon may be necessary as the inert gas for highly reactive metals, although it is much more costly and has a competitor in the form of vacuum processing.

Heat hardening or softening in inert atmospheres. The use of an inert atmosphere in handling metals is of general importance because of the reactivity of air (mostly the oxygen and moisture in air) with most metals. Even with metals which are unreactive at room temperature and modest heat, heating of metals in air to the 500°C or more needed for metal working mostly results in tarnishing due to oxide film. It is often reasonable to accept the tarnishing or remove it subsequently by polishing. However, polishing can be expensive, needing additional machinery, and may be labour intensive. It is, therefore, very useful to be able to replace the air in a furnace with another, more inert, atmosphere.

Annealing. Bright annealing is carried out in furnace atmospheres containing hydrogen. The result is a clean surface finish (no oxidation) and, with the correct temperatures and low cooling rates, typically a soft condition, ideal for most metal-working. Pipes in stainless steel and copper are bright annealed as part of their processing, between cold-drawing processes, to make them easy to bend.

Annealing is carried out on metals that have undergone any kind of cold-working. Hardening occurs when a metal undergoes a large plastic deformation as in cold-washing. It is often undesirable because, for example, it may make further cold-working more difficult or may allow cracking in use. The component cracking occurs because non work-hardened regions of the component will yield more easily, being softer, whilst the work-hardened regions will be forced to take more stress and may crack. The work-hardening itself represents energy stored in microscopic prestresses, which are relieved when heated, as the metal flows to eliminate the strains.

Annealing of magnetic iron laminations used for the core in transformers is a particularly important application of gas atmosphere annealing. Iron laminations, rather than bulk iron, are used in transformers to avoid eddy current losses in the core. The laminations are often annealed in reducing furnace atmospheres, although the object here is not to retain a bright surface; the laminations will be hidden from view and protected with varnish in service. (Silicon/iron alloys, with suitable treatment, can also form their own glassy insulating oxide surface which does not require varnish.) Annealing heat treatment of iron laminations improves their permeability and reduces their hysteresis, making them more 'soft' in a magnetic sense. Laminations are often made from cold-rolled low-carbon iron sheet and then further cold-worked, in effect, by the stamping process used to shape them. Other laminations are made from 3% silicon/iron alloy, also low carbon, which has a permeability approximately ten times higher than that of simple iron. The deformations result in hysteresis, which leads to losses in the transformer application directly since the heating of the transformer iron core is proportional to the hysteresis loop area. Deformations also reduce magnetic permeability, which means that there are higher Joule (resistive) heating losses in the conductor coils. Annealing treatment is therefore doubly valuable.

Decarburising. Decarburisation is also useful for transformer iron laminations. An iron with the lowest possible carbon content is desired because then the maximum proportion of the highly magnetic 'ferrite' (soft magnetic) iron phase is obtained with no 'pearlite'. Pearlite is a phase of steel, comprising microscopically closely spaced layers of cementite (Fe_3C, iron carbide, a hard, non-magnetic compound) and ferrite. This phase is formed when 0.9% carbon is dissolved in molten iron. Even with 0.1% carbon, a transformer lamination might be 10% undesirable pearlite. The presence of the other phases also prevents the remaining ferrite from being so effective. Gas decarburisation using a mixture of nitrogen (to reduce cost), hydrogen and steam, or possibly a nitrogen/methanol/steam mixture, is a possible solution to further reduce carbon content. With 80 or 90% of the gas being more-or-less inert nitrogen and only 10–20% of the more expensive active gas, the following reaction is principally responsible for decarburising effects

$$Fe_3C + H_2O \Rightarrow 3Fe + CO + H_2$$

By this reaction it is easy to decarburise typically thin transformer laminations (approximately 0.5 mm) and it is possible to decarburise up to 3 or 5 mm laminations with longer anneal/decarburise times. The water/hydrogen ratio for the optimum decarburisation rate is reported to be in the range 0.2–0.5. The hydrogen presumably prevents damage to the steel by restraining the oxidation possibilities of steam on red-hot iron

$$Fe + H_2O \Rightarrow FeO + H_2$$

A temperature of approximately 800°C is necessary for this reaction, with heating for a few hours followed by cooling rates in anneal of as low as 50° per hour (faster for thinner laminations). A dwell time in the iron temperature of about 450°C in an oxidising atmosphere leads to a thin oxide layer, which in certain thicknesses gives a blueish colour. This part of the process is therefore often called 'blueing'.

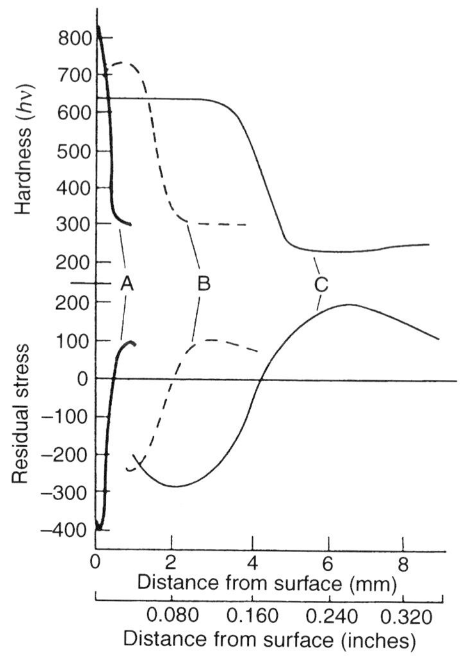

Figure 3.6 Hardness traverses (top) and residual stress distributions (bottom) through the surfaces of: A, nitrided 722M24 (80 h); B, carburised alloy steel and C, induction hardened Cr-Ni-Mo steel (Harper and Parrish, 1985).

Carburising. If many steel grades, including low carbon varieties, are heated in the presence of carbon or carbon-containing compounds to 1000°C or so, then hard martensitic steel and hard carbides will form in the surface and for up to 2 or 3 mm into the bulk of the material (depending upon time and temperature of treatment). As for nitriding, the carburised components can retain a tough core. Originally carburising simply involved packing steel parts into a charcoal mixture and then heating. Nowadays gas processing is used as it is more controllable and leaves the parts clean (Figure 3.6).

The carburising mixtures can be a blend of nitrogen and a carbonaceous gas such as propane (or methanol vapour), usually with a little adventitious air in the typical industrial furnace. If the gas exiting the furnace is flammable, it is normally burnt off using an automatic ignitor to minimise explosion risks; treated steel parts come out in an impressive gush of flame in the more old-fashioned type of furnace. Inside the furnace the gas is too rich or too diluted with inerts to catch fire. Fuel gases generally crack at the temperature of operation into H_2, CO and water vapour. Carbon dioxide can be used in carburising mixtures.

The 'carbon potential' (the percentage of carbon that will be incorporated in the steel surface) of the gas mix must be controlled. Carbon potential depends not only on the concentration of C in the furnace gas, but also on the percentage of water vapour present; more water vapour leads to a lower carbon potential. For example, a gas mix that reacts to produce 20% CO and 40% H_2 (from methanol) in N_2 would vary from 0.9% carbon potential at 0.3% water to 0.2% carbon potential at 1.2% water. An Ellingham diagram is sometimes used to predict what oxidation/reduction effects and carbon potential a particular gas mixture will achieve at operational temperature. This diagram is essentially a nomogram recording equilibrium reaction constants versus temperature.

Nitriding. Nitriding is a lower temperature process (approximately 500°C) than carburising. Originally carried out by immersing steel parts in molten salt baths based on sodium cyanide, the process is nowadays mostly based on ammonia as the active gas. The ammonia cracks to some extent on the surface of the steel, thus incorporating atomic N into it. The nitriding effectively inserts into the zone near the steel surface minute particles of nitride compounds, which strain the iron lattice and render the steel harder. The layer resists corrosion but penetrates only 1 mm or less into the bulk of the metal. Thus the steel retains its toughness qualities and is resistant to shattering, whilst having a glass-like surface hardness.

Argon and vacuum atmospheres. Vacuum is an alternative to furnace gas inerting with argon. Aluminium, magnesium, titanium, etc., can all be processed under argon. It is interesting to consider the costs involved in these two alternatives:

vacuum solution:	vacuum pumps: high capital costs strong vessel: high capital costs pump-down time needed slows down process
argon inerted solution:	thin-walled vessel: low capital costs argon supplies: high running costs purging time needed slows down process but this is a less serious defect because capital costs are low.

Clearly, if the emphasis is on minimum capital costs, then argon inerting is indicated, and vice versa. Vacuum may be most appropriate for sites with high utilisation factors where capital costs are less important.

Endothermic generators versus nitrogen-based atmospheres. Where a nitrogen-based furnace atmosphere is proposed, an alternative to simply assembling the components of the atmosphere from bought-in gases is to employ an inert gas generator to produce a nitrogen-rich atmosphere. A more sophisticated version of the inert gas generator described in section 2.2.1, an endothermic generator, burns fuel with air but restricts the oxygen supply to give a CO and H_2 content in the nitrogen-rich mixture. CO/H_2 content in a nitrogen-based mixture is normally achieved by adding methanol to nitrogen. The heat of the furnace causes cracking to CO and H_2

$$C_xH_y + O_2 + N_2 \Rightarrow CO + H_2 + N_2$$

The reaction needs the supply of external heat and yields, for example, 20%/40%/40% $CO/H_2/N_2$ if methane is used as the feedstock.

The use of relatively pure fuel supplies makes the control and output gas quality of an endothermic gas generator more acceptable than would otherwise be the case. With these higher purity feedstocks, a hot nickel catalyst bed can be used, which is heated externally. Residual fuel, nitrogen oxides, CO and sulphur compounds will all, however, be present in trace quantities and may cause process problems or corrode the pipework. Like inert gas generators, endothermic gas generators are only nowadays used in processes with minimal requirements on gas quality and a steady demand pattern, and then only when capital and running requirements must be minimised and the size of gas stream needed is not too large. The use of cryogenic nitrogen, either on-site or in bulk liquid tanks, is most often indicated. Finally, the very high CO and H_2 content means that the flammability problems of endothermic atmospheres are very serious: equivalent production is often possible with atmospheres which contain more than 90% N_2 and is much safer (carburising operations generally need the high CO content, however). Nitrogen availability also provides safety benefits in being available for purging of atmospheres prior to access; endothermic generator furnaces are at risk from fire or explosion during shut-down and start-up operations.

3.2.3 Flame welding and cutting

Flames. As noted in section 2.5.4 a flame is the reaction zone of a fuel or fuel/oxidant mixture with an oxidant. Most industrial burners use premixing of oxidant with fuel but with insufficient oxidant for complete combustion. The flame structure then comprises two parts: the central reaction zone of the premixed gases and a surrounding reaction zone where further oxidant is allowed to diffuse in and complete the combustion. An exception to this is the pure fuel 'diffusion' flame used for carbon-coating using acetylene. By contrast to the relatively dim blue optical emission of the premix flame, the diffusion flame for any hydrocarbon gas is a highly luminous yellow.

Although normally in dynamic equilibrium, a burner flame can move upstream ('lift-off') or, for premixed flames, downstream ('lightback' or 'blow-back'). Generally, a burner that runs satisfactorily on a certain pressure of a certain gas can be run satisfactorily on a range of different gases after adjustment of pressure or replacement of orifices. However, if it is desired to run a burner on a different gas without adjustment, then the replacement gas must be checked for its Wobbe index or Wobbe number. The Wobbe number is the calorific value per unit volume divided by the square root of gas density. Gases of the same Wobbe number can usually be used at the same pressure on the same burner without adjustment and without changing the heat output of the flame. This is because the flowrate, Q, of gas, density, ρ, Wobbe, W, and calorific value, C, are usually determined by a restriction of some kind, which will follow an orifice plate-type of pressure drop equation

$$Q \propto \surd(\Delta P/\rho)$$

$$H \text{ (heat output)} \simeq C\,Q, \text{ i.e. } H \propto C \surd(\Delta P/\rho) = \surd(\Delta P)\, W$$

The replacement gas will have the same pressure P, as the original gas, either because its supply pressure is the same or because pressure regulators in the system and at the burner are not changed. The flow rate Q', of a replacement gas of density ρ', Wobbe W' and calorific value C' will therefore be given by

$$Q' \propto \surd(\Delta P/\rho') \text{ i.e. } H' \simeq C'\,Q', \text{ i.e. } H' \propto C' \surd(\Delta P/\rho') = \surd(\Delta P)\, W'$$

Hence

$$H' = H\,(W'/W)$$

Note that the units in which the Wobbe number is expressed are irrelevant since the heat output change seen on changing gas to a burner is given by the ratio of Wobbe numbers. Note also, however, that even a gas of correct Wobbe index will not guarantee that lift-off or blow-back will not occur on changing gases, since the new gas may differ in other properties such as flame speed from the original gas.

Oxy/acetylene cutting. With metal thermal-cutting processes, the metal must typically first be heated to beyond its melting point over a small localised area. The molten metal must then be ejected somehow. In flame cutting, the metal is generally burnt by using an excess of oxygen in the flame. The oxidation of the metal is itself exothermic and generates additional heat, making the process a highly efficient one.

Ingenious designs of burners are employed to try to optimise the cutting speed and efficiency. The most common designs involve a mixture of oxygen and acetylene issuing from an annulus for heating the metal with a nozzle in the centre for cutting oxygen, the oxygen flow being switched on and off conveniently with a trigger on the torch (Figure 3.7). A metal mesh filter is connected in the gas supply lines to prevent flashback, i.e. the flame from the burner tries to travel back up the flexible supply hoses.

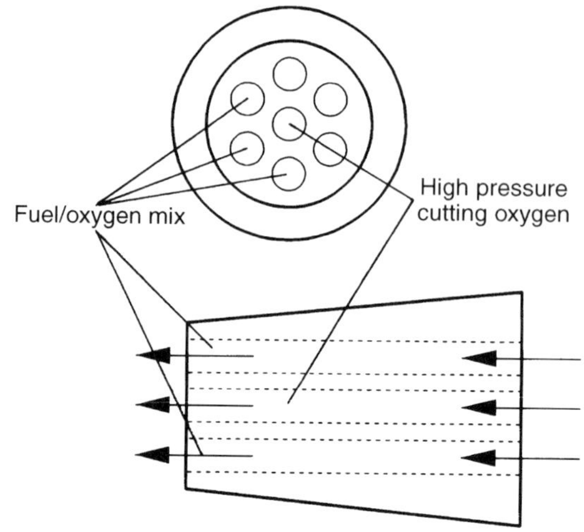

Figure 3.7 Oxy/acetylene cutting torch for cutting thick steel slabs.

The oxy/acetylene cutting process is, perhaps surprisingly, effective having:

- very high efficiency in the use of the relatively expensive acetylene fuel, although copious oxygen is still needed;
- very high speed, even on thick sections; sections up to 2 m or more are possible, whilst speed is such as to match steelrolling-mill output on specially designed high speed cutting burners;
- very narrow kerf width, down to 1 mm or so;
- a surprisingly neat surface finish with striations often below 0.5 mm along the cut edge.

There are many specialised designs of cutter, those used for 'scarfing' billets in steelworks, for example, or gouging nozzles for making channels in thick steel sections. It is even possible to inject extra oxygen via a separate lance at ultrahigh pressure, thereby burning and ejecting steel even faster. A patented device from the Messer Griesheim Corporation (Germany) employs a liquid oxygen feed right to the burner head at high pressure.

Steel plates which have been cut with oxy/acetylene do have a hardened region or 'heat-affected zone' (HAZ) which can make subsequent machining difficult. The hardening is due to the formation of an interstitial iron/carbon phase in the steel (martensite). This forms when the steel has cooled too quickly to allow the formation of stable, tougher iron/carbon phases such as pearlite. However, this hardened zone is acceptable in many mild and other steels, and can be ground out if necessary. If the next operation to be performed is welding, then the hardness of the cut zone is not such an important factor as the welding operation will introduce a HAZ of its own.

The HAZ is actually used in the process of flame-hardening, where an oxy/acetylene flame is used to rapidly heat up the surface of a steel component, whilst immediately after heating the steel is cooled with a jet of water. Special burners incorporating a water-cooling jet just behind the flame are often used. Because of the rapid heating, the depth of metal hardened is limited, leaving the core of the component still tough.

Oxygen purity effects. Oxygen purity is now so high (99.9%) that effects are not normally seen. However, in cutting of steel in the past, when oxygen purities from 97% upwards were commonly supplied, cut rates and finish quality were found to be related to the oxygen purity. The Welding Institute in Cambridge accumulated much data on this, as did shipyards in the former East Germany.

More recently work with laser cutting has shown that oxygen purity effects can be seen in the quality of cut in the more controlled conditions of a laser/oxygen cut. The effect of (typically argon nitrogen) inert impurities would appear to be the formation of an insulating blanket of gas, slightly reducing heating and reaction of the molten steel with oxygen. The blanket forms as the molten metal reacts instantly with the oxygen, leaving the inert gas fleetingly in a thin layer above the kerf.

Oxygen cutting processes in steelworks. Oxy-fuel cutting is a very large user of oxygen in the steel industry. Scrap must be cut up so that it is small enough to go in the furnaces. Large ingots will typically need cutting into smaller ones; sometimes this can be done whilst the ingots are still at red heat, sometimes on cold material. Ingots are often split up into different sections. Steel from the middle of an ingot, for example, has different properties from that at the edges and different end-applications will need those differing properties. Sheet material is often cut by multiple oxy/acetylene cutting torches – these can be arranged to cut diagonally across the metal so that a cut at right angles can be made on a continuously moving sheet.

'Scarfing', a specialised form of cutting, is used to remove surface defects from ingots prior to rolling and further operations, and is a significant oxygen application. Automatic scarfing equipment blasts high-pressure oxygen, together with propane fuel, at billets of steel as they pass at a walking pace, removing a millimetre or two of surface in a spectacular shower of sparks. In this way high surface quality can be guaranteed on steel plates and coils. This is particularly important in thin sheet.

Alternative oxidants and fuels for hot flames. The nitrous oxide/acetylene flame has been suggested and used on occasions because nitrous oxide can be stored as a liquid in relatively lightweight high-pressure cylinders, whereas oxygen must be stored as a gas in heavier higher pressure cylinders. This means that an N_2O/acetylene flame kit can be made to run for longer

Table 3.2 Adiabatic flame temperatures

Reactants	Approximate temperature of flame (°C)	Stefan–Boltzmann factor
O_2/acetylene	3070	6.1
O_2/methylacetylene	2975	5.4
O_2/propadiene(allene)	2975	5.4
O_2/ethylene	2900	4.8
O_2/propane	2815	4.3
$0_2/H_2$	2805	4.2
O_2/methane	2780	4.1
Air/acetylene	2325	2.0
Air/propane	1955	1.0
Air/methane	1850	0.8
N_2O/acetylene	2880	4.8
C_2N_2/O_2 (cyanogen)	4630	31.5
F_2/H_2	HOT!	

from the same size and weight of gas bottles than an O_2/acetylene kit, but still has a very hot flame.

Nitrous oxide flames are also employed in some flame spectrometers. The flame vaporises aqueous samples, typically mixtures of trace metals, to temperatures where they emit light in characteristic line spectra. Analysis of the spectrum emitted gives the concentration of each species down to parts per billion or lower concentrations. NO_x emissions are likely to be higher in nitrous oxide-based burners. (In coal-firing it is well-known that nitrogenous compounds in the coal fuel lead to higher NO_x levels in the exhaust gases.)

The temperature achieved in an oxy/acetylene flame is the highest of the common practical flame systems. A list of a few theoretical adiabatic flame temperatures gives a flavour for the maximum achievable (Table 3.2).

The figure in the last column of Table 3.2 raises the temperature of the flame to the fourth power and normalises on air/propane to give an idea of the power radiated by the flame if it follows the Stefan–Boltzmann radiation law.

The use of a vaporisable liquid fuel gas has clear advantages in terms of cylinder calorific capacity. The MAPP (methylacetylene/propadiene) mixture is popular in the USA, rather than acetylene, because it provides a very hot flame at lower cost and more conveniently. Propylene is also favoured in applications such as flame-spraying where a large gas capacity and usage is expected.

More exotic flames are also possible, as shown in Table 3.2. Cyanogen is a hazardous gas that is only manufactured and used as a laboratory curiosity. It is poisonous and condenses at −21°C. Fluorine is too dangerous, as is its combustion product (HF), to be considered as anything other than a curiosity with regard to flame heating.

The achieved adiabatic flame temperatures listed in Table 3.2 are functions not just of the released heat of reaction but also of the heat capacity of the combustion products and the degree to which the final products are dissociated. With air the flame is cooled because of the requirement to heat up the inert nitrogen. Similarly, dissociation of the reaction products tends to reduce the flame temperature of hotter flames. There is also an effect that the measured temperatures are reduced by the production of free radicals and incompletely reacted products in the flames. This can be easily verified by measuring the electrical resistance of a volume of gas at modest voltage (normally infinite) and repeating the experiment with a flame. Most flames show a surprisingly high conductivity, though still in the gigaOhm or more region, showing the presence of ions and free radicals. An acetylene flame, for example, is a potent source of free radicals.

The order of flame temperatures is not the correct parameter for ranking fuels for a vehicle, for example, where heat of reaction is important, and different again from the ranking for rockets where heats of reaction per unit weight and molecular weight come into the equation.

The relative effectiveness of these high temperature flames in the task for which they are most usually required, heating an already hot solid hotter still, is a very steep function of their temperature but also depends on their emissivity. The Stefan–Boltzmann law for thermal radiation of grey bodies illustrates the effects. It states that the power, P, radiated by a body's surface is related to a constant for the surface the emissivity, ϵ, and the absolute temperature, T as follows

$$P = k \epsilon T^4$$

The complete problem is, in fact, more complex than this owing to the fact that thermal emission from flames does not follow the simple Planck spectrum, which is a smooth broad peaked distribution of power. The frequency of the peak varies with temperature: white when hot, yellow, orange or red when cooler. Instead, the infrared emission is banded, with polyatomic species emitting much more effectively than diatomics or monatomics (section 2.3.5). This may account for some of the claimed advantages of the bubbler systems sold for flame enhancement. These add a vapour from a bubbler filled with liquid, consisting of highly unsaturated hydrocarbons, to the basic fuel gas stream. An enhancement of the flame temperature and effectiveness is claimed for the systems in the case of the lower performance flames such as oxy-methane.

In the radiation of visible light the particulate carbon content of the flame seems to be the most important factor. This explains why burning hydrogen is almost invisible.[48] The hydrogen flame emits light and becomes electrically conductive in exact proportion to the rate at which carbon atoms are added to it. (Advantage is taken of the latter effect in gas chromatograph instruments with flame ionisation detectors as mentioned in section 2.3.4.)

[48] The near invisibility of hydrogen flames is a real potential hazard. Hydrogen flames can occur at leaks and not be perceived until the heating effect of the flame has disastrously melted or set fire to something else. Hydrogen does show up in the most surprising places and, if ignited, can cause trouble. Caution is advised even near simple pieces of domestic apparatus such as a battery whilst it is being charged or a hot-water radiator whilst it is being bled to remove trapped gases. In both these cases small quantities of H_2 can accumulate and, because of the minute ignition energy and wide flammability limits of hydrogen, start fires.

Thermic lancing. An unusual example of a fuel oxygen flame used for cutting is the thermic lance. The thermic lance is one of the few devices that will cut through rock and concrete, even steel-reinforced concrete. The lance consists typically of a bundle of steel rods inside a steel tube of approximately 20 mm diameter. Oxygen under a few bars pressure is passed down the tube and the end is lit with the aid of an oxy/acetylene torch. The result is a spectacular white-hot firework flame with formidable penetrating powers. Note that once lit the thermic lance is running on 'solid fuel', i.e. the steel rods and tube.

The thermic lance is now falling out of favour, however, owing to advances in the art of diamond-tipped drilling for concrete; the drill makes a much neater though much slower job than the lance. However, a typical 3 m lancing rod, driven by a normal oxygen cylinder, can cut a 3 cm hole in reinforced concrete 50 cm thick in approximately 3 min. The noise, copious clouds of smoke and blast furnace heat of the lance are drawbacks, however.

It is possible that thermic lancing on a smaller scale could be more attractive and small (6 and 10 mm diameter) lances which are electrically ignited are now available. They can be started by earthing one end, passing oxygen over them and igniting with the small arc on a copper plate connected to a car battery.

3.2.4 Flame spraying/coating

Hard-facing with welding rods. It was noted early in the history of welding that the weld surface on mild steel was much harder than the parent metal (section 3.2.3). This led eventually to hard-facing softer metals with harder metals and from there to the whole industry of flame-coating and flame-spraying. In its simplest form, a coating of, say, hard Stellite alloy can be applied as a welding rod to a mild steel or other substrate exactly as if a welding operation were being carried out. Subsequently, the workpiece may need heating to consolidate and fuse the deposit, and the surface can be ground completely flat if necessary. This sort of operation has often been used as a maintenance tool, with a worn component being built back up to its original size and form with the weld metal. The repaired article, being hardened on its surface, so slowing the wearing process, will often improve on the lifetime of the original monolithic article.

Flame-sprayed powders for coatings. This technique has now been extended to include much more precisely controlled processes with a flame held a precise distance from the substrate (usually using an automatic machine) and actually keeping the substrate out of the direct flame. The coating material is added as a very fine powder to the flame gases. The red-hot minute particles of coating impact on the substrate surface and partially fuse there but do not heat the substrate enough to damage it, which is the case with 'welded-on' hard-facings. The flame-spraying technique using powders is also more versatile because the substrate is not unduly heated and the coating applied can be very thin but just as effective. The thinness of the coating is

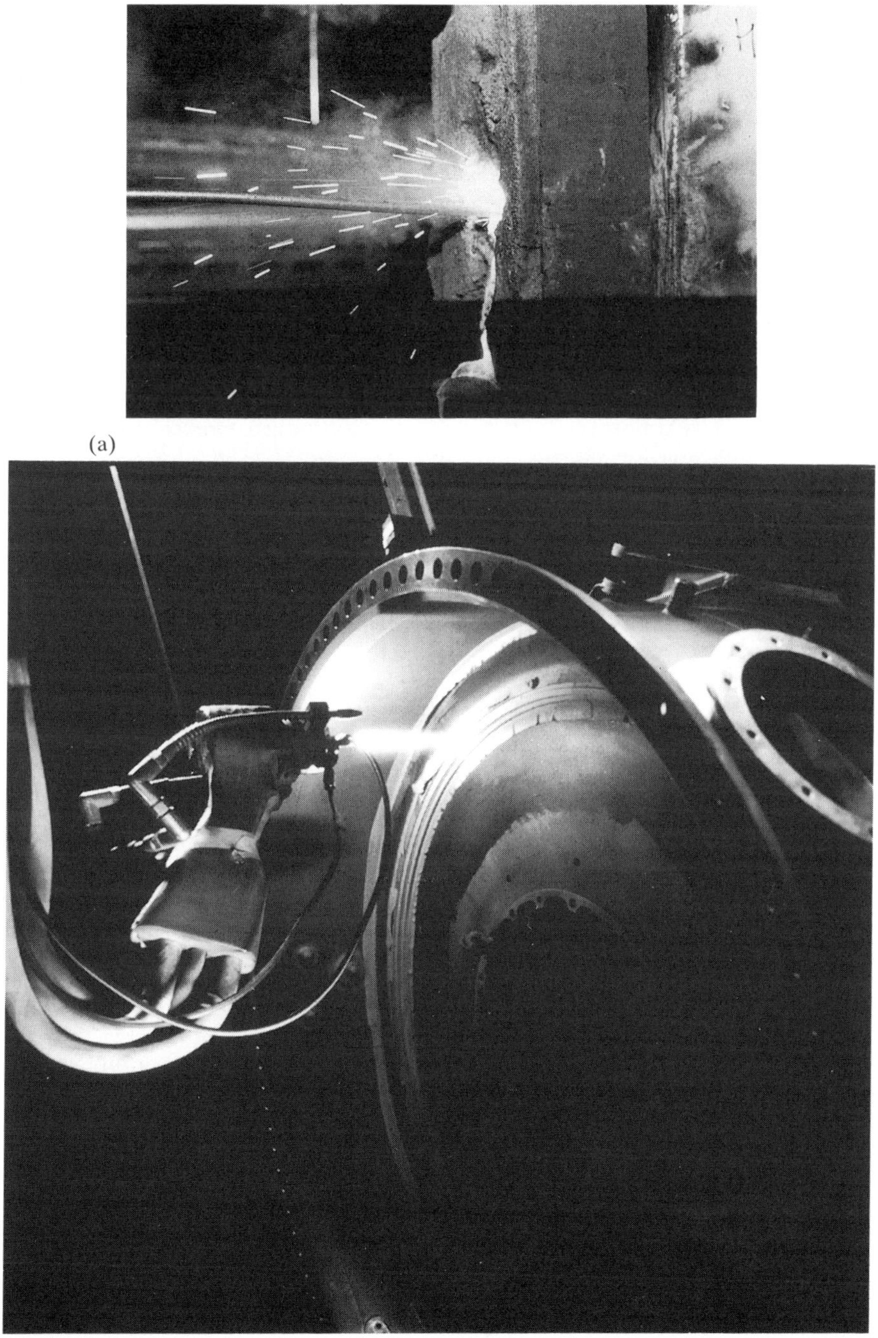

(a)

(b)

Figure 3.8 (a) Thermic lancing of masonry. Note stream of molten iron oxide and minerals running out of hole. (b) A Sulzer Metco plasma spraying system being used to spray a component of an aeroengine. The spray gun is mounted on a robotic arm which moves in five axes, the robot itself being moved linearly to provide a sixth axis (courtesy of EMMS).

often an advantage because no further grinding or machining operation need be performed.

A large number of materials can be flame-sprayed, not just metals. Ceramics such as carbides or alumina (for wear resistance), calcium minerals (for rendering metal screws and parts biocompatible) or cermets (metal/ceramic composites) can be treated. Patents covering boron nitride, zirconium nitride, chromium carbide in nickel and refractory silicides have been filed, covering applications from simple wear resistance to protection of metals from attack by other metals in the molten state.

Increasingly sophisticated apparatus is now used for flame-spraying, in which high-pressure fuel gas and oxygen are fed to a supersonic nozzle, or a 'detonation gun' (the Praxair patent process), using fuel/air mixtures (Figure 3.8). The flame nozzles are often designed to use a very high flow rate of fuel and oxygen at a relatively high pressure (up to approximately 10 bar). A liquefied high-energy gas such as propylene or MAPP is often specified because the quantities needed are inconvenient with the much less efficiently stored acetylene and because of pressure limits on the latter. However, with draw-off rates of many cubic metres per hour, it is often necessary to use a liquid withdrawal from the cylinders of fuel and a heated vaporiser unit.

The metal or ceramic powder to be coated achieves quite high temperatures and very high speed (up to 1000 ms^{-1}). Despite these extreme conditions, the substrate must not be heated unduly in order to avoid damage and differential thermal expansion problems when the coated parts cool. The maximum substrate temperature depends on substrate and coating but is often a rate-limiting step in the coating process as temperatures as low as 100°C are sometimes required. Despite very high fuel/oxygen powers (100 kW is not unusual), therefore, the coating rates are limited: (a recent paper quoted 20 μm per minute for coating a commonly required tungsten carbide/cobalt composition).

Coating rates can be improved by using forced air cooling on the substrate immediately after a layer has been coated. In other cases, liquid carbon dioxide has been sprayed onto the substrate immediately after coating. Liquid nitrogen is also being tried. When coating rotating substrates, cooling can be achieved by simply following the coating flame gun with a cooling blast. When coating flat substrates effectively, instant cooling can be achieved by raster scanning the flame gun with an adjacent cooling blast gun and thus high coating rates can be achieved.

Flame welding. Welding with an oxy/acetylene torch (flame or 'gas' welding) is still a very popular process. It is not, however, now regarded as giving the highest quality joint for many purposes. The weld produced by a gas torch is often inferior to that produced by an inerted arc process. By careful choice of welding rod, flux and practice of technique, however, perfectly satisfactory results can be obtained.

The technique of gas welding is perhaps more of a manual skill than arc welding (section 3.2.5). The problem on thinner sections is that the welding torch wants to be a cutting torch and beginners often burn right through the workpiece. This problem can be restrained by small circular or side-to-side motions of the torch but it requires skill to achieve good penetration of the weld at the same time. In an ideal situation, the metal plates to be welded are prepared with a 'V' channel between them of a little under 90°. The torch is angled up the channel and used to produce a small pool of molten steel into which the filler rod, angled down the channel, is melted. The whole process is then moved steadily up the joint.

3.2.5 Arc welding and cutting

With the development of high-current power supplies grew the realisation that extremely high temperatures, far higher than flames, could be reached by striking and maintaining an arc. The first processes used an arc struck in air between the workpiece and a rod of weld metal. Because of oxidation and nitrogen embrittlement caused by the high oxygen content of the arc, flux materials were packed around the welding rod, resulting in the familiar 'stick electrode' process used today for rough welding work. The use of shielding gases to substantially keep out the air was a later development that enables much higher quality welds to be obtained. A landmark in the development of arc welding was the large-scale promotion, beginning in 1948, of the MIG (metal inert gas) concept by the Air Reduction Corporation (now AIRCO) in the USA.

Plasma and arc cutting. Plasma cutting involves transferring a powerful stream of highly ionised gas, also containing many atoms and molecules in an excited energy state, onto an earthed metal plate. There the highly energetic neutralisation and deexcitation processes of the stream rapidly heat the earthed plate to melting point and the high-speed stream then carries the melted metal away leaving a relatively neat cut. The arc is struck between a tungsten electrode and a water-cooled copper nozzle, the whole torch being maintained about 10 mm from the workpiece. Arc cutting employs a simpler torch in which the arc struck is between the tungsten and the workpiece. Used only for rough work, arc air cutting relies on blowing air at pressure through an arc struck on a metallic workpiece.

Argon–nitrogen and argon–hydrogen mixtures are used for high-quality cutting, giving kerf widths of about 3 mm in a variety of materials including aluminium, steel, copper and other non-ferrous materials. The fact that the method does not rely on the oxidation of the material, as in oxy/acetylene cutting, gives it greater versatility. There are some safety implications for plasma cutting: the exhaust gases contain NO_x and H_2 and must be extracted, whilst noise levels are high.

Argon arc welding. Arc welding can be performed with stick electrodes which are coated with a thick layer of flux materials, such as carbonates, which evolve CO_2, thus providing a gaseous shield, or with fluxes which form a glassy slag film that isolates most of a joint from the air. However, manual metal arc welding, with stick electrodes, does not in general produce very high quality joints. The arc is difficult to stabilise, the glassy film of slag makes the finished joint messy, whilst air still penetrates the slag and causes oxide and nitride formation making joints in steel brittle.

Inert gas arc welding employs two properties of argon: the stable electric arc it forms at a few hundred amps and a few tens of volts, and its inability to react with metals at white heat. The concept of the process is very simple. A few hundred litres per hour of argon-based gas are passed up a tube around an electrode to which a suitable power supply is connected. An arc is struck on the workpiece and heats it while the gas flow forms a blanket, preventing excessive oxidation.

Careful control of the process is essential and the following points must be noted:

- metal must not drip out of the joint;
- the heat-affected zone must be minimised and have acceptable metallurgical properties;
- inclusions of oxides or nitrides will lead to weakness;
- segregation of alloy components is usually undesirable;
- excessive oxidation must be avoided;
- the weld material and HAZ should retain similar electrochemical properties to the parent metals in order to avoid joint corrosion;
- blobs of molten metals ejected from the joint area, 'spatter', is difficult to clean off afterwards and should be controlled to a low level.

Naming of arc welding processes. The names of electric welding processes are currently going through a transitional period in which new names (which more accurately describe the processes) are supplanting the older terms (Table 3.3).

Table 3.3 Acronyms for welding processes

Process	Current acronym	Old acronym
Manual metal arc	MMA	
Gas–metal arc welding	GMAW	MIG
Metal–active gas welding		MAG
Gas–tungsten arc welding	GTAW	TIG

Tungsten–inert gas (TIG) welding. An argon arc can be used simply to transfer heat to the parent metal. By ensuring a close fit between the parts being joined (a fit of a fraction of a millimetre is possible for many small components and for larger components with regular shapes) all that is needed is for the arc to provide the necessary localised heating in an inert atmosphere to melt both metal edges and fuse them together. Addition of helium to the argon is used for aluminium, whilst hydrogen is added for stainless steels. For duplex stainless steel argon with a small (less than 3%) amount of nitrogen can be used. (Duplex stainless steels are a newer series of alloys which have a two-phase structure and offer superior strength and corrosion resistance compared to standard stainless steel.)

The TIG welding 'torch' is a simple device with a tungsten tip and a ceramic nozzle to protect the end of the gas supply tube. The tungsten tip is usually made from tungsten with additions of thorium or zirconium oxides, which increase electron emission and ease starting of the arc. Depending on power level, the tip can be anything from 0.5 to 6 mm or more in diameter. The power supply is usually a versatile device capable of supplying positive, negative or a.c. voltages up to 100 V or so. Aluminium is usually welded with an a.c. electric supply, the positive half-cycle allowing positive ion bombardment of the tenacious oxide film on Al, which inhibits its welding with a d.c. supply. Most other metals are joined with a negative electrode d.c. supply to reduce loss of tungsten. Pulsed power supplies are now common in professional equipment as well as 'arc-striker' high-voltage pulse or radio frequency (RF) burst options. The power supplies are nearly always arranged to drop in voltage in a controlled way as current increases; this stabilises the arc produced.

The whole process is fairly easy to automate and small inexpensive automatic TIG welders are now common. Simple geometry welds, such as pipe butt welds, are now commonly performed only by automatic 'orbital' pipe welders. These are clamped onto the two pipes to hold them butted together. Inside the weld head are electric motors to rotate ('orbit') the head around the pipe. These can be programmed to allow for differences in welding at different ambient temperatures, orientations and with different metals. Once programmed, they produce virtually perfect welds consistently with very little fuss.

Metal–inert gas (MIG) welding. MIG welding replaces the tungsten electrode with a piece of filler wire that is automatically fed by an electric drive up to the torch head. It is now a very popular process in industry and is commonly seen in the motor trade where an oxy/acetylene burner would previously have been used. It can be carried out at high speed on thick sections but is also very good on the thin plate used in vehicles. 'Gasless' MIG welding relies on a barium carbonate-based flux in the middle of the feed wire to produce CO_2 gas (although it also evolves an unhealthy amount of barium). Gasless MIG welding suffers from quality problems similar to those of stick-electrode welding, however, and it is difficult to make the wire with enough flux to be effective and still leave it easily handleable.

MIG welding does not work well with pure argon. Early MIG processes relied on CO_2. CO_2, however, is today only used occasionally on mild steel, mainly in low-cost work, because it is more difficult to control and has increased spatter relative to the new gas mixtures. The success of most MIG welding depends on a mixture based on argon or helium. Oxygen at 1% or so produces a more stable arc and better fusion and penetration via lower surface tension in the weld pool. Increasing helium content stabilises and increases the arc heat output. CO_2 addition in the argon/CO_2 mixtures improves 'wetting' in the weld relative to argon and results in very low spatter. Argon plus 5% CO_2 or similar is now the most common welding gas mixture for MIG processes.

The rate at which the filler wire is fed to the welding head determines to a large extent the welding current (which may be anything from 100A to 600A) and the rate of welding. The filler wire is copper-plated to avoid corrosion and possible jamming in the head feed. Depending on the arc current, arc to work distance and other parameters, metal may be transferred from the consumable electrode to the work in three distinct ways:

- at short arc distances by the electrode actually dipping into the molten weld pool;
- in longer arcs by high-speed formation and transfer of droplets a few millimetres or less in diameter;
- in longer arcs by a fine spray of liquid metal from the electrode.

The voltage/current characteristic of the arc depends on the metal transfer mode. In dip-transfer large currents pass at low voltage, whilst in spray-transfer currents are lower and voltage higher. The gas type affects the process, with argon-based gases encouraging much higher metal transfer rates.

The MIG welding 'torch' is slightly more complex than that used in TIG welding. The feed wire goes down the centre of the tubular central ceramic nozzle and is surrounded by a gas supply tube. A switch on the torch starts and stops the wire feed to the unit. Wire from a motorised reel mounted on the power supply is pushed up the torch tubing, which also provides gas, electric current and, on large units, water cooling. The torch may also need to have a 'puller' unit for feed wire if the wire is too thin to be fed from the power supply end.

Noxious welding fumes. The role of gas in the production of noxious welding fumes and gases has been studied in recent years. Although fumes and gases from arc welding should be sucked away if at all possible by an exhaust hood, in practice it is still an advantage for the operator if less noxious exhaust is produced. Using lower currents for welding often reduces fume production but, of course, it also reduces productivity so the total amount of fume produced per amount of product fabricated is not decreased. Ensuring various conditions are about right, i.e. arc length, shielding gas flow, arc current/voltage control (including correct selection of a.c., d.c. and polarity), is important in restraining fume and noxious gas production but does not hold the key to really low levels. The gas companies have for many years believed that the composition of the shielding gas might hold the key and have devoted considerable amounts of R&D effort to this problem over a long period of time.

Unfortunately, this is by no means an easy problem: it seems that gas mixtures that reduce fume increase NO_x and vice versa. Ozone and UV production is reduced by the use of CO_2 or helium-based mixtures. However, fumes are in general increased in CO_2 or helium-mixture welding. As ozone, NO_x, iron and oxide particles, and intense UV light radiation are all potential hazards, the best that can be done today is a compromise.

A discovery that may eventually lead to a solution is the AGA company's research showing that NO in the welding mix restrained the production of ozone without appreciably raising levels of NO_x. The potential toxicity of a welding mix with NO, albeit at a fraction of a per cent concentration, has stopped this discovery being generally adopted. More R&D directed towards a more viable solution is in progress at several gas companies.

Arc/plasma coating. Just as flame coatings can be applied by simply welding additional metal onto the substrate, so arc welding can be used to carry out hard-facing of softer alloys. In a similar way, just as flames can be used to apply coatings formed from metal powder, so controlled arcs can be used to coat substrates using powdered materials. In fact, the variety of powders that can be applied with arc or plasma-torch processes is much larger than can be used with the comparable flame processes. Ceramic powders and high-melting compounds, such as tungsten carbide, are possible and the process can be applied not just to metal but also to ceramic or glass substrates.

Related to arc/plasma coating are processes that aim to build up bulk articles by thickly coating a mandrel and then separating the final component. In these processes, of which the 'Osprey' process of Osprey Metals is one example, a cup-like mandrel or die is sprayed with metal droplets of size as small as a few microns by an argon or nitrogen spray-gun fed with liquid metal. The mandrel or die is often rotated to ensure even deposition. After the die is full, the article is separated from the die, further heated if necessary and given a final compaction with a hydraulic press. This process is aimed at achieving high strength in difficult-to-work aero-engine components. It can produce a metal with a completely even microstructure;

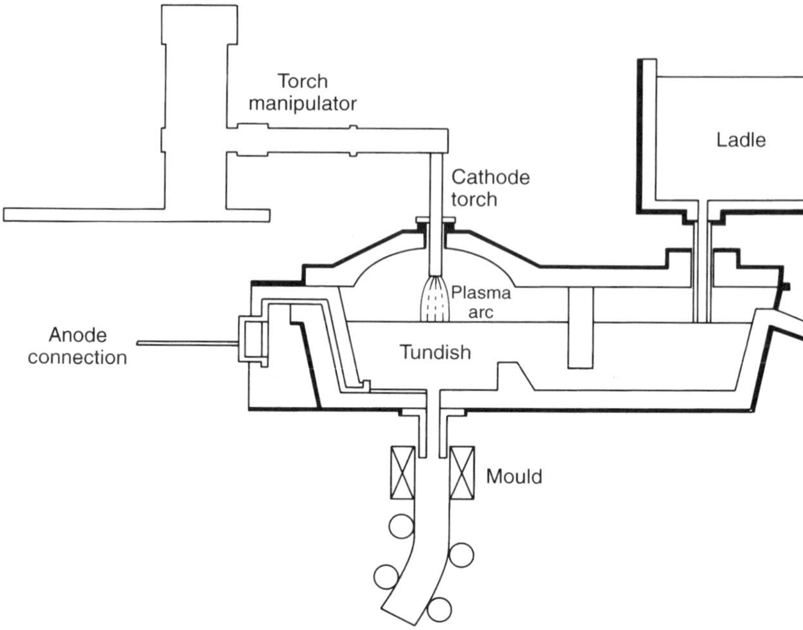

Figure 3.9 Megawatt argon plasma torch and tundish assembly.

no crystal growth patterns relating to casting or boundaries relating to powder metallurgical origin are seen. Components up to a few kilograms have been made in this way in Ni/Co/Cr alloys intended for 800°C service.

Plasma-torch heating and melting. Megawatt plasmas are used in heating metal in the steel foundry industry. Steel for casting is prepared in a ladle (section 3.1.4) and, when its composition is satisfactory, often poured directly into moulds. However, for continuous casting of strip and bar, a holding vessel above the water-cooled continuous casting mould ('tundish') is employed. This vessel can be advantageously heated with an argon plasma torch (Figure 3.9).

A typical apparatus has a lidded refractory vessel with a plasma torch which can be inserted through a relatively small aperture in the top. The torch comprises a coaxial copper probe (with water cooling in the walls) with electrical connections to supply a relatively low voltage (100–200 V), high-current d.c. arc and high melting point tungsten tips. The tips are treated with thoria to enhance electron emission and assist with arc striking. High-pressure argon at 10 m^3 per hour or more is used to blow the arc 30 cm or so out of the torch tip, the intense heat melting the scrap charge swiftly with the minimum of added impurities and exhaust stack emissions.

The big advantages of the plasma torch over other heating methods are probably convenience (with low capital cost) and controllability. Molten steel can easily be regulated in temperatures to 5°C or so and the small lightweight torch manipulated through its hole in the top dome of the tundish (steel-casting melting vessel). When steel is first poured into the tundish some heat is added to compensate for that lost to the cool refractories. As casting continues, the level in the tundish goes down and the charge begins to cool, needing further heating. The addition of a new ladle of molten steel will require further adjustments in temperature. By arranging the steel-making ladle to deliver a melt that is always a little cooler than is required by the chilled moulds, simply adjusting torch power can regulate precisely the casting temperature. Standard gas burner heating is less controllable and can cause unwanted carbon, nitrogen or oxygen additions to the steel.

There are also side-benefits from plasma-torch heating that derive from the fact that plasma torch systems can be sealed, or nearly sealed, from air ingress. Oxidation and nitridisation of the melt can be minimised as can pollutant gas and particulate emission.

Plasmas are also used in the coagulation/melting of metal dust in smelter and electric arc furnace exhaust stacks. Here fine powdered dusts, containing large amounts of volatile metals such as zinc, lead, cadmium and other potentially toxic metals and their oxides, are collected in the usual bag filters. Instead of dumping, however, the dust is melted by a plasma arc with micronised coal and fluxing agents such as calcium salts. The melt produces a relatively innocuous slag, which is tapped off and dumped, and a stream of reduced vaporised metals that can be recycled to a large extent after condensing and chemical extraction. It is worth pointing out that some of these 'by-product streams' from smelters can be extremely valuable. Companies such as Britannia Refined Metals near London, UK, for example, which carry out lead smelting, rely on the fact than many lead ores produces copious quantities of valuable silver.

In the case of electric arc scrap steelmelting, prodigious quantities of dust are emitted and these are typically dominated by zinc from galvanised scrap. The volatile zinc is evaporated from the plasma arc furnace and taken off to a zinc condenser. Zinc recovery can be further enhanced by passing the hot gas from the plasma furnace through a flame with excess air, forming yellow zinc oxide, which can be filtered (the zinc-only filter baghouse is much smaller than that required for immediate filtration of the whole dust stream and the zinc oxide is pure enough to be saleable).

Similar techniques are used to melt down spent catalysts from automobile exhaust systems. With the rise in the use of catalysts to restrict CO and NO_x emission from cars, the use of platinum or other precious metals, coated on to porous ceramics supports, has grown. Although the amount of platinum in a exhaust catalyst weighing several kilos is only a few grams, it is still essential to recover it on economic grounds.

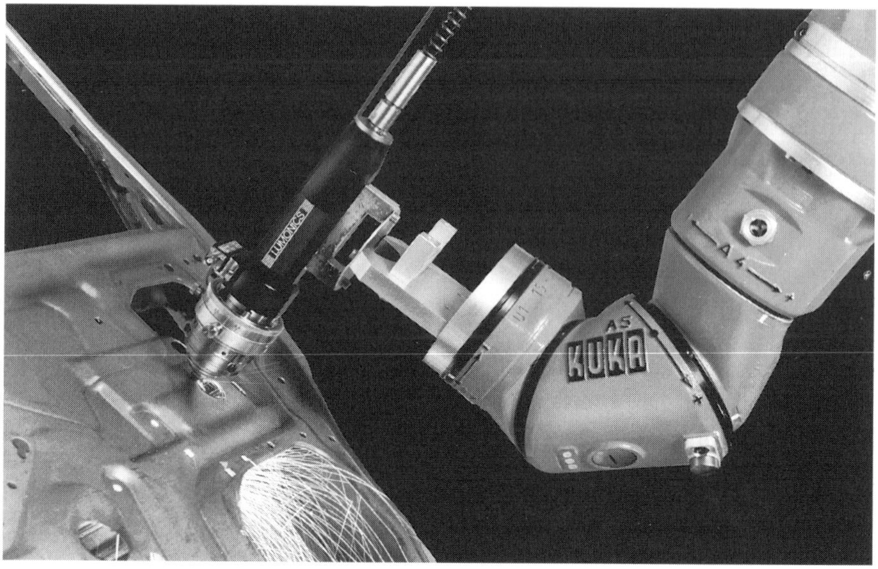

Figure 3.10 A robot manipulates a carbon dioxide laser to cut steel automotive parts using oxygen assist gas (courtesy of LUMONICS Ltd).

3.2.6 Laser welding and cutting

Cutting metals and other materials, from plastic film to refractory ceramics, is a well-established application of lasers. In industrial practice, a continous infrared carbon dioxide laser beam is the workhorse of the industry and this cuts an almost invisibly small slot in metal at a distance of a millimetre or two from the end of its nozzle.

The use of lasers for cutting relies in general on an 'assist gas', typically oxygen, which reacts with the metal melted by the laser and leaves a clean cut. Similarly, for materials which cannot or should not be burnt, a high-pressure jet of nitrogen can be employed. The assist gas is supplied to the cutting point by directing it into a coaxial nozzle assembly. This ensures the precise placing of gas pressure above the molten metal pool formed by the laser.

In operation, the first task is to start the cut, either by starting at the edge of the metal or by 'keyholing', i.e. cutting an initial hole straight down. The keyhole is often ragged so keyholing is performed on a part of the workpiece that will be scrapped. Once started, very heavy gauge and very tough steel can be cut, leaving only a tiny kerf of wasted material, often less than 0.5 mm. The finish of the cut edge is the key to the economics of laser cutting; frequently no further operation is required to smooth the edge, saving considerable finishing costs.

Carbon dioxide lasers, emitting continuous beams of a few kilowatts in the

10 μm infrared band, are the most common laser used in metal processing. Neodymium-YAG (yttrium aluminum garnet) lasers, emitting short extremely intense pulses of 1 μm radiation, are favoured only for drilling small holes in thin material. Copper vapour lasers are now being evaluated in pilot applications for drilling of small holes.

As noted for flame cutting, oxygen purity is an issue with laser cutting. In laser cutting, the effects of inert argon present in commercial oxygen at the 0.2% level can be seen. Both cutting speed and cut quality deteriorate as the argon content rises above this level. Pulsing the oxygen cutting gas pressure rapidly (up to 500 Hz) has been tried as a means of improving cut quality but it is not clear that any advantage is gained. Laser cutting is now a relatively popular process and is growing as its quality advantages become widely known.

Laser welding is still an unusual process. However, it produces very high quality welds in conjunction with a pure shielding gas, which have the advantage of no contamination by filler wire or tungsten. Even very thin sheet metal can be welded; for example, production of disposable steel cans by welding is now possible, with a joint width and depth of only a fraction of a millimetre. Helium (or a mixture that is predominantly He) is the gas favoured for laser welding as its low optical index minimises beam distortion due to gas heating near the joint and minimises ionisation of the gas, which can occur at high powers near the focus of carbon dioxide welding lasers.

3.2.7 Ceramic and metal powder forming technology

Powder metallurgy involves the formation of metal or ceramic components from a precursor powder by pressing into a mould, followed by heating to bond the particles together (sintering). The precursor powder may well be mixed with an organic binder, which is burnt out during processing. Most metals can be prepared in powdered form by spraying molten metal into a tower filled with inert cooling gas (section 3.2.10).

The sintering process is as follows. First, the mixture of metal powders, binder and other active substances, such as ceramics, is prepared and pressed into an oversize mould. The 'green' metal part is then removed from the mould to a gas-inerted furnace where it is heated to a little (20 or 30%) below the melting point of the metals used and their alloys. Interdiffusion and partial melting along the boundaries between powder grains then occurs and the part shrinks and consolidates. Porosity can be controlled so that sufficient holes are left that the part can be impregnated with a lubricant such as PTFE (polytetrafluoroethylene), oil or graphite (a favourite trick for gear wheels and bearings).

A binder may not be necessary. Small parts may simply use metal powders, which are then pressed together very hard indeed using a hydraulic press of a hundred tonnes or more capacity, even for a part weighing a hundred grams.

Gas inerting of the sintering process is essential because of the large surface area and the consequently potential disastrous effects of even slight oxidation. In fact, so serious is the oxidation problem that furnace atmospheres for sintering are typically reducing in nature, with hydrogen being used or added to the nitrogen or argon inerting. Powder metallurgy therefore works primarily with metals that do not embrittle with hydrogen.

Examples of sintering are now legion, but a few are given here to illustrate how sintering is used in industrial practice. Tungsten metal is worked by sintering. Powdered tungsten produced by chemical reduction of pure tungsten ore is very hard and brittle (although ultrapure tungsten is softer and more workable, it is too expensive and unsuitable for industrial application generally). Tungsten cannot be melted for casting in any normal fashion and cannot be cold-worked except in thin wires after extensive prior hot-working: sintering is the only route to bulk tungsten metal working. The reduced powder is first pressed into preform tubes. These are fired in air up to red heat then pure hydrogen and temperatures of up to 2000°C or more are used to sinter the powdered compact together well below tungsten's 3140°C melting point. The still rather brittle sintered rods can then be rolled, hammered and eventually drawn into wire by raising their temperature to 1000°C or more. Tungsten wire made in this way is used to fabricate filaments for the familiar household incandescent electric lamp.

High performance non-porous alumina products are now made by sintering. Spray-dried alumina pelletised powder, containing pure alumina with binders, is pressed, sometimes using isostatic gas pressure techniques (section 3.2.9) at room temperature, into compact moulds. The compact is then fired at low temperature in air to drive off binders and form a strong opaque white ceramic 'compact' very similar to a conventional alumina ceramic. Finally, in a pure hydrogen atmosphere at approximately 1300°C, the compact is sintered. It shrinks dramatically, by a factor of 30%, and becomes translucent, non-porous and very strong. A typical application for this material, which is strong, heat-resistant and oxidation resistant up to well over 1000°C, is for the plasma discharge tubes in sodium high-pressure lamps (section 3.8.10).

In each of these cases, pure hydrogen is favoured for sintering because it does not react with the product, diffuses quickly and can be used to form dense, non-porous sintered products more easily than other less diffusible gases.

3.2.8 Production of ferrites and ceramic high critical temperature superconductors

Ferrites are magnetic minerals containing the ferrite anionic group. They are unusual amongst strongly magnetic materials in that they are good electrical insulators. This enables them to be used in high-frequency circuits with low eddy current losses. Simple ferrites have the formula MFe_2O_4, where M is a metal ion or mixture of ions typically Mn, Ni or Zn and less frequently Mg, Co or Cu. Strontium and barium hexaferrites, $XFe_{12}O_{19}$, are examples of ferrites with a more complex crystal structure.

Ferrite powders are made by fine ball-milling a mixture of oxides then reacting, 'calcining', followed by re-milling of the resultant sinter. The powders can be formed into artefact shapes by 'green forming'. As with other ceramics, powder is mixed with binders and solvents until it forms a plastic mass which can be moulded. The moulded 'green' preforms can then be heated slowly to drive off solvents and burn off organic binders. They can be fired in oxygen to achieve high densities but are more commonly fired in hydrogen or inert gases when cracking due to excessive formation of iron(III) oxide (Fe_2O_3) is avoided.

The production of working copper-oxide based ceramic superconductors was worked out very recently; first results were reported only in 1986. Superconductors have a number of unique properties, the most important of which is that they can conduct direct current electricity without any loss, i.e. they have zero resistance, which is a highly desirable property in electrical engineering. Most superconductors only achieve this state of electrical nirvana at temperatures a little above absolute zero. Above this critical temperature (T_c) they revert to being ordinary electrical conductors. They thus need liquid helium as a coolant, which is an expensive technology, restricting application of superconductors almost entirely to esoteric applications where huge magnetic fields generated by huge zero resistance electromagnetic coils are needed.

Ceramic superconductors were valued at first for their high T_c, with many compounds being developed to have a T_c above the 77 K T_c of liquid nitrogen. At these temperatures refrigeration is simple and inexpensive, making them very attractive. In addition to high T_c, however, many of the cuprate superconductors also possess a property which may turn out to be more important than their high T_c; they retain their superconducting state even in ultrahigh magnetic fields.

Both the T_c and the maximum magnetic field of a high T_c (HT_c) material are very dependent on how the superconductor has been prepared. Typical HT_c superconductors are composed of copper oxide with rare earth metals, such as YBCO (yttrium/barium/copper oxide) and BISSCO (bismuth/strontium/copper oxide). A typical preparative method would be to grind the metal oxides together very finely in a ball mill, followed by a reaction firing or calcining at a high temperature. Other methods start with precursors which are water-soluble salts, carbonates or acetates, which are mixed in aqeous solution, evaporated and pyrolised in oxygen before firing.

The product of this firing is partially reacted ceramic, which is then ground up into a fine powder again and put through a forming process to achieve the required artefact shape. The forming processes employed vary from packing powder into a silver tube and detonating explosives all around it, to hot isostatic pressing (HIP) and the so-called plastic ceramic route. The latter involves mixing the powders with organic solvent and binder compounds, followed by plastic forming of the resultant black dough into the required shape. The organics are removed by a low-temperature pyrolysis step.

Next comes the vital firing process in a precisely controlled oxygen-containing atmosphere. During this firing, a complex layered crystal structure of HT_c ceramic minerals forms.

3.2.9 Hot isostatic pressing

Hot isostatic pressing of powdered preforms, whether of metal or ceramic, is now an established industrial process. Instead of pressing a powered preform between two mechanical anvils, the preform is placed in a very high pressure vessel while this is filled with ultrahigh pressure gas (up to a few thousand bars). The preform is then heated. Hot isostatic pressed materials often have improved properties; mostly incrementally improved but occasionally spectacularly better.

Many metal-working processes involve consolidating more or less porous metal artefacts. Mostly this is done by the application of hundreds of MPa (few thousand bar) of compression using a hydraulic press with a suitable die at an elevated temperature. However, it is difficult to make anything but simple components in this way and the pressure applied is essentially uniaxial in nature. With uniaxial pressure, any pores that may remain are disc-shaped and more liable to cause failure in the component. Also, the control of temperature is difficult unless the die and press jaws are heated and temperature controlled, which is difficult in a simple press.

Originally developed primarily for the processing of reactor fuel rods (diffusion bonding of cladding to uranium fuel) and then used extensively for tungsten carbide, HIP spread in the late 1970s to many other applications in the area of reducing porosity in complex castings and consolidation of components produced by powder metallurgy. A typical HIP process involves the following steps:

- encapsulation or coating of precursor object;
- placing of object in pressure vessel, evacuate or purge;
- filling of pressure vessel to 50 or 100 MPa with pure argon;
- raising temperature of object and gas to around 0.7 times its melting point at constant volume (this step causes a further pressure rise, up to hundreds of MPa).

The essential idea of HIP is to apply heat and an isotropic compression to a partially porous component. Ultrahigh pressure pure argon is a convenient means of delivering that heat and pressure. At these temperatures particles of metal or ceramic bonded only at their extremities become subject to diffusion of atoms through the bulk of the object, especially along grain boundaries and along the porosity surface, the process of sintering is taking place. In an ideal case, pores shrink because the temperature allows atoms to move and surface tension tries to shrink the pores. However, sintering at low pressure leaves pores, often spheroidal gaps which have filled with gas and have an internal pressure, preventing their complete disappearance.

Components produced by compaction and sintering of metal or ceramic powders will always tend to contain pores. Similarly, most metal castings, because of crystal grain growth and shrinkage problems, are slightly porous. High levels of trapped gases such as hydrogen or carbon monoxide, often present in casting metals, further exacerbate these tendencies. Only by the thorough removal of this residual porosity can full strength be attained.

The application of pressure in addition to sintering temperatures enables the complete removal of porosity. The pressure applied will completely swamp any residual gas pressure and enable the gas from remaining pores to be dissolved.

The HIP process is not a metallurgical panacea, however, as it has a number of problems:

- the vessel needed is expensive and potentially dangerous;
- the compressor needed, generally a diaphragm-isolated reciprocating type, is expensive to buy and run;
- the cycle time is necessarily fairly long (usually a few hours but the vessel size, a metre or so diameter, limits the batch size);
- surface-connected porosity is not compacted but is simply consolidated as if it were part of the component design.

The latter point was a serious objection to the application of HIP to castings with surface defects and powder components until suitable coatings and encapsulation were developed. Thin sheet metal, usually steel, is often used for powder metallurgical components. Ideally the sheet covering is pressed into shape, the component sealed inside under vacuum and removed after HIP by acid dissolution or machining. A potentially simpler alternative is to coat with a slurry of ground glass. The glass type is chosen to soften at HIP temperature; it must be fused before HIP pressure is applied. There is much research into other surface-connected pore treatments, such as metal deposition by plating or sputtering.

The argon gas needed for HIP is usually recycled on frequently used systems but must be carefully purified before reuse; a typical specification might be 99.99% purity. This is because even small partial pressures of impurities such as oxygen can lead to serious oxidation in the components or furnace. The diaphragm compressor is preferred because there is then no question of the compressor adding impurities to the gas stream. The argon gas should ideally be checked for common contaminants, such as H_2O, O_2, H_2, CO_x and CH_4, with trace gas analysers.

Argon is used above nitrogen or other gases because it does not dissolve to any appreciable extent in most materials and hence there is little danger of argon bubbles forming on depressurisation. Helium can also be used, although at some penalty in cost. At the conditions used argon gas has a density comparable with that of water. At these unusual densities, the gas is an excellent heat transfer medium and HIP processes normally rely mostly on this convective heat transfer rather than radiation.

HIP of materials which dissociate to some extent in an inert atmosphere at elevated temperatures can be aided by mixing some of the dissociation product with the argon. For example, addition of small amounts of oxygen or nitrogen can aid in the production of oxide or nitride ceramics.

A typical HIP installation comprises a pressure vessel with heaters and a circulator with safety systems for leaks and pressure vessel cracks. A diaphragm-isolated reciprocating compressor is used to pump the argon, which is often repurified and recycled after use. Facilities for the preparation of samples, such as an atmospheric sintering furnace, powder compaction and encapsulation, are also required.

3.2.10 Metal powders by spraying

A lot of metal powder is used as a powder, for example in metallic paints, as a filler for plastics, for catalysts and so forth. However, a major application for metal powders is for applications where bulk metal is the ultimate object, such as the preparation of sintered metal objects. For sintering, it is a big advantage to have a free-flowing and easily compacted metal powder, and this can be achieved by using metal powder comprised of more or less spherical particles.

There is a long history of manufacture of spherical metal particles of larger size for use in shot for sporting ammunition. Lead or lead/arsenic alloys were poured or sprayed into the air inside a 'shot tower' a few tens of metres above a quenching pond. Surface tension forces ensures that approximately spherical particles of a few millimetres in diameter are formed, which cool sufficiently in the air of the tower that they will not be distorted when they hit the cooling water. The formation of spheroidal particles of metal for powder metallurgy follows similar lines but uses an inert gas for the cooling medium and no pond.

A typical installation pumps a liquid metal stream into a jet of nitrogen or argon at up to 40 bar breaking up the metal stream into small droplets. The atomised metal, from a few microns up to 100 μm particles, can be allowed to cool in a free fall through stationary inert gas or in recirculating gas blast, from which the particles are separated from the gas. The result is small spheroidal particles quite unlike the angular particles produced by simple crushing and aqueous atomisation followed by evaporation or oxide reduction. A large atomisation unit is a big user of inert gas; figures up to several hundred cubic metres per tonne of metal powder are quoted for some installations.

Atomised metal powders in finer grades are often 'passivated'. Unpassivated fine metal powders, even of such relatively stable metals as nickel, ignite very easily in air and may even be pyrophoric, i.e. they spontaneously ignite on being poured through air, such as iron filings emitted from a grinding wheel working on steel ignite. Gas mixtures containing small amounts of oxygen are often used to 'passivate' pyrophoric powders. Such fine powders are most often used as catalysts; the hydrogenation of vegetable oils, for example, is mostly based on finely divided nickel powders. One way of avoiding the handling dangers that this entails is to slightly oxidise the surface of the powder, by drying it in nitrogen or similar and then blowing nitrogen or argon with 1 or 2% oxygen added through it so that a thin oxide film can form with no danger of combustion. The resultant powder is thus rendered non-pyrophoric.

Finally, it is not always necessary to isolate the powder production from solid object production. In some processes, typified by the Osprey process, an aerosolised liquid metal stream made with argon or nitrogen is formed directly into a solid object of preform by spraying the particles, while they are still hot, onto a rotating or otherwise moving mandrel.

3.3 Industrial gases and chemical manufacture

3.3.1 Oxygen for catalytic converters

The use of oxygen in many petroleum processes is now routine. Most large refineries have a large on-site air-separation unit, producing both nitrogen and oxygen. Within the many potential applications for air gases around the refinery, selection is based on economic criteria.

One often selected is the addition of oxygen to catalytic 'cracker' (cat-cracker) units. These units are designed to convert much of the heavy hydrocarbons in crude oil into lighter fractions suitable for motor fuel.

Cat-cracker units vary in design but one type functions as follows. Zeolite-type silicate catalyst, ground to 100 μm or less, is mixed with heavy crude oil and the resultant slurry heated to several hundred °C. Lighter fractions are produced, as required for automotive petrol and for liquid petroleum gas (LPG), along with very heavy hydrocarbons stuck firmly to the particles of zeolite. The zeolite is then regenerated by removing as much volatile material as possible by steam stripping, followed by combustion in a pressurised air stream in the catalyst regenerator unit. Waste heat from this process is used to preheat the new slurry whilst the burnt catalyst is regenerated and can be mixed with more crude and reused.

If cat-cracker units are already working to their design capacity, a relatively simple way to increase their capacity is to 'debottleneck' them by boosting their throughput with oxygen injection. The oxygen enrichment of the burn-off air reduces the amount of coke deposited on the catalyst during the cracking operation. The large volumes, up to hundreds of tonnes per day, used in this application mean that it is one of the larger uses in the chemical industry.

3.3.2 Gasification of coal

The most plentiful fossil fuels on the earth are the various sorts of coal. However, coals are inconvenient sources of energy because they:

- are buried deep underground;
- produce copious waste (noxious sulphur dioxide, unpleasant ash) when combusted;
- are highly variable in quality;
- are solids and therefore are expensive and difficult to handle both in transport and inside engines and boilers.

The efficient production of a gaseous fuel from coal is, therefore, a desirable objective, eliminating, at least for the user, all the problems of using coal and leaving only the problem of mining it. This fact was also plain in the nineteenth century when gasworks were installed at most industrialised towns of any size. However, the gas they generated (a mixture of easily ignitable hydrogen and toxic CO) is no longer acceptable for domestic consumption.[49]

The production of high-grade methane fuel gas from coal is not an obvious use for oxygen. The basic reaction desired is, after all, the gasification of coal with water

$$2C + 2\,H_2O \Rightarrow CO_2 + CH_4$$

However, most processes for substitute natural gas (SNG) from coal do involve oxygen, with most processes, perhaps surprisingly, needing more than a tonne of oxygen for every tonne of methane produced. The apparent anomaly is because this direct process is not in general possible, the above reaction is endothermic and heat must be obtained from further coal burning, so the overall reaction scheme is approximately

$$3C + 2H_2O + O_2 \Rightarrow 2CO_2 + CH_4$$

The additional combustion is best carried out with oxygen because it does

[49] That said, early gas appliances, including Bunsen burners and the like in the laboratory, were still running on 'coal gas' in the late 1960s and 'safer' methane would also have caused problems in many situations. When methane first replaced coal gas in UK, there were a number of incidents where burners failed to ignite, owing to the much higher ignition energy of methane, causing potentially dangerous accumulations of fuel gas.

not introduce further unwanted gas (N_2) into the reactant stream. Alternatively, the combustion process can be separate from the gasification step, the heat being transferred by conduction or transfer of hot uncombusted coal, for example, so that nitrogen from the air is not present in the gasification step. However, a typical process does use oxygen.

A typical gasifier process, in outline, is as follows. Pellets of coal are added steadily to the top of a reactor running at tens of bar pressure, the pellets reacting and sinking down whilst a counter-current of hot gas flows upwards. Reactions are zoned, with combustion at the bottom heating the coal to as much as 2000°C in a stream of steam and oxygen. In the cooler middle region CO is formed by the reaction

$$C + H_2O \Rightarrow CO + H_2$$

Also in this region some methane is formed by the endothermic gasification reaction. Finally, in the top zone more methane is formed by hydrogenation of tar in the coal whilst the new pellets of coal are preheated. The overall reaction in a typical gasifier falls short of the ideal in that the calorific value of the outlet stream is rarely over 50% methane. Some additional upgrading process, e.g. a catalytic reactor, is required to shift CO and H_2 into more methane.

Considerable variations in SNG processes are possible, with temperatures from 1200 to 2000°C in the combustion zone, for example, and the coal used varies from soft, highly impure lignites to the almost pure carbon of anthracitic coals. This latter variability is a problem, as is the current low price of natural methane. However, if SNG is required as a real alternative fuel, then very large oxygen plants may be needed whatever the type of process and coal. 5000 tonne per day plants are indicated for commercial scale SNG.

3.3.3 Nitrogen and CO_2 injection in oil wells

The 'gusher' is still, in the minds of most people, the stereotype of an oil well. A deep hole is bored in the right part of Texas or Arabia and when the drill finally penetrates the oil-bearing rock oil magically squirts out of the hole, soaking the rig operators in their 'black gold'.

In the early days of petroleum the large hydraulic and gas pressures present in many oil wells were commonly relied on for initial extraction. As a well began to run out of its own natural pressure, pumps, often the classic 'nodding donkey' pumps, were applied. All oil used to be produced by gushers and pumping. Today, however, when it is necessary to ensure a much higher percentage of the oil in a reservoir is recovered, injection is common.

Injection into an oil well, by pumping into neighbouring boreholes, was first carried out mostly with water, often treating the water with chemicals to avoid plugging of the well. Gas injection is also now common. Often hydrocarbon gases are injected. However, it is cheaper to inject CO_2 or nitrogen into a formation. Nitrogen has the advantage that it is only slightly soluble in oil and in rock, and so smaller volumes can be injected than would be the case with CO_2 and hydrocarbons. A non-oxidising gas is needed for injection; air injection, at the high pressures and temperatures involved, would lead to partial combustion down the drill-hole. The latter has been tried for *in situ* gasification of petroleum, tar and coal but is experimental compared to the straightforward gas injection to enhance production that is now standard in the industry. PSA equipment is often used for making the large quantities of nitrogen needed. The low cost of gas produced on-site by PSA is the key, whilst the typically low purity of the PSA plant N_2 is not important.

3.3.4 Oxygen for oxidation of chemical intermediates

The preparation of many oxygenated solvent and intermediate products relies today on the catalytic reaction of oxygen with unsaturated hydrocarbons in the presence of a transition metal or precious metal catalyst at a few hundred °C. Oxidation occurs via an intermediate oxidant whose state is restored by reaction with gaseous oxygen. Acrylic acid, ethylene oxide, formaldehyde, phthalic anhydride, vinyl acetate and vinyl chloride are all susceptible to manufacture by catalytic partial oxidation.

The use of oxygen, as opposed to air, for these oxidation reactions depends primarily on economic factors such as size and therefore capital size of plant. With the more intensive reactions achieved with oxygen, smaller plant is needed with a lower capital cost. Where more production is needed from an existing plant, the use of oxygen enrichment is often a possibility, with minimal capital investment. Provided the cost of the oxygen is exceeded by the value of the extra product less its other marginal costs, therefore, the use of oxygen is justified.

Other economic factors are the cost of heating or cooling the unwanted nitrogen stream, even if this can partly be recovered by heat exchangers. If a process needs compressed air at a few bar, then there is again wasted energy involved in the compression of the unwanted nitrogen. Indeed, the compression of air to a few bar is all that is required for the operation of an air-separation plant producing pure oxygen. Similarly, the running costs of scrubbing plant to extract the product and recycle the gases are reduced with oxygen feed.

There may be strong technical, rather than simple commercial, factors favouring the use of oxygen, however. Classic cases of this kind occur when a process does not carry a reaction close to completion. In order to use the unreacted gases effectively, they must be recycled back to the input. However, if air has been used, nitrogen and argon will build up in the system and will need to be vented or extracted. This would cost more valuable energy or materials through the need for venting or separation of the nitrogen, as otherwise the reactants would be diluted, with a resultant loss in plant productivity. The BOC company has patented several PETROX processes which combine a pressure-swing absorption unit with oxygen enrichment to carry out reactions such as butane to maleic anhydride (oxidation) or propane to acrylonitrile (propane dehydrogenation then ammoxidation using O_2 and NH_3). These constitute an efficient process by recycling the unreacted alkane back as input feedstock to obtain high yields. Similarly, nitrogen oxide production occurs in all high-temperature plant using air feed; this can be avoided by using oxygen.

Some processes are unsuitable for use with pure oxygen because they proceed too vigorously, whilst for others the plant size is already small. The wider explosive limits for gas/oxygen mixtures also restrict the applications in which oxygen can replace air in gas-phase reactions. It may even be necessary to dilute air with nitrogen to reduce the oxygen content of the reactant stream; another potential application for air-separation gases. Reactions in aqueous media do not generally, of course, have this problem.

An example of a process which is simpler and cheaper with pure oxygen is the Hoechst–Wacker ethylene to acetaldehyde conversion. This uses a mixture of aqueous palladium and copper chlorides to effect the direct oxidation

$$2C_2H_4 + O_2 \Rightarrow 2CH_3CHO$$

The conversion is very efficient at a few bars pressure and 100°C. The simplest apparatus involves a reactor followed by a water scrubber. The liquid from the scrubber is distilled for the product. In a similar way, it is economic to use oxygen for the production of ethylene dichloride, ethylene and propylene oxides and vinyl acetate.

There are further economic factors involved in the full use of all the products of air separation. If a chemical complex has many uses for nitrogen, for inert blanketing, flammable feedstock and product tanks, for example, then the use of oxygen in a process may be favoured.

3.3.5 Oxygen-enhanced sulphur recovery

The Claus process is one example of several patented processes for reducing the sulphur content of petrochemical streams. It uses oxygen injection to both remove sulphur from the process stream and to recover tolerably pure sulphur. The oxygen oxidises the hydrogen sulphide in a controlled manner so that elemental sulphur is produced. If H_2S is burned in the normal way with oxygen the result is sulphur dioxide. Early Claus plants simply took one third of the input stream and combusted it completely then mixed it with the original stream where the reaction

$$2H_2S + SO_2 \Rightarrow 3S\downarrow + 2H_2O$$

produced elemental sulphur. Claus recovery plants typically work on gases containing N_2, CO_2, H_2S. This is the sort of mixture which results from treatment of fuels such as natural gas for acidic gas impurities. A liquid absorber takes up the CO_2 and H_2S, leaving, in the case of natural gas, pure methane. The absorber liquid is then transferred to another vessel where it is heated and depressurised to desorb the gases after which it can be recycled. The Claus plant passes this mixture of desorbed gases through a catalytic reactor. Recovery factors for sulphur content vary up to about 97%. Arguments for the use of oxygen rather than air for the oxidation revolve mainly around the larger size and therefore higher capital cost of air-based plant versus the cost of oxygen.

3.3.6 Ammonia and fertilisers

The production of ammonia often proceeds industrially by the most direct route possible, i.e. the Haber process. Nitrogen from large on-site air-distillation plants is added to hydrogen from large steam reformers. This reaction is exothermic but requires extraordinary pressures of approximately 200 bar. After multiple stages of compression via enormous banks of axial and centrifugal compressors, the gas mixture is reacted at 200 bar or more. Argon is one of the by-products of the ammonia business. This arises as follows. Nitrogen must be produced in large quantities, along with hydrogen, in the ratio 1:3, according to stoichiometry of the reaction

$$N_2 + 3H_2 \Rightarrow 2NH_3$$

The hydrogen is generally produced by reacting steam and methane in the 'steam reformer' reaction

$$CH_4 + H_2O \Rightarrow CO + 3H_2$$

However, this reaction needs high temperatures and is endothermic. The heat and nitrogen needed can be supplied by adding air to the reaction stream, oxidising further methane to CO exothermically. This is ingenious as it is not necessary to separate the nitrogen from the air

$$CH_4 + \text{air} \Rightarrow CO + H_2 + CO_2 + N_2$$

A catalyst, e.g. iron oxide at 500°C, will convert the CO to CO_2 if more steam is added, leaving only CO_2, which is easily scrubbed out with an absorber tower, N_2 and H_2. The H_2 stream is supplemented as appropriate from the steam reformer reaction to yield a 1:3 N_2/H_2 mixture.

There is a snag, however, to the cunning use of air in this way. The resultant N_2/H_2 stream contains a fraction of one per cent of argon. This would not perhaps be too much of problem were it not for the fact that the

N_2/H_2 mixture is not reacted in one go into ammonia but is recycled several times, depending on process conditions and catalyst, each time scrubbing out more ammonia but leaving increasing amounts of argon. Eventually, the argon builds up to tens of percent and N_2/H_2 must be wasted to wash out the inert argon. Alternatively, if the process is operated to maintain the argon at a compromise level in the N_2/H_2, then Ar can be commercially extracted in a cryogenic plant. Too high a level of Ar will inhibit the ammonia synthesis, too low a level will make the Ar extraction too expensive.

Most ammonia is made into fertiliser. After reaction with nitric or sulphuric acid, and the admixture of phosphates or other components, the aqueous fertiliser is sprayed into a 'prill tower'. Here an updraught of warm air carries the droplets upwards. They rapidly dry not into crystals but into small 'prills' i.e. free-flowing rounded granules a few millimetres across. They are collected from the top of the tower and then bagged for customers. Ammonium nitrate is also important in the manufacture of nitrous oxide gas.

3.3.7 Inerting chemicals

Inerting pipework. As noted above (section 2.5.10), before a pipeline is filled with a flammable or reactive gas, it must be purged. To prevent fire and explosions in large pipes carrying inflammable gas or vapour, pipes are purged with several volumes of nitrogen before and after use, ideally with pressure cycling. Reactive gases and pipelines needing high purity may require much more purging than this using all the purging tricks detailed above.

The purging is often carried out far beyond safety requirements to ensure quality. Silane in electronics applications, for example, is useless if even a few ppm of oxygen creeps into it; the O_2 reacts to produce particulate silica (i.e. a rather fine sand), which will deposit on the microcircuit being fabricated and ruin it. Similarly, large stores of resins and paints are often much more carefully inerted and controlled than fire safety would dictate because even small amounts of oxygen from the air lead to oxidative polymerisation, i.e. the resin thickens and skins, leading to problems later.

Inert blanketing of tanks. The ullage space above tanks containing liquids which are flammable or reactive to oxygen or other components in air, such as moisture or carbon dioxide, is best maintained with an inert gas. As noted, many paints will react to set prematurely if oxygen reaches them, whilst dessicating agents and cyanoacrylate glues must be kept in an anhydrous condition to be saleable. Still other chemicals can usefully be inerted because of materials considerations. When maintained oxygen-free and anhydrous, many chemicals, even acids, are much less reactive towards inexpensive constructional materials such as mild steel.

With liquids forming a flammable vapour, the breathing of a tank can result in vapour being lost via vents during filling or temperature increases, with the risk of a flame at the vent. Worse still, on emptying or cooling, the tank can inhale air to form an explosive mixture in a confined space, which is clearly extremely dangerous.

The simplest way of ensuring blanketing is to simply connect a steady stream of inert gas (almost invariably nitrogen) to the top of the tank. This stream is then vented continuously. Provided the product withdrawal rate from the tank does not exceed the inert gas input rate, the tank will remain under fully inert conditions. A more sophisticated variant is to add another valve which switches to a higher flow rate when the tank is being drained and otherwise uses a low inerting flow. This method, although simple, leads to higher gas usage and loses a small amount of product vapour continuously (see Figure 3.11a).

Alternatively, sensitive gas regulators can be employed so that when liquid is pumped in or out of the tank a slight positive pressure of gas, a few millibars on a water gauge, can be maintained above it by venting or adding gas. The slight positive pressure ensures that any small leaks will be outwards in direction while putting minimum strain on the tank. Product vapour is only lost during product filling. Even this loss can be recovered to some extent by the use of carbon absorbers. These absorb product vapour when the tank is venting and can be stripped at least partially of that vapour by backflushing into the tank. A more sophisticated and more effective vapour recovery system uses cryogenically cooled absorbers to further increase efficiency. At least for very large tanks, in this kind of vapour recovery system inert gas is only used during product withdrawal. Inert gas can be shared between tanks on large-tank farms, provided slight admixtures of vapours can be accepted, further reducing the quantity of inert gas needed.

A scheme for implementing tank inerting in this way is shown in Figure 3.11b. Two ordinary regulators reduce the house nitrogen line pressure down to a few hundred millibars or less when a sensitive and accurate low-range final regulator is used. Check valves are included to avoid the possibility of solvent entering the nitrogen system and pressure-relief valves are supplemented by a vacuum-relief valve to prevent the tank collapsing. Flame arrestors on the vent pipes are good practice that avoids a flame spreading to the ullage.

During commissioning, the nitrogen regulator is set artificially high so that the large flow rate can purge out the ullage. In operation, the nitrogen regulator is set a few millibars lower than the two relief valves, giving the system some hysteresis, which will avoid unnecessary hunting of the system control and excess venting losses.

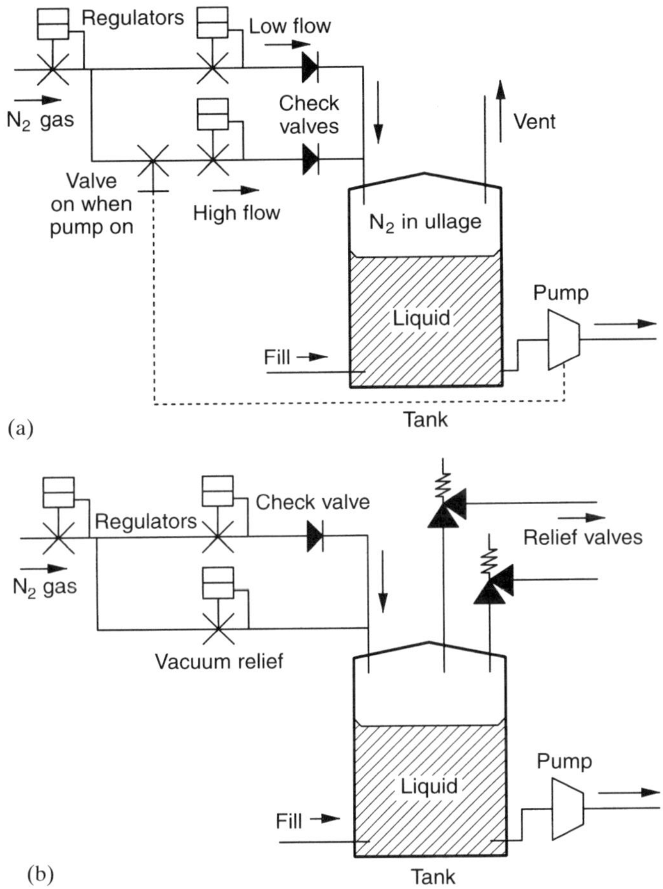

Figure 3.11 Balanced pressure tank inerting system: (a) tank blanketing with continuous nitrogen purge; (b) tank blanketing system with semi-closed circuit.

Nitrogen foam inerting. Nitrogen can be used to generate inert foams, which are especially good fire suppressants and extinguishants. Such foam has often been employed when modifying or demolishing chemical plant that has, or used to have, flammable materials inside it. Specialist liquid nitrogen trucks are provided with a propane-fired vaporiser along with a pump for water and a tank of medium- or high-expansion foam concentrate. A foam 'gun' combines these ingredients and can be used to make up to 50 m^3 of foam per minute.

The typical job for such a foam generator is during demolition or repair of tank farms and other chemical plant. Tanks that contain small amounts of residual flammable liquid will be more or less full of flammable vapour, which, if air is allowed to enter, constitutes a severe explosion risk. During demolition, tanks can be filled with foam and then cut up to leave a wide-open top. An oxy/acetylene torch can be used and foam leaking out of the cut will not extinguish the cutting flame. Once the top is off the tank it can be left open and demolition completed after the foam has drained away and freely circulating air has removed any vapour. For repairs, the tank is drained (although sometimes this is not necessary) and foam fed in. The hole or crack can then be repaired by electric or oxy/acetylene welding in the usual way and the foam drained off. After that the tank is washed and refilled ready for service.

There may be problems with some solvents which react with and disrupt the foam stability but often choosing a different detergent or soap can handle this. Most solvents and oils are less dense than water and it is therefore important to drain off excess liquid from the bottom.

3.3.8 Miscellaneous chemical applications

Pressure transfer and pneumatic lift. Highly corrosive and solvent liquids and, in the semiconductor industry, ultrapure liquids can be difficult to pump. The rotating or reciprocating seals in many pumps add particulates to ultrapure liquids and even the minutest leaks of corrosive materials past seals can cause disaster. Modern diaphragm pumps, often fabricated almost entirely from Teflon and related perfluorinated polymers, have overcome many of these problems but are expensive, give a pulsating flow and often require regular maintenance, which can be awkward in confined locations. Using gravity is often possible but placing tanks high enough to give suitable dispense pressures is often difficult and undesirable for large quantities of hazardous liquids.

Two solutions which take the radical approach of removing pump problems by removing the pump, are pressure transfer and pneumatic lift. 'Pressure transfer' refers to the method of dispensing liquid from a tank by applying gas pressure to the ullage space. 'Pneumatic lift' is where gas is bubbled into the bottom of a liquid pipe that is taken well below supply and outlet level. The gas bubble/liquid mixture column has in effect a lower density than a liquid column and thus the liquid can be carried up to considerable heights.

Pressure transfer requires a vessel capable of handling a reasonable pressure (often dubbed a 'pressure egg'). Incoming liquid is delivered into the pressure egg by first venting the vessel then admitting it. Once the correct level is reached, the vent and inlet valve are closed and nitrogen pressure applied. Whenever the dispense valve is opened, the product is dispensed under pressure. With good pressure control and electronic timing of the valves, precise quantities can be dispensed into a downstream process. The level in the pressure egg can be measured in the usual way, e.g. by a

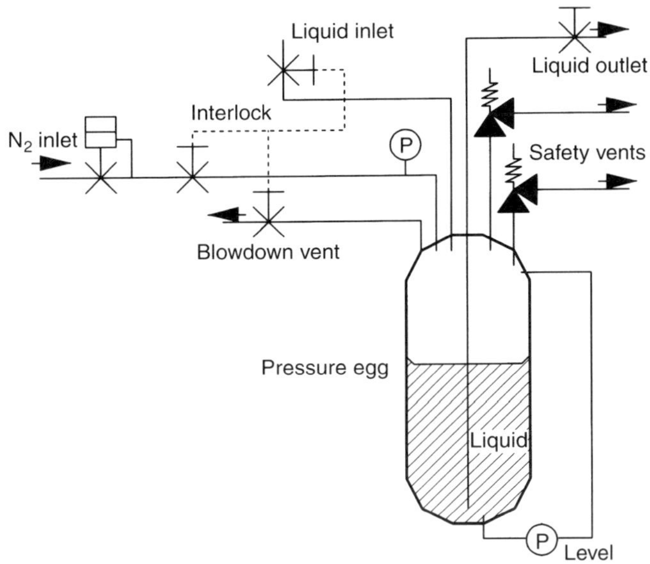

Figure 3.12 Pressure transfer system.

differential pressure indicator or transmitter, or simple on/off level indicators can be used.

Some applications lend themselves to pressure transfer. Highly viscous materials, such as glues and molten sulphur, cause problems in pumps, particularly as it may not be possible to heat the pump to the same temperature as the tanks and pipeline. In these cases a pressure transfer system is especially appropriate.

Pneumatic lift systems are usually designed to transfer liquids from one atmospheric pressure tank to another, the latter being higher than the supply tank. They rely on a pipework system which goes considerably below the supply tank. Gas is injected into the pipework below the supply tank at a position (1 m or so into the up-going pipe) where bubbles are sure to rise towards the receiving tank. The gas and liquid in the up-going pipe then constitute a column of material of lower density than that in the down-going pipe. The pipe is chosen to be of sufficiently small bore that the gas cannot simply bubble up the column, but the liquid and gas form a genuine mixture (although there will always be some 'slippage', i.e. gas flowing faster than liquid). A hydrostatic equilibrium is soon established with the lower density column displaced upwards so far that liquid overflows out of the up-going pipe and is delivered to the receiving tank.

The gas/liquid mixture in the up-going pipe, at reasonably high rates of gas injection, shows the bubble amalgamation effect seen in many bubbling apparatus (section 3.6.1) and the flow pattern soon settles down to a 'slug flow'. In this, slugs of water, interspersed with large bubbles that fill all the pipe except for a thin film of liquid, flow upwards. At higher gas flows the slugs are smaller and more irregular, and the term 'churn flow' is sometimes used. Unstable 'hunting' of the flow rate is often seen, related to acceleration of liquid in the pipes. This can be controlled to a large extent by incorporating a check valve somewhere in the pipework, thus preventing backflow.

The depth at which the gas must be injected can be calculated as follows. At hydrostatic equilibrium the pressure supporting the supply and receiver vessel fluid columns of density ρ and ρ', respectively, must be equal, so

$$g \rho' (\Delta H + D) = g \rho \Delta H$$

or

$$D_{min} = \Delta H \phi / (1 - \phi)$$

where D_{min} is the minimum 'submergence' (distance of gas injection point below supply tank), ΔH is the difference in tank heights, ϕ is the density ratio ρ'/ρ and g is acceleration due to gravity. Clearly, due to frictional losses and other non-ideal effects, D must be made substantially more than D_{min}. It can be seen that if ϕ is only, say, 0.8, only a little gas is injected. In this case a pit or borehole may have to be dug to allow D to be large enough ($D_{min} = 4\Delta H$). On the other hand, if it is permissible to use a large amount of gas injection, then $\phi = 0.4$ might be usable, giving $D_{min} = 0.7\ \Delta H$.

The mean density in the slug flowing pipe is the key factor in the calculation of whether a particular configuration is workable. This can be estimated for water and other common fluids by scaling from empirical measurements in the literature. For unusual fluids (the effect of dissolved surfactants or viscosity modifiers are of course important) there is no substitute for a small trial. However, it will often be found that a mean density of 0.5 or so can be achieved (assuming the gas density is negligible compared to that of the liquid). A warning should be sounded, however, that achieved density varies with flow rate of liquid as well as that of gas and, perhaps surprisingly, at low liquid rates, because of the change of flow conditions in the upgoing pipe, density may not fall with increasing gas rate.

Older standard textbooks (see, for example, Coulson and Richardson, 1965, 1970) often give the full theory of pneumatic lift pumps. It is sufficient to say that they can be made very efficient (50% in terms of usage of the energy of the compressed gas) provided a deeply submerged pipework system can be managed and very high flows are not required.

Both pressure transfer and pneumatic lift can be carried out with many substances in air. However, the very reactivity of many chemicals for which the methods are useful and the maintenance of ultracleanliness in semiconductor work, mean that using inert nitrogen gas is often the method of choice. Nitrogen avoids problems with oxidation of product, fire and explosive vapour hazards, and is not contaminated with other potential

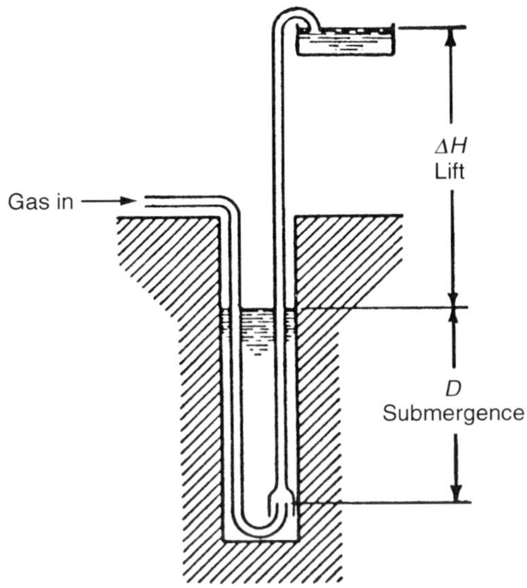

Figure 3.13 Pneumatic lift system.

reactants such as carbon dioxide and water vapour. Triboelectric charging phenomena often occur, causing large potential differences between insulated vessels containing non-conductive liquids. Sparks produced by these voltages can cause ignition in an unexpected way, so nitrogen inerting can be particularly valuable.

3.4 Gases in the food industry

3.4.1 Freezing and chilling

The freezing and chilling of food using liquid nitrogen is now very big business. In many countries it is the biggest user of tonnage liquid (as opposed to gaseous) nitrogen. The total world market for frozen food is now of the order of $100bn. At first, in the 1880s, crude techniques such as brine immersion were used. Business pioneers such as Clarence Birdseye developed economic freezing systems in the 1930s using cold-plate freezing. The much more rapid mechanical air-blast and the even faster cryogenic techniques followed soon afterwards, along with a shift towards higher quality IQF (individual quick frozen) products. IQF foods are preferred by consumers as the quick freezing gives a tastier, better textured product and the separate pieces of produce means just the right amount of product can be defrosted for use.

To preserve it, food must be stored well below the freezing point of water, around −18°C giving a reasonable lifetime. Simply placing food in a cold store at −18°C and hoping for the best, however, is not in general sufficient. Under these circumstances it could take a day or more even for a small (e.g. a few pounds) package of typical food to freeze completely and cool to near −18°C with natural convection alone. This is so slow that some changes will occur in the food. There would be time for some bacterial growth or other spoiling effects and therefore food that is frozen slowly does not taste fresh when thawed. Slow freezing may allow ice crystals to form inside the food, ruining its texture when thawed. Thus there arises a need for a specific food freezing process prior to storage or distribution as frozen produce.

The freezing process. Normally freezing of foods is done by use of large-scale, but otherwise rather conventional, refrigeration plant. A working fluid such as a CFC (or HCFC for newer plant) or anhydrous ammonia is compressed and then cooled by an air or water-cooled heat exchanger back to room temperature, at which point it liquefies. The liquid refrigerant is then allowed to expand at a nozzle then cools, absorbing heat from the cooling plates inside the plant and boiling-off. The gaseous refrigerant is then recycled back to the compressor.

In order to effect rapid freezing of foods, the conventional plant is run at a very much lower temperature (−30 or −40°C, or lower) than a normal cold store and large fans are employed to blast the air rapidly past the food being frozen. This is the more or less standard 'air-blast' freezing machinery used in many freezing plants.

The time taken to freeze food with this sort of modern freezing machinery is, however, still quite lengthy. A 10 kg package, for example, might take 4 to 8 h to freeze and cool down acceptably. Some acceleration of this is possible by processing smaller packages and by using a higher speed air blast; air speeds of up to 10 m s^{-1} are used. With mechanical freezing, even with these powerful icy blasts, there is still a distinct pause phase in which ice crystals form. The initial cooling to a degree or two below 0°C is rapid, as is the final cool-down to −40°C, but there is a long plateau. It is during this plateau phase that growing ice crystals can cause damage and evaporating water from the product surface can leave it dehydrated. The freezing plateau can also lead to 'drip loss'. This is an effect caused by osmosis from the plant or animal cells into the intercellular space. As the ice crystals form in the intercellular space, the liquid remaining there becomes concentrated and an osmotic pressure differential is built up which causes the cells to leak water. Subsequently, both cells and space are frozen and further osmosis stopped. On thawing, however, there is more water (up to 5% more) in the intercellular space, some of which simply runs out of the product as drip loss. Drip loss is the percentage of the produce lost in watery fluid as it is thawed before consumption.

Liquid nitrogen (LN) offers the potential for very rapid freezing. This is

Table 3.4 Advantages and disadvantages of food freezing with liquid nitrogen

Advantages	Disadvantages
Low capital cost	High running cost
High-quality produce due to very low freezing temperature	Regular truck deliveries needed
Small plant size	
Low maintenance	
CFC inventory (typically) or other refrigerant not needed	

important because it means that many foodstuffs which contain a lot of free water can be frozen. If frozen slowly, the water will crystallise into long spicules which break the cell walls within the foodstuff. When this frozen produce is then warmed up to room temperature it will be mushy because of the destruction of the cellular structure which normally gives it texture. Frozen produce with this kind of problem will also suffer excessive drip loss. Frozen raspberries are a classic example of this problem.

The freezing speed of LN systems varies, as with air blast, with package size but is typically only a few minutes, i.e. 20 times or more quicker than standard air-blast systems. The speed of freezing is also important because it enables a much smaller and cheaper set of equipment to process the same throughput of frozen produce when compared to a conventional air-blast system. Finally, the very quick initial cool-down of produce minimises staling reactions and the chances for pathogenic bacteria or other spoilage effects to grow.

A typical LN system comprises a cryogenic storage tank, usually rented by the gas supplier, and a freezing tunnel. The tunnel has a moving stainless steel mesh belt on which produce is placed. The product traverses first a cool nitrogen gas zone, then a cold N_2 zone and finally is sprayed with LN from sets of nozzles above the belt. Fans of moderate power within each zone ensure a good convective thermal contact between N_2 and the produce. The tunnel can be quite small, as small as 3 to 6 m long, as opposed to the 30 or 50 m lengths used in air blast, even though the belt is moving at a metre or more per minute.

Advantages and disadvantages of freezing with LN. Although expensive in running costs relative to mechanical air-blast freezing, LN is the best choice in a substantial minority of applications, at least in well-developed economies with low LN costs (Table 3.4).

Cryogenic air blast freezing. For very large installations used fairly continuously, it is now possible to obtain the advantages of fast cryogenic freezing by using expansion turbine air-blast systems. These operate a little like a cryogenic air-separation unit, compressing and then cooling air back to room temperature, drying the air and then expanding the compressed air with a large and efficient turbine, where it can cool adiabatically to −150°C (123 K) or less.

The resultant ultracold air blast has many of the fast-freezing quality advantages of liquid nitrogen freezing but avoids the use of liquid refrigerants. The use of a heat exchanger on the exhaust from the freezing tunnel or spiral means that wasted cooling power can be minimised. (A spiral belt freezer is preferred because of its lower heat losses.) However, the laws of thermodynamics predict that such systems will show significantly higher power requirements than a conventional blast freezer; with Carnot efficiencies, operating at 123 K will cost around 4.5 times as much power as operating at 223 K. As well as this, the plant required is substantial and has to be engineered for lower temperatures than are normal in the freezing business so capital costs are expected to be high until systems of this type become more common. By using the cooling power inherent in the air blast right up to room temperature, the comparison can be made rather more favourable (see next section).

LN belt freezing efficiency. To be efficient, nitrogen freezing must use not just the latent heat of boiling of liquid N_2 but also the specific heat of cold gaseous N_2. This is achieved by employing the counter-current principle. The specific heat of nitrogen gas from 77 K to room temperature (293 K) and the latent heat of nitrogen as it boils amount to the following thermal capacity. The theoretical heat absorbed per kilogram of LN at 1 bara up to −18°C is:

$$\text{specific heat} \times (255-77) + \text{latent heat} = 1 \times (178) + 200 = 378 \text{ kJ kg}^{-1}$$

A figure reflecting achieved practice might be 300 kJ kg^{-1}. The mass of the produce frozen (in water equivalent to −18°C) per kilogram of LN is

$$\text{heat absorption} = \text{water specific heat} \times (293-273) + \text{latent heat} + \text{ice specific heat} \times (273-255) \times \text{mass of produce}$$

N_2 has a latent heat of vaporisation of 200 kJ kg^{-1} and its gas specific heat is close to 1 kJ kg^{-1} $°C^{-1}$ so

$$\text{mass of produce frozen} \simeq 0.83 \text{ kg per kilogram of LN}$$

If the cost of LN is \$100/tonne, frozen food should cost about \$0.12 per kilogram more than fresh food; an acceptable premium for all but the lowest value frozen produce. A comparative figure can be estimated for an air blast freezer only if thermal efficiency and electricity prices can be assumed, along with cost of capital for the much larger capital sum needed for the large air-blast system and the real estate needed to house it. However, the overall

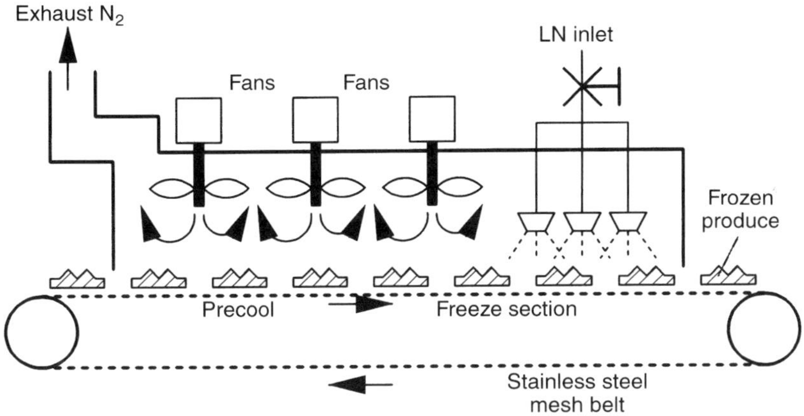

Figure 3.14 Principle of belt freezer using liquid nitrogen.

thermal efficiencies of LN versus air-blast freezing can easily be compared on the assumption of ideal Carnot efficiency as follows. If freezing power were delivered wholly at the cold-end temperature of the system

Energy needed (theoretical Carnot cycle) per joule cooling for air blast, $-50°C = 0.3$ J

Energy needed (theoretical Carnot cycle) per joule cooling for LN, $-195°C = 2.8$ J

On the face of it air blast requires only 11% or so of the energy needed by LN. However, this ignores the cooling effect of the sensible heat of LN and the effect of this sensible heat being absorbed in the ASU plant. In effect, only half the ASU plant's cooling is at −195°C; nearly half is at intermediate temperatures. As a result of this consideration, a fairer comparison would be that the air-blast technique only requires approximately 20% of the LN energy input.

On this basis there would appear to be little to commend LN as a freezing medium. Other factors, such as the possibility of freezing in a few minutes, however, and the low capital costs of LN freezing mean that it is, in fact, the most attractive option for many freezing plants. For very large plants one option is to take advantage of the higher thermodynamic efficiencies of large cryogenic plants. Very large freezing plants in a few locations have actually employed an on-site LN plant. The 1970s operation of King Harry's Pizza in Thame, UK, is an example.

The most common equipment for LN freezing is the straightforward belt/tunnel freezer (Figure 3.14). Food is placed on a relatively slow-moving stainless steel mesh belt. Occasionally a thin stainless steel foil is used which presents a polished surface to the food being frozen. This can be useful to avoid sticking and marking of the product base. Produce is normally taken via zones of an intermediate cold until it reaches LN sprays and complete freezing. In some belt freezers provision is made to achieve a rapid initial freeze so that a frozen crust forms which will keep produce, such as tarts, ice cream or piped cream, in shape until freezing is complete. Crust-freezing can also be used to temporarily strengthen products which are not to be totally frozen but are subsequently simply kept chilled or at ambient temperature. The strengthened crust allows handling and package of the produce to be achieved more easily with less damage.

Crust-freezing can also help with frozen or chilled prepared goods that are distributed in vacuum display packs. Vacuum-packing is carried out by wrapping goods on both sides with a rectangular piece of impermeable plastic film, heat-sealing the plastic films together around three edges and then, using the fourth side, putting a vacuum on the inside of the package, removing the air and squeezing the product between the two films of plastic. The crust-freezing of a prepared food such as pizza allows it to withstand the crushing force of atmospheric pressure and avoid flash-off of moisture from the surface when the vacuum is applied.

In some plants spiral-belt freezers are used. These have a continuous belt onto which food is placed. It is then bent around into a spiral before being brought to the exit again. They have a much smaller 'footprint' on the factory floor and have lower thermal losses, their compact shape having a much lower surface area for a given volume. Strawberries, raspberries, prawns, mushrooms and tomatoes are examples of produce commonly LN-processed on spiral-belt freezers.

For more robust products, simpler and higher capacity systems can be used for LN freezing. A simple cupboard can be used for batch freezing, whilst for freezing of minced meats and similar diced products a slowly rotating drum with screw-drive suffices and will achieve large throughputs in a small space.

One important feature of frozen granulated or diced food is the requirement that the frozen produce should not form large lumps but should be formed as separately frozen individual pieces which are more convenient for the purchaser. LN freezing techniques are more easily adapted for this than conventional blast freezing, in that they achieve very high rates of freezing, allowing crust-freezing to occur. Once this has happened further cooling can occur without individual pieces sticking together.

For freezing of liquids into solid granules, a droplet freezer can be employed. Although immersion droplet freezers are possible, a more common system is to exude droplets of materials, such as fresh frozen egg yolk (used in the catering trade), onto the outside of a rotating drum that is cooled on the inside with LN. Individual frozen droplets are formed that are scraped off the drum and packaged as free-flowing granules.

Immersion freezers. Before modern air-blast equipment became efficient enough, freezing by immersion in ultracold (down to −40°C) brine was common. An immersion concept that has carried on into modern cryogenic freezing practice is the use of an ice glaze to protect valuable and delicate foodstuffs such as prawns. After initial freezing, produce is taken on the moving belt briefly through a bath of water at a controlled low temperature. Here it acquires a coating of ice, which is hard frozen by an additional freezing stage after exiting the water bath.

Early attempts at ultrafast freezing relied on immersion in LN to achieve the ultimate chill rate. However, there were often several practical problems with this apparently simple approach and this method is still rather unusual, although it has now become commercial practice in a few factories.

Freezing and chilling on the move. Once frozen, produce must be kept that way. Similarly, there is a requirement to keep fresh produce at around 5°C to improve its keeping qualities. Large cold and frozen stores are now common, some the size of a football pitch inside, with surprisingly small refrigerator plants maintaining their interiors cold. A more difficult problem is distribution of the produce while keeping it at a low temperature.

The most common solution adopted is the use of a separate diesel-driven mechanical refrigerator plant on each semi-trailer or truck. However, these units have some problems. If economically sized, to deal with the average thermal loading, they are very slow to cool down after warming during loading or unloading and the interior fluctuates in temperature considerably. An alternative is to use a small tank of liquid nitrogen on the truck. Using LN, almost instant cooldown and good subsequent controllability for separate chilled and frozen compartments can be achieved in a truck or semi-trailer unit.

Some high-quality fresh and frozen produce retailers have adopted this ingenious LN-powered system for delivery to their shops. Goods are loaded into the delivery truck, which is equipped with LN spray heads connected to an onboard LN tank and control system. Once the doors are closed the LN system comes into play, cooling the goods within a few minutes to the required temperature. Moveable partitions between goods needing different temperatures are available so that with a few additional components trucks can deliver mixed loads of deep frozen, heavily refrigerated and slightly cooled goods. An alarm is necessary to ensure that when the doors are opened air mixes with the nitrogen atmosphere and an asphyxia accident is avoided.

Freeze-grinding. For many flavouring foodstuffs, such as peppers, spices or coffee, grinding is an essential part of the processing. Without grinding, the products would not release their desirable flavours. However, grinding is not easy for some products as they are not hard enough and they tend to clog

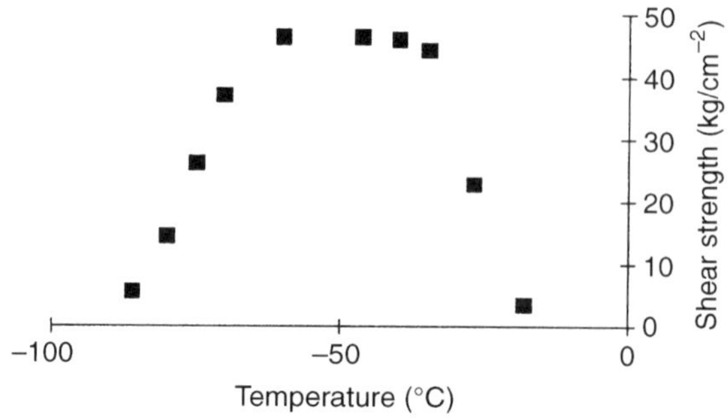

Figure 3.15 Ice adhesive shear strength for 10% glucose solution on a copper plate versus copper temperature (courtesy of Air Products Corporation).

the grinding machinery. Worse, the grinding process tends to heat the flavourings up, releasing some of the volatile flavouring essences which are the material's *raison d'être*. The addition of a little liquid nitrogen to the grinding machine avoids these problems.

The LN makes the foodstuff hard and less liable to clog, and keeps the temperature so low that loss of essential flavouring oils and esters is minimised. Nitrogen, in suitable systems, can provide an inert atmosphere, which avoids the problem of oxidation that occurs particularly in grinding to fine particle size. Fine particles in air can present a fire or explosion hazard, which is eliminated if much of the air is displaced by nitrogen (see also section 3.7.3). The LN cost can often be recovered in a premium price for the product (as in freeze-ground coffee) or in increased throughput of grinding machinery.

Non-stick freezing. By utilising to the full the ultrarapid freezing properties of liquid nitrogen, it is possible to freeze foodstuffs so fast that their surfaces become 'non-sticky'. Recent accurate measurements have confirmed a curious effect that the adhesion between a polymer surface and ice formed on that surface is very strong when the surface is cooled to −20 or −30°C but, surprisingly, is lower by a factor of up to 10 or so when the surface is at −80°C. Free-flowing minced products can be produced, as can small fruits and small or sliced vegetables, without sticking. The latest application of this ultrarapid freezing technique is in the production of moulded fruit ice/ice-cream novelties; the Air Products Corporation offer proprietary techniques in this area (Figure 3.15).

The science of why ultracold surfaces should offer such poor adhesion still appears to be unknown. Although the phenomenon has been known for some years in the preparation of frozen sections for microscopy using LN-cooled copper surfaces, there is no comprehensive theory. It is thought that ice formed on a cold surface bonds by first wetting the surface, then freezing, hence the greater adhesion to wettable surfaces and the lower (although still very substantial) bonding to non-wettable surfaces such as PTFE. If a surface is not adequately wetted when it is very cold (a sort of reverse Leidenfrost phenomenon), because the water freezes too rapidly, then this could account for adhesion reduction. The other more obvious phenomenon that comes into play is the difference in thermal contraction between ice, water and the cold surface. Ice has double the thermal contraction versus temperature of typical metals and shear forces can build up at the ice/surface interface which, as the bulk of the ice forms, can cause debonding after initial bonding has occurred.

The best guess about the reduction in adhesion strength at cryogenic temperatures would appear to be a combination of freezing before wetting and debonding following thermal contraction of the ice formed. There might also be an 'unzipping' or 'boundary crack' effect, which is seen in other adhesive bonds. Once a bond begins to peel off, stress is concentrated and the debonding will propagate along the interface between two materials. Perhaps thermal contraction effects can drive along such a debonding phenomenon.

3.4.2 Dry-ice freezing

Where there are copious supplies of inexpensive carbon dioxide, this can be a very reasonable cryogen for freezing food. It has the disadvantage relative to LN that it can impart a taste to the food, particularly non-acidic food, owing to its mildly acid reactive nature. Its use in delicately flavoured non-acid produce is therefore sometimes not recommended. It is also not capable of such a rapid freeze as LN, subliming at 194 K as opposed to the 77 K of LN. In addition, CO_2 at atmospheric pressure is a solid and cannot form as intimate a contact with the food being frozen as LN does.

CO_2 has a latent heat of vaporisation of 570 kJ kg^{-1} (from the solid at 1 atmosphere but rather less from liquid, around 250–300 kJ kg^{-1}, depending on the pressure), as opposed to the effective 380 kJ kg^{-1} of LN. This means that CO_2 can be more economical in use than LN, provided it is available at a low enough price. In the USA, the price in many regions is low, so CO_2 is used for refrigeration of freight cars on railways and truck trailers. It is sometimes used in solid form as CO_2 snow, which is loaded as liquid into the roof of the freight compartment initially but allowed to flash-off *in situ*. This is simpler than shovelling CO_2 snow as a solid into the roof of the compartment and also avoids the need for a control system for dispensing CO_2.

There are some handling problems with CO_2. In bulk applications, like food freezing, it must be stored in refrigerated tanks. These typically hold liquid CO_2 at around −30 or −40°C and approximately 10 bar. The tank does not need to be an expensive vacuum-insulated vessel, although insulation is needed. A modest refrigerator (a few kW) suffices for most installations with a few tens of tonnes of CO_2 stored. As noted above (section 2.4.7), precautions should be taken so that the CO_2 does not solidify in the tank.

Liquid CO_2 stored in this way does not contain the freezing capacity of solid CO_2. When it is sprayed out of a nozzle into 1 bar air, half of the liquid sprayed immediately 'flashes-off' as gaseous CO_2. The resultant solid CO_2 snow can be mixed with a mixable product like minced beef or small particulate foods, such as mushrooms, to increase contact between cryogen and food. Freezing rates can be greatly increased by this. This amounts to co-current heat exchange, however, and is less efficient than the effectively counter-current techniques used in best LN practice.

The cooling of airline in-flight catering foodstuffs or portable ice-cream sellers' displays is carried out not by low density CO_2 snow, however, but by means of compressed bricks or slabs of solid CO_2 of much higher density, close to the 2 g cm^{-3} theoretical density (section 3.7.2) made by compressing CO_2 snow with a hydraulic ram.

3.4.3 Gases as preservative chemicals

The use of gases for food preservation purposes other than cooling or freezing food is now routine, although the volumes of gas used are much smaller than in the freezing business. The preserving effects of gases such as oxygen or carbon dioxide are modest but particularly valuable because they are intrinsically safe and healthy, and do not affect food taste. Traditional salt and sugar preservatives are unhealthy in excess and can affect taste adversely, whilst many modern chemical stabilisers, although carefully chosen and tested, will continue to have a question mark over their absolute safety, at least in the minds of the general public.

The appearance and freshness of food in sealed plastic packages can be enhanced by the correct selection of gas to be included in the package. For example, oxygen will help to preserve the taste and the red colour of fresh meat. Some further examples are shown in Table 3.5.

Table 3.5 Preservative gases for foodstuffs

Produce	Gas
Red meat	Oxygen
Fish	CO_2
Milk	CO_2
Mushrooms	15% CO_2 + 2% O_2
Apples	N_2 + 2% CO_2 + 2% O_2
Strawberries	N_2 + 20% CO_2 + 20% O_2

A lot of produce undergoes radical maturing changes over the period of time after picking or other preparation and its consumption. Meat is usually 'hung' or kept for a few days, when it is subject to reactions which render it less tough; game meats are too tough unless hung for a week or so. Oxyhaemoglobin gives meat its red colour, but slowly gives up its oxygen during hanging, giving the meat a livid blue/grey tinge. Oxygen helps preserve the haemoglobin and hence the red colour a little longer. Fruits are still to some extent 'living'; they respire, taking in oxygen and giving off CO_2. Exclusion of oxygen from fruits can help restrain respiration and lengthen storage, whilst avoiding the unwanted oxidation reactions (which give rise to browning) which occur with imperfect or slightly bruised fruit. The appropriate gas atmosphere will also help restrain pathogenic bacteria and moulds in some cases; growth of *Clostridium botulinum*, for example, is retarded in high levels of oxygen.[50]

The storage and transit of fruit and other produce is often now enhanced by systems which either seal in the appropriate atmosphere, as above, or, in more sophisticated systems, actively manage the atmosphere by adding appropriate amounts of nitrogen and CO_2.

CO_2 has been proposed for the preservation of milk but the proposition has failed to find widespread acceptance. A milk with CO_2 fizzes and has a slightly acid taste rather than its usual more alkaline and creamy effect. Consumer objections to the curious taste of the product might be expected in simple liquid milk; 'too much magic' to be acceptable would probably be the advice of a food marketing expert. However, CO_2 in milk destined for cheese, butter, yoghurt and other dairy products, or as additions to primarily non-dairy recipes, might be perfectly acceptable.

Ozone sterilisation of food equipment. The ready availability of inexpensive, electronically controlled ozonisers begs the question of whether traditional liquid chemical sterilants such as bleach, sulphur dioxide solutions or organic phenolic/resorcinolics/cresolics, used in the food industry could advantageously be replaced to some extent by ozone. Ozone, being gaseous, reaches all the internals of pipes and vessels more effectively than a liquid, which may leave the tops of vessels and pipe loops ineffectively sterilised. Furthermore it decays away to harmless oxygen, avoiding any question of pollution, and is intrinsically safe and taste/odour free if any food is accidentally contaminated.

Pumping. The mundane business of pumping beer or lemonade from a tank or barrel to another tank, or to the customer's glass, can be carried out with a pump. However, it is often convenient, and more reliable, to employ

[50] Carbon dioxide is not just used to preserve plant matter. CO_2 can enhance the growth rate of living plants (see section 3.4.5).

gas pressure to the pump. In essence this is the 'pressure transfer' system which is used in chemical works (section 3.3.8). The use of gas pressure is on the increase, both on small and large scales. In large breweries, PSA nitrogen plants are being installed for beer pumping. On a smaller scale, although a gas cylinder is usual, small PSA plants, smaller than a gas cylinder, are being considered for use in bars and pubs.

In the case of nitrogen, pressure transfer is used, i.e. gas pressure on top of the liquid is used to transfer it from the storage tank to the point of use. Carbon dioxide can also be used but it dissolves in drinks, a process that has been pursued for centuries to enhance the taste. Carbon dioxide is used in an ale-house to supply beer with not two but three effects : not only does it pump the beer and add fizziness but the gas can be expanded from the cylinder via a heat exchanger, or added to the beer in the line just before the dispensing tap, in order to give a cooling effect. A 10°C or more drop in temperature is easily achieved, giving a pleasantly cool drink, provided the gas used is sufficient and the cellar temperature is already fairly low.

Sparging. Sparging is the injection of a gas into a stream of liquid to remove an unwanted gas dissolved in that liquid. The most common example is the removal of oxygen from water during the preparation of beverages. Excessive oxygen leads to undesirable oxidation reactions in the drink and limits shelf-life.

The process works because bubbles of nitrogen equilibriate with the oxygen in the water until the nitrogen reaches around 10% or so oxygen content. Thus, by blowing enough bubbles of pure nitrogen through the water in a counter-current arrangement, most of the oxygen can be removed.

Two patterns of sparging are in use: the in-line sparging bar and the packed column sparger. The former is less effective but is cheap and very small, whilst the latter effectively exploits the counter-current principle, and can be made very efficient, but is high in capital cost and will typically need a large space to put it in.

3.4.4 Carbonated drinks

Pressurised carbonated fizzy wines, such as champagne, and fizzy beers have been made for at least 500 years. The fermentation process that produces alchohol from sugar also produces carbon dioxide, which improves the taste of the drink giving the beverage a fizzy acidic taste. However, a simple cork seal in a glass bottle is incapable of retaining much overpressure of CO_2, limiting the fizziness factor to a slight tingling on the tongue. The practice of

Figure 3.16 Early Victorian lemonade bottle. The ball visible in the neck seals against the narrower neck orifice to retain CO_2 pressure. To drink the contents, the user pushes the ball down. It is trapped by the crimp in the bottle from sinking to the bottom and prevented from blocking the neck during pouring by the two lugs above the crimp.

adding CO_2 has been employed since about 1800[51] although efficient and inexpensive seals for the bottles for such beverages was a limiting factor for many years (Figure 3.16). With the advent of effective seals for bottles, such as the multilayer champagne cork and the internal glass bead check valve seal, it became practical to retain a large amount of CO_2 in a drink. Today, common fizzy drinks, such as Coke, Pepsi or sparkling mineral water, employ around 1% CO_2, giving a pressure of a few barg in the worst case of a bottle left in bright sunshine. One might expect that with this considerable overpressure the opening of a lemonade bottle might invariably lead to an instantaneous foam of drink and gas showering the unfortunate consumer. This does not usually happen, however, because the CO_2 is normally released from solution rather slowly. If, however, nucleation centres are

[51] Joseph Priestley, as well as discovering oxygen, made carbonated drinks and proposed in the eighteenth century that the British Royal Navy could use fizzy drinks on long voyages with the intention of reducing the vitamin-deficiency disease scurvy.

added to encourage bubble formation, or mechanical agitation applied, the equilibrium is reached more quickly, resulting in a mass of foam.[52]

Not all drinks require a high fizziness. Nevertheless, it is useful to have some internal pressure of gas and exclude oxygen (to prevent undesirable oxidation leading to rancidity or vinegar forming). The presence of a small internal pressure of nitrogen enables the use of much thinner aluminium cans for the canning of uncarbonated or low-carbonated beers and juices; a practice called 'can rigidification'. Special dosing systems which add just a few drops of liquid nitrogen to an open can a fraction of second before its lid is sealed on have been designed for can rigidification.

Guinness, the Irish black 'stout' beer, is sometimes canned in special nitrogen-pressurised containers; cans with a N_2 capsule in the base dubbed the 'widget'. The widget's function is to present the beer in a form as close as possible to the 'real' draught Guinness. The widget, containing nitrogen, is a small chamber located in the bottom of the can which has a small hole in its top. During filling, the widget becomes filled with beer nitrogen under pressure. On opening the can and releasing the overpressure in the main body of the can, the nitrogen is released and bubbles through the beer causing nucleation of small bubbles of CO_2 as well, the resultant very small bubbles forming a satisfying fine frothy 'head' on top of the beer.

There are still problems today with maintaining fizziness and excluding oxygen over an extended period in plastic containers. Early plastic fizzy drinks tended to lose CO_2 via permeation through the plastic and impermeable coatings were at first thought to be necessary. However, the advent of the PET bottle, with its low permeation characteristic and high strength enabled the production of thin, lightweight, inexpensive, high capacity carbonated drink bottles.

Occasionally today, as in the past, lemonade bottles explode. The stresses on bottles are quite high. The modern PET bottle is very thin and has to be strengthened by bi-axial stretching of the polyester polymer to achieve the necessary strength. The rupture of bottles can be due to a number of factors. However, if the bottle was intact and correctly filled, then rupture is usually due to excessive temperature rise in bottles exposed to full sunlight. Interestingly, the most violent detonations are not from bottles which are full but from those which are partly consumed. A little thought will show that, as with hydraulic testing of gas cylinders, the energy release is greater in the second case.

Ripening of fruit. Ethylene is used in small quantities during the transport of bananas by ship. During their sea voyage, the bananas are subjected to a small concentration of ethylene which helps to bring out the bananas'

[52] The reader is encouraged to observe the effect of adding finely milled salt to mineral water.

flavour and change the colour from the bright green of their fresh-picked state to their familiar yellow colour.

The banana-ripening reaction using ethylene is essentially an enhancement/speeding up of the natural process. It is catalytic or hormonal in action, as can be judged from the effective amounts; a 50 g dose is sufficient to induce ripening in tens of tonnes of bananas in an enclosed cold store. Special small gas cylinders ('ethylene tubes') are sold which contain around this amount of gas.

The ripening effect of small amounts of ethylene is by no means confined to bananas. Leaks at ethylene plants, although typically small and producing sub-ppm concentrations of gas, have often been noticed to have effects on local vegetation; for example, premature dropping of leaves from deciduous shrubs and trees. Other fruits are also ripened by ethylene. Kiwi fruits are very sensitive to ethylene and ripen more quickly in the presence of a few parts per billion of ethylene.

As well as responding to ethylene, many fruits also evolve trace levels of ethylene during ripening. As ethylene is evolved, it causes more ripening, releasing more ethylene and so on. This autocatalytic action serves to accelerate ripening and to ensure that ripening within a batch of fruit is synchronised as the ethylene diffuses through the air in a store. Clearly, the old saying 'it takes only one bad apple to make a barrel rotten' is not just a parable but actually contains a piece of science.

Ripening is delayed for many fruits if these traces of ethylene are removed from the atmosphere surrounding the stored fruit. For some years now, ethylene-removal systems sometimes as simple as a large bubbler with a dilute aqueous reducing agent, for example, potassium permanganate, have been used to reduce ethylene in large stores. Containers for shipping fruit are now also being made which remove ethylene, although regenerated absorber beds are favoured in these more high-technology devices. Shipping containers, as well as removing ethylene, can also inject CO_2 and/or nitrogen using gas cylinders and PSA units to control the basic constituents of the storage atmospheres. The BOC Group INTAC system is typical of the most sophisticated of these systems, seeking to control O_2, CO_2, ethylene and water vapour levels by adding air, PSA nitrogen, cylinder CO_2 or water, whilst scrubbing ethylene or water, as well as governing the temperature given by the refrigeration system.

3.4.5 Carbon dioxide in intensive agriculture

The basic action of a green plant is to convert water, sunshine, carbon dioxide and a little nitrogen into more of that plant's carbohydrates and protein. In order to boost plant production, therefore, farmers have for millenia added more water and sought to grow their crops in the sunniest spots available, consistent with that water being available. It is only recently, however, that farmers have sought to add the gaseous component, carbon dioxide, in efforts to increase yield. (Small concentrations of another gas, ammonia, have also occasionally been applied to intensive agriculture to boost nitrogen.)

If a greenhouse can be more or less effectively sealed, it can be economic to add CO_2 to the atmosphere, boosting the natural 350 ppm to up to 1500 ppm. The acidulating effect of CO_2 limits the amount that can be added, which depends on crop type. Strawberries, cucumbers, table grapes and tomatoes are all treated successfully with CO_2, with yield per acre or growth rate enhancements of the order of 20%. More controversially, CO_2 has been added to irrigation water with cotton and open-air tomato crops in fairly arid areas. It is claimed that in alkaline soils acidulating the irrigation water leads to enhanced dissolution of phosphorus and trace elements and better crop growth. Although mineral acids can be used, CO_2 has advantages in cost and safety and, like mineral acids, helps to limit scaling on irrigation pipework with alkaline waters.

Just as plants can benefit from carbon dioxide, brewing yeasts can benefit from treatment with air enriched with oxygen,[53] whilst oxygen can be used to enhance dough-making in the bread industry. Oxygen can be used to replace the potassium bromate oxidising agent often added to bread, which is now being banned in some countries as a potential carcinogen.

3.4.6 Gases in pest control

Pesticides are prepared in various forms but only a very few are actually gaseous. The selective action usually required means that a complex molecule, which is inevitably large and thus non-volatile, is indicated. The use of liquid CO_2 in high-pressure cylinders as a solvent for organic plant and animal pesticides has now become a standard means of delivering these materials. The liquid CO_2 acts as a reasonable solvent for these materials, although a co-solvent such as methanol or ethanol is often used. An example of these pesticides is pyrethrum, the environmentally benign extract from daisy flowers used for killing insects such as aphids. It dissolves easily in liquid CO_2 and is effectively aerosolised when the mixture is allowed to spray out into the air.

Similarly, tea tree oil can be dissolved to quite large volumes in liquid CO_2 and such a mixture, on release, aerosolises the oil very effectively. Tea tree oil is used to prevent the growth of bacteria and moulds inside apparatus; a

[53] Oxygen is used by the yeast for its respiration during its normal growth. Yeast, like higher animals, converts its food, sugars in this case, into energy by oxidation with O_2, producing CO_2 as a by-product. It is the CO_2 which is relied on to give a beer its natural fizziness and to cause bread to rise. It is only when deprived of oxygen that yeast starts to produce alcohol along with the CO_2 from the sugars it feeds on. However, the growth of yeast cell mass is enhanced by O_2.

typical use might be in bacteriostasis in air conditioning. (The Australia-based Commonwealth Industrial Gases is developing the use of a tea tree oil product in air-conditioning systems.)

Air-conditioning systems are known potential sources of bacteria and other microbes. Damp parts of the system are particularly prone to developing colonies of microorganisms, hence the outbreak of fatal *Legionella pneumophilia* infections in buildings where air-conditioning cooling-tower water became contaminated. Although air filters are in general effective, old filters, particularly if they have become wet for some reason, can let through numbers of organisms, which can actually grow through filter material under some circumstances. Fortunately, the powerful defence mechanisms of the human body, lysosomes in the nasal and tear duct areas, for example, swiftly kill all but a few microorganisms. However, some of the survivors will be pathogenic and may cause a disease directly or exacerbate another condition such as asthma. Tea tree oil, dispersed into 10 μm or less droplets by CO_2 aerosol flash-off, has been shown to be active against Legionella and several other important bacteria.

Very few pesticides can be used directly on food for human consumption. Even if they are not actually toxic to humans (as was arguably the case for DDT, for example), their subsequent removal to low levels in the foodstuff is usually required.

Simple carbon dioxide is one material that can safely be used. It is harmless in any quantities in foodstuffs but can affect the growth and respiration of fungi and plants. Hence CO_2 can be used on its own for fungus inhibition in grain silos. CO_2 can also be used on its own as an insecticide in grain silos. Because of its physiological effect on respiration at percentage levels (it causes hyperventilation) it is effectively toxic in concentrations in the 10% or so range and it is not necessary to achieve the 50% or more levels needed if a simple inert asphyxiant gas such as nitrogen were used. Its low toxicity in low concentration and its acceptability as a foodstuff ingredient mean that there need be few fears about its safety.

Experiments in the US have shown that in grain treated with as little as 1 kg of CO_2 per 200 kg of grain, insects are killed immediately following hyperventilation and dehydration, whilst their eggs and pupae larvae are also killed after a few days if the concentration of CO_2 is high enough (approximately 60%). Systems have been installed which blow CO_2 into grain silos at the top using a high initial flow rate for purging and then a lower rate. The system then measures achieved gas-space concentration at different points down the silo using a simple analyser (the CO_2 sinks through the grain).

A more potent agent for the destruction of insects in grain siloes is phosphine. Although phosphine is highly toxic it is also highly reactive and unstable, so it can be guaranteed that none will remain a few days after treatment. Under normal conditions of moisture, the small amount of phosphine used is swiftly hydrolysed and oxidised to a few parts per million of harmless phosphoric acid

$$2PH_3 + 3H_2O + 2.5O_2 \Rightarrow 2H_3PO_4$$

However, the high acute toxicity of phosphine means that stringent precautions are necessary.

3.4.7 Margarine and edible oils

The margarine industry is said to have originated out of a published government requirement for a substitute for butter during the economic slump in Paris during the Napoleonic period when wars were exhausting the normal means of production. However, animal fats have been substituted for each other for time immemorial, so this could be viewed as simply the first time that an attempt was made to make the substitute simulate butter.

Surprisingly perhaps, it was nearly 100 years later that chemical modification of edible oils was carried out to extend this process of substitution. The first practical process was patented in Germany in 1902 by Wilhelm Normann, following the process still largely used today employing nickel and hydrogen. This was only a useful process at that point, however, because fairly pure hydrogen was becoming available from the burgeoning coal gas and chemical industry. The fat-hardening industry expanded rapidly, into baking fats and fats for soap makers as well as butter, and is today a major consumer of merchant hydrogen gas.

The chemistry of hardening oils with H_2. Natural oils and fats are, chemically speaking, esters, being the result of combining three carboxylic (organic or fatty) acids with the trihydric alcohol, glycerol. The fatty acids involved may be saturated (no C=C double bonds), unsaturated (one or more C=C bonds) or polyunsaturated (two or more C=C bonds). An example of an unsaturated vegetable oil is glyceryl linoleate (Figure 3.17).

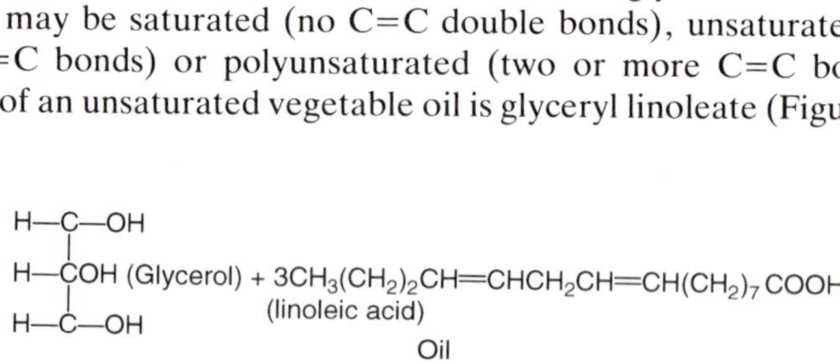

Figure 3.17 Glyceryl linoleate.

In the presence of nickel catalysts, high-pressure hydrogen gas saturates double bonds in oils, reducing them to 'harder' (higher melting point) fats. An example of a fat is glyceryl tristearate (Figure 3.18).

$$
\begin{array}{ll}
CH_2OCH_3(CH_2)_{16}COOH & \\
| & \\
CHO{-}CH_3(CH_2)_{16}COOH & \quad \text{Fat} \\
| & \\
CH_2OCH_3(CH_2)_{16}COOH &
\end{array}
$$

Figure 3.18 Glyceryl tristearate.

Many vegetable oils are unstable, reacting with oxygen in the air to form rancid or off-flavours of fatty aldehydes and ketones, for example, as well as discolouring. Hydrogenation generally reduces these problems by saturating some of the reactive double bonds both in the pure oil and its contaminants, making the oil more stable.

Some polyunsaturated carboxylic acids, such as linoleic acid, are regarded as essential components of the human diet. They are used by the human body to make hormones and the like and are known as essential fatty acids (EFAs). Saturated fatty acids, by contrast, are generally regarded as not just inessential but possibly harmful, being implicated in coronary arterial blockage heart disease. It is therefore an aim of most processes not to fully saturate the linoleic and other polyunsaturates but to saturate only part of the linoleic content and other acids whilst still achieving a higher melting point.

There are problems in many oils with the linolenic acid content. Linolenic acid has the formula $CH_3(CH_2CH = CH_2)_3(CH_2)_7COOH$. Hydrogenated, linolenic acid yields some valuable and some undesirable isomers of linoleic acid. Some of these isomers cause a nauseous sweet odour to arise. However, if unreacted, a substantial content of this acid will cause the oil to polymerise. Linseed oil, which is mainly composed of linolenic acid, was up until the 1930s the principal component of most oil paints as it reacts over a few days in the air to form a tough skin.

There are many transition metal and noble metal catalysts that if finely divided, will catalyse the hydrogenation. Copper has been extensively researched and is valued for its removal of linolenic acid content whilst not saturating all the linoleic content of an oil. However, problems with poisoning and the necessity for virtually 100% removal of copper from the final product have meant that nickel is still the mainstay catalyst.

Oil hardening in practice. The process is as follows. First, oil is prepurified to remove air, particulates and catalyst poisons. It is then loaded into the reactor vessel and vacuum applied, avoiding safety problems with H_2/air mixtures in the ullage space. The oil is then warmed to around 120–180°C and agitated in the reactor vessel whilst hydrogen is bubbled through at 3 barg or so.

The nickel catalyst is one of the most ingenious aspects of the whole hydrogenation process. Although fixed-bed catalysts are used, the most popular system is the addition of finely divided (10 nm particle size) nickel on kieselguhr silica support powder (*c*. 10 μm) to the reaction mixture, followed by subsequent filtration. The filtration must be very effective; the resultant mixture from the reactor is jet black in colour. The approach usually adopted is to recycle the first few minutes of filtrate. The filter bed accumulates caked catalyst, which acts as the true filter.

Commercial hydrogen purity is now well within the purity requirements for alimentary fat hardening. Less pure sources of hydrogen are occasionally used, particularly for less demanding applications such as soap feedstock hydrogenation. Mercury contamination of electrolytic hydrogen is reduced to infinitesimal levels by iodised carbon beds and oils made with electrolytic hydrogen contain mercury levels much lower than many natural foodstuffs.

3.4.8 Decaffeinated coffee and detarred tobacco

Both nitrogen and carbon dioxide have had some impact in the business of coffee drinks. Nitrogen is used both in grinding and in roasting while carbon dioxide can be used for decaffeinating. Coffee beans ground with the assistance of liquid nitrogen retain more of the flavour of the bean (section 3.4.1). Coffee roasted in a nitrogen atmosphere avoids the excessive oxidation of some of the flavour compounds in coffee that can occur in air roasting. Perhaps more ingenious than these applications, however, is the use of carbon dioxide.

After roasting, coffee beans, and infusions made from them, contain considerable amounts of the drug caffeine, which has both stimulant and diuretic effects. A large market has been developed for a beverage with the taste of coffee but without these side-effects. One process uses dichloromethane as the caffeine solvent. This works tolerably well but has snags in the cost and toxicity of dichloromethane, which must be very thoroughly removed (an earlier process used benzene but this is now completely unacceptable as even small amounts of benzene are considered carcinogenic). Dicholoromethane's high volatility, it boils at 40°C, makes this relatively simple, although it is also highly inflammable on this account. A more recent process using CO_2 has many advantages over the dichloromethane process.

CO_2 supercritical extraction of caffeine. Supercritical carbon dioxide (section 1.3.15) is capable of considerable solvent powers, including organic substances such as caffeine, a fact which has been put to good use in the processing of coffee beans. High-density supercritical CO_2 is percolated through the roasted coffee beans, dissolving a substantial proportion of the caffeine. The caffeine is then removed in the CO_2 and is absorbed in water or carbon, allowing the CO_2 to be recycled. The whole process is carried out at

high pressure, over 70 bar, which leads to the only economic problem in the process: the cost of the pressure vessels required.

Tobacco tar-extraction and fluffing. Mass-produced tobacco products, such as cigarettes, have now reached a pinnacle of sophistication. As well as chemicals to keep the tobacco from going out and ultrathin clean-burning paper, some gas science is also applied. In low-tar cigarettes, air is drawn in through holes in the paper near to the filter end to dilute the inhaled smoke stream. Only by laser drilling these holes accurately in each cigarette is it possible to be sure that a consistent effect is achieved. Filter tips are now carefully made too. Particular grades of triacetate fibre tow, compressed and chopped up, are used. The fibres are parallel to the direction of flow and act as diffusion filters, having little impact or sieving filter effect (section 2.4.3). These offer reasonable retention of tar droplets and particles with low pressure drop.

More sophisticated still is the use of supercritical fluid carbon dioxide to process tobacco leaf prior to incorporation in cigarettes. Liquid CO_2 treatment of tobacco has several effects which tobacco companies value:

- the removal of some harmful tars via its solvent effect;
- the removal of the addictive nicotine via solvation;
- residual CO_2 left in the tobacco can be allowed to vaporise quickly, leading to expansion of the tobacco leaf ('fluffing').

The expanded tobacco is, of course, less dense than standard leaves and thus less is needed to fill cigarettes, enabling still lower tar content ratings.

3.5 Gases in the semiconductor industry

The semiconductor industry continuously pioneers new processes for processing materials. Every three years or so it doubles the capability of its technology. Memory circuits, currently (1996) 4 or 16 Mbits in capacity, will be 16 or 64 Mbits in capacity in less than six years. Processors, currently running at 66 MHz will be running at 120 MHz with the same data in three years' time. The industry is on a technology treadmill in which it relies on constantly convincing its customers that their current information technology, although only half-way through its life, needs replacement by technology containing new semiconductors. The replacement cycle can only be sustained by offering a substantial improvement, at least a factor of two, at each approach to customers.

The semiconductor industry's need for constant technical improvements and its high (albeit erratic) profitability, means that it is a hotbed for innovation. Techniques used today for semiconductors are quite likely to spin-off into other industries. Chemical vapour deposition (CVD), developed largely for the semiconductor industry, is now, for example, used for optical fibre manufacture. The ultraclean gas techniques described in this section seem particularly likely to spin off in the future into other industrial sectors. For this reason, and because it is also a pioneer for processes using gases, the semiconductor industry is particularly interesting and appropriate for this book.

An integrated circuit (IC), is simply a large number of simultaneously manufactured electronic circuits. Each individual circuit is made of conventional electronic components, i.e. resistors (made of semiconductor), capacitors (made of two metal plates insulated from each other with insulating oxide), conductors (made of metal) and, of course, the 'active components', i.e. the transistors used as amplifiers and switches (made of semiconductor). There was a conceptual leap from circuits made from laboriously hand-wired individual components to the IC in which conductor, insulator and semiconductor materials are 'printed' onto a substrate to form the same circuit function. The IC, although it has a totally different scale and physical appearance to the hand-wired circuit, has the same topology and performs the same electrical functions. The real innovation of the IC is that it can be mass-produced to a degree which dwarfs the productivity improvements of organisational innovations such as the factory assembly line. Productivity in terms of circuits produced per manhour has already been multiplied a million-fold.

The IC plant or 'wafer fabrication' plant, often shortened to 'wafer fab' or just 'fab' essentially prints ('fabricates') active microelectronic circuit layers onto thin round wafers of silicon semiconductor, 0.5 mm in thickness and mostly 100–200 mm diameter.[54] The single crystal silicon substrate, or a silicon layer contiguous with it, is used for fabricating the transistors, whilst upper deposited silicon layers can be used for resistors. Metal layers are used for conductive interconnections, whilst insulating layers provide isolation from conductors and the dielectric portion of capacitors. The patterning for each layer is defined by UV exposure of a mask pattern onto a photosensitive plastic 'resist' layer applied to the silicon wafer. The resist is polymerised only in the exposed areas and the unexposed resist removed in the development process. Other processes can then be used, such as etching, which will only affect the exposed wafer, whilst leaving the resist-protected parts unaffected. This simple principle can be used up to 20 or 30 times (20 or 30 'mask layers') to form the necessary circuit topology.

[54] The term 'wafer' came about because originally the largest single crystals of semiconductor that could be obtained were only a few millimetres or so in diameter and the 0.5 mm slices from these used for early integrated circuits were indeed very similar to a Christian communion wafer. With future fabs planning to use 300 mm diameter silicon, a new term is perhaps called for.

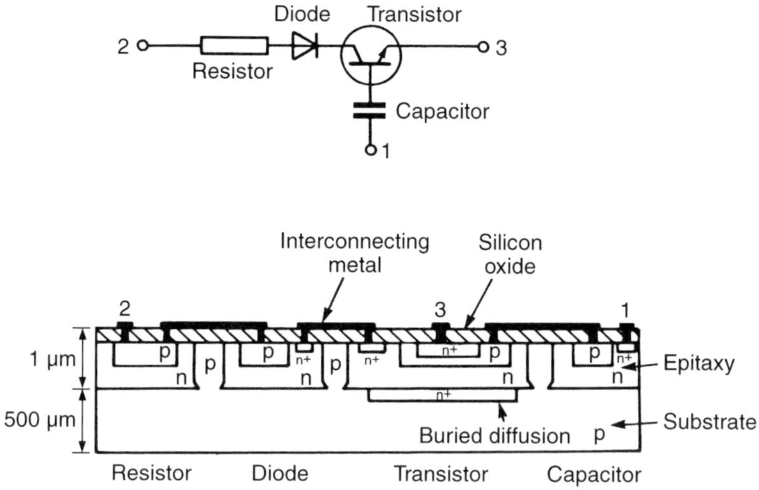

Figure 3.19 An integrated circuit with diode isolation; (a) electronic circuit; (b) integrated circuit layout.

Throughout a wafer fab, industrial gases are taken for granted as an integral part of nearly every procedure carried out. It is often the case that when all the major applications for gases inside a wafer fab are added up, they total much less than the gas consumed by the site. This reflects that in fact there are a myriad of minor applications which together consume large volumes of gas. Semiconductor manufacture is arguably more dependent on industrial gases than any other manufacturing process today.

3.5.1 The need for high purity in semiconductors

Gases are used in many parts of the processes in the manufacture of semiconductor integrated circuits. Nitrogen is ubiquitous in the wafer fab. It is used for all sorts of processes, from drying wet silicon wafers to diluting process gas. Other bulk gases used are oxygen, argon and hydrogen, in much smaller amounts. Only cylinder quantities are needed of most of the hundred or so other gases used but these are the essential reactive gases, from the silane used for epitaxy and glass insulator deposition to chlorine- and fluorine-based etching mixtures and even more exotic compounds used in quantities of a few grams for 'doping'.

The requirements for gases for making semiconductor devices are distinguished from most other applications for gases in that ultimate purities are generally needed. The silicon on which most ICs are made is itself an ultrahigh purity product. The 150 or 200 mm diameter silicon wafers processed in wafer fabs today are thin slices from giant single crystals, the biggest commercially used monolithic crystals, and are pure at the level of 10 parts per billion.

The wafer fab is an ultrahigh purity manufacturing environment which is cleaner by far than, for example, a hospital operating theatre. Special non-dust-generating cleanroom 'bunny' suits and face masks are required in the cleanroom. An ordinary room environment has millions of tiny particles of dust per cubic foot whilst in the cleanroom there are fewer than ten particles; less than a bacterium in each cubic foot of air. Particulates as small as virus particles, 0.1 μm, are effectively monitored and controlled in the latest cleanrooms. Still higher levels of cleanliness are achieved near the silicon wafers themselves by the use of 'mini-environments'. These are the suites of equipment now used where wafers do not contact even the ultraclean cleanroom air but only see air or nitrogen inside the equipment or inside the SMIF (standard mechanical interface) boxes used to transport wafers between different parts of the cleanroom.

Many of the processes in the wafer fab have a special purity requirement, as a glance down the list in Table 3.6 will confirm. Ultraclean gas technology has therefore grown up largely to meet semiconductor needs. Only in a few cases is there is good quantitative evidence for the effect of impurities on a semiconductor process. It is well known, for example, that oxygen and moisture impurities in gases used for CVD silicon nitride growth give a predictable proportion of oxygen in the nitride layer grown (CVD nitride is often used to passivate circuits, protecting them from reaction with the atmosphere). However, in many other processes theoretical predictions are clear-cut or qualitative results are available so that a picture has been built up, albeit incomplete, of the most critical impurities (Table 3.6).

Particulate contamination. A particle of larger than 0.1 μm may ruin an integrated circuit using 1 μm design rules, as 0.1 μm is representative of the thickness of some these IC structures. It is possible for even smaller particles to cause problems if they are composed of 'poisons' such as alkali metals. The latter can diffuse freely through the semiconductor structure, poisoning a volume of silicon much larger than their own volume. They cause unpredictable and time-varying doping, off-spec device parameters or, worse, drift of device characteristics over time.

For any particular scale of circuit, the maximum allowable particle size can probably be set at 10% of design rule for 'ordinary' particulates and perhaps between 1% and 3% of design rule for poisons. As far as the number of particles above this limit that are allowed, a typical standard calls for less than 10 or less than 100 per cubic foot. As this is in excess of what could be typically guaranteed for industrial gases in bulk, most semiconductor processes will require filters on all critical processes.

Particulates are of little or no importance once the wafers have been passivated as the active devices are then largely protected against any

Table 3.6 Semiconductor fabrication processes and critical gas impurities

Process	Gas	Critical Impurities
Crystal growth	Argon	Oxygen, water, hydrocarbon
Blow dry, blow clean (gun)	Nitrogen	Particulates
Carbon dioxide snow cleaning	Carbon dioxide	Particulates, hydrocarbons
Epitaxial deposition	Hydrogen	Dopants: Group III, V elements, oxygen, water, hydrocarbon
Field oxidation	Oxygen, Hydrogen	Hydrocarbon
Purging	Nitrogen, argon	Oxygen, water
Nitriding	Nitrogen, hydrogen	Oxygen
Diffusion	Argon	Oxygen, water
Gate oxidation	Oxygen	Water in dry oxidation, hydrocarbons
Chemical vapour deposition (CVD)	Hydrogen, nitrogen	Oxygen, water, dopants
Polysilicon	Silane	Oxygen, water
CVD oxide	Oxygen, nitrogen	Water, metals, dopants
Annealing	Nitrogen, argon	Oxygen, water
Metallisation, alloying	Nitrogen	Oxygen, hydrogen
Silicides	Argon	Oxygen, water, nitrogen
Sputtering, glass passivation	Argon, nitrogen, oxygen	Water
Pneumatics	Nitrogen	Particulates
Encapsulation	Nitrogen, hydrogen	Oxygen, water

foreign body intrusion. Higher levels of particulates can also, perhaps surprisingly, be tolerated in some of the vacuum processes, such as low-pressure CVD, plasma etching or sputtering. This is because at the low pressure of operation the particles can settle, just like sand in air, and because the processes are frequently performed with wafers upside down or sideways, reducing the possibility of particles settling on active surfaces. In addition, some processes are always followed by a wet chemical cleaning step that will also remove particles. (CVD and etching can create large numbers of particles themselves.)

Reduction of particulates is not easily achieved by clean-up after particles have been deposited. This is because the electrostatic forces developed between the wafer and the particle are very large compared to the particle weight. If the force between a spherical particle and a conducting surface is calculated on the basis that it adheres with an electrostatic force (due to a small charge on the particle), for example, is easy to show that the force of adherence, expressed in terms of the gravitational strength required to dislodge it, varies as $1/R^5$, where R is the particle diameter

particle weight, $W \sim R^3$

contact electrostatic force, $F \sim 1/R^2$

electrostatic force in gravities $\sim F/W \sim 1/R^5$

A similar effect, i.e. very high adherence forces for small particles, is also seen if the weaker van der Waal's forces are assumed between particle and surface. Given the difficulty of removing particles once deposited, it is important to avoid deposition in the first place. A reduction in particles derived from gases can be achieved by filtering gas before the point of use. However, care must be taken that the filters used do not add particles derived from their own structure to the gas stream. This is a typical characteristic of, for example, fibrous filters.

Removing particulates from wafers usually requires a wet process. At entry to the wafer fab, wafers are often wet-scrubbed with rotary polymer sponge brushes (which removes particles). Wet processes are good at removing particulates because a liquid, by acting as a dielectric between particle and surface, can reduce the adhesion forces. In addition, a liquid can coat the particle and cause it to be absorbed into the body of the liquid as a colloidal particle. At various points in its progress through the wafer fab, therefore, wafers are subjected to wet cleans, though not with brushes after the initial scrubbing. Wet cleans often employ an acid/oxidising bath and an alkali/oxidising bath (section 3.5.10). Oxidation can help remove particles by oxidising the particle and/or the wafer surface at the point where the two are adhering, enabling the particle to be detached.

Gaseous contamination. Generally, the standard industrial gas specification, in the low parts per million for most impurities, for bulk gases is adequate for much semiconductor use as far as dissolved impurities go. However, there are many critical applications where this is not the case. Silicon and metals of ICs will react with even tiny amounts of water and oxygen in high-temperature inert gas processes such as annealing. A typical wafer fab plant may require these amounts to be below, say, 50 ppb and 10 ppb respectively, for all its high-temperature argon and nitrogen processes and these specs are constantly being forced to higher purities. Even greater purities are desired for the silane, hydrogen, nitrogen or ammonia used in nitriding. These must contain sub-ppb oxygen as silicon dioxide is thermodynamically favoured over silicon nitride and the latter will only form stoichiometrically in the complete absence of oxygen. Frequently, layers are seen which correspond to silicon oxynitride caused by very slight oxygen contamination during the reaction.

For many processes the common impurities in oxygen, water and hydrocarbons, are regarded as unimportant. However, some companies have insisted on low hydrocarbon content for Mbit and above circuits and a number use purifiers to lower the water content to the 1 ppb region. These are mostly using a dry oxidation process for the critical gate oxide, which small quantities of water can render uncontrollable by accelerating oxide growth in an unpredictable way. Carbon contamination can cause gate oxide problems.

Table 3.7 Gas purity specifications for semiconductor wafer fabrication processes

Gas	Purity level
Nitrogen (cryogenic on-site plant)	1 ppb O_2, H_2O 0.1 ppb CH_4, CO_2; 200 ppb CO 500 ppb H_2 (depending on plant design)
Nitrogen (via bulk on-site purifier)	0.1 ppb O_2 0.5 ppb H_2O; 0.1 ppb CH_4 0.1 ppb CO_2; <0.3 ppb CO <1 ppb H_2
Argon (delivered high purity liquid argon)	40 ppb N_2
Oxygen (delivered high purity liquid oxygen)	50 ppb N_2 20 ppb hydrocarbons
Argon (via bulk on-site purifier)	1 ppb H_2O 1 ppb N_2 0.2 ppb O_2 all other impurities (save inerts) <0.05 ppb

It is sometimes assumed that purge gases can be of low quality. This is not the case. HCl and the other corrosives etchants will react with moisture and attack metal parts, producing metallic contamination if the purge gas contains moisture. Silane and the other pyrophoric gases will react with residual oxygen to produce oxide particulate contaminants. Purge gases in general must be of similar quality to those used in the process if they are to be effective in maintaining the gas lines of the critical processes.

Table 3.7 lists the level of purity now achieved (and expected) in state of the art wafer fabs. All these figures should only be taken as example achieved purities. Plant and purifier performance vary widely with respect to these minute trace levels of contaminants. Note that the moisture figures are probably high because of the use of Teflon polymer filters. At the moment metal or ceramic filters are still exceptional but when these become widespread these figures may well fall to the 0.1 ppb or better achieved in ultraclean laboratories. At 0.1ppb there is a consensus that bulk gas purities are probably good enough for 99.9% of all wafer fab operations and that emphasis should now shift to purity of vacuum, surfaces and special or process gases.

3.5.2 Making ultraclean gases

Ultrapure gases, it turns out, are not in general too difficult to produce. Physical distillation increases purity in principle by geometric steps. After each distillation the gas contaminants decrease by the same multiplying factor. If one pass up the purifying column, or one distillation plate, decreases contaminants by a factor of 10, then two columns or plates will increase contaminants by a factor of 100. Ultrapure grades of most gases can be created by physical separation; air gases can simply be drawn off the standard liquid air column and then redistilled in a small-diameter side column. Gases such as silanes are also fairly simple to distil as part of the production process.

It is now generally accepted that nitrogen generation on-site is a more economical way of generating nitrogen at ppb levels of purity. By adding distillation trays and small modifications to ancillary parts of the plant (adding extra beds to the air prepurifier units, for example) levels of around 1ppb can be achieved with the exception of CO, which is not effectively separated in air distillation. Depending on plant siting (CO mostly derives from vehicle exhaust in urban areas), this could be as high as a few hundred ppb CO with occasional peaks up to ppm level in some sites. The addition of a purifier(s) is justified if ppb or lower levels of CO are needed.

Chemical purifiers are capable of even more spectacular decreases in contaminants in one pass, even more so if they can use more or less irreversible chemical reactions (section 2.2.11). Purifiers offer the convenience that it is only necessary to purify gases used in critical processes and offer an 'insurance policy' in case a batch of off-specification gas comes through.

With gases produced in bulk by the general chemical industry, ammonia or hydrogen chloride, for example, there is more of a problem. After attempting to purchase the best grade of gas in the best container available, the gas company supplying the semiconductor industry may be able to select batches. Batches not up to semiconductor standard can be sold for less critical applications. Failing this, there is no alternative but to build a special purification rig. This may be as simple as a blowdown facility to remove volatile gaseous impurities such as oxygen on top of a liquefied gas such as HCl. Similarly, vaporising and filtering the product, leaving heavier particulates at the bottom of the cylinder, can also be carried out. A chemical absorber rig can also be very simple, comprising an absorber material on inert packing. The chemical chosen needs to match both the substrate gas and the impurity gas, however. Ultimately, if none of these simple solutions suffices, a distillation rig for upgrading gases is used.

One elegant solution for distillation purification employs the cryogenic storage tank itself as a still (Figure 3.20). Liquid hydrogen tanks are available which allow vapour withdrawal, the lost heat of vaporisation being added back to the tank via an external ambient heat exchanger on the withdrawn gas and an internal coil heat exchanger in the tank. A regulator ensures that tank pressure is not raised too far, when direct vapour withdrawal without back-addition of heat occurs, until the tank pressure has fallen far enough. In the bottom of the tank, solid contaminant gases can build up; with a hydrogen/oxygen slurry a few grams could explode and rupture the tank if ignited. A periodic withdrawal, perhaps only every few months or so, of LH from the bottom of the tank avoids the problem. This

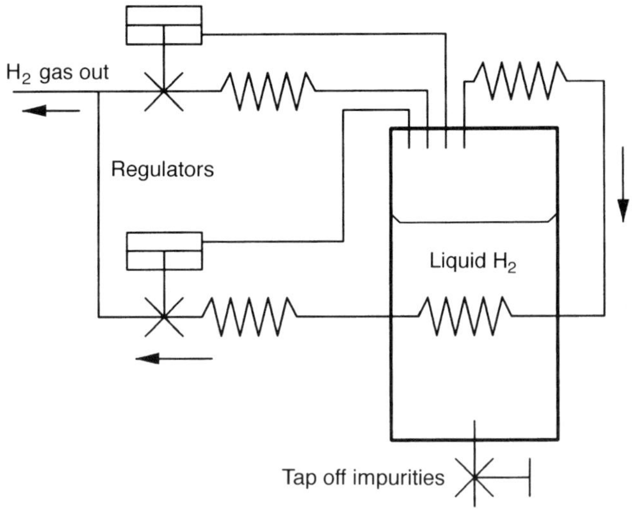

Figure 3.20 Principle of the self-distilling liquid hydrogen tank.

system works well for liquid hydrogen, partly because the heat required to boil H_2 is so small that a small heat exchanger in the tank suffices. Perhaps, despite problems with the larger heat flows needed, self-distilling tanks will be devised for other gases in the future.

3.5.3 Keeping ultraclean gases ultraclean

Keeping gases clean once made is as big a problem in ultraclean technology as making them in the first place. Basic ultraclean pipework and vessels are now made as much as possible from 316L stainless steel, often electropolished. The 316L grade of stainless steel has 16% Cr and 10% Ni. It is completely rust proof under normal conditions and highly resistant to most chemical attack, even chlorine under dry conditions. Although copper tubing is still encountered in wafer fabs, and can give perfectly acceptable in service with more benign gases, it is no longer used in new installations. Copper tubing takes longer to dry than electropolished stainless and is subject to corrosion if accidentally exposed to contamination. Mill-finish (or 'as-drawn', i.e. not polished) stainless steel still has its devotees, however. Although mill-finish tubing does take longer to 'dry-down' to low moisture levels, at least when sub-ppb levels are needed, the presence of polymer filters or process chambers which contain moisture anyway may mean that this is not important. In fact maintaining continous flow in a gas line may be just as important as the tubing finish, since tubing with little or no flow will always show higher levels of contaminants than tubing with a good flow.

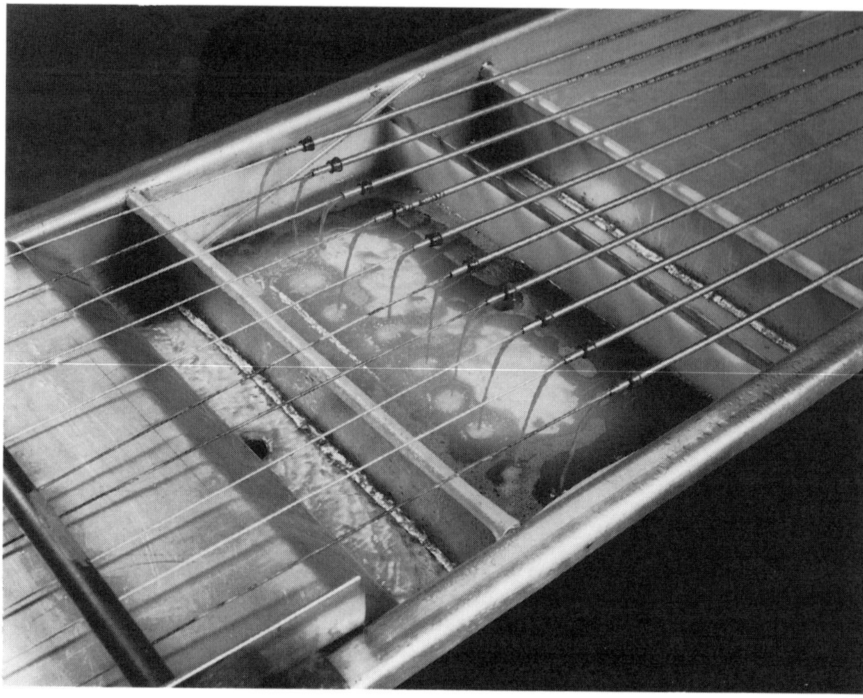

Figure 3.21 Electropolishing of stainless steel tubing showing polishing electrode inside tubes along which is flowing the electropolishing solution (courtesy of Accles & Pollock Ltd).

Electropolishing. The electropolishing process, as the name suggests, gives a mirror-like polished surface to the stainless steel. Electropolishing also has the effect of increasing the surface concentration of chromium and especially nickel, which are the elements that lend stainless steel its thin but adherent and unreactive oxide surface. Electropolishing is carried out by making the surface to be polished an electrode in a proprietary aqueous solution of salts, inserting an inert electrode, then applying an electrolysis current. More current flows from sharp points on the stainless steel surface than from the 'lands' in between. In this way, high spots are removed and the polishing action is achieved. At the same time, iron atoms are removed more rapidly than chromium or nickel atoms, which enrich the Cr/Ni concentration. A carefully electropolished surface has a much smaller tendency to trap particulates than the standard stainless steel smooth finish and it also retains much smaller amounts of moisture.

Passivation. Fluorine passivation is used in installations for corrosive fluorine-containing gases and has been used on other gas supplies, although there do not seem to be many numbers available on its performance. It is carried out using a gas mix with a small amount of fluorine in it, which forms a thin but stable fluoride film in a slow deliberate manner. Subsequently, the gas line will be safer to use with fluorine or HF gas and will give lower levels of contamination.

Stainless steels of the standard 316L variety are normally passivated by passing a mixture of at least 2% F_2 in N_2 down the pipework at ambient temperature. A fluoride layer about 6 nm thick forms, basically overlying the pre-existing 'native' oxide, which is 2 or 3 nm thick. This fluoride layer, mainly chromium and iron fluorides, is passive. It will not react further with fluorine nor with agressive fluorides such as WF_6 (which otherwise tend to react, causing corrosion and plating out on the inside of the pipework). Presumably the presence of involatile iron fluorides stops migration of metal ions to the surface, preventing further reaction.

Unfortunately, standard fluorine-passivated surfaces are spoilt by exposure to ambient moisture and a passivation procedure must be repeated each time a gas distribution line is exposed to air or other moisture-bearing gas. Professor Ohmi's team at Tohoku University, Japan, have published an improvement to fluorine passivation, which relies on exposing pipeline components to 100% fluorine at 220°C for an hour followed by 320°C annealing in nitrogen (Ohmi *et al.*, 1988). This process is clearly much more hazardous than 2% F_2 treatment and cannot be done *in situ* because of the need for strong heating. However, this treatment produces a thicker passivation film and one that is resistant to moisture attack (see also section 3.5.5).

Pipework joints. Electropolishing, high-grade stainless steels and passivation are all capable of improving gas quality for a wafer fab but there are problems. The first problem is in jointing stainless steel. Plates can only effectively be joined by welding and this is also the most common means of joining pipes. Pipes are conveniently joined with the automatic orbital TIG welders which are now quite common in the industry. The choice of the low carbon 316L grade of stainless steel is explained here. The low carbon means that TIG or MIG welding will lead to a HAZ with little unwanted hardness and a good quality joint. Unfortunately, welding does lead to the generation of fine fumes of submicron metal particles, though this can be controlled to some extent by inerting with argon inside the vessel or pipe as well as outside.

It is obviously not possible to produce an all-welded system for gases; it is always necessary to have demountable joints. It is also convenient not to have to weld in confined areas such as in small underfloor ductwork and inside equipment. Although many systems still use Swagelok-type crimped ferrule compression joints (section 2.4.1), standard practice is now VCR™-type demountable joints.

Good though they are, Swagelok type joints have small narrow crevices for particles to accumulate in, and expose part of the outside of the pipe to the gas contained. A better mechanical joint for applications needing the ultimate in cleanliness and gas purity, such as wafer fab plant gases, is a face-seal joint such as the VCR. These special end fittings, which are welded onto tube ends or machined onto components, have a face with a ring-shaped ridge. These annular ridges cut into a flat soft washer placed between the two faces as the union nut clamps the two ends together. The washer is replaced each time the joint is put together. The joints are similar in many ways to the all-metal joints used in vacuum work where knife-edge rings on the faces are forced to seal on a soft copper washer. These fittings are more complex to apply than Swagelok fittings because of the requirement to weld the fittings on either end of a tube, so they are only used when ultraclean gas is definitely required.

MCG face-seal fittings take the principle of face-sealing one stage further by adding a ball-bearing to the union nut so that there is virtually no tendency for the pipe to rotate during the assembly of a joint and a more highly leak-tight and contamination-free joint can be guaranteed, giving a welcome bonus of a joint that needs only minimal force to tighten it. MCG fittings are used with 'C'-ring hollow metal seals, giving improved freedom from particulate contamination.

Ultraclean valves. The next weak point which must be attacked in an ultrapure gas installation is where some kind of flow control fitting is fitted. A conventional control valve, with its packed stem, its soft polymer seat, springs and other metal parts not made in the same 316L stainless steel, is clearly not going to be satisfactory. The polymer seat may outgas moisture and oxygen, the packing may leak air, as well as being a source of contamination itself, and the sealing surfaces will in general release particles on operating the valve, especially if the valve is neither fully open nor fully closed but in a throttling condition.

An ultraclean valve should ideally be all metal with a hard metal valve seat and a stainless steel diaphragm or bellows to separate stem and spring from the gas stream (Figure 3.22).

Other special features may be needed in control elements. For example, the use of tied-diaphragm regulators. Instead of relying on the inlet gas pressure to push the valve stem onto the diaphragm, these have the stem welded to the diaphragm. This avoids erratic operation when small particles form, for example, when silane or BCl_3 contacts residual oxygen or moisture. The oxygen or moisture produces particulate silica or borax in the system which can get on the valve seat and prevent correct operation.

Ultraclean gas control panels. Ultraclean gas control panels and ultraclean gas cabinets for toxic or corrosive cylinder gases follow the same principles as standard equipment (and indeed, some standard equipment now follows

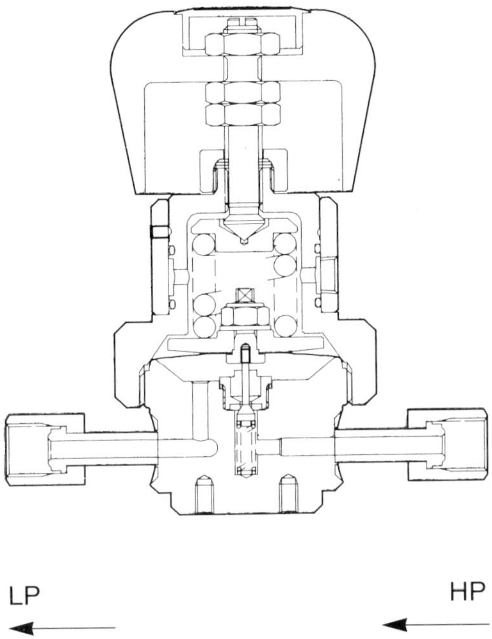

Figure 3.22 Pressure regulator for ultrapure gases. Fabricated in 316L stainless steel with electropolishing on inner surfaces and a fluoropolymer seat material. The diaphragm is tied to the valve to give a positive shut-off for toxic gases (courtesy of Hale Hamilton (Valves) Ltd).

semiconductor industry practice). However, apart from the selection of ultraclean-compatible components such as diaphragm valves and perhaps electropolished parts, there are a number of features which differentiate them:

- they consist of an all stainless steel (316L) largely welded assembly;
- the lengths of connections between components are optimised to minimise lengths of pipe that will be dead-volumes in operation;
- access to vacuum source for the pressure purge cycle is essential (ideally a rotary vacuum pump at a fraction of a millibar pressure);
- automatic purge operation is usual;
- they can be remotely controlled by computer (with local override) for complex installations via pneumatically operated diaphragm actuators on many or all valves;
- they have electronic monitoring of cylinder contents (weighscales or pressure);
- electronic pressure gauges are used to avoid the dead-volume of a Bourdon tube gauge;
- a relief valve is usually necessary after the regulator to ensure that

Figure 3.23 Manually operated ultraclean gas cabinet using electropolished VCR components and vacuum-assisted purging, designed for one-purge cylinder (right) and two-process gas cylinders (courtesy of Cambridge Fluid Systems).

sensitive downstream equipment cannot be overpressurised in the event of a regulator failure, as well as for safety reasons;

- an excess flow valve or orifice is desirable, again to protect downstream equipment as well as for safety reasons;
- check valves are required to prevent back-contamination of purge gas;
- a separate cylinder of gas for purging is needed to totally eliminate the possibility of back-flow of process gases into the site main gas supplies;
- a pneumatically-operated cylinder valve is desirable for safety reasons;
- the whole assembly is put together in cleanroom conditions.

Note that both the connection to the cylinder as well as downstream of the regulator, can be purged in the gas panel shown in Figure 3.23. The purge cycle may be automatically repeated up to 10 times, which is why computer control is very convenient, releasing the operator from this tedious job and reducing the possibility of error.

Ultrapure gases in wafer fab equipment. Much effort goes into achieving ultraclean gases up to the gas supplier's definition of 'point-of-use', i.e. where the pipe carrying the gas enters a piece of equipment or the process chamber where the silicon wafer will react with the gas. However, the conduction of the gas to this point in ultraclean condition is futile, of course, if an ultraclean reactor is not then used to process the actual wafer. The gas must be kept ultraclean right up to the 'point-of-reaction' (POR).

In well-built facilities it is not so much in the gas pipework or storage system but more often closer to the POR, inside the process reactor, that many wafer fab cleanliness problems seem to occur. When apparently ultraclean reactors are opened up, rubber 'O'-ring seals (potential outgassers), plastic pipes (these can diffuse moisture and oxygen into the system), cracks in glassware and leaking valves are often found. The latest generation of semiconductor equipment does seem to have eliminated most of these weak points when new, although this equipment will still be subject to degradation via cracking and valve failure with time.

Some of these problems are difficult to avoid. A recent short study from the ultraclean specialist research group of Marc Heyns of IMEC, Belgium, for example, showed TiO_2 flaking from the inside of a chamber used for Ti sputtering. Just 250 ppb of O_2 in the nitrogen or argon gas admitted to the chamber had led to the formation of the oxide, which caused particulate problems in the process.[55]

There are two general trends in wafer fab equipment which will tend to improve gas purity at the POR. There is a trend towards performing wafer processing in 'cluster tools', i.e. several related operations are performed on a wafer without that wafer being removed. The wafer is simply shunted around by a robot, operating in a vacuum, from one process module to another. A cluster tool with four chambers for etching might, for example, carry out a tungsten etch in SF_6, followed by an aluminium etch in BCl_3, then a titanium/tungsten barrier layer etch in SF_6+ HBr, followed by a photoresist stripping operation in oxygen plasma, before the wafer is returned to the cleanroom environment.

[55] Other problems may seem so huge as to be easy to avoid. The author has measured up to 20 or 30 ppm O_2 during the ramp-up of a rapid thermal processors, although once up to temperature this O_2 level could be swiftly purged out down to ppb levels again. Similar levels of H_2O were seen in rapid thermal processor experiments recently during the testing of a fast moisture analysis technique. However, high levels of moisture in this case may well relate to moisture taken into the reactor absorbed onto the surface of the wafer. A monolayer of moisture would give 10^{-5} moles on a 200 mm wafer, for example, enough to give 10 ppm for a minute or so at 5 l per minute purge gas flow.

The other important new ultraclean processing development is the move towards 'mini-environment' processing. In this, each piece of processing equipment, whether ion implanter, oxidation furnace or plasma etcher, has its own filtration system for clean air (or nitrogen) and is otherwise sealed off from the cleanroom except for its SMIF port. Wafers to be processed are brought in the SMIF sealed box to the port, where the machine can unload the wafers in an loading airlock ('loadlock') and then move them into the machine to be processed; the wafers will not contact cleanroom air. Newer SMIF-type systems such as the INCAM (individual cassette manufacturing) system, from the CEA-LETI group in Grenoble, France, may eventually take over from SMIF to achieve even higher levels of performance, even with the huge 300 mm wafers expected soon. INCAM is being designed to handle single wafers or wafers in batches of five or so.

3.5.4 Ultrapure gas pipework simulation

There is little doubt that the best practice for ultraclean pipework layout can be encapsulated in two principles: avoid dead-legs and ensure continuous flow. Within these broad principles, however, there are many detailed compromises to be made and computer modelling can help this process.

The behaviour of trace contaminants in a gas system is not readily predictable, especially when the system is complex. A relatively simple computer model, simulating the effect of dead volumes, dead or low flow pipework legs, under different assumptions of flow rate, input purity and leak rates, can be very useful in pointing to the parts of a system critical to purity. It can be useful both in the original design of systems and in the commissioning and improvement of existing pipework. The simulation of the effects of moisture retention, although less tractable analytically, is also useful. Ideally, such a system works both for steady flow conditions and during transients (e.g. start-up or gas-line switching).

Such a pipework simulator was first put under development and offered to semiconductor users by the L'Air Liquide company, although many of the other gas companies have now followed suit. The models have the following features:

- modelling of dead volumes of different shape;
- modelling of contaminated dead branches;
- modelling of different (but constant) outgassing rates.

They require the input of a pipework connectivity and flow map, and output a predicted purity map. These inputs are then applied to a model of the gas volumes and absorbing surfaces. The latter are particularly important for predicting moisture. Moisture on stainless steel surfaces follows an absorption principle a little different from the Langmuir isotherm (described in section 2.1.6) for absorption on PSA zeolite absorbers. Rather than having many sites for absorption that have the same binding or absorption energy, moisture on stainless steel is absorbed strongly (high binding energy) for small surface coverage and weakly (low binding energy) at high surface coverage. This follows from the irregular polycrystalline nature of the steel's surface where there will be a variety of different sites, of varying binding energy, the high binding energy ones (more stable) being filled by moisture molecules first. With this kind of isotherm and algorithms linking volumes, surfaces and flows together, fairly precise modelling of pipework behaviour can be achieved.

There are other phenomena which impinge on purity achieved in pipework systems, i.e. back-diffusion and leaking into the pipeline. Back-diffusion (see also section 1.3.10) occurs where impurities from a process chamber leak back into a pipe supplying them with an ultrapure gas against the stream of gas issuing from the leak. The most sensitive method of leak detection available should always be employed in the fabrication of semiconductor gas lines. Generally, gas lines and most vessels can be evacuated, allowing the use of the mass-spectrometer/helium technique described in section 2.4.4.

3.5.5 Further developments in ultraclean gas technology

In addition to the basic ultraclean technology outlined above, gas technology for semiconductor gases has come to encompass hundreds of technical improvements, of which the following are some of the more important.

Heated filters. When a new PTFE-based filter is installed in a gas line, the gas flowing down the line will typically be contaminated to a ppm level for a few hours, depending on the size and type of filter, and the gas flow rate, by outgassing from the huge surface area of the filter medium. It will still have ppb contamination due to the new filter after days of operation. Many wafer fabs expect to use heated filters in the future so that outgassing can be carried out after installation or replacement. With conventional PTFE filters, however, heating can only easily be achieved by heating the purge gas used and the maximum usable temperature is only approximately 200°C as the PTFE-based materials will degrade at 250°C.

Metal and ceramic filters. The Teflon membrane filter is the mainstay of the semiconductor industry filter. It is relatively inexpensive, very efficient in filtration and chemically highly resistant. However, it has the considerable disadvantage that it can absorb moisture and other contaminants from the atmosphere, slowly releasing these into the gas stream being filtered. Advances in sintering technology mean that alumina-based ceramic and metals such as stainless steel or nickel can now be prepared as sintered blocks offering fine pore size and good filtration properties. The latest filters are based on technology developed for ceramic membranes for use in uranium hexafluoride diffusion isotope separation (supplied by, among others US Filter and SCT). A strong (2 or 3 mm or more) support of rather macroporous (20 μm and larger) ceramic alumina is coated with several layers of a fine alumina 'slip' ceramic that, when fired, forms a thin layer of microporous ceramic with pore size from 10 μm down to a small fraction of a micron and just a few tens of microns thick. Several layers of slip are used so that pinholes can be avoided and a gradated transition from coarse to fine texture managed. In recent filters, filters with three layers of successively finer alumina are deposited. The result is a robust membrane filter with a texture of the order of 0.1 μm that can efficiently filter gases to a very high standard without blocking or blinding. Gas flow is into the macroporous alumina, through the microporous layers and out on to the clean side. Nickel and ceramic filters have the advantage that they can be used with highly reactive gases such as HCl, HBr, WF_6 and dichlorosilane.

Sintered filters typically have to be 10 or 20 times thicker than equivalent PTFE membrane filters. The most penetrating size (section 2.4.3) for a PTFE filter is around 0.04 μm with a theoretical penetration factor of $<10^{-22}$ for a 100 μm thick membrane. The most penetrating size for a recently tested sintered 316L stainless steel semiconductor filter was 0.1 μm with a penetration factor estimated at $<10^{-9}$ for a 1.7 mm thickness.

Metal filters, and to some extent ceramic filters, can simply be heated from the outside with a small purge gas flow to carry off outgassing. Caution should be taken to heat ceramic filters slowly, however, as the ceramic filter medium can be cracked by thermal shock. These new filters can be heated to quite high temperatures with a purge gas ('baked out') to achieve ultrahigh vacuum levels of contaminant gas and moisture retention. Even without bakeout, the metal and ceramic filters show much less outgassing than PTFE. However, their high cost means that they will, for the moment, be used mainly in small units near the point of use where they may well be baked out in order to be a good match in performance to a high quality point-of-use purifier.

Low-turbulence components and slow-opening valves. There is an expectation that components will become available with lower turbulence factors than hitherto, for example via smoother internal flow paths. These will cause less entrainment of particles in cleanroom gases.

Putting restrictions or small orifices in pneumatic control lines to slow the rate of opening or closing of valves has been shown to reduce the number of particulates added to gas lines by valves. Some electric and pneumatic actuated valves can be fitted with glycol-filled dampers to increase opening and closing times. Replacing single valves with systems comprising two valves and a laminar flow restriction element would also seem to be a promising technique for achieving a slow opening characteristic effectively.

Ultralow dead volume and integrated valves. Dead volume must be avoided in ultraclean systems because it can accumulate particulates, make purging by gas flow impossible and pressure-cycle purging slower. It can be reduced by careful valve design and by incorporating the valves in a gas panel design that minimises connection dead volumes. A major contribution to the latter can be made by integrating several valves into one metal body with several control orifices connected by small-bored passages. Such integrated valves are now becoming more common on gas panels, despite the limitations that they place on design flexibility.

Flow restrictors in pyrophoric and toxic gases. Flow restrictors reduce the impact of a cylinder gas accidental release, allowing the accident to be localised. Simple orifices are frequently used for this purpose. The only alternative here is an excess flow valve of some description but this may not be as intrinsically safe as a simple orifice flow restrictor. Flow restrictors do lead to particles being released and injected into the gas used, however. Measurements by USA gas company Matheson have put the contribution (with inert gas) from a typical cylinder gas restrictor at *c.* 250 particles >0.1 μm per cubic foot.

Reactive purifiers. These specialist purifiers are now being sold in larger numbers, especially to the semiconductor industry. They mostly rely on a microporous solid coated with a reactive chemical and are effective absorbers of ppb water or oxygen. They do have problems, however. With early models, accidental exposure to air would always result in almost instant loss of capacity and could result in a disastrous overheating as the reactive coating reacted exothermically with air.

More work needs to be done to test the long-term effectiveness of these purifiers in different situations. For example, experiments with some early models showed that for every molecule of water absorbed around 0.01 molecule of butane or other hydrocarbon is emitted from the purifier material during impurity challenge tests. Some of the emitted hydrocarbon may be bonded/dissolved further downstream within the resin of a purifier unit but temperature cycling or simply a long time period could result in it being emitted to the process. Particles of reactive material may be released into the process gas stream, too, but these are easily stopped in the filters normally downstream of the purifiers.

Claims are made for the latest generation of reactive purifiers (e.g. Millipore Corporation Reactive MicroMatrix) that would seem to make them the perfect device for very small volumes of gas (they are very expensive). For example, they will not undergo ‘melt-down’ if exposed to ır. The purifier formulations designed for inert gas/H_2 need no pretreatment or heating elements. This is an advantage compared to getters, which need an initial heating cycle to activate or require continuous heating. They are based on a metal carrier linked to a reactive metal such as lithium.

Purifiers are now available to purify some of the reactive gases such as Cl_2, $SiCl_2H_2$, SiH_4, AsH_3, PH_3, NH_3, $GeCl_4$, $SiCl_4$ and WF_6. They are based on a large surface area inert support coated with an organometallic reactive compound. The organometallic is chosen to provide a suitable coating capability and to react with the unwanted impurities water and oxygen to produce an involatile oxide or hydroxide of the metal. Note that these purifiers will not remove hydrocarbons. The purifier is first activated or 'conditioned' by flowing the gas to be purified through it. A typical example is lithium triphenylmethane. This purifies, arsine (AsH_3), for example, as follows

$$\text{TPMLi} + AsH_3 \Rightarrow \text{TPMLiAsH}_2 \text{ (conditioning)}$$

$$\text{TPMLiAsH}_2 + H_2O \Rightarrow \text{TPMLiOH (involatile)} + AsH_3 \text{ (purifying)}$$

Note that a small additional amount of AsH_3 is produced in the purifying reaction, adding to the AsH_3 flowing in. In a similar way HCl can be purified with anhydrous $MgCl_2$-coated support, inerts and hydrogen can be purified with MgH_2, while AlF_3 can be used for fluoride gases such as NF_3.

In the case of other hydrides, the pretreatment coats the purifier active sites with bonded gas, e.g. NH_2 groups for NH_3, which are then released by displacement by the preferentially absorbed impurities forming, e.g., NH_3 again and absorbing H_2O.

Purifier end-point detection. Several point-of-use purifier manufacturers are now considering whether they can offer a simple analyser to give a signal indicating the exhaustion of the purifier material. Currently, only large purifiers include analysis of output purity. POU purifiers are simply replaced on a schedule, normally ensuring they are only fractionally consumed. This is both wasteful of purifiers and risky; an unforeseen air contamination can easily result in premature exhaustion of the purifier. One concept is a very simple quartz microbalance assembly, not based on reversible water absorption as with the DuPont water analysers described in section 2.3.7 but a cumulative instrument based on barium-coated quartz crystals; more than a few ppb for a few minutes is capable of producing a frequency shift measurable with low-cost electronics.

Oxygen passivation. Although it now seems that it really does work, it is unclear that oxygen passivation is necessary for the current generation of chip production. Oxygen passivation has been much more extensively reported on recently and is offered by a number of, though not all, gas equipment suppliers. The oxide film is formed by heating tubing or components in a furnace at a few hundred °C, while passing argon-diluted oxygen through the tube or inside the components. The result after a few hours is an oxide film which retains little moisture and other contaminants. Although the oxide film is very thin and not particularly tough, it does seem to resist the addition of monolayers of other molecules such as moisture. Oxygen passivated tubing has been shown to 'dry down' much more quickly than standard electropolished tubing. On purging, moisture levels in passivated tubing fall quickly to tens of ppt (1 ppt = 1 part in 10^{12}); levels that can only be measured with sophisticated APIMS techniques.

Ultraclean cylinders and aluminium. Aluminium continues to be the best material for moisture performance. It is interesting to see that despite heroic efforts with compound electropolished and passivated stainless steel cylinders the ultraclean research team at Tohoku University have found the same to date. Their heavy and expensive stainless steel cylinders cannot really outperform a relatively ordinary high-quality aluminium cylinder. Aluminium cylinders are typically specially treated, often by proprietary anodising methods, which give them an unusually stable, pin-hole free, unreactive and hydrophobic surface.

A consortium in Japan has been promoting and researching the use of cleaned/passivated aluminium for vacuum chambers and pipework, evaluating aluminium components against stainless steel. There are now some test results with hardened aluminium VCR-style joints for Al pipe available for use with softened aluminium gaskets. There are, however, still problems to be solved. It is possible that future ultrapure systems, as well as vacuum systems, could use aluminium.

Cylinder coatings. Coatings have already been mentioned (sections 2.2.12 and 2.4.7) in connection with low-ppm gas mixtures and cylinder manufacture. Semiconductor gases are always packaged in cylinders which have a controlled inner surface preparation. Aluminium cylinders are anodised with proprietary techniques whilst steel cylinders may also receive a proprietary coating, often based on nickel. Chemical or electropolishing can also be employed along with coatings.

Bags made from Teflon or a similar unreactive polymer could be installed in cylinders to reduce contamination. Such bags would not save the cylinder from corrosion necessarily (their permeability would be too high) but particulates might be effectively reduced by this kind of measure. Although not current technology, cylinders based on this concept were tried a few years ago but without commercial success.

Continuous flow. Simply by sitting inside a gas line, for a variety of reasons from dead-legs to outgassing, reacton or back-diffusion, gas becomes less pure over time. When gas begins to flow from a stopped line into a process it will not become pure for a considerable period. This is not satisfactory and is one of the factors making the first batch of wafers different from subsequent batches, which is always a headache in wafer fab plants. The best purity gas lines are those which are in continuous use, the process gas itself purging out any residual trace impurities.

Another argument for continuous flow is that cross-contamination can occur because of back-diffusion when lines are inadvertently left open to atmosphere at their downstream ends. The latter is often the cause of the contaminants which are sometimes found on the downstream side of filters in semiconductor process equipment. When the equipment is open to the air, in order to admit a new batch of wafers, for example, or for routine maintenance, the gas lines are switched off. Reactive materials such as silane or chlorosilanes then have the opportunity to see small amounts of O_2 and moisture with which they form particles. When the machine is restarted the particles are carried into the process chamber. If the lines are switched from a reactive to a purge gas, instead of switched off, this can be avoided.

3.5.6 Semiconductor deposition processes

Ultrapure single crystals of semiconductor. Before semiconductor device fabrication can begin, the substrate, ultrapure monocrystalline silicon, must be made. The bulk polycrystalline silicon from which IC grade single crystals are grown is produced at a very few sites by companies such as Huels (Monsanto, Dynamit Nobel), Union Carbide, Wacker, Shin-Etsu, Osaka Titanium and Dow Corning.

The process starts by reacting HCl with crude metallurgical silicon to produce a mixture of chlorosilanes. Distillation of these produces the feedstock for the silane (SiH_4) plants and also the starting material for the pure grades of chlorosilanes used in the wafer fab. For bulk silicon, these chlorosilanes are then reacted with hydrogen. The Siemens process involves heating a polysilicon rod, held in graphite connector chucks, to well above 950°C to ensure reaction, depending on the silane or silane mixture used. The starter rod, in the form of a U-shape in the classic process, is soon coated with a thick layer of ultrapure polysilicon, which can be easily be broken into chunks for crystal growing. The volumes of gases consumed are considerable but these processes are carried out at only a very few sites worldwide.

Impressively pure polysilicon from the above process is melted and then grown into huge monolithic single crystals in an inerted furnace. The Czochralski process pulls the single crystal out of the melt and effectively carries out zone-refining during the growth. The growing of silicon crystals takes place at very high temperature (the melting point of silicon is 1400°C) and a high quality inert atmosphere is essential. Argon is the gas of choice, with helium being chosen occasionally.

Other semiconductors also need to be grown into single crystals. These include gallium arsenide, gallium phosphide, mercury cadmium telluride, indium phosphide and many alloys of these. The rule is that the ensemble atoms in the crystal structure should have an average valence of exactly four. Gallium nitride has recently attracted more widespread interest in optical devices, whilst silicon carbide, until a few years ago only made for use as an industrial abrasive, is showing promise for circuits, such as those used in car engines, which must withstand high ambient temperatures. Most recently, the element used in the first generation of semiconductor transistors, germanium, looks like it may be able to stage a comeback in compound layers with silicon. SiGe analogue circuits were recently demonstrated by the IBM company and showed exceedingly good high-frequency performance. The use of a mainly silicon silicon–germanium alloy means that circuits can be processed in a similar way to silicon and fabrication is much easier than for compound semiconductors such as gallium arsenide.

There are greater difficulties in growing compound semiconductor crystals because of the problem of one of the elements escaping and spoiling the necessary average valence. In gallium arsenide, for example, the arsenic can sublime at 613°C, whilst gallium, although melting at 30°C, does not boil until 2000°C so heating up melts of GaAs tends to result in loss of arsenic. Sometimes this problem can be avoided by crystal growth inside an encapsulation of molten glass or by the use of high pressure gaseous blanketing.

Epitaxial semiconductor layers. An epitaxial layer is a continuation of the native single crystal wafer substrate into a deposited layer, typically of higher purity than the substrate. Typical gases used are silane or chlorosilanes for silicon and mixtures of organo-gallium volatile compounds with arsine for gallium arsenide. Generally these gases are applied by being flowed over wafers in a high-temperature furnace, dilution of the active gases being employed. Silane is used for the lowest temperature deposition processes (around 500°C or so) with chlorosilanes requiring higher temperatures, rising with their chlorine content up to 1250°C for $SiCl_4$.

Exceptional purity is required in silane gas for epitaxial deposition. The epitaxial layer has to have a low intrinsic conductivity so that different parts of the integrated circuit are not short-circuited. However, this low conductivity can only be achieved if doping levels are extraordinarily low and this requires ultrapure gas. Often silane gas is specified in terms of the resistivity of the epitaxial silicon it is capable of forming. This is more sensitive (ppt) than any gas analysis method.

The diluting carrier gas may be hydrogen (for chlorosilanes), occasionally an inert gas such as argon or sometimes a ternary mixture. Vapours carrying

Group III or Group V elements such as boron and arsenic, respectively, are added to achieve phosphorus or nitrogen doping in the epitaxial layer. Reactors for epitaxial silicon, frequently used in advanced bipolar circuits, are formidably large consumers of hydrogen and (chloro)silane. A wide range of pressures can be used, from a few millibars up to atmospheric pressure.

Today, single-wafer thermal reactors (RTP reactors, see below) are available for epitaxial deposition, some with rotation of the wafer to assure greater uniformity of the deposited epitaxial layer.

CVD reactor processes. Epitaxial growth from a silicon-containing vapour is just one example of a chemical vapour deposition (CVD) process. A CVD process is in general a gas-phase deposition of any compound onto a wafer surface. CVD processes are used for depositing polysilicon (poly) used for interconnection for circuits and for silicon nitride insulator layers. For silicon nitride, mixtures with ammonia are employed. Although much silicon dioxide produced is grown on native silicon (below), some stages of processes require deposited oxide films, often created mixtures of silane, nitrous oxide, oxygen and dilutant. CVD performed at a few millibars of pressure is called low pressure CVD (LPCVD). These processes are more modern in origin than the atmospheric pressure CVD processes and involve the addition of large vacuum pumping systems to the reactor vessel. Many CVD processes cannot guarantee good adhesion of the deposited film unless they include a mild etch as an initial step. The mild etch step is often a brief exposure to HCl gas.

CVD reactors can be either ‘cold wall’ or ‘hot wall’. The latter type are simply a type of vacuum furnace with a controlled atmosphere. They have the snag that the CVD reaction happening on the wafer is happening on the walls of the reactor, too, so that the latter will need frequent cleaning. Cold-wall reactors employ RF inductive or radiant lamp heaters to heat the silicon wafer and its carrier far hotter than the walls of the reactor, which may be water-cooled to keep them reasonably near room temperature. Deposition of the CVD compound on the walls is thus largely eliminated in cold-wall reactors.

Common CVD reactions are those to deposit silicon dioxide (insulator) or polysilicon (transistor gate) layers. Polysilicon is deposited using silane or possibly chlorosilane (SiH_3Cl), although the latter does need a higher temperature to deposit, typically mixed with hydrogen. The addition of diborane, arsine or phosphine gas to the silane allows the production of heavily doped polysilicon, which has a high electrical conductivity and is used for polysilicon gate conductors.

Silicon dioxide-based insulator layers are typically laid down by mixing silane with small amounts of oxygen near the heated wafer surface. This sort of process can be carried out at pressures from high vacuum (a few millibars) right up to atmospheric pressure. CVD glassy layers are often 'reflowed' or 'planarised', i.e. melted into compact layers with a smooth top surface, by heating in an Ar annealing furnace if necessary.

Mixtures such as silane and oxygen, or silane and nitrous oxide, can be doped with diborane or phosphine to deposit silicon dioxide based glass dielectric material. If both are added, the dielectric is described as BPSG (borophosphosilicate glass). These dopants make a glassy layer which can be reflowed at low temperature. Glassy dielectrics are often used as insulator layers, especially between metallisation, and are also used, laid down very thick, in the final encapsulation ('passivation' or 'glassivation') layer that protects the finished integrated circuit from environmental contamination. These thicker layers are often put down in an atmospheric pressure system. The pyrophoric nature of the gases used leads to the term 'pyro reactor' for some of the CVD systems used for this.

Vaporised liquid sources are now popular for dielectric CVD. Compounds such as tetraethylorthosilicate (TEOS) are mixed with an inert stream such as He or nitrogen in a heated 'bubbler'. The TEOS stream meets an ozone stream near the wafer surface and silicon dioxide is deposited on the silicon wafer, which is heated, typically to 400°C. A 500 nm layer can be grown in 5 min. TEOS is much less dangerous to handle than silane, being only a moderately reactive liquid. Dopants to make the glass reflow at low temperature are vaporisable liquid sources of phosphorus and boron such as trimethyl phosphite and trimethylborate.

More and more exotic materials are being used in CVD in order to utilise the fast deposition, high-purity characteristics possible with good CVD. Starting with the less, and working toward the more exotic, some of the more well-known compounds are:

- tungsten hexafluoride (WF_6): for tungsten conductor 'via plugs' forming connections between different conductor layers in the integrated circuit;
- tetrakis(diethylamido)titanium/ammonia: for titanium nitride barrier layers as an alternative to nitriding a sputtered titanium layer;
- tripropyl aluminium: for Al conductors, instead of sputtered Al films;
- various copper metallorganics (e.g. copper(II)beta-diketonates using argon or hydrogen as a carrier gas for copper conductors): replace Al conductors in future chips needing the ultimate electrical conductivity of copper;
- tantalum pentaethoxide: for producing films of tantalum pentoxide, a high dielectric constant material, useful for increasing the data storage capacitor values in DRAM (dynamic random access memory) circuits.

Most of these compounds are vaporisable liquids, not gases. They are delivered to the wafer surface by their carrier gas, often inert (argon or, for lower temperatures, nitrogen) but often, too, a reducing carrier (hydrogen). The carrier gas is bubbled through the liquid compound with some precautions to avoid too much liquid droplet carry-over and temperature control of the bubbler to ensure a constant partial vapour pressure of active compound. Further carrier gas is then mixed in downstream until the desired concentration is attained. This additional dilution also avoids condensation problems.

Some of the compounds need to be in relatively high concentration in the carrier gas and this can lead to problems with liquid condensing out in the distribution pipeline. In this case, the gas lines (and supply cylinder, too, perhaps) will need to be heated, with a positive temperature gradient towards the point of use. Low-temperature trace heating by heating tapes is a suitable technique, if combined with suitable lagging so that cold spots are not formed. (It should also be noted that chlorosilanes can also condense in cool tubing under pressure and similar precautions are occasionally necessary even with 'gases'.) Alternatively, the supply bubbler can be cooled or the gas can be delivered at subatmospheric pressure. Special regulators are made with large orifices which will work down to a few millibars of absolute pressure.

Selective CVD. CVD reactions are typically 'blanket' reactions, that is, they form a layer, or blanket, right across the whole wafer. This does not have to be the case and reactions such as WF_6 tungsten deposition can be operated so that they take place selectively; for example, depositing on Al or silicon areas on the wafer but not on SiO_2 areas. Sometimes reacting the wafer surface in some way, by etching it mildly, for example, can render an otherwise non-selective process selective. These selective processes are more difficult to operate, however, and are mostly slower than the blanket reactions. Thus tungsten deposition for via plugs is most often achieved today by blanket tungsten deposition followed by etch back until only the plugs are left.

Diffusion doping of semiconductors. Semiconductors in the pure state are almost insulators. To make them conduct electricity, they must be 'doped' with small amounts of elements which replace some of the silicon atoms but also donate electrons or remove electrons from the crystalline structure. The donated electrons render the semiconductor conductive with negatively charged carriers (n-type). Less obviously, where electrons are removed the other electrons in the crystal behave in such a way that they simulate mobile positive charges, called 'holes', in p-type semiconductors. A typical p-type doping element is boron, the atoms of which have three electrons in their outermost electron orbitals. Boron replaces a silicon atom while removing an electron. A typical n-type doping element is phosphorus, which donates an extra electron to the crystal, having five electrons instead of four in the outermost orbital.

Doping gases are thus those containing elements from the III and V columns of the Periodic Table for silicon, and the II and VI columns for gallium arsenide. Phosphine and diborane are typical, being supplied mostly in dilute mixtures with hydrogen. Only tiny amounts of the elements are required as the doping concentrations are in the parts per billion range. Purity of the gas mixture is therefore of strong importance, otherwise inadvertent doping or undesirable dopants can be introduced. The doping gases are often used in a diffusion furnace, in which atoms of dopant are deposited on the semiconductor surface and then driven into it by being heated so that solid state diffusion takes place. Dopant atoms can also be incorporated by addition of doping gases to the silane-based gas mixture used to deposit polycrystalline or epitaxial silicon.

Ion implantation doping of semiconductors. A more modern alternative technique for doping, which uses very tiny quantities of dopant, not always as gas, is ion implantation. Here beams of dopant ions are created by a plasma ion source and then accelerated into a high-vacuum chamber by electrostatic fields. The beam of ions, which may have energy anywhere in the range of a few tens of keV to 2 MeV or more, is then impacted into the target silicon wafer. An electron beam gun ensures that the wafer surface does not become charged up. The plasma ion source is typically fed by tiny flow rates of hydride gases such as arsine, diborane or phosphine. The ion beam currents used commercially can be up to 100 mA or more, with limits on beam current depending on ion repulsion defocusing effects at low ion energy and on heating effects at high ion energy.

In principle, the purities needed are not high since ion implanters normally transport the ion source output via a magnetic focusing and separation beamline to the target. The beam transport is in fact a mass spectrometer, allowing only ions of the correct charge/mass ratio through onto the wafer surface.

Oxide growth. When a layer of silicon oxide is required on top of silicon, it is produced by oxidation of the wafer silicon or by addition of silicon dioxide. The former process is preferred for high quality oxide. There are a number of variations on the silicon oxidation reaction used in the industry:

- wet oxidation: uses steam + oxygen
- dry oxidation: uses dry oxygen
- HIPOX (high pressure oxidation): uses oxygen at high pressure (although this process is currently largely obsolete).

HCl is often added to accelerate the rate of oxidation and improve the oxide quality. These processes are all fairly modest users of oxygen, a small LO tank supply being the norm. One variant of the wet oxidation process is a more substantial user of gases, however. This is the variant 'pyrogenic steam' in which hydrogen is burned with excess oxygen to provide high purity steam plus oxygen for the furnace oxidation atmosphere. The consumption of oxygen and hydrogen can be considerable.

CVD-deposited oxide can be formed at temperatures as low as 200°C. The grown oxide process and its resultant silicon dioxide film is often referred to as 'thermal oxide' because it is formed only at very high temperatures (above 800°C). The wafer is heated to an accurately known temperature in the oxidising gas stream for a precise period of time and then cooled until further reaction is not possible. Oxidation reactions can take many hours or just a minute or two because the speed of oxidation is a sensitive function of temperature and oxide growth is not linear. Thick oxides are particularly slow because oxygen must diffuse through a thick intervening layer of oxide to reach silicon to oxidise towards the end of the process. The oxidation reaction is so important to silicon semiconductors that it has been extensively studied.

The reaction is governed by two thermally activated processes:

1. diffusion of oxygen atoms through the already-formed silicon dioxide;
2. reaction of oxygen atoms at the silicon/silicon dioxide interface.

If E_d is the activation for the first process and E_r is the activation for the second, the oxidation thickness X takes the following course versus time, t

$$X = D/K\ (\sqrt{(1 + 2N_s\, K^2(t + \tau)/(DN_m))} - 1)$$

Where D is the diffusion constant for O in Si, $D = D_o \exp(-E_d/RT)$ (where D_o is a constant), K is the reaction rate constant for oxidation ($K = K_o \exp(-E_r/RT)$, the constant K_o is 5 or 10 times greater with H_2O or HCl than for dry oxidation), N_s is the surface concentration of O_2 (related to O_2 partial pressure), N_m is a constant (O atoms in unit volume of SiO_2) and τ is a constant related to initial oxide thickness.

For small diffusion times or low temperatures oxide growth is 'reaction-rate limited' and X follows a linear law

$$X \propto K\,(t + \tau)$$

For longer diffusion times or higher temperatures oxide growth is 'diffusion-rate limited' and X follows a square root law

$$X \propto \sqrt{(Dt)}$$

(This kind of model also applies to many other solid–gas reactions, such as the formation of oxide on metals in air. Metals which oxidise slowly in air generally form stable, continuous oxide layers, e.g. silicon, which have a low diffusion constant. These metals, such as aluminium or stainless steel, reach the diffusion-limited regime in seconds and further oxide growth is almost imperceptible. Metals which oxidise rapidly are generally those which do not form such a diffusion barrier. The oxide formed is lost by flaking or some other process and they oxidise continuously in the reaction-rate limited regime. The picture is more complex than with silicon because of catalytic effects, electrochemical cells and other corrosion reactions in the presence of moisture and other atmospheric contaminants such as SO_2.)

A typical industrial process in a conventional quartz tube diffusion furnace might operate at 1100°C for two or three hours to obtain 1000 nm field oxide, using wet oxidation. With single wafer rapid thermal processor dry oxidation, a much shorter time in a hotter furnace is used, e.g. 1200°C for 200 s for 40 nm of gate oxide.

The purity of oxygen used in these processes, formerly not regarded as important, has recently become an issue. One of the problems is that even cryogenic oxygen typically has 20–40 ppm of hydrocarbon impurities, which can lead to harmful carbon in the oxide; carbon typically lowers oxide breakdown voltage, important in the ultrathin gate oxide used today. Oxygen purifiers are now used in new wafer fabs.

Rapid thermal processing. One technique that is slowly growing is that of using high speed single wafer furnaces, i.e. rapid thermal processing (RTP). Here a single wafer is loaded into a (typically) quartz furnace, purged for 30 s, heated in approximately 30 s up to operational temperatures (up to 1100°C), and then cooled in a flow of inert gas back to 200°C, at which temperature handling is possible, in a further 30 or 60 s or less (see Figure 3.24). This rapidity of processing is possible only by using banks of tungsten/halogen lamps for heating, with pyrometric temperature measurements providing feedback with which to control the heating rate, evenness of heating and final temperature. The speed at which processing can be carried out, with only 2 min or so per wafer, means that RTP can be competitive with multiwafer furnaces taking 25 or 50 wafers but needing hours to process.

RTP can potentially be used for many processes normally carried out in large multiwafer vertical or horizontal tube furnaces, e.g. silicon epitaxy, gate oxidation, CVD silicon dioxide, annealing or CVD tungsten. However, only a small number of operations routinely use RTP today. An example is titanium silicide layers, formed by reacting sputtered titanium with wafer silicon at 700°C, where the controllability of the conditions in a single wafer reactor have allowed it to take over from more traditional furnace treatment.

There are a number of problems with RTP; some fundamental, some more in the nature of teething troubles for a technique that is still not quite a standard production technique. In the first category is temperature uniformity, especially during ramp-up and ramp-down. Evenness of lamp illumination is never quite perfect. Different parts of a part-processed (patterned) wafer absorb IR energy from the lamps at a different rate and heat up differentially, an effect so pronounced that some wafers will 'curl up' visibly during the heating cycle. However, the latest RTP designs claim to

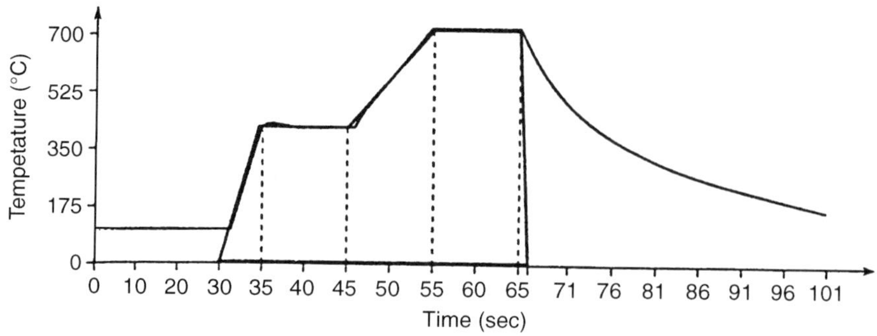

Figure 3.24 RTP cycle.

have solved this problem, getting around ±2°C uniformity across a wafer. Another furnace avoids the problem of evenness of lamp-heating by simply pushing the wafer into a constantly hot furnace section at the top of a vertical tube furnace, achieving wafer processing with a cycle time only a few times slower than tungsten/halogen RTP.

Another fundamental problem is that of purging of moisture. The constraints of the short cycle time are such that it is difficult to purge out moisture completely from the system down to the ppb levels possible in larger multiwafer systems. As noted above, wafers themselves have water absorbed onto their surface when they are first put into the RTP at room temperature.

Sputtering for metallisation. The sputtering process basically involves the bombarding of a metal target surface with argon ions, releasing metal atoms which can then be deposited in a layer on the IC surface, the whole process taking place at lower pressure than CVD ($\sim 10^{-3}$– 0.1 mbar). Sputtering is popular because it yields higher quality films than other processes, such as vacuum evaporation. It is a user of argon gas, which is the favoured gas for generating a stable glow discharge.

As well as straightforwardly striking a glow discharge between the metal target and the wafer, it is possible to extract positive ions from the argon to direct at the target, an arrangement often used at lower pressures, higher powers and for dielectrics. The glow discharge can be made very much more powerful by using RF, with 10 kW and more power supplies being in use to speed up the process. The wafer being coated is often heated to a few hundred °C. This gives a better film, presumably because it allows some thermal relaxation of the initial landing position of the sputtered atoms – at lower temperatures the atoms freeze where they land. However, IC manufacturers try to minimise the amount of thermal exposure that a wafer is subject to late in its processing, so sputtering at 200–300°C or less is now normal.

Sputtering is a relatively slow deposition process, although speeds have increased dramatically with the advent of high power RF plasmas, and is mainly used for metal films (Al and Ti are typical requirements). The sputtering process, being largely a physical phenomenon, also works on metal alloys and on dielectrics by using a target made of these and adjusting process parameters. Typical sputtered thicknesses are low, e.g. 20–500 nm, with deposition rates of the order of 0.5 nm s^{-1} in powerful machines being typical of the maximum rates attainable on a metal like Ti; some materials are, however, much slower to sputter.

Problems with sputtering are few. A high quality vacuum is necessary as otherwise the vacuum residual gas will be incorporated in the sputtered film. High-purity argon is indicated, backfilled into a high achieved quality of vacuum. Sputtered material covers the entire inside of the reactor vessel, not just the wafer being processed, so the machine must be regularly dismantled and the deposits removed, normally by suitable chemical processing, or replacements substituted for the coated parts. By careful design, inexpensive shield parts can intercept nearly all the surplus coating. These shield parts are quickly replaced, being either disposable or easily cleaned on a wet bench. Flaking off of the vessel coating is a major potential source of particulate contaminants. Some sputtered films on the reactor vessel are so firmly attached that removal is not necessary. Other films are all right in place provided they are not oxidised. Small amounts of oxygen impurity in the gas can lead to oxidation and flaking of the oxide formed; the use of ultrapure gases and clean vacuum can avoid this problem.

Reactive sputtering is sputtering carried out in an atmosphere which can react with the newly sputtered film. Titanium nitride, for example, can be laid down by sputtering titanium on to a wafer in an atmosphere of argon with a few tens of percent nitrogen at 10^{-2} mbar pressure.

3.5.7 Plasma etching: a key technology

The art and science of plasma etching has been one of the most important innovations in the semiconductor industry. Plasma etching (sometimes called 'reactive ion etching' or loosely dubbed 'dry etching') is simply a means of removing a thin layer material from a wafer surface. It can be non-selective, that is, removing the surface whatever the material there, or it can be selective, removing metal while leaving insulator or vice versa. Selectivity can be defined as the rate of desired etching divided by the rate of unwanted etching, minus one (although the one is often not subtracted). Etching can also be either isotropic or anisotropic, terms which warrant some explanation.

The etching of patterns on a silicon wafer in general is achieved by using an etch process which removes substrate but not the photolithographically patterned polymer 'resist' material. An isotropic etch simply etches the

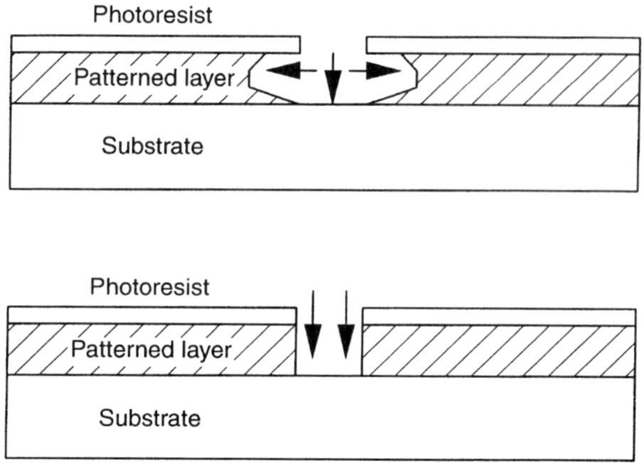

Figure 3.25 Isotropic (top) and anisotropic (bottom) etching.

substrate (but not the resist) in all directions at the same rate. This means that the resist will be undercut about 1 μm sideways for every 1 μm of downward etch. Anisotropy can be defined as the rate of vertical etch divided by the rate of sideways etch, minus one, though again the latter is often omitted. Anisotropic etching occurs where several effects, such as plasma bombardment and the lifetime of excited species are significant and areas which can 'hide behind' the polymer resist will not be etched much, if at all, so that, for example, a 1 μm etch downwards might only undercut the resist (if at all) by a small fraction of a micron (Figure 3.25). Anisotropic etching has big advantages over isotropic etching, particularly with the small size of circuitry now being sought by integrated circuit manufacturers. Anisotropy allows structures such as deep but narrow conductors and allows much denser packing of circuits. High anisotropy is one the key achievements of modern plasma etching processes.

Isotropic etching. Simple dry etching in which a plasma is not used, using an anhydrous reactant gas, cannot possess the advantages of anisotropy but can be cleaner and more controllable for small etch rates than wet etching. Etchants include HCl, HF and chlorine. These processes are sometimes called thermal etching, as it is usually necessary to raise temperatures. The term 'dry etching' is often used more loosely to refer to plasma etching, which is also a dry process.

Dry-etching processes, such as plasma etching, are popular anyway because they avoid the handling and purity problems of wet chemicals. Plasma processes are even more popular because they open up the possibility of anisotropic etching, which is increasingly important with the

Table 3.8 Plasma etch chemistries

Material	Etch species	Possible gas mixtures
Silicon	Br, Cl or F atoms (+ oxidant)	$BC1_3/CC1_4$
		HBr/O_2
		SF_6
Al	Cl (+ oxidant)	Cl_2
		BCl_3
W	F (+ oxidant)	SF_6/Cl_2
		NF_3/Cl_2
SiO_2	F	C_2F_6
		CF_4

shrinking geometries now evolving. The use of an electrical discharge to create excited atoms or ions is the basis of the plasma etch technique, the excited species being derived from gas mixtures (Table 3.8).

Dry-etch gases include fluorine and chlorine compounds such as the CFCs, chlorine, mixtures of these with oxygen, sulphur hexafluoride and more exotic materials such as silicon tetrafluoride or boron trichloride. Reactive ion etching (RIE) cmploys lower pressures at which these gases are more substantially ionised and at which more anisotropic etches can be obtained. Argon is frequently employed as a chemically inert diluent gas, and, as usual, as a gas in which plasma formation can be easily obtained. Hydrogen may be added in order to help form volatile hydrides of etch products.

The simplest plasma reactors place a silicon wafer on a platform consisting of one electrode facing a flat-plate electrode. Plasma conditions, a highly ionised gas, are produced in plasma reactors by using low pressures (from 0.1 mbar or less to tens of mbars) and an RF (e.g. 13.56 MHz) or microwave (e.g. 2.5 GHz) alternating electric field. Efficiency of plasma production is raised by carefully tuning the RF circuits or, in the case of microwaves, by adding a d.c. magnetic field to give an electron cyclotron resonance enhancement. The electron cyclotron resonance (ECR) mechanism deflects the electrons by using a magnetic field such that one (or more) orbit of an electron in the magnetic field takes place at about the same frequency as the applied electric field. The cyclotron effect occurs because the orbital time of the electrons is constant, no matter what their energy, because as their velocity increases, the radius of curvature of their orbit increases proportionately (until relativistic velocities are reached).

Etchers can be designed to etch boats of 25 wafers, but it is difficult to ensure etch uniformity. Normal designs either etch one wafer at a time against a parallel planar counter-electrode or etch up to a dozen or so clamped on a rotating polygonal prism ('hexode') or on a dodecahedron with stationary counter-electrodes around.

The electric fields accelerate natural ions and an avalanche of (mainly positive) ions and electrons quickly forms an equilibrium state where 1 in $\sim 10^6$ of the gas molecules are ionised. The ions and electrons are fast-moving, much more so than the gas temperature would suggest, and this is the key to part of their activity. They collide with the other gas molecules, breaking apart the molecules and releasing free radicals and free atoms as well as further ions. A space charge of ions builds up which is positive, repelling ions from itself so that an RF plasma tends to have ions flowing away from its middle.

Anisotropy, selectivity and etch rate are the key parameters of plasma etch processes. Selectivity of 100:1 of polysilicon on SiO_2 is possible, whilst etching at just 4° to the vertical is possible for sidewalls (corresponding to a high degree of anisotropy). Etch rates vary enormously, depending on the layers being processed on the wafer. Suffice to say that plasma-etched trenches of 50 or 100 μm are possible for special applications such as microengineered devices, although normal etch rates might be only 1 μm per minute or less. In addition to these primary parameters, the effects on photoresist coating, the deposition of particles or polymer (the latter derived from the materials being etched and the photoresist), the surface damage and the surface smoothness of surfaces after etch are also important.

The 'ideal' plasma is one which has a very high density of low energy ions while operating at low gas pressure. The low energy of the ions will avoid disrupting the atomic structure of the semiconductor surface whilst the high density assures a high etch rate and anisotropy. However, designers strive to achieve this using different designs or plasma generation. They have the problematic compromise that high power, needed for high density, tends to produce high energy ions, causing damage unless gas pressure is increased, when anisotropy decreases. Etch residues are also a concern since, if the products of the etch reaction are not volatile and speedily removed, there will be problems.

The process variables which may be manipulated are the gas composition and pressure, the RF field, the d.c. voltage on the semiconductor wafer, the wafer temperature and the reactor geometry. Plasma processes are, in industrial practice, largely empirically determined simply because the science is complex. The effect on anisotropy, selectivity and etch rates of a change of process variables cannot be predicted in general. Plasma etching is of such industrial importance that a huge amount of effort has been put into producing workable processes. Processes and equipment are now available to carry out almost every conceivable etch job, although with some compromises on performance, despite a largely empirical basis. Much work over the years has extracted the underlying mechanisms, however, and these are used as a guide to predict the process parameters at least approximately. Plasma etching uses several basic physical effects as follows.

- **Sputtering**: the mechanical dislodging of surface atoms by impinging high speed ions and neutrals (section 3.5.6). Sputtering is highly directional, so is important in anisotropic etches but is normally very slow. It fails to account for the speed of plasma etching by many orders of magnitude.
- **Chemical etch** by species such as free atoms and radicals produced in the plasma. Straightforwardly, for example, F atoms produced in a plasma can attach to silicon atoms and form SiF_4, which is volatile and lost from the surface. Chemical etch is isotropic, at least for longer lifetime species, so the chemical etch mechanism must be less important in highly anisotropic etch processes. In downstream etching, such as stripping processes, presumably chemical etch by excited species is the only mechanism.
- **Collision-energy induced chemical reactions** are reactions which occur only because the bombarding molecule or atom is moving fast enough to 'damage' the etched wafer. The surface of the semiconductor wafer being etched is excited by the collision in some way; it may have a 'dangling bond', for example. The surface at that site will then react readily with the gas present or with another bombarding ion. This mechanism leads to high anisotropy; only surfaces bombarded are etched in this way, surfaces hidden by photoresist are not excited and not etched.
- **Sidewall protection** occurs when the natural chemical etch reaction produces an involatile species which forms a protective layer on the wafer being etched, stopping further etch occurring. The tough protective oxide layers which protect stainless steel and aluminium from oxidation in the air are familiar everyday examples of this effect. However, in the plasma etch reactor the protective film can be removed by the action of bombardment by the sputtering process or by ion-collision induced chemical reactions. This mechanism does not affect the sidewalls, whose protective film stays largely intact. Sidewall protection also leads to high anisotropy. Examination of sidewall films sometimes reveals this mechanism at work; for example, fluorocarbon-based polymers are sometimes found after etches using C_2F_6.

Choice of etch gases is determined partly by product volatility. For example, for etching Al interconnection layers, chlorine-containing Cl_2, CFCs or BCl_3 are used because aluminium chloride is relatively volatile. When etching 'blanket' tungsten (used for interconnect vias between layers), F-containing CFCs, PFCs or NF_3, for example, are used because WF_6 is volatile. Oxygen is frequently added to mixtures because the basic reaction is an oxidation where, for example, metals are being etched. Gases such as chlorine are themselves oxidising agents and mixtures with chlorine will not require oxygen addition.

Problems with plasma etching. Although plasma etching is now ubiquitous, it has had to overcome many problems. A few are described below.

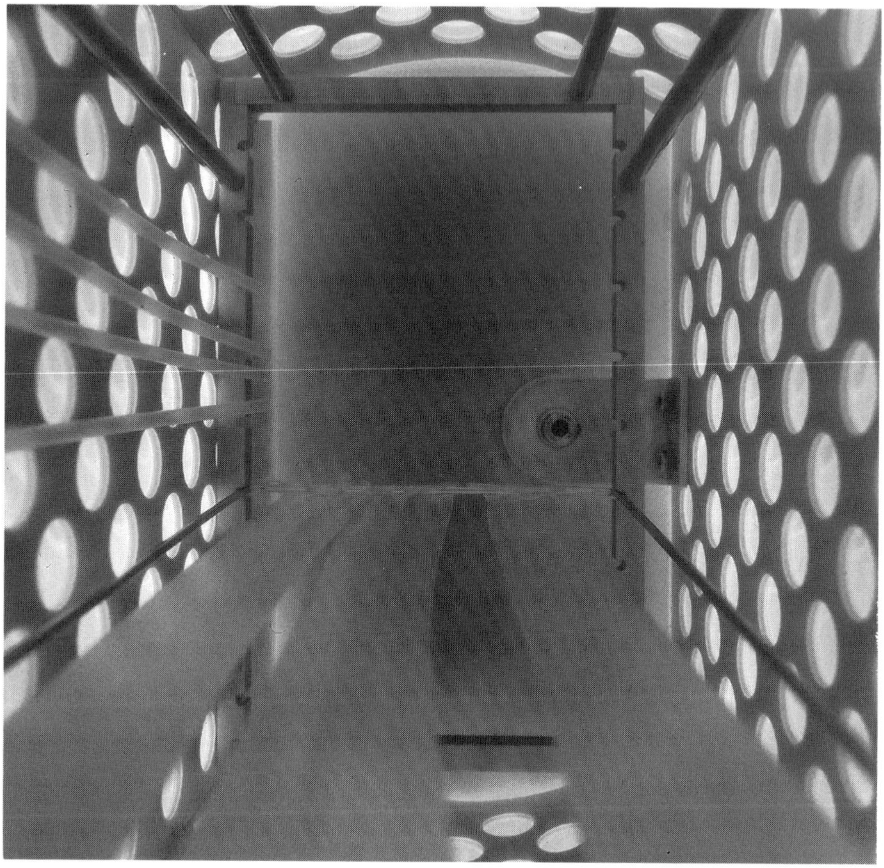

Figure 3.26 Plastic film strips being plasma-treated for surface modification (courtesy of Oxford Plasma Technology).

Etch start is sometimes inhibited on silicon and metals by the presence of 'native oxide' film, i.e. the oxide formed while the wafer was sitting around in the cleanroom air. This can be avoided by adding a minute amount of an oxide-removing gas to mixtures; fluorine-containing molecules would work for the native oxide which forms on silicon surfaces preventing silicon etch, for example.

Particulate formation can be a big problem in plasma reactors. As particulate contamination specifications for integrated circuits move ever downwards, even small numbers of minute 0.2 or 0.1 μm particles on the processed wafer may be unacceptable. Reactions in the plasma may create particulates, which is bad enough but might be acceptable if they could be immediately swept away by the gas flow. However, it is frequently not so simple as this. The plasma charges up all particulates by adding electrons to them and the charges generated may then allow the particles to become trapped in the plasma, and be lost only slowly by removal mechanisms such as gas flow, deposition or evaporation.

First observed by astronauts near their spacecraft in orbit, particles trapped by plasma can easily be seen in a low power plasma by, for example, shining a laser beam through the chamber during operation and observing with a narrow-band red filter. Video sequences were first reported by Selwyn's IBM group in the USA and showed this curious trapping phenomenon starting up. Small ($\sim$1 μm) particles accumulate, suspended in patches characteristic of the reactor geometry and RF fields. Small clouds are often found near changes in the reactor electrode surface, such as the edge of the wafer. Often flakes or needles form, each made up of thousands of particles agglomerated together. Deposition of trapped particles on the wafer being processed may occur when the plasma is switched off and the charged particles are dumped on the wafer either by gravity, migration or by electrostatic forces.

Monitoring of particle densities is an important diagnostic in any vacuum reactor. Equipment is available (e.g. the USA-based HYT Corporation's *in situ* particle monitor) which uses right angle scattering from a laser beam in the exhaust line from a chamber. These particle monitors are rather like an atmospheric OPC (optical particle counter). The detection chamber uses mirrors to create a 'net' of laser beams across a wide (50 or 100 mm) aperture, however, rather than forcing the sampled gas stream through a narrow constriction illuminated by the laser.

Excessive polymer deposits can often be caused by polymerisation of fluorocarbons. This can be controlled by reducing the fluorocarbons, using another fluorine source, or adding oxidant. High flow rates can carry away polymer precursors and hence suppress deposits.

The ions bombarding the target, i.e. the wafer being processed, cause charge build-up on that wafer. The resulting voltage on the wafer is called 'bias' and can affect the etch effect markedly. Ion implantation can also occur, causing sub-surface damage to the wafer. At lower pressures the bias voltage can be considerable, resulting in ions being accelerated into the wafer leaving trails of ionisation and disruption of the silicon crystal below the level being etched. This damage can be minimised by operating at a lower bias voltage condition with plasma conditions that minimise the ion energy. The latest designs achieve 20 eV ion energies, whilst maintaining ion densities in the plasma of 10^{11} or 10^{12} near the wafer surface in a low pressure (0.01–0.05 mbar) gas.

Next to particulate contamination, non-uniformity is probably the biggest problem of many semiconductor processes and leads to many rejected ICs from a wafer. An ideal process etches every wafer, and all of that wafer's surface, to the same degree. For processes like photoresist strip this may not be critical in that the etch process will be largely stopped once the resist has cleaned off and over-etch can be generous. With processes with a less effective etch-stop, however, over-etch must be minimal and the uniformity of the process must be as high as possible. Clearly, single-wafer diode and hexode etchers will have important advantages over boat etchers here and the boat equipment is only used for non-critical stripping processes. Improvements to the flow patterns inside reactors can improve uniformity, as can use of 'guard' areas on flat-plate etchers and rotation of the wafer holder. Temperature control of the wafer is important. Many etchers now use a cooling system, such as flowing helium, on the wafer backside. Uniformity is generally enhanced at lower plasma power levels, an effect at least partly due to the more uniform temperatures achieved. Roughening is also a kind of non-uniformity. Roughening of an etched surface is often seen and needs to be controlled and limited if it is not to cause problems.

End-point detection. It is very important to be able to stop a plasma process when its etching is finished. Commonly, the plasma discharge will change its electrical characteristics near to the end-point. The effect of the plasma becoming less conductive, for example, modifies the load seen by the RF power supply; sensing changes in the latter enables an end-point to be declared. With increasingly difficult processes using plasma, and with less selective processes being used more, however, such a simple approach is now much less useful.

A more sophisticated, but expensive, approach, is to attach a small quadrupole mass analyser to the plasma rig and look for the presence of tell-tale ions which indicate the end of an etch, perhaps simply by seeing a large increase in ions from the substrate at the etch bottom. However, the addition of a mass spectrometer system requires a large intrusion into the vacuum system in order to be able to sample the plasma effectively. The fact that plasma emits a typically blue–mauve glow, which changes slightly with the material being etched provides an easier means of monitoring. Optical emission from the plasma can be used as an end-point detection system, which is more easily retrofittable. It is now common to monitor plasma via its optical emission, sometimes using two emission frequencies: one a wavelength indicating end-point, the other a wavelength providing normalisation by giving a measure of the overall plasma power.

3.5.8 Other plasma processes

Resist dry stripping. 'Resist dry-stripping', often called 'ashing', is a specialised kind of isotropic plasma etching that is in many cases replacing wet solvent methods. The polymer-based photoresists used during silicon processing must be removed after the main etching process. They are often surprisingly difficult to remove. After coating, exposure and development they have to be 'hard-baked' on by heating strongly, perhaps also with UV light, converting them into the very tough polymers needed to resist powerful etch processes. Plasma etching may leave polymeric deposits mixed with silicon, metals and minerals along the sidewalls, and sputtering can also occur, further modifying the resist. Other processing, such as ion implantation, can cause a toughened, even carbonised, skin on the resist surface. Even with mixtures of powerful solvents such as N-methylpyrrolidone with water and ethanolamines, and even with heated solvent mixtures, there will often be residual resist adhering to the silicon wafer after wet-stripping.

Dry-stripping is a common solution to these problems. Oxygen atoms are generated and these combine vigorously with all the resist components while etching the wanted layers only slowly. An oxygen-based low pressure (1–20 mbar) plasma is employed, pumped by RF or microwave energy, usually with a fluorine-containing gas such as NF_3 or a CFC. As well as processing systems with the plasma generated immediately above the silicon wafer, 'downstream' plasma generation is possible.

Alternative schemes have employed ozone in oxygen at atmospheric pressure or nitrous oxide as the oxidant in a low-pressure plasma system. A variant of the ozone generator system normally used in water treatment (section 2.2.7), silent electrical discharge, can be used. Illuminating the wafer surface with UV light to further activate the ozone by breaking it down into oxygen atoms *in situ* can also be carried out. Often, resist dry stripping is followed by a relatively mild wet-chemical etch, sometimes just a deionised water rinse, to remove any mineral and other non-organic deposits which the oxygen atoms have not removed.

Plasma-enhanced CVD (PECVD). PECVD uses the plasma excitation as an energy source to drive deposition of materials from a gas mixture on to the silicon wafer surface. A large number of traditionally CVD processes can be moved to a PECVD process, often with the advantage that a less aggressively reactive gas or vapour can be used. PECVD is a process which is growing steadily in importance in the semiconductor industry.

One recent example of this is the use of PECVD for creating low dielectric constant insulator layers made from fluorinated silicon dioxide (these speed up circuit functions by reducing the electrical capacitance of interconnections). A silicon source, either tetraethoxysilane or silicon tetrafluoride, with oxygen and, if necessary, a fluorine source such as nitrogen trifluoride or perfluoropropane is used, at a few millibars pressure. In this way, when the plasma is struck, layers of 1 μm of insulator can be deposited. The insulator typically has 5–10% fluorine content and a dielectric constant of 3.5 or so, giving a useful advantage over pure silicon dioxide whose dielectric constant is typically 4 or more, depending on density and doping.

Dry cleaning processes for silicon wafers. Many of the process steps that a semiconductor wafer must pass through during fabrication are cleaning steps and most of these are still performed with 'wet benches'. Wet benches are sets of baths of ultraclean chemicals and deionised water, sometimes automated with the addition of a robot handler for boats of wafers. Broadly, the cleaning steps carried out on wet benches amount to breaking the electrostatic adhesion of particles with an aqueous liquid, oxidising the impurities or etching back the wafer slightly, releasing surface particles and removing native oxide (thin oxide formed on exposure to air). Popular cleaning baths are named Piranha™ (a hot sulphuric acid oxidising process), SC-1 (an ammonia akaline oxidising bath) and SC-2 (a hydrochloric acid oxidising bath). Others are known by their chemistry, as in 'buffered HF' (HF plus ammonium fluoride) or 'HF dip'.

Wet benches consume large amounts of chemicals even though recirculation of the bath chemicals through filters reduces consumption. The use of hydrogen peroxide as an oxidiser is inconvenient as H_2O_2 is relatively dangerous, expensive in high-purity grades and decomposes easily. Spray cleaners, which are economical with chemicals, are still trailing wet benches in performance for most cleaning applications and are only used for less stringent cleaning requirements. As a final disadvantage, silicon wafers must be dried after each wet-cleaning process. The drying process causes delays and has the potential to concentrate any contaminants from the last rinse bath onto the wafer surface.

Supplanting wet-cleaning steps might improve yields and would make the wafer fab safer and less expensive to run. No generally applicable and accepted process is available currently to replace wet cleaning but it is the goal of a number of research projects worldwide to produce an alternative dry-cleaning process, i.e. a process for cleaning wafers without using the usual mixtures of unpleasant wet chemicals.

A few proposals simply use a reactive gas mixture. HF gas has been shown to contribute a cleaning action and, like hydrofluoric acid, leaves the wafer passivated with F atoms, preventing oxidation in the cleanroom air. The problem of having to handle HF, however, is significant. Gas companies, as well as vacuum equipment and semiconductor equipment companies, have joined these research efforts, such is the importance of such a process. UV is being tried as the source of excitation for oxygen mixtures, as well as plasma sources, and many different combinations of gases are being researched.

Hydrogen and argon downstream plasma arrangements are being tried in which the gas mixture passes through a plasma discharge and then over the wafers, avoiding problems with ions from the plasma and allowing only relatively long-lived excited species to reach the wafer surface. Both excited Ar and atomic hydrogen are known metastable species with high reactivity. A demonstration of the downstream plasma process is occasionally seen in undergraduate chemistry laboratories. If an organometallic with a high vapour pressure, such as lead tetraethyl, is pyrolysed at a section of a quartz tube by heating strongly, a shiny metallic film is laid down. The pyrolysis yields free radicals. The deposited film can be cleaned away by heating the tube again but this time upstream, exposing it to a stream of vapour containing free radicals from the decomposition.

Cleaning with liquid CO_2 blast. One unusual idea that seems to be inapplicable to silicon wafers but might well become popular in other applications, is the use of a blast of CO_2 gas/liquid/solid. Ultraclean liquid CO_2, under its vapour pressure of 50 barg or so, is directed at the surface to be cleaned via a small carbide/teflon nozzle assembly. The resultant high-speed stream of CO_2 gas, containing fine CO_2 snow particles and probably also (fleetingly) liquid, is very effective at removing contamination from smooth surfaces, even down to very small particles. The technique is in use in the manufacture of photoconductive drums for photocopiers and laser printers, and has been proposed for cleaning the inside of gas cylinders for ultrapure gases.

Fingerprints and oily marks as well as particles are removed by CO_2. However, there are problems. The CO_2 used must be of a special high purity; standard CO_2 is of a much lower purity than other cryogens and a special grade is needed. Another problem is that the substrate being cleaned is cooled down considerably. In air, water will be condensed onto the surface almost immediately, presenting another potential contamination problem. Cleaning of parts inside a 'dry box' filled with dry air or nitrogen avoids this problem but throws away some of the technique's convenience.

There is a further proposal to use CO_2 under pressure as a supercritical fluid for the development of photoresist. It turns out that certain substituted polymeric silanes, such as poly(methyl phenyl silane), are depolymerised to siloxane by UV light and this exposed resist material can be removed by a solution of the siloxanes in supercritical CO_2.

A dry-cleaning process that is now an accepted technique is the use of a glow discharge or low-power plasma for cleaning the inside surfaces of vacuum equipment. Depending on the contamination, mixtures containing oxygen and fluorine or chlorine compounds are the most popular. For hydrocarbon contamination, such as might be seen in a plasma etch system working on photoresist, a mixture of SF_6 with oxygen is common, although with etch chambers that are contaminated with aluminium a chlorine-containing etch mixture is required. A mixture of SF_6 with chlorine can both oxidise hydrocarbons and form volatile aluminium chloride, so solving this problem.

3.5.9 Semiconductor effluent gases

CFCs, PFCs and toxic emissions. In recent years processes have been developed which largely avoid using CFC gases which can damage the earth's ozone layer (section 2.5.12). The Montreal Convention will reduce CFC production and emission; to zero for some compounds and to low levels for others. This will have the effect of increasing the prices of the CFCs that are still available and processes which avoid their use are expected to take over almost completely in the next few years. The majority of the new plasma processes use a perfluorinated compound to replace the CFC. CF_4, C_2F_6, SF_6 and NF_3 are the most commonly used.

The perfluorinated compounds (PFCs) all have negligible effects on atmospheric ozone but they are long-lived in the upper atmosphere, from hundreds (NF_3) to tens of thousands of years (CF_4 estimate), and have strong infrared absorption bands. The latter give rise to global warming via the greenhouse effect (section 2.5.12). However, emissions from semiconductor etch processes worldwide is modest; only a few per cent of total emissions. The worst polluters for perfluoropropane are electrolytic aluminium processing plants. Only in the event of aluminium smelting cleaning up considerably would strong pressure fall on the semiconductor industry to eliminate PFCs.

Despite this, there are concerns about PFCs and many wafer fabrication plants are seeking to reduce PFC emissions, either simply by more careful gas control, using lower pressures and shorter etches, or by actively removing PFCs (and their breakdown products) from the exhaust gas. Wet and dry scrubbers (section 2.5.14) can both be employed and give good results with care. Dry absorber scrubbers can be effective but need correct incineration disposal after use. Some of the other scrubber problems are that, for wet scrubbing to work, a high-temperature flame 'burn-box' is necessary to ensure that the PFCs (or allowed CFCs) are broken down to more reactive radicals and compounds that will be effectively scrubbed by the wet-spray chamber. The integrated burner/scrubber 'submerged flame' technique is also claimed to be good here, as are dry scrubbers which are heated, like the Edwards GRC system.

An alternative that is now edging up towards acceptable performance is dry treatment at low pressure, i.e. before the vacuum backing pump. This employs a plasma to excite the exhaust gas mixture so that it will react, yielding mineral deposits which are caught in a chamber before the backing pump. This kind of scrubber provides a useful protection for oil-sealed vacuum pumps which soon wear out under such conditions. (Even with the oil circulation and filtration and frequent oil changes a particulate load will reduce pump and oil lifetime considerably.) The newer Northey claw backing pumps also suffer from wear and occasional overloads when operated with particulates. The pre-pump filter (necessarily a coarse filter to maintain flow rates) or cyclone chamber will intercept many of the particulates but will leave them as an almost unagglomerated powder, which is easily re-entrained. The striking of a plasma causes the deposits to become much stronger, more like eggshell in texture. Large-area electrodes allow several kilogrammes of material to be collected by such plasma traps.

3.5.10 Other processes in the wafer fab

Annealing processes for silicon wafers. After a number of processes to which the wafers are subjected, point defects, small amounts of damage done to the perfection of the single crystal, will arise that will impair the device operation, for example by reducing carrier lifetime. These defects can be repaired by a controlled heating process or 'annealing'. Ion implantation, for example, causes a lot of sub-surface damage such as interstitial atoms and vacancies, and is invariably followed by annealing.[56] A heating cycle may also be necessary after implantation or diffusion to carry out 'drive in', that is, to diffuse dopants from near the surface into deeper layers of the device. Similarly, the last traces of impurity from a wafer can be removed by a controlled heating process called 'gettering'. The heating in this case causes mobile impurities to diffuse until they reach oxygen-induced stacking faults in the silicon or microcracks on the back side of the wafer. The impurities are then trapped by the faults or cracks.

Defect damage can be repaired by controlled heating at high temperature, typically done in an inert atmosphere of nitrogen or argon. A tube furnace can be used for annealing a whole boat of 25 or more wafers or single wafers may be annealed in an RTP furnace.

Purging of reactors. Before a furnace or reactor operation can be performed, the ambient air, which would react vigorously with silicon at high temperature and with the active ingredients of process gases, must be expelled. Frequently there is a vacuum step but this is only possible with a vacuum-qualified reactor, i.e. one with good seals and strong enough to withstand atmospheric pressure. This is, however, not always sufficient. Hence there is a huge demand for purging gases, typically nitrogen. The purging gas requirements are difficult to estimate since the purging process itself is not always predictable and the effectiveness of the purging required will vary with equipment and the type of ICs being made.

For ordinary (and now increasingly old-fashioned) tube 'diffusion' furnaces, purging is at a rate of a few litres per minute or so, rising during loading and unloading to provide additional shielding of the wafers from atmospheric contamination and to provide additional cooling. While not in use, purging must be maintained at some level to keep the furnace ready for

[56] Note that annealing of silicon is not done to soften the metal, as in metallurgical annealing of steels, aluminium or copper. Annealed silicon remains the same as the original cut wafer, i.e. a bright silvery metallic-looking material that is, in fact, brittle.

use. Once air has been let in, the system will need to be purged for hours and baked at temperature before it is ready again.

Gas ballast for vacuum pumps. Pyrophoric gases such as silane require dilution with nitrogen before they can be diluted with air and expelled up the factory chimney. Similarly, vacuum pumps used on many processes are subject to problems because of the condensible nature of the gas load they are evacuating. The recommended solution is the addition of a small flow of nitrogen to the final stage of a multistage rotary vacuum pump. This is known as 'gas ballast'. Manufacturers will advise on the flow rates needed and the (small) loss of ultimate vacuum this imposes.

Note that the new types of 'dry' vacuum pumps can take larger flows of ballast than oil-immersed types. Dry pumps are those recommended for applications where there are corrosive condensible vapours. A gas ballast will almost certainly be essential with these gases.

Blow-drying. The blow-drying of wafers after a wet clean or wet etch process is a common application of nitrogen. In principle, high-quality clean dry compressed air (CDA) could be used instead and this is the practice in some wafer fabs. The use of a blast of dry nitrogen to evaporate solvents from newly coated photoresist is also common for both convenience and quality assurance. The inert nitrogen may avoid possible problems with resists that are more sensitive to oxidation.

Vacuum wands, blow guns and pneumatics. Vacuum wands are venturi-operated 'paddles' which are used whenever it is necessary to handle wafers manually. They allow handling of wafers via the back surface only and avoid touching of the delicate edge or patterned surfaces of the wafer. Blow guns are handheld valves supplied with nitrogen via a filter and used to blow-dry or clean-off visible dust particles from wafers, again where manual handling is necessary. Pneumatics, which would normally be operated by oil-mist lubricated compressed air, are often operated in a wafer fab by nitrogen; certainly the oil-mist in standard compressed air is totally unacceptable in a wafer fab. These are all applications for a clean dry gas and nitrogen is frequently employed, again with some companies using CDA where this is available.

Gas inerting for IC storage. For storage and transport of the chips, they need to be kept in a clean, dry and preferably inert atmosphere. Nitrogen is the gas of choice and considerable amounts can be used where chips are stored in large banks of cabinets, all of which need to be kept freshly purged. Practically all glovebox handling cabinets for semiconductors (sometimes called 'dry boxes') are purged with nitrogen.

Chemical handling. Even the use of liquid rather than gaseous processes does not mean that no gas is used. Frequently nitrogen is used for propelling corrosive liquids without the use of pumps (pressure transfer or pneumatic lift pumping) and for agitating the reaction baths. Even where centrifugal spray processing is used, employing liquid reagents, considerable quantities of nitrogen may be consumed.

Point-of-use chemical generation. Aqueous chemicals, such as hydrochloric acid, ammonium hydroxide, and hydrofluoric acid, are difficult to prepare in ultrapure form. They tend to pick up particulates and metallic ions during their preparation and more contaminants are added during transfilling into transport containers and pumping to the point of use from those containers. The very fact that they are used to clean-off metallics and particulates from wafers should tell us that this is to some extent inevitable. These tendencies can be almost, but not completely, suppressed by the use of special materials. Containers and pipelines made of PVDF (polyvinylidene fluoride), PFA (perfluoralkoxy) and high-purity PE (polyethylene) polymers are, for example, favoured. However, even these highly inert polymers contain residual catalyst, reactive end-groups, etc., which can result in slight contamination.

The ultimate way to control all this might be the preparation of aqueous chemicals by dissolving the appropriate ultrapure gas in ultrapure water at the point-of-use chemical generation (POUCG). The ultrapure water needed is already available at a wafer fab where water is treated by ion exchangers followed by reverse osmosis to achieve very high levels of purities. Resistivity very close to the theoretical maximum of around $18.26 \times 10^6\ \Omega\ cm^{-1}$ is normal in wafer fabs, indicating that the ionic content of cleanroom deionised water is very good (just 10 ppb salt will reduce the resistivity of ultraclean deionised water by half). The gases used, HF, HCl and NH_3, can all be filtered to remove particulates down to 0.1 μm or better and cannot intrinsically harbour metal ions so chemicals generated in this way have lower levels of metal ions and particulates.

Several companies are now offering this approach using equipment which will prepare small batches of chemical near the wet bench or other point-of-use inside the wafer fab. An ultraclean oxidative aqueous solution to supplant hydrogen peroxide can also be provided by this approach, by bubbling clean ozone, made from ultrapure oxygen, through water. Ozone solutions are problematical , however, as they have a very short lifetime and concentrated solutions are not possible.

Although the technique is in its infancy, a typical installation is already becoming evident. In this a tank containing a few litres of deionised water is gassed with ultrapure anhydrous gas, which has been filtered to remove particulates. Two or more gases can be added to form ready-to-use mixtures used in wafer cleaning and etching processes. An example would be the 'buffered oxide etch' (BOE) used on silicon dioxide layers, consisting of aqueous ammonia and HF. Another example would be an aqueous ammonia/ozone mixture (with hydrogen peroxide (H_2O_2) instead of ozone this would be SC1 or RCA1). The same apparatus can then be used to generate the mixture which often follows, i.e. an HCl/ozone mixture (with H_2O_2 this is would be SC2 or RCA2). Ozone mixtures are not yet used routinely, however, because ozone cannot be delivered in simple gas cylinders but must be generated by a relatively expensive electrical discharge generator (section 2.2.7).

With an increasing move towards more dilute cleaning solutions, this approach is going to look increasingly attractive, despite generally falling prices for ultrapure chemicals delivered. Already, POUCG equipment is claiming <50 ppt levels or better for all metal contamination, comparing favourably with the ppb levels typical in many high-quality delivered chemicals.

3.5.11 Gases in IC assembly and hybrid circuits

It is not simply the wafer fab sector of the electronics industry that uses industrial gases today. The IC assembly process, in which silicon dice are cut from wafers and put into their protective ceramic or plastic jackets, is also using industrial gases, although not with the same order of purity requirements. The manufacture of hybrid circuits also uses gases in modest quantities.

Many applications in the manufacture of hybrids and packaging use nitrogen as an inert atmosphere. Great care must be taken in these applications to ensure a low oxygen/humidity environment throughout, as levels of only a few hundred vpm (volume parts per million) can cause problems that are notoriously difficult to trace. The nitrogen itself is pure enough for any of these applications. However, problems may arise with inefficiencies in the purging process, perhaps due to 'backwaters' in the flow path or simply due to the purging flow rate being too small.

Once wafer processing is complete, hundreds of IC circuits are formed in the surface of each wafer. Subsequent operations to make them usable in actual operational equipment are required; these are collectively known as 'packaging' or 'back-end'. The first operation is 'wafer probing'. Clusters of tiny spring-loaded probes are lowered onto each circuit on the wafer and a low bandwidth basic test carried out. The second operation is to saw the wafer into individual circuits using a high-speed diamond circular saw. The individual naked circuits (known as 'dice' or, colloquially, 'chips') are then sorted and the dice which were bad at probe are rejected. Dice may then be further tested, although it is by no means easy to connect them to and disconnect them from a high bandwidth testing circuit on a high speed

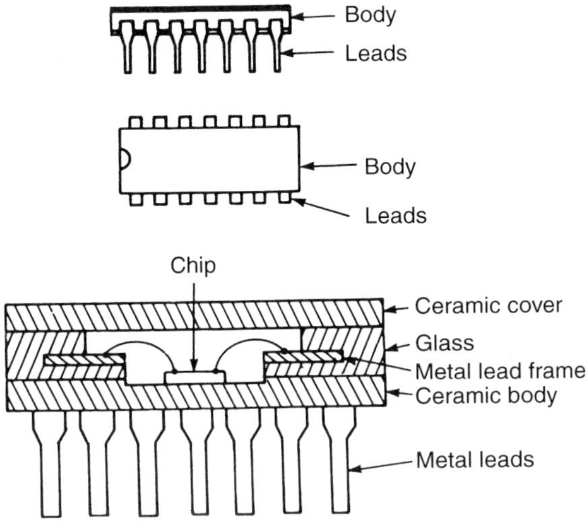

Figure 3.27 Dual in-line packaging: (a) side view; (b) top view; (c) cross-section of ceramic package.

production line. Dice which pass these more stringent tests are known as 'known good dice' or KGD. KGD are often required in hybrid circuits, where the failure of just one die can lead to an expensive rejection of the whole hybrid with all its dice.

IC packaging. Dice may then proceed either by the 'plastic packaging' or the 'hermetic packaging' route. In the former, the naked ICs are bonded (the 'die-bonding process') to a metal frame which will become the completed package's leads, i.e. a lead-frame. Die-bonding may be by silicon–gold eutectic solder or conductive epoxy. Individual parts of the IC are then connected to the metal leads of the frame with fine gold wires ('wire-bonds') attached by a microwelding process. The whole assembly, leaving the leads protruding, is then encapsulated in epoxy in a moulding machine and the complete package cropped to separate it from the lead-frame edges. In hermetic packaging, the ICs are bonded down to a metal or ceramic package then wire-bonded to leads sealed into the package by the package manufacturer (see Figure 3.27). A lid is then added under moisture-free atmospheric conditions, either by welding or by reflow of low-melting-point glass. By warming up the assembly to 150 or 200°C during packaging, much moisture can be driven off, ensuring freedom from corrosion effects. Many of these packaging processes are carried out in 'dry boxes' in nitrogen gas to ensure that moisture is not spuriously included in the packages, to avoid discoloration of welded joints and to avoid oxidation damage to the ICs caused by hot processes. Silicon chips are susceptible to oxidation in room air but oxidise especially quickly above about 100°C.

Hybrid circuits. Manufacturers of hybrid circuits are also users of gases. Hybrid circuits are so-called because they are a 'hybrid' between integrated circuit technology and printed circuit boards. They also lie in between monolithic silicon integrated circuits and printed circuit boards in their efficiency at packing in circuits. They typically comprise a circuit board, several times smaller than a conventional glass-epoxy PC board, made of a ceramic such as alumina. (Other hybrid systems are based on silicon substrates using techniques more closely related to the monolithic silicon IC industry.) In alumina-based hybrids, the alumina is the substrate onto which simple electronic passive components such as capacitors or resistors, as well as conductor patterns, can be printed using special 'inks' which can be fired at high temperature. The inks are composed of a 'frit' or powder-glass-type material, together with palladium or gold for conductive inks, ruthenium dioxide for resistors and high dielectric constant glass for capacitors. Once printed on the substrate, the inks are fired at temperatures from 500 to 1000°C, when the frit component flows, binding the active layer together and bonding it to the substrate. After several printing and firing cycles, the board part of the circuit is complete and it remains only to screen solder paste onto the board, add active components such as ICs and transistors in special miniature 'surface mount' (SM) packages, and reflow the solder. The whole circuit will often then be encapsulated in epoxy. This is essential if instead of SM ICs, unpackaged naked silicon dice are used. These are wire-bonded onto the hybrid or inverted and joined by heating, causing reflow of solder balls deposited on them to effect electrical connections. The latter process is called 'flip chip' because the solder balls are on the active surface of the silicon so the silicon chip must be turned upside down to bond them on.

The inks used for alumina hybrids have to contain organic binders to make them flow and handle in the screen-printing machinery typically used for patterning. The ink must flow sufficiently well to be easily squeezed through the mesh of the screen but not so runny that it 'bleeds', i.e. flows away from the desired patterned areas, after deposition. The organic binder is burnt away early in firing leaving the glass frit to bond the ink to the substrate and the metal or oxide to form the active part of the layer. Many hybrid circuits use air-firing ceramic inks, such as palladium–silver conductor ink and ruthenium oxide-based resistor ink. However, many circuits are now being switched to more highly conductive inks based on copper, which cannot be air-fired. To go with these, new resistor and insulator formulations have been devised so that the whole hybrid can be fired in nitrogen. Copper-based hybrid inks will also oxidise far below firing temperatures and many should be heated in nitrogen even at the low temperatures first used to burn out the organic binders.

Fault detection and thermal testing. Intermittent faults due to faulty contacts or semiconductors in circuits are a serious form of failure and can be frustratingly difficult to locate for repair. It is standard practice in electronic fault-finding to spray suspect components with a liquefied gas. On cooling to −50°C or so an intermittent bad connection will often fail, allowing it to be traced. A non-toxic, non-inflammable, non-conductive liquefied gas boiling at −50°C or so, which can be stored at room temperature in a cheap disposable spray can, is needed. With CFCs such as CFC-12 banned for applications of this sort, the choice revolves around materials such as HCFC-22 ($CHClF_2$) or HFC-134a ($C_2H_2F_4$) or mixtures of these.

To find faults early, and to assure functioning at all temperatures (for military-grade ICs), new ICs and complete circuit boards or other subassemblies are often tested in environmental test chambers, typically in the range −25°C±125°C. These chambers are required to be able to reach and then maintain any desired temperature within minutes. Often thermal cycling tests are needed, in which many cycles of heating and cooling must be completed in the minimum of time, and the powerful cooling effect of a cryogen injection system is often used to achieve this (section 3.7.3).

Lid brazing. Many ICs and hybrids are packaged in metal-plated ceramic packages designed for metal lids. Metal lids, typically gold-plated Kovar, are then fixed to the ceramic or metal header by a brazing or welding operation. Despite the plating, the packages and their lids would be damaged by air, as would the IC connection pads and wire bonds if they got hot enough, and it is necessary to maintain a nitrogen or argon atmosphere. Ceramic lids are melted down onto a low-melting-point glass layer on the package body. A nitrogen atmosphere is maintained in order to ensure that the headspace above the chip will not contain moisture or oxygen, which would attack the chip over a period of time. Inert atmospheres are not just used in ceramic IC sealing; there are also advantages in using nitrogen in some plastic encapsulation processes.

Helium leak testing. The hermetic seal of packages may be checked by a helium leak test. There are many ways of doing this but one simple way, requiring no particularly expensive equipment, is as follows. Packages are placed in a chamber and pressurised with helium at a few bar pressure. They are then removed and placed in a bath of solvent. Packages with leaks will have taken up He and become pressurised, and on placing them in the bath at 1 bara, small bubbles of He will be seen, even in the case of quite small leaks. More sophisticated systems employ mass spectrometric detection of He.

Nitrogen–hydrogen mixtures. These can be employed wherever a high-temperature operation may cause undue oxidation of silicon or other metal surfaces, even with nitrogen inerting. These 'forming gas' mixtures are especially helpful where small amounts of oxygen cannot be avoided in the high-temperature zone. Although often used in the wafer fab itself, these mixtures are more often encountered in electronics assembly; for example in heating ceramic packages to avoid oxidation of the metallised surfaces prior to lid seal or die-bonding.

3.5.12 Gases in compound semiconductor manufacture

Most semiconductor devices are made of silicon. However, devices for RF and microwave regions and devices such as light-emitting diodes, use other semiconductors. A few per cent of the semiconductor industry is based on the alternative III–V and II–VI compound semiconductors such as gallium arsenide and indium phosphide.

GaAs is a much more difficult material to work with than silicon and other compound semiconductors are more difficult still. One of the biggest difficulties is that there is no simple process for an insulator. With silicon, silicon dioxide is an insulator that can be grown on or added to silicon and is a reasonable thermal match to it. No such convenient insulating oxide can be grown on GaAs and deposited layers are also fraught with difficulties.

Another difficulty is that single GaAs wafers are difficult to make with the necessary levels of purity and freedom from faults. During growth, the more volatile arsenic tends to be lost from the melt, giving rise to vast numbers of faults if uncorrected. Epitaxial layers are often therefore needed in GaAs but they are more difficult than for silicon; a mixed gas atmosphere is needed with gaseous sources for both arsenic and gallium which will deposit pure GaAs. Trimethyl gallium and arsine (AsH_3) are the standard gases for carrying this metallorganic vapour phase epitaxy (MOVPE) process. The process is sufficiently tricky that some device manufacturers do not use MOVPE themselves but buy in wafers with MOVPE layers already added to their specification. There are companies which specialise in carrying out GaAs MOVPE as a service in Europe and the USA.

Doping and alloying of GaAs and other compound semiconductors is accomplished with unusual compounds, such as dimethyl zinc, dimethyl cadmium, disilane (Si_2H_6), germane (GeH_4), hydrogen selenide, hydrogen telluride, tripropyl aluminium, trimethyl aluminium, trimethyl indium, tetramethyl tin, stibine and other metallorganics. Only minute amounts of these compounds are needed; the entire world market for all these compounds is only a few hundred kilograms.

Most of these gases are thermally unstable, that is, they are endothermic and release heat on decomposition. Instability leading to loss of product can be avoided by dilution with, for example, hydrogen; this is standard practice for diborane. Another problem of endothermic gases is that they can actually undergo an unwanted CVD process inside the gas cylinders they are supplied in. Occasionally a cylinder valve on these gases will fail because deposition has occurred on the valve seat and they cannot be turned off completely, giving rise to a dangerous leak. An additional technique which is being looked at seriously is that of using the deuterium isotope to form hydrides. Fully deuterated diborane has enhanced stability, for example. As noted above (section 2.2.10), deuterium is the least expensive of the separated isotopes and is not an unreasonable cost for gases for GaAs use, where quantities used are in tens or hundreds of grams.

Compounds such as germane and hydrogen telluride are sufficiently endothermic that a spark or other source of ignition can start a runaway chain reaction in which the heat released from the decomposition is sufficient to make surrounding molecules to decompose, too. To avoid disasatrous detonation, gases like germane have therefore to be stored at relatively low pressure gas (10 bar or less), as well as being diluted with hydrogen in many cases.

Recently, the preparation of these exotic compounds at hitherto unheard levels of purity has begun to be carried out commercially by companies such as Voltaix and Epichem. The higher purities available will enable the preparation by CVD and MOVPE of exotic semiconductors with properties (defect densities, carrier lifetimes and so forth) which are closer to the near-perfect properties now achieved only with silicon. Impurities down to < 1 ppm of metals are now routinely achieved, whilst analysis with ICP-MS down to ppb levels can be carried out for critical impurities. Even more sophisticated analytical chemistry tools are now being brought to bear on occasion. For example, 250 MHz NMR can see oxygenated compounds in trimethyl aluminium; this is important in final devices.

The Epichem company (Bromboro, UK) is an afficionado of the 'adduct' purification route. Adduct purification is only justified for purer grades; up to 99.5% purity alkyl compounds are straightforwardly prepared using standard materials and techniques. It is only beyond this purity level that another technique is needed. An example of adduct purification is the preparation of ultrapure trimethyl aluminium.

Bis-diphenylphosphenoethane ($PPh_2CH_2CH_2PPh_2$, DIPHOS, in which Ph is C_6H_5) is first reacted with trimethyl aluminium (TMA). The resulting adduct compound, DIPHOS-TMA, is an involatile solid that will contain only adducts with trivalent compounds, the adduct bond being a donation of a pair of electrons from each of the phosphorus atoms in the DIPHOS. Impurity compounds will not form adducts and can be extracted by vacuum and gentle heating as they are mostly volatile. Heating to 100°C will decompose the adduct compound, releasing the TMA and allowing a further stage of distillation. Similar reactions are used to purify the other trivalent organometallics such as trimethyl gallium and trimethyl indium.

These compounds are now used in MOVPE growth. Wafers up to 6" in diameter, based on GaAs and InP, can now be made with high-purity epitaxial layers and precise levels of doping. Lasers and high efficiency LEDs are now made from gases prepared in this way, as they show much better performance. GaInP LEDs made by MOVPE seem likely to replace liquid-phase epitaxy devices as they are up to a thousand times brighter with higher electrical efficiencies.

3.5.13 Gases for optical fibres

Optical fibres are an example to rival silicon semiconductors of an industrial development which depends on an ultrahigh purity material, albeit on a smaller scale. Optical fibres are the principal means used to conduct long-distance telecommunication traffic around the world. Electric signals from telephones or computers are converted into pulses of light and these light pulses are directed along the centre of a solid fibre of quartz glass which acts as a 'light pipe'.

Light approaching the interface of a piece of glass with a medium of lower refractive index from within at a shallow angle is reflected. In fact it undergoes total internal reflection; 100% of the energy is reflected back into the glass. This principle is the reason why a piece of glass fibre can be used as an efficient light pipe. Every 50 or 100 km of optical glass fibre there is a repeater station, which amplifies the weakening light pulses and passes them on. At the receiving end, the light pulses are reconstituted into electric signals for the receiving telephone or computer.

Ordinary glass even a couple of feet thick noticeably tints and dims light passing though it. Workable optical fibres are thus wholly dependent on achieving the utmost clarity (and therefore purity) in the glass used.

In fact, many of the processes and gases pioneered in the semiconductor industry were taken over and modified to suit the manufacture of optical fibres. Helium, nitrogen, oxygen, silane, HCl, BCl_3, $GeCl_4$, $POCl_3$ and chlorine are all used in the industry.

The process of producing an optical glass fibre bears some resemblance to the process used since the 1800s for the preparation of 'rock' sweets (sticks of mint candy with a name or picture running through their cross-section). The required fibre has a high optical density core of $\sim 10\ \mu m$ in a $100\ \mu m$ body of lower optical density. To achieve this radial variation of optical refractive index, a large, short, fat boule or 'preform' is produced that has a much larger (several inches) diameter. The boule then heated and drawn down from a few inches to $100\ \mu m$ in diameter by heating until it reaches the

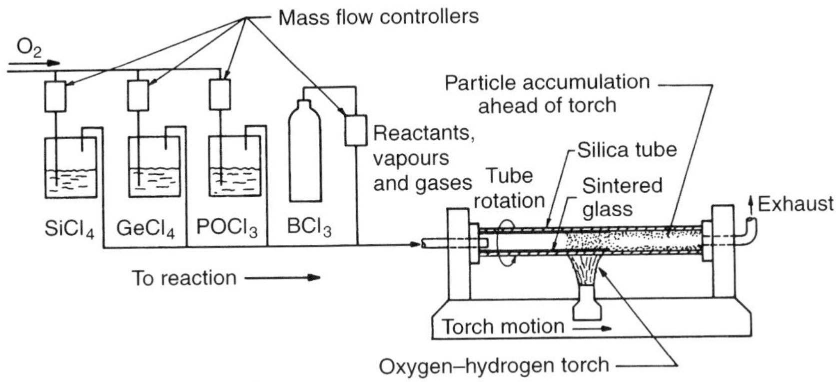

Figure 3.28 Glass preform lathe arrangement.

consistency of sticky toffee and then pulling out a sticky string of glass on to a carefully controlled winding drum (Figure 3.29).

The preform can be produced in many ways but must maintain extraordinary levels of purity. One process uses a CVD technique to deposit a 'soot' of porous glass onto a central silica rod. The central 'bait' rod is rotated in a lathe machine whilst a burner is played onto it, moving laterally to ensure an even distribution of deposit along the axis. The burner is fed with methane, oxygen, $SiCl_4$ and dopant chlorides; first heavy dopants such as $GeCl_4$, then lighter ones such as BCl_3. The radial variation of composition and refractive index of the fibre will be determined by the programmed changes in the flame dopants during the deposition of the preform (see Figure 3.28).

The reactions in the flame are analogous to the CVD reactions used in the semiconductor industry to lay down the final insulating 'glassivation' layers on silicon but also have some similarity to flame-coating processes since a fine particulate soot of hot glassy material is formed in the flame before deposition.

The presence of any moisture in the process would leave OH radicals in the glass and ruin the fibre so conditions of high cleanliness are needed and gas inerting is used in subsequent processing. However, the flame process adds a lot of moisture to the boule and this must be eliminated. First the central rod is removed. The resultant hollow porous dry glass rod is then run through a tube zone-sintering furnace to convert it into a solid boule. The use of a helium atmosphere in the sintering furnace enables bubbles to be more easily eliminated. Heating in hydrogen chloride or chlorine gases, sometimes in mixtures with oxygen, is then used and results in the elimination of residual moisture and OH.

Other preform methods are closer to semiconductor CVD as they form a

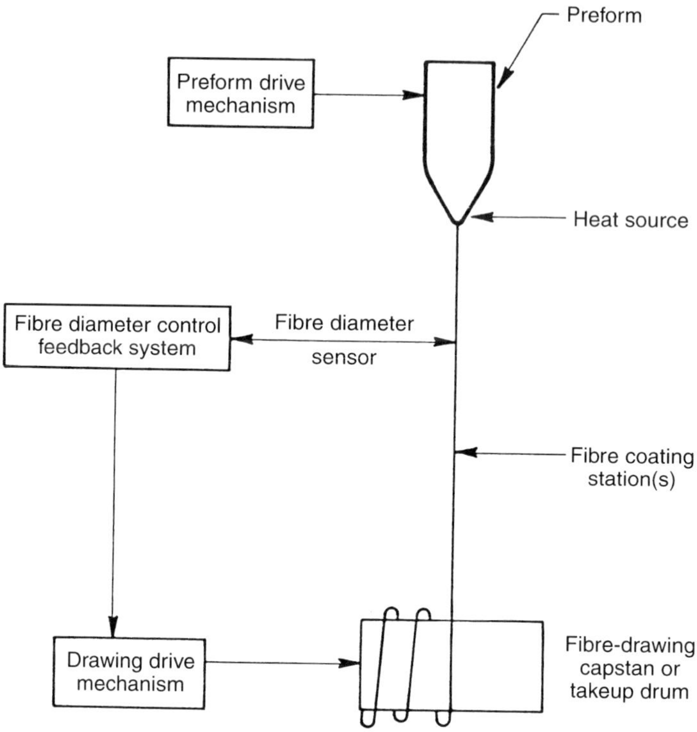

Figure 3.29 Fibre preform melting and pulling machine.

deposit on the inside of a glass tube that is heated by a flame from the outside and do not need to add methane fuel gas to the deposition mixture, with consequent moisture problems. As in the above method, the tube must be furnace-treated to eliminate the central hole and convert it into a solid boule.

The production of fibre is done in a pulling tower by pulling a 'gob' of molten glass from a hot boule, suspending it in the furnace, then stretching this to form the initial fibre (Figure 3.29). The fibre is wound onto a take-up spool, whose speed is controlled to control fibre diameter, and continuous production can begin. Once the machine is running, at up to a few metres per second, the fibre is cooled within a fraction of a second of its production and coated with a plastic coating which protects it from crack-forming or surface contamination. Coated fibres are very strong, despite being only 100 μm or so in diameter, which makes handling easier.

The pulling furnace may be a small oxy-hydrogen torch furnace, a graphite tube furnace or another type but is invariably a vertically mounted tube furnace. Ideally an inert atmosphere is maintained around the fibre in the furnace. Cooling the newly formed fibre in a pure gas preserves its perfect surface and increases the potential production rate of the fibre. An inert gas also means that a higher quality UV setting lacquer (without additives for counteracting air oxidation) can be used for coating. Even with thermoplastic coatings, an inert gas for the coating section can avoid some of the thermal degradation of the plastic that may occur in air.

3.5.14 Gases in general electronics

Soldering. Although traditionally done in air, soldering of electronic components to printed circuit boards, using nitrogen or other gases, is now becoming more common. Solder is usually a joint between an electronic component, lead-wire (made of copper, often coated with tin or solder), and a circuit board (also copper, often tinned). The solder used is usually a simple 62/38 tin/lead eutectic composition, which melts at 183°C, although small amounts of additional alloying elements such as antimony can give solder melting at a little higher or lower temperature, allowing multistage soldering. Multistage soldering requires that subsequent soldering does not loosen previously soldered components. A little flux is employed to aid wetting of the joint by the molten solder. The flux is a mildly acid material such as ammonium chloride (a powerful flux), adipic acid (mild) or a rosin (a mixture of natural diterpene acids with a milder action, favoured today). The flux removes a slight oxide film and protects the soldered surfaces from further oxidation during the heating that immediately precedes the molten metal.

The two most common techniques for soldering (other than by hand) are reflow and wave soldering (Figure 3.30). After placement of the components, sometimes using glue to retain them, the latter process passes the circuit boards through a wave of molten solder. In the reflow process, the components are placed/glued after screening a pattern of solder paste onto all the places where the component will be. The boards are then passed through an oven where the small patches of solder paste fuse into soldered joints. (Another gas soldering technique uses a hot blast of nitrogen for fusing solder paste to fix small, single components.)

Nitrogen, used in wave soldering, reduces the amount of 'dross' and improves quality, although it is still unusual to see gas inerting on wave-solder machines. Dross is a mixture of solder and oxides of lead and tin that accumulates on the surface of molten solder after a few minutes' exposure to air. Dross is a considerable problem, causing bridging and wetting problems on circuit boards, and often requiring manual intervention in the automatic soldering process to remove it (it can be reprocessed back into solder). A large wave-soldering machine may form more than 1000 g per hour of dross, on a solder wave just 20 cm across. With small reductions

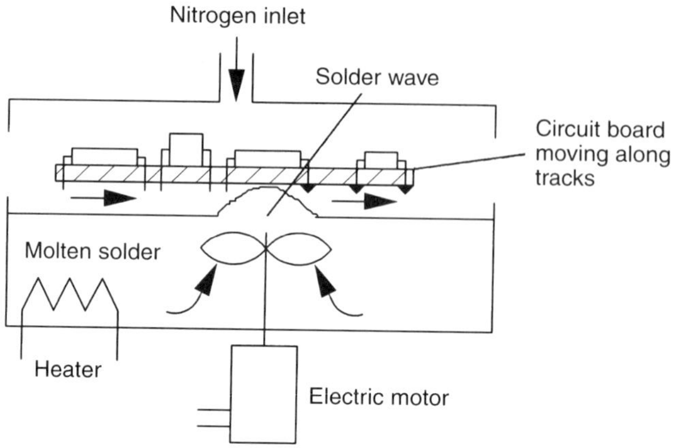

Figure 3.30 Wave soldering with gas inerting.

in oxygen content, dross is produced as copiously as ever. However, reducing the oxygen below about 2% in the atmosphere makes improvements and dross formation is minimal at 0.1% or less oxygen.

Gases have not hitherto been much used in reflow soldering but this may eventually change. Traditionally, circuit boards that have been soldered go through a cleaning bath based on CFC liquid solvents to remove residues of the flux used and ensure wetting of the metal parts by the solder. This is no longer environmentally acceptable. New approaches vary from the use of different solvents (one 'new' solution is a soap and water system), to avoiding the creation of a flux residue in the first place.

The solder paste used in reflow is a mixture of powdered solder, solvents and flux. The use of nitrogen inerting in the reflow oven enables the use of milder fluxes, in smaller amounts, allowing circuit boards to be used in less exacting applications without cleaning. One process was developed using a flux which reacts with the gas used in the reflow oven, leaving an extremely small residue. 'No-clean' fluxes are now available from several suppliers that have extremely low residual deposits and do not need cleaning-off. No-clean fluxes are based on very mildly acidic compounds, such as adipic acid, and have a small solids content, relying on the use of nitrogen to largely protect the copper or tinned surface in the hot atmosphere until solder wetting occurs. In air, they would not be able to stop excessive oxidation and inadequate wetting occurring.

Occasionally, references are made to 'vapour-phase' soldering. This is a form of reflow soldering in which vapour from a boiling liquid, contained in a deep tank, condenses on a circuit board lowered into it and heats it up to near the liquid's boiling point. Inert fluorocarbon liquids that boil at 300°C are used for this process. As a technique it is now regarded as obsolescent, although it does provide some of the advantages of an inerted soldering technique.

Gases in printed circuit board manufacture. The preparation of multilayer printed circuit boards (PCBs) is now a sophisticated technology. The use of nitrogen-filled autoclaves for laminating them has become a standard procedure.

The laminae, 'green' epoxy/glass fibre composite sheets a fraction of a millimetre thick and patterned with copper tracks, are placed together in stacks and placed inside a vacuum bag. Hundreds of such bags are then connected to a vacuum pipe inside a large heated pressure vessel (an autoclave). Whilst a vacuum is drawn inside the bag, a 10 bar nitrogen atmosphere is applied and the whole apparatus heated. After a few hours, the epoxy resins are set and the boards can be removed, the laminae efficiently bonded together. The use of nitrogen avoids risks of fire and explosion that the use of high-pressure air with the large volumes of resin would entail.

PCBs, after lamination, are drilled and the holes are plated with copper using an 'electroless copper' process followed by electroplating, providing a connection between each side of the board. After drilling the PCBs, it is often the case that the sides of the holes are 'smeared' with resin, preventing the correct plating of plated-through holes. The holes can be 'desmeared' by processing in an oxygen/argon plasma (section 3.5.7), which quickly oxidises the resin blobs away while leaving the bulk board more or less unaffected.

3.5.15 Thermionic valves

Although semiconductors now account for 99% of all electronics sold, there are some applications for which the thermionic valve is still the only solution. The cathode-ray tube (CRT) used in television sets and in computers is probably the most common thermionic valve today. In the CRT, the display seen is simply the result of fluorescence caused by electrons from the hot filament striking the screen. Magnetrons for microwave ovens are probably the second most common thermionic device in use. Similar but larger thermionic valves, up to 2 m long, are used for television and radio broadcasting, and many other high-power RF applications. All these devices employ a similar range of gases in their manufacture but magnetrons, still made in large quantities today, are a good example. The quantities of gas used are not necessarily as small as might be thought; one magetron plant, for example, specified 100 000 m^3 H_2, 50 tonnes of Ar, and 1000 tonnes of LN per annum.

Magnetrons consist of a glass envelope with a filament and various metal electrodes inside, and a powerful magnet around the outside (Figure 3.31).

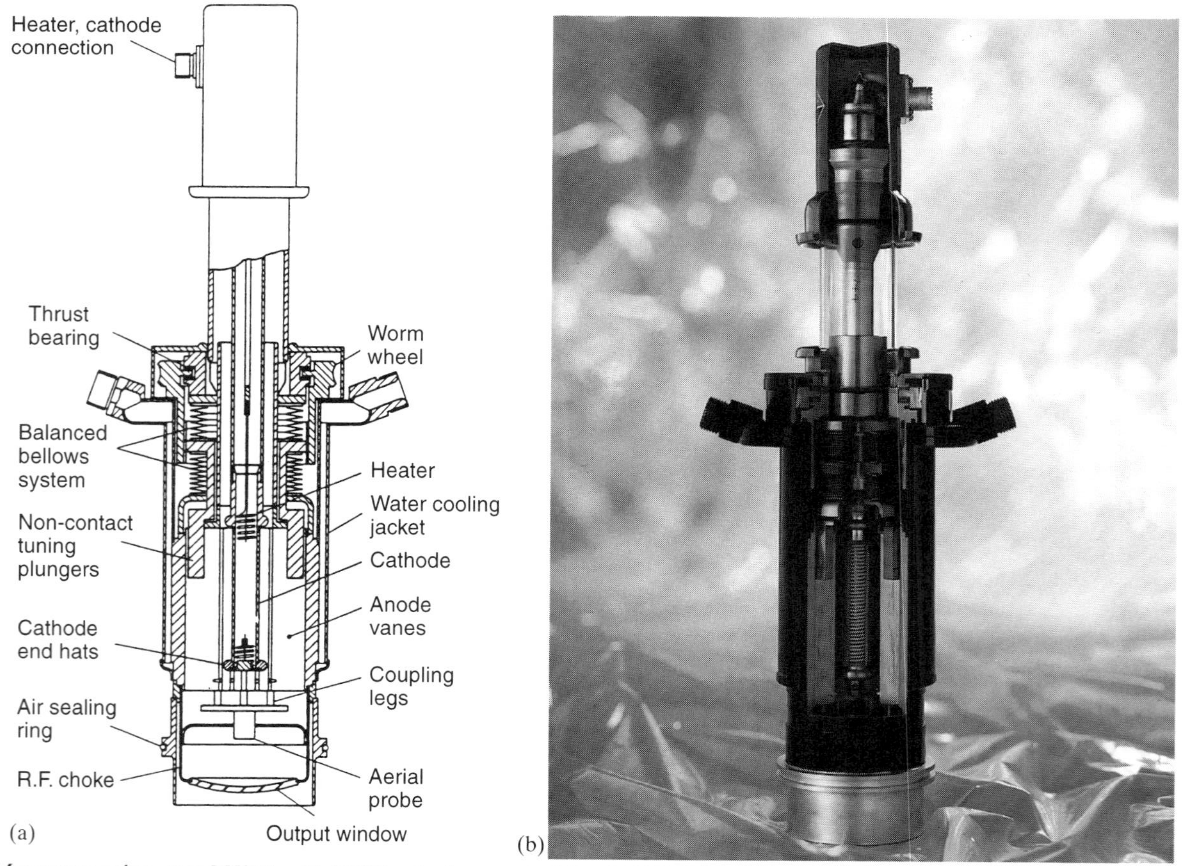

Figure 3.31 Magnetron valve, type M5028, with 5 MW pulsed output in S-band (2850 MHz): one of the most powerful such devices made. The output window (base) may be cooled by being in contact with sulphur hexafluoride. The many ceramic-to-metal seals in the 50 cm long device are made by heating in hydrogen gas mixtures (courtesy of EEV Ltd).

They create powerful microwave electromagnetic waves in the 1 to 10 GHz range, typically 2.4 GHz, which can penetrate several centimetres into non-conducting objects such as food and heat them up; hence the application for microwave cooking.

The anode (outer electrode) is typically made from copper, whilst the inner electrode is typically a thorium (and other reactive metal) oxide-coated shell with heater wires inside. The glass and ceramic envelope must be very carefully sealed to the metal components. This often involves furnace processes to form a controlled oxide film on the metal prior to sealing. The sealing process itself can be performed in a nitrogen/hydrogen atmosphere belt furnace.

Glass-to-metal and ceramic-to-metal seals are in principle difficult but have long been studied and many solutions are now available. Molten glass bonds well to ceramic and to low expansion metals such as Kovar (Fe/NiCo 54/29/17) when the metal has a controlled thin oxide film on its surface. Metal powder mixtures such as Mo/Mn can form a sintered bond between ceramic or glass and metal when heated to 1400°C in a hydrogen/steam atmosphere. Ceramics may be sealed to metals by brazing or soldering. In order to do this, the ceramic must first be coated with a metal. The metal plate may be achieved by sputter-coating and then electroplating to form a thick, well-bonded metal surface. Alternatively, a paint composed of a hard metal powder in a nitrocellulose lacquer can be painted onto the ceramic parts. If fired in a hydrogen/steam atmosphere, paints of this sort will form a solid metal layer for brazing.

The magnet of a magnetron is typically a permanent type made from a dispersion type of FeCo/NiAl (Alnico) cast alloy, quenched and then annealed at 500–700°C. Sometimes clever techniques involving cooling (under a controlled atmosphere) from a molten mixture in a magnetic field, which allows the full potential of high performance alloys, such as Ticonal™ and Alcomax™, to be realised. Alternatively, barium hexaferrites may be employed (section 3.2.8). These are made by sintering a magnetic oxide ceramic mixture in a controlled atmosphere following pressing in a magnetic field.

3.6 Gases in water treatment

Wastewater treatment is commonly classified into primary, secondary and tertiary treatment.

Primary treatment. This is the removal of particulate matter. Following a screen to eliminate pebbles and other gross particulate above 20 mm or so, settling tanks are employed in which much of the suspended solids simply settle out to the bottom. Aiding this process is very commonly a flocculation system. This adds a highly charged ion, such as are given by iron(III) chloride or sulphate or aluminium sulphate, which acts as a centre for the agglomeration of the unfilterably small colloidal particles into macroscopic 'floc' particles which will settle or are filterable. Sometimes specially designed polyelectrolytic chemicals (charged organix polymers) are added instead to achieve the same effect. Occasionally, lime or another alkali is also added to ensure that the pH, which should not be too low, is correct. The flocculating chemical added is often added at a constant rate. It is only needed in very small quantities.[57] Primary treatment reduces the biochemical oxygen demand (BOD), the amount of (undesirable) biologically oxidisable material in the water, by 50% or more.

Secondary treatment. This is today considered an essential part of the wastewater treatment process. In this, colloids not removed in primary treatment, dissolved organics and other oxidisable contaminants are oxidised by processing with colonies of microorganisms to harmless molecules such as CO_2 and water. In this way, BOD can be reduced to 10% or so of input values. Two systems, activated sludge and trickling filter, are in use, with the former tending to become more popular. Activated sludge simply blows air or oxygen through the pre-treated wastewater, to which some of the sludge extracted previously has been added (this contains the necessary dose of micro-organisms). Trickling filters are beds of rock chippings over which water is trickled, the microorganism colonies adhering to the chippings. Secondary treatment, finishing with filtration, largely eliminates organisms harmful to humans and animals, although a further sterilisation step with chlorine or ozone is used for drinking water. An alternative to the aerobic secondary treatment is anaerobic treatment, which, in addition to purifying the water, can produce useful amounts of methane fuel gas. This can be used to run a gas turbine electrical generator to power sewage plant operations.

Tertiary treatment. This is aimed at reducing non-BOD contaminants, inorganic phosphorus, nitrogen, etc., which, although not pathogenic, cause problems in water courses by allowing excessive growth of algae and water plants that can clog or pollute rivers and streams just as effectively as quarrying or detergent discharges. Addition of lime (calcium hydroxide) or aluminium hydroxide is now used at many plants to reduce ionic content.

Water purification. This refers to operations downstream of wastewater processing. Wastewater processing aims simply at producing water fit to put back into rivers and streams. Water purification is aimed at producing purified piped water for human purposes, either domestic or industrial. For piped mains drinking water, little extra processing is done except additional filtration if necessary and sterilisation of any remaining pathogenic

[57] There is now a patented system (Matchfloc Ltd) that will regulate the amount added.

microorganisms by chemical oxidation. Chlorine, chloramine (NH_2Cl, produced by injecting ammonia into previously chlorinated water), chlorine dioxide and ozone are all used for sterilisation. However, ozone may well become the dominant sterilant in the future (section 4.3).

Water softening. This refers to the removal of metal ions. For many industrial purposes, it is necessary to ensure that water is 'soft', that is, it does not contain more than a minimal amount of calcium and magnesium salts, which deposit as hard, irremovable scale when water is evaporated or heated. Hard water seriously limits the effectiveness of washing with soaps and detergents, and so soft water may also be regarded as desirable for domestic purposes. However, soft water is now regarded as undesirable for drinking, being associated with increased rates of coronary heart disease. In the older developed world lead pipes have been installed for drinking water, only rather recently being replaced by copper and plastic. There are thousands of kilometres of lead pipe still in existence, which release lead ions slowly into soft water, whilst transmitting hard water uncontaminated. In areas having naturally hard water, therefore, those where water comes via limestone or chalk rocks, for example, there is a strong move away from municipal water softening and towards leaving hardness in water pipelines (or even adding it, see section 3.6.2). Water softening apparatus is now usually employed at the point of use. Water softening is carried out in general by ion exchange resins, which replace calcium and magnesium ions by sodium ions taken from brine.

Beyond water softening, further purification processes may be carried out, e.g. deionisation (in which ion exchange treatment is carried further to remove both anions as well as cations), distillation (this is rare nowadays, except on a laboratory scale or when it is carried out under reduced pressure with a low-grade waste heat source), reverse osmosis (to remove salts completely), UV treatment (to remove residual organics) and degassing or gas stripping.

3.6.1 Oxygen for controlling aerobic sewage treatment

After use in domestic processes water is contaminated, mainly with suspended fine mineral particles, salts and organic compounds ranging from animal and human excreta to soap emulsions (from washing processes), and often including appreciable concentrations of bacteria and other micro-organisms, some of which are harmful pathogens. Industrial water uses may add to these contaminants water of too high or too low pH, dissolved heavy metals and non-biological organics such as solvents. The most serious problems in terms of volume derive from domestic biological contaminants, however, and it is to these that the most effort of wastewater treatment is aimed, principally through aerobic biological treatment.

The primary measure of success of secondary wastewater treatment is usually the reduction achieved of the BOD of the water being treated. BOD expresses the level of undesirable organics in the water in terms of the oxygen that will be needed to oxidise it completely. Typical BOD figures are difficult to specify, being highly variable between sites and seasons, but figures up to around 1–10 kg m^{-3} are often encountered.

Aerobic sewage treatment requires a high level of oxygen gas to be dissolved in the wastewater being treated. However, oxygen has low solubility in water and this sets a large minimum size for sewage plant. This minimum size is made even bigger if air must be relied on as the source of oxygen. The partial pressure of oxygen in air is only 0.21 bara, giving a maximum equilibrium O_2 concentration of about 10 g m^{-2}, i.e. 1 part in 10^5. With pure oxygen the maximum concentration achievable in shallow tanks is $\sim$ 50 g m^{-2}, simply because of the higher partial pressure. This would in principle allow pure oxygen sewage plant to be five times smaller in volume relative to air plant.

Although air dissolution is improved by using a deep or pressurised injection point, it is still a fundamental limitation that the concentration achievable is very low. There is another problem, the 'gas stripping' effect. The effect arises because the nitrogen content of the air acts as a sparging stream, stripping malodorous volatile materials from the wastewater. If pure oxygen is used and carefully regulated, then virtually all the oxygen is dissolved, with the result that no gas-stripping of malodours occurs. Measurements at plants with bad odour are occasionally made and levels of up to 20 ppm H_2S, for example, can often be found. At these levels H_2S is not simply unpleasant, it could be dangerous to operators of the plant.

Using oxygen injection, oxygen levels higher than saturation can be achieved for a few minutes; 200% supersaturated water will only slowly generate microbubbles and these take some time to coalesce and rise to the surface. This effect can be useful where reaction or dilution will allow the oxygen to react or dissolve within a short time and can substantially increase the amount of oxygen it is possible to deliver over the maximum amount calculated using oxygen equlibrium values.

The systems for injecting oxygen into wastewater basically fall into three categories:

1. inject fine bubbles into the bottom of the processing tank;
2. pump a stream of sewage from the tank into a pressure vessel where oxygen is added under pressure, then return the stream to the processing tank;
3. add a stream of oxygen to the sewage whilst it is inside a high-pressure pipeline.

Method 3 can be very low in capital but depends on a suitable injection point. If the injection point is low in a system, and thus at a high hydrostatic pressure, or is after a pressure-raising pump, simple injection into the water can be very effective. The high pressure of the water means that the oxygen solubility is increased roughly in proportion to the absolute pressure, and ensures rapid and complete dissolution of the oxygen. Patented treatment processes such as Vertech (section 3.6.3) capitalise on this effect even further. A cautionary note needs to be sounded about the effect of oxygen injection on high sulphur content wastewater in pipes. The addition of oxygen can lead to conversion of sulphides and reduced sulphur forms to acidic sulphate solution, which can cause severe corrosion of pipes (even, surprisingly, those made of concrete).

Method 2 is very efficient in terms of gas consumption, achieving high levels of dissolved oxygen efficiently but at some capital cost. It is becoming more popular, being promoted by several of the larger gas companies. The addition of oxygen to a sidestream allows conditions to be carefully optimised to produce more or less full dissolution, whilst a helpful mixing effect is achieved on pumping the oxygenated water stream back into the tank. The use of a pressurised side-stream allows efficient oxygen injection with shallow tanks; something that is difficult with method 1, as oxygen is lost to the atmosphere if high levels of dissolved oxygen are attempted.

Method 1 typically involves some loss of oxygen gas but is the most common method because it is low on capital cost and because it can almost always be retrofitted to existing sewage plants. An ingenious touch in method 1 is to use slots in a flexible plastic hose for the gas injection; when gas pressure is applied these open up and allow gas into the sewage but on release of pressure the tiny slots seal up again. This gives greater freedom from clogging; small bored holes in rigid pipes tend to block easily when used intermittently. A typical installation involves simply laying dozens of perforated plastic hoses across a tank. The hoses are connected to a manifold pipe on the side of the tank.

All dissolution methods, but especially method 1, have limitations relating to the behaviour of bubbles. Anyone who has visited a major science museum recently may well have seen an educational toy in which bubbles of air are sent up a large column of a viscous clear liquid. The equipment has viscous flow and demonstrates that large bubbles rise faster than small ones according to Stoke's law for drag on a spherical object. However, even more obvious than this is that the bubbles have a tendency to combine together after formation but the converse, bubbles splitting up, never happens. Bubble coalescence happens because the excess free energy due to the surface tension possessed by two bubbles is less than that of one bubble of equivalent volume. Taking the case of two identical spherical bubbles of volume V (radius R) and overpressure ΔP coalescing, for example, the additional gas pressure energy shrinks by 20%

$$E = 1/2\Delta PV + 1/2\Delta PV = \Delta PV = 2S/RV \text{ (two bubbles)}$$

where S is the surface tension and ΔP is the pressure inside the bubble above the local hydrostatic pressure.

On coalescence, one bubble, pressure $\Delta P'$, radius R' and volume V' is formed and

$$V' = 2V, R' = R2^{0.33}, \Delta P' = 2S/R'$$

Hence

$$E \text{ (one bubble)} = 1/2\, \Delta P'V' = 0.79(2S/RV) = 79\% \times E \text{ (two bubbles)}$$

This effect limits the maximum flowrate through diffusers: if the flowrate is too high, then bubbles appear too close together and they begin to coalesce. The large coalesced bubbles rise more quickly, tend to coalesce with other smaller, slow bubbles as they rise and coalesce further until the neat regular streams of small bubbles seen at lower flowrates, which dissolve very efficiently, are completely disrupted by irregular bubbles with large gouts of bubbles rising to the surface incompletely dissolved.

With all these methods, a dissolved oxygen electrode suspended in the tank or pipe being treated can feedback achieved oxygen level to a controlling valve, thus economising on oxygen usage. Typically oxygen is only required only for a few hours a day to handle peaks in BOD. However, oxygen electrodes are subject to fouling and other failures in services in water treatment and some care is necessary in choosing and locating the electrode. The type chosen will typically be isolated from the water stream by a plastic membrane that allows only small molecules, such as oxygen, to pass, as described in section 2.1.7.

For continuous supply of oxygen to wastewater treatment, bulk liquid pure oxygen supplies are regarded as rather expensive. More popular is to use ~ 90% pure oxygen produced by pressure swing absorption (PSA). These units can be made to operate with long intervals between maintenance and are steadily growing in energy efficiency. PSA oxygen injection is popular, for example, for oxygen treatment of tanks where there is a foul odour problem. A conventional tank aeration system needs a similar power for its air compressor or other agitation system to that needed for the PSA oxygen system. However, with conventional aeration, nitrogen bubbling out of the surface of the tank carries an aerosol into the air, efficiently vaporising any foul-smelling molecules from the wastewater. Other aeration systems which pump a fountain of water through the air have a similar problem. With oxygen injection, bubbling at the surface can be reduced virtually to zero, minimising odour problems.

Pure oxygen for peak demand. Even with gas from an efficient PSA, the cost of continuous oxygen supply is still generally regarded as too high in most cases and aerobic sewage treatment using air is the norm. Pure oxygen treatment is becoming very common, however, for 'peak shaving'. The volume and BOD per m^3 of wastewater in many areas varies between summer and winter in addition to shorter period variations. Although some very short period variations are averaged out anyway by the natural watercourse, and by lagoons and settling tanks at the water treatment plant, long period variations are more problematical. One attractive alternative is to meet the base-load oxygen demand of the sewage stream with air, while during peaks of high BOD or high flow oxygen is injected. This approach optimises the compromise between running and capital costs; although the oxygen is expensive, it is used only occasionally, whilst the main wastewater plant can be sized only a little above average demand, rather than having to be sized to meet peak demand. A typical liquid oxygen installation of this type uses an annual average of only a few tonnes per week of oxygen but is capable of high rates of O_2 addition when necessary.

In many peak-shaving plants, oxygen injection is necessary only in summer. In summer, water volumes are reduced so that even where there is a similar absolute BOD there will be a higher BOD per unit volume. This is just when the lower solubility of oxygen at higher temperatures makes it more difficult to achieve adequate oxygenation. Furthermore, in summer aerobic processes in water are faster and therefore have a higher oxygen demand. Such considerations, along with rainfall and water-usage patterns, determine the peak and average oxygen demands. The cost reduction achievable with an LO peak system depends on this peak/average demand ratio but can be a factor of two or more.

Oxygenation of rivers and lakes. The River Thames in London, UK, flows through hundreds of towns and industrial areas, and receives run-off water from intensively farmed crop-growing areas. Despite this, the river has a large population of fish, including even the occasional salmon. This is only possible because oxygen is injected in tonnage quantities whenever the BOD of the river rises too high. Just as oxygen injection can be used to increased oxygenation in sewage works, it can also be employed to provide respiratory oxygen to natural river microbes and fish.

In lakes with a relatively small throughput of water, small amounts of organic contaminants, from sewage, food processing or paper-making, can have serious consequences. These cause turbidity and excess BOD in the water. Turbidity inhibits photosynthesis of oxygen. It is possible for BOD to go up, dissolved oxygen levels to fall and living organisms in the lake to die. This creates a whole new mass of organic material with its own BOD, further exacerbating the problem. This classic vicious circle quickly results in a dead lake. Conversely, if BOD falls to below the critical level, plants and animals can recolonise the lake, resulting in the consumption of the organic chemicals and the production of further oxygen, further reducing BOD. Inland lakes with small flows are thus 'bistable'. Small lakes can be revived by tipping them out of the more or less dead anaerobic state into the living aerobic state by a steady injection of oxygen over some weeks or months.

3.6.2 Drinking water treatment

Water from most sources other than rainwater or direct from springs, clean lakes and reservoirs needs some treatment before it can be guaranteed to be healthy for drinking. Algae, bacteria and other suspended matter are removed to a large extent by filtration, often in simple sand bed filters. Addition of alum or similar salts in part per million quantities helps in the coagulation of suspended matter and aids filtration. However, the water may still contain some pathogenic bacteria. These can be killed by adding a very small amount of chlorine. Increasingly today, however, consumers find the taste of even a small amount of residual chlorine unacceptable. Also, there are worries about the formation (in almost undetectably small quantities) of potentially carcinogenic organochlorine compounds. Trihalomethanes are one example of this. Chlorine dioxide and chloramine treatment would also appear to reduce the production of organochlorine compounds, although as these treatments both use the element chlorine there are still some worries.

In the future ozone will probably be the most important alternative to chlorine in the killing of pathogens in water. Ozone water treatment is already the largest commercial application for ozone. Oxygen or air is passed through a silent high-voltage discharge, producing a gas mixture with a small percentage of ozone. When this is bubbled through water in small amounts it easily and quickly kills residual bacteria.

Ozone has the fundamental advantage that it is unstable and will always have disappeared completely before delivery from the tap. In any case catalytic reactors are available which complete ozone decay very rapidly. Ozone-treated water is therefore completely pure, although ironically this means that to retain bacteriocidal action along pipelines an amount of chlorine (a small fraction of the normal amount) must typically be added to ozone-treated water.

Ozone also has the advantage of being a more powerful oxidation agent than chlorine. This means that any residual organic matter, that not oxidised by the microbial treatment of sewage, is decolorised and deodorised more completely than with chlorine. With ozone there are also fewer handling risks relative to chlorine. Ozone treatment is, however, a more expensive and more complex technology than chlorine treatment.

Hardening/stabilisation of drinking water. In many parts of the world, depending on rock formations in the rain catchment area, water is naturally 'hard', containing soluble calcium and magnesium compounds. Other areas have the opposite problem: an acidic water without any calcium or magnesium to buffer and prevent the pH from falling (much of Scandinavia and Scotland, for example). Paradoxically, although hardness is generally regarded as a nuisance, a small amount of hardness is actually beneficial in that it suppresses pipe corrosion and reduces pick-up of potentially harmful pipe and tank materials such as copper, lead and zinc. New European water standards now insist on a certain low minimum of hardness so plants in northern England, Scotland and Scandinavia will be adding lime ($Ca(OH)_2$) to attain hardness targets. However, lime is minimally soluble and practical systems in fact add a lime slurry and inject CO_2, producing one of the principal constituents of naturally hard water, calcium bicarbonate ($Ca(HCO_3)_2$), which remains in solution. Typical treatment levels are of the order of 10 mg l^{-1}. Good mixing is essential to ensure that CO_2 is not lost before it has time to react.

Stabilisation of lime-softened water. Although ion exchange resins are used on a relatively small scale for softening water, this would be impossible on economic grounds for municipal water supplies. They are, however, used to treat water for hardness economically by using cruder processes such as lime softening, discovered in 1766 by Cavendish. Lime softening is achieved by adding CaO or its hydrate, $Ca(OH)_2$, the cheapest industrial alkali, to waters containing magnesium and calcium bicarbonates. Calcitic or dolomitic deposits are precipitated and the water is softened

$$Mg(HCO_3)_2 + Ca(OH)_2 \Rightarrow CaCO_3/MgCO_3 \downarrow + 2\,H_2O$$

$$Ca(HCO_3)_2 + Ca(OH)_2 \Rightarrow 2CaCO_3 \downarrow + 2H_2O$$

However, these sparingly soluble compounds are not totally precipitated and are carried on into the remainder of a water treatment works, where they may cause severe problems by precipitating out in filters, pipes and pumps. Stabilisation prevents this unwanted deposition by adding CO_2 to the water stream after a settling tank has accumulated nearly all the particulate calcite and dolomite. The injected CO_2 reacts with dissolved and microparticulate carbonates and produces the much more soluble bicarbonates,[58] although in much smaller quantities than the input water, and also neutralises any excess lime that may have been used

$$CaCO_3 + CO_2 + H_2O \Rightarrow Ca(HCO_3)_2$$

$$Ca(OH)_2 + 2CO_2 \Rightarrow Ca(HCO_3)_2$$

If a modest excess of lime is always added, the CO_2 addition can be controlled by a downstream pH meter since the limed effluent will have a pH of, say, 10 to 14, dependent on the amount of excess, and the CO_2 can be added to bring the pH back to, say, 6 to 9.

One 'wastewater' which is used in probably the largest quantities anywhere in industry is the cooling water used in power stations. The difference here is that most power stations recycle most of their cooling water. Addition of CO_2 to acidulate the water slightly is common practice as it helps to reduce scale, as in the lime-softening of water.

[58] This reaction is the reason why limewater turns milky with CO_2 then turns clear again if the CO_2 stream is continued.

3.6.3 Industrial wastewater treatment

Some industrial wastewater treatment is biologically based, although an intensification of the treatment process is often necessary. For example, pure oxygen injection systems are sometimes seen in the treatment of industrial wastewater with high BOD. The steady tightening of regulations on the pollution of water in many countries means that occasional high BOD discharges, which used to be acceptable, must now be avoided. Many industrial concerns, breweries and abattoirs for example, with wastewater problems are therefore now installing pure oxygen equipment. Oxygen injection is typically used on an occasional peak-shaving basis with the wastewater normally being allowed straight out to sewers.

Industrial wastewater treatment more commonly uses chemical rather than biological methods. The high toxicity and wide variations of pH and pollutant concentrations mean that conventional biological treatment is difficult or impossible. Very often, however, industrial chemical treatment of wastewater is used near the point of generation of the wastewater simply to reduce toxicity and pH to reasonable bounds. The water can then be safely and economically piped away for further treatment in the conventional municipal sewage works. It is this chemical primary effluent treatment in which gases have their most important role.

One technique for aerobic treatment of industrial wastes which may be important in the future is the use of hydrostatic pressure. Without microbial action the high hydrostatic pressure down a deep shaft can be used to dissolve large amounts of oxygen in water. The Vertech process involves pumping sewage down a shaft more than a kilometre deep, adding oxygen at a few hundred metres down, and pumping it back up again. High pressures and high temperatures are obtained without expensive pressure vessels by natural hydrostatic pressure and the increase of temperature with depth. This very efficient dissolution of oxygen at high temperatures allows an increased rate of reaction. The first commercial-scale pilot plant in Apeldoorn, Netherlands, is now operating, treating 100 tonnes per day of 5% sludge water with considerable oxygen demand. Similar non-biological oxygen treatment using surface equipment is used here and there for industrial wastewater but has never caught on on a large scale.

Ozone is also now used in industrial wastewater processing, in order to achieve limits imposed by local authorities for discharge to sewers and water courses. The ease with which ozone can convert potentially harmful organic substances to CO_2 and water, and even highly toxic ions such as cyanide to the relatively harmless cyanates, makes it very attractive as a small-scale treatment for industrial discharge.

3.6.4 Control of pH with CO_2

One environmentally friendly way of controlling the pH of wastewater streams is to add CO_2. Many waste streams have too high a pH, i.e. alkali pollution problems. In addition, the pH varies widely, depending on the stage of the process. For example, a cleaning process might typically involve caustic soda followed by rinsing.

Carbon dioxide forms the weak acid, carbonic acid, when dissolved in water. Chemical equilibrium principles mean that addition of a weak acid to an alkali-contaminated stream will both reduce and stabilise the pH. Even low-pH wastewater streams are sometimes treated with CO_2 by predosing with trace amounts of alkali, followed by precise control of pH with CO_2 addition. Carbon dioxide is convenient because it can be easily controlled, is non-toxic, inexpensive and readily available. A simple pH control rig might consist of little more than a control valve supplying CO_2 gas to a bubbler in the waste stream, with a pH electrode downstream giving a feedback signal to the control valve.

Wherever weakly acidic salts, such as ammonium chloride or ammonium sulphate, are used to control pH, CO_2 may be a possible candidate as an environmentally sounder substitute. For example, CO_2 can be used instead of ammonium chloride to reduce pH in leather treatment. After removing hair and then pickling in alkali (liming), leather has to be gradually neutralised (delimed) before tanning under acidic conditions with sulphuric acid/sulphates followed by dichromate. CO_2 can sometimes replace the more traditional ammoniacal solutions by reacting with the $Ca(OH)_2$ and lowering the pH from 12 to about 7, when the acidic treatment can begin. The ammonium salts are normally discharged into water courses where they may cause significant smell and pollution problems. CO_2 avoids these problems totally and is less expensive.

CO_2 can also, in one application, replace chlorine. Chlorine is used in swimming pools, in much the same way as it is used in drinking water. However, here again, the taste of chlorine, and its effect on the sensitive eye membranes, is undesirable. The level of chlorine is normally maintained by injection of small amounts of chlorine at intervals, the level being maintained to control the pH (acidity) of the pool water. A more recent scheme has been devised whereby the pool pH can be controlled by means of CO_2 with bacteriostatic properties being maintained by addition of liquid and solid chemicals. This relatively new scheme means that gas cylinders of chlorine need not be handled and the small risks of excessive chlorine leaks or excessive addition are avoided.

3.6.5 Gas stripping

As mentioned above, certain industrial uses of water require processing beyond the achievement of drinking water standards. The removal of dissolved gases, particularly oxygen, is one example of this. The removal of volatile materials and dissolved gases from water can sometimes be achieved by the elementary expedient of bubbling an inert gas through the water. Gas stripping or 'sparging' is used especially for aqueous product streams (section 3.4.3). Often air is sufficiently inert that it can be used in gas stripping; unless, of course, the gas to be stripped is oxygen.

Gas stripping is most effectively carried out in counter-current packed columns but the capital cost of this kind of installation often means that less efficient simple bubbling tanks are used. An alternative method is to add inert gas to the low pipework somewhere in an installation and then allow the sparging gas to escape at the top of the system, perhaps from the headspace of a tank.

An adjustment of pH upwards will enable dissolved alkalis, such as ammonia, to be stripped out by nitrogen sparging, whilst addition of acid to lower pH will enable removal of dissolved volatile acids, such as hydrogen sulphide or cyanide. The use of nitrogen for the stripping operation is justified where oxidation, often of other contents of the wastewater, must be avoided.

Gas stripping is seen in semiconductor water purification, in the manufacture of soft drinks and in the preparation of boiler feedwater. In semiconductors (section 3.5) absolute purity is important, whilst in boilers the use of water which will not cause oxidation/corrosion of the boiler tubes is very important. In semiconductor and boiler applications an alternative to sparging is the addition of a reducing chemical; perhaps after an initial gas-stripping operation. A typical chemical dosing is with hydrate of hydrazine, the reaction producing simply nitrogen and water with dissolved oxygen. Clearly, however, sparging is superior in that it does not involve a hazardous chemical and any carry over of the sparging gas will be harmless, whereas hydrazine carryover could be highly undesirable.

3.6.6 Oxidation treatment

Even with pure oxygen instead of air, oxidation treatment of industrial effluent can be a slow reaction. The exception is the oxidation of iron and manganese in which oxidation followed by precipitation is a possible treatment. Interestingly, iron and manganese ions can be deliberately added to organic-contaminated streams where they act as catalysts to allow the oxidation of oxygenated organics, alcohols, aldehydes and ketones, which otherwise react much too slowly. Similarly, transition metal catalysts can be used to catalyse oxidation of sulphides in wastewater.

Ozone treatment. Except with biological treatment, and with the exceptions noted above, oxygen is often ineffective in industrial wastewater treatment. It is insufficiently active as an oxidant and it is too difficult to dissolve in water. If the oxygen is converted into ozone, however, oxidation is almost invariably successful. However, ozone has all the problems seen in drinking water treatment. It is expensive to make, whether direct from air or from oxygen. Furthermore it is difficult to dissolve in water, though more soluble than oxygen, and it rapidly decomposes back, in a matter of minutes, to unreactive oxygen again.

An example of where ozone is used is in phenol and cyanide oxidation. Ozone neatly converts many toxic and persistent phenol products, such as occur in phenolic glue manufacture and use, into harmless and readily biodegradable smaller molecules. Similarly, cyanides are converted to much less toxic cyanates. Ozone system suppliers now claim reductions of serious phenol pollution down to ppb levels with ozone injection.

Paper-making and wastewater. Another area in which ozone and oxygen may make inroads in the future is in the paper-making industry. The manufacture of paper is associated with the consumption and potential pollution of very large quantities of water. With all paper-making processes there would appear to be good opportunities for much greater use of industrial gases. Large amounts of water end up containing suspended organic matter and the waste stream has a very high BOD. This effluent can be treated by conventional aerobic biological plant or with pure oxygen biological plant

The basic paper-making process is the dissolution of mashed heartwood from trees (the bark is discarded). Many variations on the process exist but the following is representative. In Kraft pulping, wood chips are 'cooked' with aqueous sodium hydroxide/sodium sulphate under pressure at $\sim 150°C$ for a few hours. The cooking process destroys the lignin, the resin that binds the cellulose fibres of the wood together. The released fibres are more or less pure cellulose fibres but they are mixed with chemical by-products: the alkali that was added, sulphide formed from the sulphate, and evil-smelling mercaptans and complex alcohols with a dark colour known as 'black liquor'. The processed pulp is washed and the black liquor is recycled by evaporation followed by separation into cooking chemicals, waste lime sludge and 'tall oil', the latter containing resins and fatty acids from the wood, including turpentine. A more recent variation on Kraft pulping uses a sulphite mix, blowing SO_2 through the mix, after adding magnesium hydroxide and/or lime.

After the initial chemical cooking process has extracted cellulose from the wood-pulp, purified cellulose is usually bleached white with an oxidising agent, commonly chlorine dioxide. In many cases, chlorine-based industrial treatment processes are being frowned upon today because minute amounts of potentially harmful chlorinated organic compounds can be formed. Dioxins, for example, have been found (probably in harmless trace amounts, it must be admitted) in the effluent from paper mills and in the paper itself.

Oxygen can sometimes be used and the bleaching of lignin by oxygen/caustic soda is a now a proven process, replacing either hypochlorite or chlorine dioxide usage. Where oxygen is an insufficiently powerful oxidant, ozone can often be effective. Ozone is an effective bleaching agent and can be used both in paper fibre and textile fibre bleaching, replacing chlorine or hypochlorite. Recently, babies' disposable nappies came under the spotlight because of fears of organochlorine contamination due to chlorine-based bleaching. Ozone or oxygen would guarantee that organochlorine compounds would not be present even at exceedingly low levels but would be entirely absent.

3.7 Industrial cryogenics

3.7.1 Properties of liquefied gases

The properties of common cryogens are important parameters for the design of cryogenic apparatus. There is in fact only a rather small set of acceptable cryogens. Even in the short list given in Table 3.9, hydrogen and methane are flammable, neon is rather expensive and CO_2 is barely cryogenic (195 K is, after all, only −78°C).

There are a number of things which Table 3.9 emphasises. One is the wide variation in latent heat of the materials, although the exceptionally low density of helium and hydrogen mean that their latent heats are actually lower per unit volume. Even after the addition of more exotic molecules to this short table, it is noticeable that there are a number of 'gaps' in the spectrum of boiling points. There are in fact numerous slightly less ordinary gases which can be used to fill these gaps from 77 K upwards. However, if a controlled temperature bath is needed at 12 or 50 K, for example, then the convenience of boiling cryogen is not available.

3.7.2 Cryogenic phenomena

Remarkable effects occur at temperatures just a few degrees below those normal on earth. Below 0°C a large number of things happen related to the freezing of water. The vast bulk of university cryogenics research is devoted to the fascinating, but mainly esoteric, pheonomena that occur at temperatures below the 4 K boiling point of liquid helium. However, in industry cryogenics is almost entirely confined to the region above the boiling point of liquid nitrogen, i.e. above 77 K.

A number of important phenomena happen at industrial cryogenic temperatures:

Table 3.9 Properties of liquefied gases

Gas	Boiling point (K)	Density (Kg m^{-3})	Heat capacity (kJ kg^{-1} K^{-1})	Vaporisation latent heat (kJ kg^{-1})	Liquid viscosity (μN s^{-1} m^{-2})
He	4	125	5.20	20	3.3
H_2	20	71	14.4	446	13.3
Ne	27	1205	1.03	86	124
N_2	77	809	1.04	199	152
O_2	90	1141	0.92	213	195
Ar	87	1393	0.52	161	260
CH_4	112	423	2.23	512	119
CF_4	145	1317	0.71	136	20
CO_2	195	1980	0.84	570 (sublimes)	
(H_2O	373	1000	1.86 (at 298 K)	2257	278)

- many ordinarily flexible or tough materials become brittle,[59] although some non-ferrous metals, such as aluminium, become stronger at cryogenic temperatures;
- most gases, except the permanent[60] gases, either condense or freeze;
- ceramic copper oxide (HT_c) materials become superconductors;
- high rates of cooling from room temperature are possible because of the very low temperature relative to room temperature (−195°C) and high heat capacity of LN;
- most materials shrink quite markedly in linear dimensions relative to room temperature.

In a few research groups, such as that at Southampton University, UK, phenomena at 77 K are studied in detail. Recent work in this area has been directed towards topics such as HT_c superconductor cooling and enhanced boiling surfaces. For example, special precautions are necessary during the cool-down of the fragile ceramic HT_c materials because differential thermal expansion can lead to cracking. Enhanced boiling surfaces, i.e. surfaces capable of even higher rates of cooling than expected hitherto, may be useful in enhancing current cryocooling equipment, as well as for future yet-to-be-determined uses (section 4.3).

[59] One of the less familiar of ice-related phenomena relates to the extraordinary strength achieved by freezing quite dilute suspensions of fibrous materials in water. Although it is more normal to utilise the weakening of materials entailed by LN freezing, it is worth remembering that some materials increase in strength when deep frozen. Some frozen food, for example, seems to be every bit as strong as wood. The British 1940 wartime 'Habbakuk' project relied on 'pykrete', a 3% wood-pulp in ice composite, which was shown to melt more slowly than ice, have many times the strength and to resist explosive detonations to a large extent. The project was intended to produce million tonne ships/icebergs to act as temporary invasion craft and aircraft carriers.

[60] Permanent gases are nitrogen, oxygen, argon, helium, hydrogen and neon.

Boiling. LN will boil vigorously when heated and the boiling process is what allows the spectacular rates of cooling that are the principal reason for using LN. However, boiling is, in detail, a subtle process. For example, the bubbles that occur in heated water are not necessarily due to boiling. Actually, the smallest of the bubbles, when they first appear, are dissolved gas from the water being evolved. It is difficult to see exactly what is happening when a saucepan is really boiling because of the violent disturbance of the surface if the heat is at all strong. For boiling to occur, molecules of a liquid have to retreat away from a cavity that is sufficiently large that surface tension forces on the forming bubble cannot immediately collapse it. The overpressure required to form a bubble is smaller for larger bubbles. As the bubble gets bigger, it becomes more stable. The vapour pressure of liquid inside a bubble depends on the bubble radius, tending to the equilibrium vapour pressure at large radii and tending to zero as the bubble shrinks. Small bubbles are unstable, tending to collapse back into liquid.

Superheating is explained by reference to what happens as the vapour pressure begins to exceed the ambient pressure. Bubbles, once formed, will grow if above a certain critical size. However, if nothing happens to form these microbubbles, then nothing happens at all except the liquid becomes hotter, i.e. the liquid is superheated. The addition of a surface on which the minimum size bubbles can be formed with lower energy than in the free liquid causes 'nucleation', i.e. a stream of bubbles originating at particular points on that surface. Nucleation around the rows of ions left behind by a high-energy ionising particle is the basis for the cryogenic 'bubble chamber'. Generally filled with liquid hydrogen, these were used in many of the 1960s and 1970s fundamental particle physics discoveries on which current quark/gluon theory is based. Particle collisions can be tracked by arranging for flash photographs to be taken of bubbles in the superheated liquid hydrogen just after a collision occurs inside it but before general boiling occurs (Figure 1.6).

Boiling proceeds in a number of different regimes depending on the heat flux applied to a liquid: convection, nucleate boiling and film boiling. At low heat flux inputs 'boiling' is indistinguishable from normal gentle heating. Heat is applied at the bottom of a bath, convection occurs, carrying the heated liquid from the bottom to the top surface, where evaporation takes place enhanced by the increased temperature.

The nucleate boiling regime exists because of the very high overpressure required to cause a very small bubble of vapour to exist against surface tension. If a nucleation surface exists, this lowers the overpressure required so boiling occurs selectively on surfaces, around points on surfaces or on suspended particles. To see precisely why this is so, consider the equilibrium of forces on a bubble cut in half once it has formed

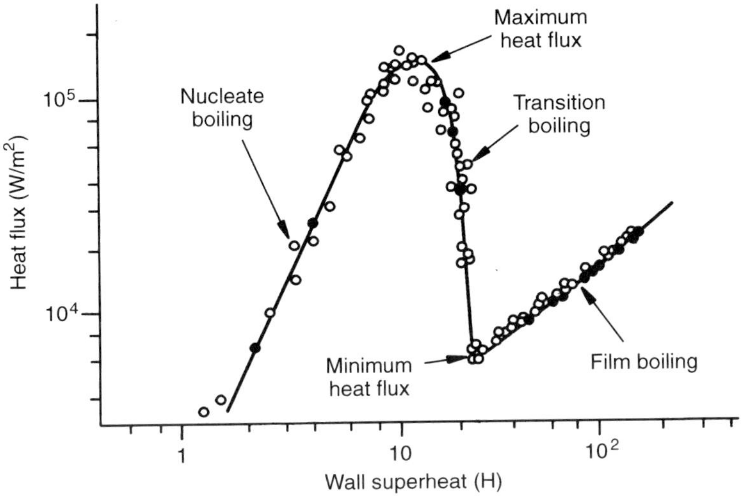

Figure 3.32 Typical experimental pool boiling curve for nitrogen (Hands, 1986).

Force on half a spherical bubble due to excess vapour pressure inside = $P\pi R^2$

Force on half bubble due to surface tension, S, per unit length = $S2\pi R$

where R is the diameter of the bubble. At equilibrium the excess pressure required to form a bubble is simply $\Delta P = 2S/R$ so for very tiny bubbles a large overpressure is needed, implying a necessity for strong superheating. If an overpressure of 0.1 atm is assumed to be needed, then for LN, despite its low surface tension (6 mN m^{-1}) the minimum bubble size is already 0.6 mm. However, if a surface can prefabricate or 'nucleate' a microbubble of smaller minimum radius, then boiling can occur closer to the theoretical boiling point.

Enhanced boiling surfaces seek to achieve exactly this, in the most efficient way possible, by applying (typically) a porous surface, which is easily wetted, will not easily show the Leidenfrost phenomenon and provides the maximum nucleation effect. Enhanced boiling surfaces have applications in improving the performance of heat exchangers used with LN in cryogenic cooling, perhaps in improving the performance of plates and packing in distillation columns, and perhaps in new devices such as HT_c superconductor fault current limiters, which seek to control cryogenic temperatures very accurately by boiling liquids (section 4.3). One way of making enhanced boiling surfaces is to create a spongy metal surface on a metal.

At very high heat fluxes the heated surface no longer contacts the liquid directly at all. A thin film of vapour forms which isolates the surface and partially insulates it. This is, in effect, the Leidenfrost phenomenon. The

Table 3.10 Boiling point of nitrogen at elevated pressures

Pressure (bara)	Boiling point (K)
1	77
2	84
3	88
4	91
5	94
7	99
10	103.8

transition to 'film boiling', as it is known, is marked by a sharp decrease in heat transfer (Figure 3.32).

Vapour pressure and boiling point. As with all liquids, cryogen vapour pressures increase with temperature approximately exponentially (section 1.3.14). This means that their boiling points depend on pressure. For example, for liquid nitrogen boiling point varies with pressure as shown in Table 3.10.

It is possible to control chemical reaction temperature to a fraction of a degree using pressure control of a boiling liquid at elevated temperature (for instance, in organic chemical reactions). Subatmospheric pressures can also be used to control boiling point. By vacuum pumping, LN can reach 65 K, instead of 77 K, whilst LHe can reach 1 K, rather than 4.2 K. Controlling pressure above a cryogen bath via regulators and pumps, whilst maintaining boiling by means of enhanced boiling surfaces, is a possible means of accurate temperature control.

Flash loss. The change of boiling point with pressure is also important because of the loss of LN that occurs when LN from pressurised tanks is decanted into open vessels. The resultant losses are known as 'flash loss' because violent flash boiling occurs when the liquid is first decanted into another vessel at atmospheric pressure.

A liquid cryogen is generally stored at a high pressure; liquid nitrogen is frequently stored at 15 barg or more, for example. This may well be convenient for supplying gaseous nitrogen to a process at that kind of pressure. Storage vessels are almost always well above atmospheric pressure because handling problems are eased in any case; the gas pressure in the ullage space can be used to drive liquid transfer rather than having to rely on cryogenic pumps.

However, when a pressurised tank is needed to supply liquid nitrogen at atmospheric pressure, there will be a considerable amount of boiling when pressure is released from the liquid stream, and a sizeable fraction of the nitrogen extracted will be lost, until the remaining liquid has cooled to the boiling point at 1 bar pressure.

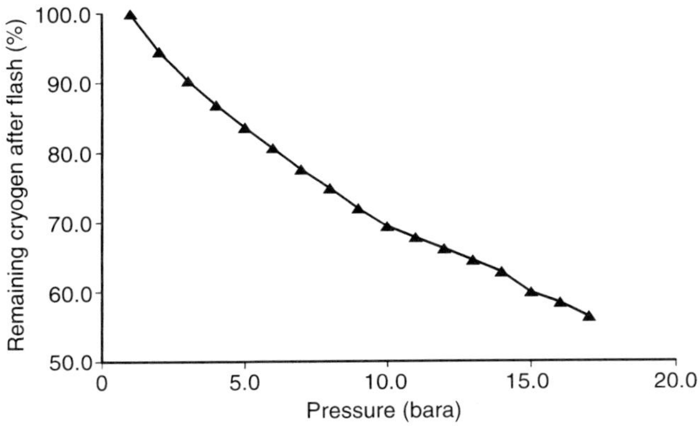

Figure 3.33 Flash losses in liquid nitorgen decant from pressurised tanks.

The flash loss can be theoretically calculated and these figures are often seen tabulated (Figure 3.33). However, the theory is only accurate if fairly large amounts of LN are withdrawn, heat-leaks in the LN decant pipework are small, the receiving vessel is pre-cooled to cryogenic temperatures and precautions are taken to ensure that droplets of LN are caught and drained back into the receiving vessel. The conditions under which cryogens are normally withdrawn from highly pressurised vessels rarely meet these criteria. In general, therefore, much larger amounts of cryogen are wasted than Table 3.11 indicates.

A caution should also be sounded about the accuracy of the figures in Table 3.11. Several major data sources consulted for data disagree with one another by 5 or 10%. The data in Figure 3.33 differ slightly from those in Table 3.11, for example. However, given the other uncertainties, greater accuracy is probably unnecessary. The lesson should be that, whatever the losses, they are surprisingly high. 'Decant' systems often try to restrain losses by using insulation and some sort of counter-current system for cooling incoming 'warm' liquid with boiling LN vapour.

Table 3.11 Flash evaporation losses from liquid nitrogen when depressurised

Pressure of LN (bara)	Flash loss (theoretical minimum) (%)
1	0
2	6
4	13
8	25
16	42

Flash losses in CO_2. Carbon dioxide sublimes at 195 K at 1 bara pressure. The cryogenic use of carbon dioxide is to some extent limited by its lack of a liquid form at atmospheric pressure. A considerable pressure on the gas (50–60 bar) is required to retain it in liquid form (section 2.4.7). The property of sublimation is useful, though, in small applications such as keeping small amounts of food cool simply and without mess. No mechanical refrigerating device is yet available which will work without an external power supply in the size range from a fraction of a kilogram to a few tens of kilograms. Alternatives that do not involve sublimation are invariably more complex to apply. Blocks of ice have the snag of needing some way of handling the water produced, whilst cryogenic liquids such as LN cannot simply be installed in a small block in an insulated product package.

Liquid CO_2 has considerably less cooling capacity than solid CO_2. For example, liquid CO_2 stored under its own vapour pressure at room temperature will suffer a flash loss of 70% (theoretical minimum) by mass on release of pressure. Even from a refrigerated storage tank at 10 bar or so, a large flash loss is expected. A relatively large volume of liquid is therefore needed to create a fairly small block of dry ice. This is not immediately apparent as the dry ice snow produced by simple expansion in a ‘snow horn’ has a very low density.

A snow horn is used to produce CO_2 snow *in situ* on a material or device to be cooled. Many CO_2 applications employ liquid CO_2 because of the ease of handling the liquid but use a snow horn to apply it. The snow horn is an expansion nozzle with a shield to prevent too much splash back from the nozzle. The nozzle itself is arranged so that there are no constrictions that could get blocked with solid CO_2 and is sized to restrict flow to a suitable value. The CO_2 snow produced has a fluffy texture and is of low density, always less than 50% of theoretical density.

Small-scale applications can use the compressed form of solid CO_2 (dry ice) which is much more compact than CO_2 . Solid of near-theoretical density (around 2g cm^{-3}) can be produced by compressing CO_2 snow in a hydraulically-driven piston and cylinder arrangement. The resultant large block of dry ice is then ejected from the cylinder and sawn up into slabs 25 or 50 mm thick, which are then individually wrapped in paper so that they will not weld together in transit. Solid CO_2 in slabs is, as a result, rather expensive but does achieve the maximum possible refrigeration capacity in the minimum space. Less laborious is the production of dry ice in pellets. These are also produced by compression of dry ice snow but the pellets, typically a few millimetres in dimension, are simply ejected directly into a bag inside an insulated cardboard box for delivery. The gas space between the pellets means that this form is not as space-efficient as solid slabs of CO_2. Small-size pellets are used in dry ice bead blasting (section 3.7.4).

Embrittlement. Many tough plastics and some tough metals become very brittle below room temperatures. Plastics generally have a glass transition temperature below which they are glassy rather than tough or rubbery. The glass transition temperature reflects the fact that atoms in the polymer molecules are no longer free to slide over each other (see Figure 2.67). Ordinary carbon steels with a high degree of toughness at room temperature are severely embrittled at low temperatures, whilst cast irons and similar materials are rendered more brittle by cryogenic immersion.

It is also worth remarking that differential thermal contraction caused by sudden immersion in cryogens (thermal shock) is a frequent cause of materials failing. As noted above, many normally tough materials become less able to withstand this kind of internal stress because of brittleness. At the same time, the thermal conductivity of materials decreases rapidly as temperature falls so that relief of stress by conduction is slow.

The modification of many plastic materials by the addition of large amounts of fillers, such as sand or glass fibre, is an important technique for rendering plastics more useful at cryogenic temperatures. These fillers often increase the thermal conductivity and always decrease the thermal expansivity of the plastic, decreasing the effects of thermal shock. Typical values of 2% contraction from room temperature to 77 K can be reduced by a factor of 10 or so by adding large percentages of filler.

Turbulent flow. As noted in section 1.3.7, many cryogens have a rather low viscosity. LN is one seventh of the viscosity of room temperature water, for example, whilst liquid helium is 300 times less viscous than water, although in this case the low density of LHe means that the viscous forces still have some importance. These extremely low viscosities mean that cryogens only behave as viscous liquids at low speeds and it is very common to have turbulent flow of cryogens in pipes and vessels, which has a number of consequences. Not only does this lead to a requirement for larger than expected pumps and pipe sizes but it may mean that some allowance will be needed for cryogen losses due to the additional heat input caused by the turbulence.

Vacuum-insulated pipework and phase separators. In order to deliver liquid nitrogen or other cryogens to the point of use without unnecessary heat loss, it is fairly obvious that vacuum-insulated pipework is advantageous. It is perhaps less obvious that vacuum-insulated pipework may actually be essential. If an uninsulated pipe is used for a steady flow, the amount of boiling along its length may well be such as to cause very unsteady turbulent flow with plug and 'churning' flow. Such an unsteady two-phase flow is difficult to measure and control effectively. Furthermore, if an uninsulated pipe is used for unsteady demand, then control problems may be insurmountable. Every time the pipe flow is stopped, it will boil dry and begin to warm up. After a few minutes, if flow is demanded, liquid will flow into a warm pipe and instantly vaporise so huge amounts of gas will pass down the line. The blast of gas cools steadily until droplets of liquid cryogen appear in it. It will often be several minutes before liquid cryogen flows out of the end. These great blasts of gas are often totally unacceptable in operation. In an operation such as can rigidifying, for example, a precision-measured few drops of liquid cryogen must be guaranteed every time the dispense valve opens and therefore any gas flow is unacceptable.

Vacuum-insulated pipework is made in a similar way to vacuum-insulated vessels. It has inner and outer pipes of (typically) stainless steel, the void being evacuated, sometimes with a powder or metal-foil fill. Long sections are welded up and then evacuated, either in the supplier's facilities or *in situ*. At the end of a run of any length a bellows section is normally included on the inner pipe, allowing the inner pipe to contract relative to the outer pipe when the cryogen is admitted. Sometimes as with cryogenic tanks, points are left where a vacuum pump can be reattached to restore vacuum quality.

Vacuum-insulated pipework is more difficult to install than ordinary pipework. Tees and elbows are often executed in ordinary pipework with plastic foam insulation, the vacuum insulation resuming wherever there is a straight run. Fairly gentle bends can be fabricated by filling both pipes with water and freezing, prior to bending, or by using double pipe with a polymer string spiral wrapping around the inner pipe as a quasi-continuous space. Demountable joints are possible in vacuum-insulated pipe. A demountable joint in an ordinary pipe costs only a few dollars. In vacuum-insulated pipes, however, these are complex assemblies costing many hundreds of dollars. Installations are only effectively carried out by a few specialist firms and major gas suppliers.

Even with vacuum-insulated pipework there is still typically a certain amount of two-phase flow, not pure liquid flow, at the point of use. Many installations will benefit from a phase separator near the point of use. This is often no more than a small insulated tank with liquid inlet and outlet and a vapour vent, although designs with plates and baffles can behave better in bumpy flow.

3.7.3 Cryogenic practice

Environmental cabinets. It is frequently the case in manufacturing that a system or component must be cooled down or heated up and tested to check that it will function when the ambient temperature is low. An insulated cabinet, fitted with a circulating fan and powerful heater, serves for the higher temperatures but lower temperatures are less easily achieved with a simple refrigerator set-up. Either the refrigerator will take a long time to cool or it will be difficult to regulate the temperature precisely. An LN-fed cabinet can be cooled very swiftly indeed and controlled accurately. In the electronics industry it is often a requirement for military circuits to be tested from −55°C to 125°C and it is important to be able to carry out this testing swiftly. LN is often used in small environmental cabinets for this purpose. Thermal cycling tests, with hundreds or thousands or more of cycles from cold to hot temperatures, are often specified during the development of electronics. A small cabinet with LN cooling and powerful electric heating is typically used. For small cabinets liquid CO_2 is an alternative that is most suited to cabinets which are used only intermittently.

LN also has the advantage of a low capital cost for large cabinets, which is often appropriate as large-scale cold test facilities are only occasionally required in most companies. The largest LN cabinets are big enough to contain a medium-sized aircraft. Testing of aircraft in different ambients is very important as they encounter very low temperatures in the upper atmosphere, even above the equator. Typically, aircraft testing has a highly intermittent use pattern which makes an LN system very appropriate.

Related to LN environmental cabinets for aircraft testing are the small supersonic wind-tunnel test and experimental facilities run by universities and institutes. An inexpensive way to construct a supersonic tunnel is to use a large capacity vaporiser and a high-pressure pump, if necessary, to achieve the bars or tens of bars pressure required for supersonic flow. Wind tunnels a few centimetres across are possible using LN so that even the smallest research laboratories and universities can now afford supersonic facilities (see section 4.3.2).

LN vacuum cold traps. The use of a cryogen to so cool a gas that it condenses and creates a vacuum is a very obvious idea that has been used, at least in research, ever since cryogens became available. The cold trap is the simplest embodiment of the principle.

Cold traps are still a very low-cost means of pumping away unwanted vapours out of a high-vacuum chamber. Installed before a high-vacuum pump, the trap consists of a liquid nitrogen reservoir thermally linked to a set of metal baffles. It prevents backstreaming of oil vapour from the pump to the chamber and condenses vapours from the chamber onto its baffles. In this way, the pump needs to extract only 'permanent' gases and can be much smaller than would otherwise be the case in a vacuum system. Periodically, when high vacuum can be interrupted, the trap is warmed to room temperature, the absorbed vapours evaporate and are pumped away easily at relatively high pressures. Depending on the pumping load, it is possible to accumulate a dangerous mix of gases, which could explode or react dangerously when warmed up for regeneration. This would happen, for example, if the pumping load were air, followed by slow outgassing of hydrocarbons. Regeneration cycles, therefore, should be carried out as frequently as is reasonably convenient and precautions, such as a flow of nitrogen purge gas, taken if necessary.

Cryopumps. Proper cryopumps are more sophisticated in two respects compared to simple cold traps. First, they employ not just metal baffles but also high surface area absorbers. Second, they normally use lower temperatures than the 77 K of liquid nitrogen, either by using liquid helium or by means of a Stirling cycle refrigerator.

Some ingenuity is employed in the design of cryopumps to ensure that the heat load on the refrigerator is almost entirely that due to the condensation of gases. To achieve this, radiation shields (optical baffles) are placed between the cold surface and the (warm) volume being pumped. These radiation shields are cooled to intermediate temperatures by LN or by an additional Stirling stage (see Figure 3.34). In many pumping jobs the gas load is dominated by water vapour. This is condensed on the first intermediate temperature baffles, reducing the heat load on the cold panels and thus increasing the thermodynamic efficiency of the system. The absorbers are generally coated onto the cold absorber panels and are usually made from some kind of microporous carbon. Further advances are now being made in the thermodynamic efficiency of cryopumps, with many now using the Gifford–McMahon cycle for the helium refrigerator.

The cryopump is capable of pumping millions of litres per second and as such is probably the most powerful high-vacuum pump available, provided the connecting pipe bores and paths are such as to allow it to display its potential. Large vacuum chambers, such as outer space simulators for spacecraft testing, for example, use cryopumps. The power of the cryopump concept is illustrated by the looking at the vapour pressure of gases. At liquid helium temperatures only helium and hydrogen have any vapour pressure that is measurable and LH has only 10^{-7} mbars.

Stirling cycle cryorefrigerators. The liquid helium temperatures needed for cryopumps are easily achieved with merchant-delivered liquid helium and this is ideal for intermittently used apparatus. However, for more or less continuous vacuum pumping a Stirling cycle engine using helium as the working fluid is the appropriate solution (Figure 3.35). These Stirling engines are also used as a more or less consumable-free variable temperature cryostat at all temperatures below 77 K, down to 20 K at reasonable heat loads (a few watts or so) and to a few degrees above helium boiling point (4.2 K). To achieve these low temperatures 2 or 3 stages are needed, with the colder stage rejecting heat to the hotter stage. Although their efficiency is generally low and they are mechanical devices, Stirling cycle cryorefrigerators can be made quite reliable and they are very popular in scientific research where heat loads are very small.

A Stirling engine works by compressing the working gas isothermally, rejecting heat to the hot side, then forcing this gas at constant volume through a regenerator, a simple macroporous metal mass, often fine wire windings or similar, which cools the gas. The cool compressed gas is then allowed to absorb heat from the cold side and is thereby expanded

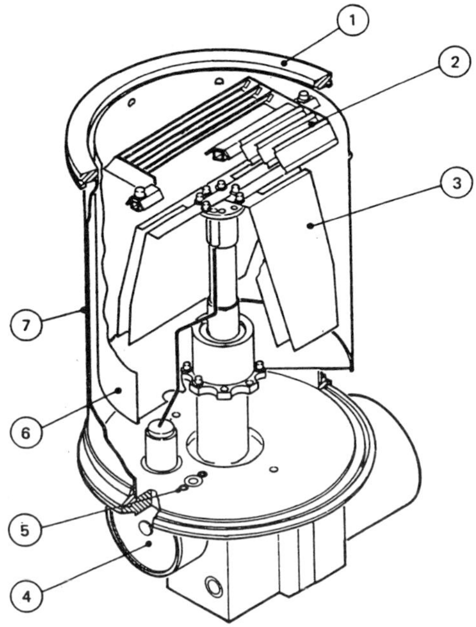

(a)

(b)

Figure 3.34 (a) Cut-away view of a cryopump: 1, inlet flange; 2, inlet louvre (80 K); 3, second stage cryopanel leaflets (15 K); 4, hydrogen vapour pressure thermometer; 5, purge accessory port; 6, first stage cryopanel (80 K); 7, vacuum shroud (courtesy of Edwards High Vacuum). (b) Two cryopumps (left, with helium compressors bottom left) evacuate a large process chamber. Note the gate valves for isolation above upper pump and below lower pump (courtesy of Oxford Plasma Technology).

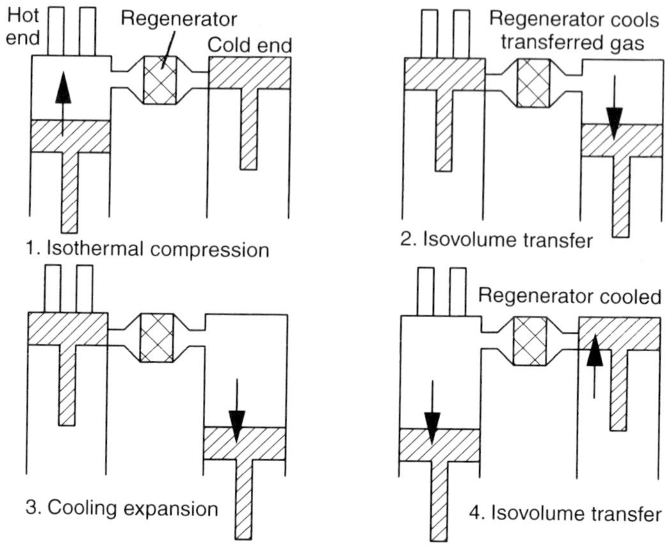

Figure 3.35 Stirling cycle principle.

isothermally. The expanded gas is then forced at constant volume back through the regenerator, cooling this back down and restoring the initial starting position ready for another cycle. A simple implementation of the principle can involve the use of a moving piston and a moving regenerator operated by the same drive shaft and a very simple mechanical arrangement.

Pipe-freezing. This is an important maintenance procedure useful for many pipework systems, from the tiniest 6 mm line to oil lines 1 m or more in diameter. Pipe-freezing enables modifications to be made to pipelines without shutdown or draining off the system. First, the system must be full of process liquid or water. The liquid flow is then stopped and liquid nitrogen poured into a box surrounding the pipeline. Within a few minutes or hours (depending on the pipeline size), a plug of frozen liquid forms and is typically sufficient to resist a minimum of a few bars overpressure. On long freezing times the freezing box must be topped up with LN from time to time. Once frozen, the LN in the box will cease to boil so vigorously, the pipeline can be broken into and a small amount of process liquid drained off. Necessary operations, replacing valves, welding on new pipeline connections and so forth, can then be carried out. The LN box is then drained and removed. The frozen slug melts, freeing flow and the system can once again be operated as normal.

There are a few potential pitfalls for the unwary, however. Although a 6 mm line containing water can be frozen and blocked in under a minute, a large 300 mm line may take some hours because there is roughly a square law relation between freeze time and pipe diameter, as expected from theoretical calculations from simple assumptions. This can be seen by assuming the growth of a cylindrical plug of length L in a radial direction, with outer pipe wall at T_o (radius R_o) and inner bore at 273 K (radius R_i), and integrating to find the heat flux Q' through a thick-walled pipe. Further integration gives the time for the ice to fill the pipe bore. The first integration yields

$$Q' = 2\pi LK(T_o - T_i)/\ln(R_o/R_i)$$

where K is the thermal conductivity of ice. On integration to measure how long (time, t) it takes an ice plug to form in a water pipe at 0°C, the following result is obtained

$$t \simeq R_o^2 / (4Ks\rho(T_i - T_o))$$

where ρ is the ice density and s is its specific heat of fusion. The cold temperature T_o might be taken to be 77 K, and T_i to be 273 K for an illustrative calculation, giving 2 h for a 60 mm inner diameter pipe. This formula, and ones like it, tend to underestimate freezing times. More accurate calculations take account of residual flow, cnd effects (only in the middle of a long bath would the above strictly apply), pipe wall conductivity and convection effects, all of which make freezing somewhat slower. Prudence dictates that a further safety margin of a factor of 10 or more might be wise where possible or a trial conducted first.

Tables giving freeze time are available in a limited number of cases. The line pressure that the frozen plug can withstand increases as the plug length increases. Rules of thumb are used in which nominal pressures of 8 bar (in oil, which is more slippery and does not expand on freezing) or 16 bar (in water) can be withstood by a plug only two diameters long. At higher pressures, it would be prudent to allow longer plugs on a pro rata basis. Freezing data assumes absolutely zero flow but often flow will be present; perhaps because a leaking valve is the reason for the need for the pipe-freezing for repair.

Table 3.12 Time taken by liquid nitrogen to freeze water in pipes

Pipe diameter (mm)	Time to freeze (min)
12	0.7
25	2.5
50	7.5
75	15
100	25
150	45
200	70
300	225

Table 3.12 provides a guide to the time required for freezing water in pipes.

It is possible to damage a pipeline by freezing. Steel pipelines are embrittled and can be shattered with a careless blow with a hammer or excessive torque from a spanner. Some plastic pipelines can be frozen but some will shatter. Ice expansion is not normally a problem, provided the ice plug has room to expand sideways. Forming an iceplug too close (less than 10 diameters) to a closed valve may cause a pipe to split as the ice is then confined sideways. Finally, once the LN box is in place to freeze the pipe, access to the trench in which the pipe is situated should be restricted, as the cold evaporated nitrogen gas will lie in the trench and could cause a fatal accident by asphyxia. Plenty of time should be allowed for the gas to clear before carrying out work on calm days or when working indoors.

The group at the University of Southampton, UK, amongst others, has carried out both theoretical modelling and practical experiments on pipe-freezing with a variety of pipeline substances and pipe sizes, including situations where there is some residual flow.

Ground-freezing. Occasionally, it is required to tunnel in soil or rock of very low strength. Alternatively, a deep hole or trench is needed and there is not room to excavate a large hole with gently sloping sides. There are a number of alternatives in this situation. However, one straightforward alternative is to insert small bore pipes into the soil to be tunnelled or dug and inject liquid nitrogen for hours or days. Eventually, the water content of the ground is fully frozen and binds together the loose soil and rock so that fairly conventional tunnelling, with little danger of roof or wall collapse, can be carried out. The process is quicker than using brine, although typically more explosive.

Cryogenic shrink-fitting. Many engineering assemblies rely on an 'interference fit'. Briefly, this means that the surfaces of two components nominally overlap by a few microns when those components are assembled, i.e. an axle of 10 mm (10 000 μm) diameter will have an interference fit of 10 μm with a wheel with a hole of diameter 9990 μm that is 10 mm thick. An interference fit is often achieved by force-fitting, i.e. using a hydraulic press of many tonnes to force the components together. However, force-fitting is more likely to result in scrap. The standard hot process for shrink-fitting is to heat the housing, whilst keeping the insert cool. Once the housing is hot, the insert is pushed into place and the whole assembly is allowed to cool until the insert is held in place by the shrinking of the housing around it.

Cryogenic shrink-fitting is a gentler alternative to the standard process and one which is particularly economical for inserting a small component into a larger housing. Rather than warming up the large housing, the small component is immersed in LN and allowed to shrink. The size of the smaller component is calculated so that when it expands to normal size it will fit immoveably in the larger housing.

An example is easily calculated to show the efficacy of the method. If a 10 mm shaft made of brass is cooled to LN temperatures, it will shrink 34 μm at LN temperatures. It will thus be 9966 μm when cold, leaving 24 μm to allow reasonably easy assembly of the shaft into the wheel by hand. To know how immoveable the shaft will be, it is necessary to know the moduli of elasticity of the components, as well as their geometry, and the interfacial coefficient of friction.

Taking this example further, suppose that the wheel is much more rigid than the brass and that the coefficient of friction, C_f, is 0.2, then, given a brass modulus of E and ignoring Poisson ratio effects for simplicity, the shaft will resist a 600 N or 60 kg axial force before it can be moved within the wheel.

A detailed calculation for the case of a cylindrical shaft radius R_i being fitted with a ring of length L of outer radius R_o gives the following result for the required interference distance I (in metres) to give an axial grip force of F, allowing for Poisson ρ_o and ρ_i for the wheel and shaft materials

$$I\pi C_f L/F = (1-\rho_i)\,/\,E_i + ((R_o^2 + R_i^2)/(R_o^2 - R_i^2) + \rho_o)\,/\,E_o$$

Detailed calculations of interference fits are given in standard engineering texts for simple geometry. There are effects due to the Poisson ratio (generally, when a body is squeezed in in one direction, it squeezes out in another direction) and effects due to anisotropy in elastic moduli. The force with which an insert will grip also depends on how hollow the insert is and on the inserted component's thickness and elastic moduli. In some cases, shrink-fitting will result in the material's elastic limit being exceeded, reducing the tightness of fit and further complicating calculations as the component will permanently distort slightly. In difficult shapes, numerical modelling using a computer software package is possible, whilst for small components there is no substitute for experiment. Finally, in production lines, where it is required to fit many sets of components together without individual measurement, allowance must be made for the worst cases of smallest insert and largest housing (which may not grip well enough) and largest insert with smallest housing (which may not be possible or may overstress the components in the finished assembly).

Shrink-fitting is certainly convenient (LN-cooled components can easily be handled with gloves, for example) and very inexpensive (a polystyrene insulated pot or bath is the only item of capital equipment needed). Also, LN rarely leads to any degradation of the surface finish of components. Heating to a few hundred °C is difficult to accomplish while keeping paint finish intact and stronger heating may lead to oxidation staining as well as giving a danger of unintentional heat-treatment effects. However, there are some problems. Embrittlement of steels at low temperature may lead to chipping or failure if a cold component is inadvertently dropped[61] or there may be failure during warm up. Note also that some austenitic steels are in effect heat-treated by LN exposure, which will usually not be desirable.

Cryotumbling and cryoblasting of rubber components. Here the embrittling effect of rubber is employed to good effect. Small moulded components normally have 'flash' and 'sprues': superfluous material that is typically removed by tumbling many components simultaneously in a large drum, perhaps with the addition of ball-bearings, abrasives and water or other liquids. After a few minutes or hours tumbling, the sharp edges, and all the flash and sprues, have been removed and a smooth overall finish is obtained. However, this process is difficult with rubbery materials unless liquid nitrogen is added to the tumbling drum. With LN added, the physical properties of rubber, hardness and brittleness, become similar to those of other hard plastic and metal components, and tumbling of rubber becomes straightforward.

In a similar way, other processes can clean-up sprues and flash from plastic components without the often undesirable rounding of edges that tumbling entails. One such is the shot blasting of trays of LN-cooled components. It should also be added that components from materials other than rubber can usefully be deflashed with LN pretreatment. Small zinc die-castings and small thermoplastic components can benefit from this method.

Concrete cooling. Concrete is made by binding together fine and coarse aggregates (sand and stones, in other words) with Portland cement. Portland cement is a mixture of silicon, magnesium and aluminium oxides that, when water is added, reacts to produce an interlocking mesh of hydrated mineral crystals forming a strong rock-like material. Unfortunately, the reaction is exothermic and will not proceed correctly except near ambient temperatures. In temperate climates, and where the thickness is not too great, ambient cooling is sufficient. The correct choice of cement (exothermicity varies with the cement oxide fractions) enables satisfactory concrete.

In hot countries the setting of concrete is seriously impaired, however, because the achieved strength is lowered as temperatures are raised. Cracking tends to occur in thicker sections because of the differential thermal expansion of the mix during setting and the slow cooling rates of the centre of thick-piece concrete. In addition, water tends to evaporate from the surface of the concrete during setting before it has had time to react and set.

Structures needing high performance from concrete, therefore, are made by cooling the concrete by adding liquid nitrogen to the mixing machine during mixing. The cooled concrete is then laid or fed to the concrete pump for placement. Cooled in this way, the concrete and its added water have sufficient heat capacity to set correctly before the exothermic reactions

[61] The author has found smaller steel parts still remarkably robust after cryocooling.

warm them up too much. Cracking and surface problems are avoided and strength enhanced; by as much as 20% or more in some experiments.

Experiments have also shown that fears that the concrete will be seriously weakened by freezing if there is a slight overdose of liquid nitrogen in the mixer are unfounded. Freezing of concrete during setting generally causes problems because the growth of ice crystals causes voids to be formed. However, the freezing caused by LN injection is both fast and slight, so the concrete setting is unaffected.

Cryogenic solvent/vapour recovery. Vapour recovery is where volatile organic solvents, such as xylene or methyl cellusolve, used in cleaning or in paints and coatings, are recovered for reuse. A vapour stream from dip tanks, spray booths or drying kilns, following a coating or cleaning process, is processed to recover the solvent and reduce environmental discharges. The two drivers for the adoption of recovery systems are the increasing environmental demands put on companies to limit solvent emissions and the cost of solvent, particularly for large or continuously-operating plant.

A common system for recovery is to force air extracted from the vapour source through carbon-based absorbers. The solvent may then be recovered, or at least a sizeable fraction of it, by warming the absorber beds and passing air (backwards) through them and then over a condenser. However, this kind of simple bolt-on system needs stringent safety precautions, is usually not very efficient and may be rather expensive to run. More satisfactory installations can be achieved using a cryogenic coolant system and inert gas.

A simple installation would take the extracted air stream from a plastic tape drying kiln, and cool it, first using water-cooled surfaces, then using a liquid nitrogen spray, typically onto a set of high surface area inert metal-contacting surfaces. The result is a mixture of solvent and water, and a much cleaner air stream to exhaust to atmosphere. The solvent and water are separated, by distillation if necessary, and the solvent reused.

Instead of exhausting to the atmosphere, some coating processes can be performed in a nitrogen atmosphere in a semi-closed cycle. An inert nitrogen stream from the solvent exhaust cooler can often be used to provide point-of-use inerting for a coating or cleaning process, reducing the effects of atmospheric water and oxygen contamination (oxygen is undesirable for some UV-set coatings, for example, and water causes problems with many coatings) and eliminating fire and explosion hazards. Nitrogen is bled from the system at a point, such as just before the point-of-use, where solvent concentration is low and the nitrogen temperature relatively high. Such systems are especially applicable to tape-coating processes in which the process oven can be most effectively sealed.

The main cost savings attached to a solvent recovery system are related to the ability to recycle the solvent. However, where a coating must be rapidly

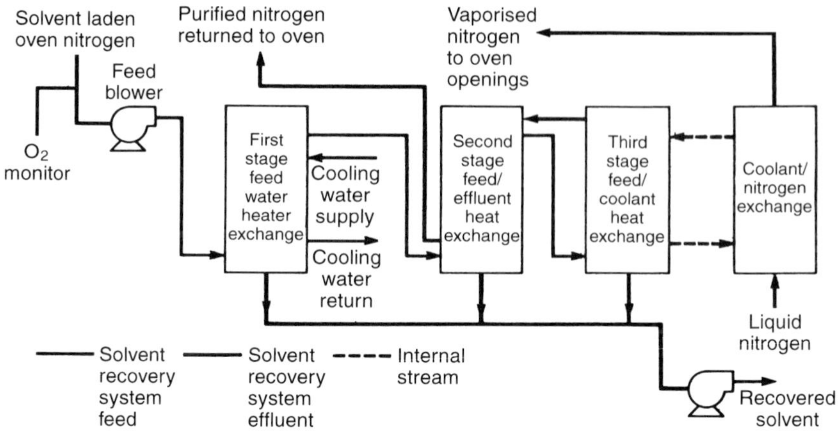

Figure 3.36 Semi-closed cycle vapour recovery using liquid nitrogen injection.

evaporated by heating strongly (known as 'stoving') there is an additional substantial saving to be made when using an inerted system. There is no need to maintain a low concentration of vapour in the coating unit, so there is no need to heat a large amount of air to keep the vapour concentration low. Because there is no danger of ignition in a coater which is inerted, vapour concentration can be much higher. In some cases (e.g. xylene) solvent fraction can be raised from 0.5% or less (to stay below 30% of the lower explosive limit) to as much as 30% by volume. If an oven maintained at, say, 150°C is converted from low vapour concentration in air to high vapour in nitrogen, these savings can outweigh the cost of the liquid nitrogen used and the solvent recovered is pure profit.

A semi-closed cycle system may allow higher temperatures to be used safely, allowing for quicker drying of solvent from the workpiece and greater production rates from a given size of production unit (Figure 3.36). Some semi-closed systems achieve almost 100% solvent recovery as moisture contamination is minimised, reducing undesirable solvent emissions dramatically. The potential for recovery is limited by the cooling achieved versus the oven temperature. This is, in the absence of other effects, the proportion represented by the vapour pressure at the cool temperature over the vapour pressure at the oven temperature is the proportion recoverable, i.e.

$$\text{recovered (\%)}\ R = 1 - VP_{\text{cold}}/VP_{\text{hot}}$$

On a typical exponential vapour pressure curve, $VP = \exp\,(E/kT)$ and the recovery R is a function of the temperature difference $T_{\text{cold}} - T_{\text{hot}}$ and the latent heat of evaporation (to which E is proportional)

$$R = 1 - \exp\left(\frac{E}{k}\left(\frac{T_{\text{cold}} - T_{\text{hot}}}{T_{\text{cold}}\ \ T_{\text{hot}}}\right)\right)$$

With an oven temperature of 100°C and an achieved cold temperature of −100°C, for example, about 99.6% of octane will be recovered. With a vapour content of only 0.4%, the vented gas resulting will be acceptable on both environmental (VOC emissions) and safety grounds (at 0.4% by volume, the mixture will not inflame in air). With achievement of temperatures even approaching LN boiling point, −196°C, even the 0.4% of vapour escaping at −100°C will be eliminated, giving an even more desirable result.

A similar system for vapour recovery can also be installed as part of a tank ullage inerting system in the large-scale storage of volatile materials such as petrol (section 3.3.7). A system as shown in Figure 3.36 can be used on the vent line from a tank ullage space. With petrol vapour at summer temperatures constituting 5 or 10% of the gas in the nitrogen inert blanket, the vent line recovery can be highly economic, with LN costs exceeded several times by the value of the recovered volatiles. Vapour recovery systems are now common in chemical tank farms both for recovery from tank to tank pumping and also where tanks are off-loaded to or loaded from haulage vehicles such as road or rail trucks, or transport containers.

With the intermittent nature of tank venting, it is rarely economic to install a mechanical refrigeration plant of a large enough cooling capacity and low enough temperature to carry out vapour recovery in tanks. A system of low capital cost with a small LN tank but a high peak capacity at low temperature is a much better match to tank venting vapour recovery. The inexpensive carbon-bed-based systems are, however, competitive with LN in many cases.

The nitrogen evolved in the recovery system is not directly usable in the inerting of the tank for which the recovery system is operating since the recovery process is needed when the tank is venting, not when it is intaking. With a little ingenuity, however, where there are multiple tanks of similar products, some of the evaporated LN can be used as (slightly impure) nitrogen inert blanketing for other tanks. In this case, however, check valves, flame arrestors and/or slam-shut valves triggered in the event of a flame may be necessary to prevent a fire in a tank ullage space spreading via the pipework system.

Cryomilling. As noted in section 3.4.1, spices and other foodstuffs tend to heat up and clog the machinery in a high-speed bulk grinding operation. The heat also drives off wanted volatile oils in spices and leads to other degradation. The addition of LN to the grinding jaws removes unwanted heat, adds an inert atmosphere and embrittles the product, leading to better quality and higher grinding speeds.

The grinding of plastics and rubber is also now commonly done with liquid nitrogen as these products are exceptionally difficult to shred or grind at ambient temperatures, particularly where smaller particle sizes are needed.

Even where the particle sizes needed are large, as in shredding, there may be big advantages in using LN grinding. Old tyres, for example, even those with steel reinforcement, can be ground up. Their steel reinforcement causes big problems for conventional shredding machinery but the cryogenic process allows the steel reinforcement to be easily separated from the rubber itself.

In grinding tasks such as tyre shredding, there is an additional effect that comes into play with the use of cryogenic embrittlement: differential embrittlement. Although two materials may both be embrittled at LN temperatures, one material will be affected more than the other. On crushing or hammering, one material is broken easily into small crumbs, whilst the other only breaks to a limited extent. This can be a big help with separation and hence with recycling.[62] The effects of differential embrittlement are used in commercial practice, for example in removing plastic insulation from cables whose copper or aluminium scrap content is to be reused.

Although energy is needed to create the LN, the energy needed in the shredding machinery may be high compared to that required for crumbing in a cryogenic process. Overall, therefore, cryogrinding and crumbing can be more energy efficient. The availability of LN grinding has even led to new products. In the surfacing on children's playgrounds, for example, small chips of rubber are now common. They provide a desirable degree of resilience beneath play apparatus where a fall on to concrete for a child could be serious. Made by crumbing with LN assist in many cases, the chips can be left loose or bonded with a small amount of adhesive to provide a resilient equivalent of tar macadam road surfacing. A similar process adds crumbed rubber to stone chippings in road surfacing, changing the road's grip characteristics.

Plastics are now commonly crumbed cryogenically to provide a convenient, free-flowing feedstock for moulding and other plastic processing operations. Plastics crumbed with ambient shredders often have crumbs with a tail which cause the aggregate plastic to clog and clump. Plastic with pigments, as used in, for example, xerographic toner, can conveniently be prepared by cryogrinding of bulk material, with powdered plastic of almost arbitrarily small particle size down to a few microns being available. This is impossible by other methods. Although cryogrinding on a small scale often employs the simplest equipment, simply adding a fine spray of LN to a more or less normal grinder, other systems employ a counter-current screw feeder to maximise the efficient use of the cold potential in the LN and achieve a good level of inerting.

[62] The plastic goes below its glass transition temperature and embrittles, whilst the copper or aluminium remains ductile. Zuwala *et al.* (1994) describe the ribbed rollers needed to cause crumbing of the plastic while preserving metal wires.

Throughput of a product in grinder is often greatly increased by the addition of LN cooling and, as noted in section 3.4.1 above, problems with overheating (partial melting of plastic chips, for example), oxidation damage and explosion/fire hazards can be avoided. Grinders of types which are normally useless on soft material, but are desirable in that they have low energy consumption (such as centrifugal hammer-mills or ball-mills, for example), are often usable with materials embrittled by LN.

Freeze-drying. Freeze-drying or 'lyophilisation' is a technique for the preservation of organic compounds susceptible to degradation if simply left in the ambient atmosphere. Lyophilisation is mainly applied to pharmaceuticals and other high-value products, including some high-value foods, although food materials such as coffee and soup are also freeze-dried in some quantities but using a simpler and less expensive process. Whereas normally aqueous or hydrated products are dried by heating gently and blowing air copiously through or across the medium, freeze-drying removes water by first freezing and then removing the ice formed by sublimation. Roughly speaking, whereas normal evaporation results in a product shrinking by an amount equal to the water loss and maintaining its density, freeze-dried products have about the same volume as the original product since the ice crystals keep the structure of the product intact, leaving interstices as the ice sublimes. The interstices allow the rapid rehydration of the product when it is ready to use and this rehydration largely reconstitutes the original structure. Evaporated products often rehydrate poorly since the water cannot penetrate the shrunken, shrivelled layers of solid product and the heating during evaporation often alters the chemical constitution of the product. Freeze-dried products made by freeze-drying of soluble compounds form granules composed of loosely bound microcrystals that are more readily dissolved than normally evaporated crystals, which are much larger and cemented together.

Rapid initial freezing of the product can be helpful as the freezing process itself can cause destruction of some products, for example plant cell walls, because of the growth of large pointed ice crystals. On the other hand, a slower freezing with larger crystals leads to a speedier drying process since there will be a more open structure in the product with better vapour paths. Most lyophilisation systems use mechanical refrigeration at approximately −50°C, often by simply placing the products on cooled trays in the vacuum chamber before applying vacuum. However, with thin products where vapour transit is not an issue, or where an extended drying cycle can be allowed, liquid nitrogen freezing is advantageous. The vacuum pressure applied is 1 mbar or less, typically for a period of several hours or more whilst no further cooling is applied but heat is allowed to enter the vacuum chamber so that the temperature does not fall too much.

After initial removal of water, more water can be driven off by gently warming the product back towards room temperature over several hours or more whilst maintaining the vacuum. Once the product has warmed up back to room temperature it will be found to comprise a porous, rather friable solid, typically hygroscopic but containing just a small percentage of water. At this level of water most materials will, if kept in sealed containers under dry nitrogen or air, be stable against chemical degradation and, more especially, be kept free from fungal, bacterial or other spoiling action. Large batch lyophilisers will accept up to a tonne of wet hydrated product for processing, in thicknesses up to approximately 1 cm.

Plastic-moulding cooling. Here liquid nitrogen is employed to cool a freshly moulded thermoplastic item so that it can be removed from the injection moulding machine more quickly. In this way, the productivity of these machines can be enhanced, sometimes by as much as 100%.

Both blow moulding and injection moulding can be speeded up. In a typical blow-moulding application, a gob of plastic is inflated by air pressure until it fills up the mould, which might be as as small as a 10 g plastic bottle or as big as a 100 l fuel tank. As soon as the 'parison', the inflated gob, has contacted the walls of the mould, its outer surface begins to cool on the water-cooled mould walls. At this point a brief but precise stream of LN is sprayed inside, cooling the inside of the parison. With cooling mostly complete, the parison is solid enough for it to be ejected from the opened mould much more quickly than would otherwise be the case.

Early plastic-moulding cooling with LN was somewhat hit and miss. Rejects occurred frequently, particularly during start-up of production, because LN in the cryogenic feed line filled with vapour and insufficient LN was admitted when the cryogen valve opened. Ways around this include careful designs of valves, use of a cool-down sequence, so that the components reach a steady-state temperature, and the installation of a phase-separator immediately above the moulding machine. AIRCO in the USA is trying the use of a 'dipping pipe' in place of a valve. The dipping pipe lies normally in the vapour space of the phase separator tank. When LN is demanded, the short dipping pipe is immersed in the LN and pressure feed forces a known volume of gas followed by cryogen into the mould. At the end of cooling, the dipping pipe is raised to drain it of LN before shutting off the cryogen (Figure 3.37).

Extrusion cooling. Just as moulding can be speeded up by cryogenic cooling, extrusion processes for plastics (and some metals such as aluminium) can be speeded up, and often quality improved, by the use of LN cooling. Particularly with higher temperature extrusions, the inertness of the nitrogen atmosphere can improve surface quality. The extrusion process is speeded up because a higher temperature can be used in the feed material,

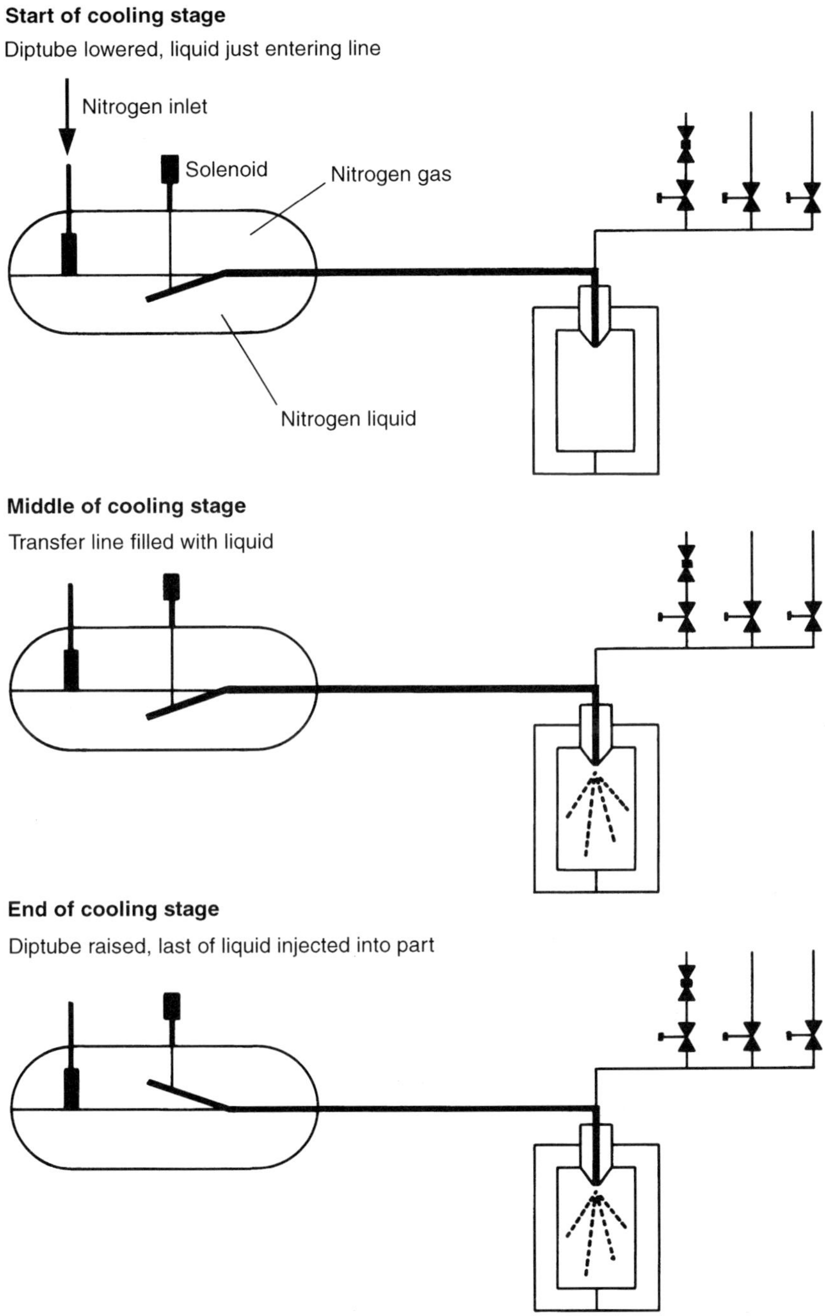

Figure 3.37 Dipping-pipe technique for accurate dosing of liquid nitrogen (courtesy of AIRCO).

giving a lower viscosity and hence higher throughput for a given die and extrusion pressure. In other installations, production is not limited by the extrusion screw but by the rate at which the product can be cooled sufficiently to be stored or further processed. Again, LN cooling can reduce this bottleneck.

LN cooling can be added to existing equipment for minimal capital outlay. This sort of installation is often encountered as an extra to inexpensively speed up the output of a previously installed piece of capital plant. However, if careful calculations are done, LN often proves to be economic from the outset.

Blown-film extrusion is to some extent a combination of blow moulding and extrusion is commonly used to make thin polyethylene film (Figure 3.38). In this process, a plastic tube is extruded and, while still hot, inflated with controlled internal air pressure to twice or more its starting diameter. The inflation pressure is very small, centimetres of water gauge or less, and must be carefully controlled. It should be varied with the extrusion speed and take-up roll speed to provide the correct amount of inflation and thus a film of the desired thickness. The film cools in the air sufficiently that it may then be led to a V-shaped funnel with a pair of take-up rollers pulling the film through the slot in the V-funnel. The result is a roll of film in the form of a wide, thin-walled tube, neatly folded at each edge. When slit up one side, the film roll is thus in length only half the width of the opened out film, making it a convenient form for handling and selling. The film extrusion rate can be increased by circulating cool air on the inside of the film (in the normal system, the only cooling is on the outside of the film). Systems using LN to make cold nitrogen to cool a partial closed-circuit air-cooling system on the inside of the film have been shown to boost production rates by approximately 50%.

Cryogenic fog and theatrical effects. There are serious uses for large flows of liquid nitrogen to simulate cold upper atmosphere conditions. The combination of liquid nitrogen and water vapour can be controlled with suitable apparatus to yield thick, cold fogs, which are also useful in testing instruments and equipment for use in foggy conditions.

Liquid nitrogen can also be used for making 'steam jets' or 'fountains' as a stage effect. The pressure of the liquid itself, boiling, can be used to propel it through small jets. The liquid stream issuing from the jets quickly boils into cold gas which condenses moisture in the ambient air, creating a spectacular but reasonably safe fountain. Similarly, by allowing the LN to boil before a larger orifice, a jet of steam effect can be created.

CO_2 foggers are a more standard stage effect. Water and dry ice are mixed, resulting in the copious evolution of cold CO_2 and water droplets. The mixture, even when diluted with much air, is considerably heavier than

Figure 3.38 Blown film plastic extrusion equipment. Molten plastic extruding from a ring (centre) is inflated and rapidly cooled by cold air to form a continuous tube. The tubing is thoroughly cooled and then wound onto reels in a continuous process. Cryogenic nitrogen cooling in such equipment allows increased productivity and the use of polymers normally not suitable for the process (courtesy of Battenfield Gloenco Extrusion Systems Ltd).

ambient air and forms a fog layer which flows over the floor. A more long-lasting fog effect can be achieved by using dry ice or LN to cool the air/smoke mixture from a glycol-based stage smoke machine so that the smoke/air stays in a layer near the ground.

3.7.4 Dry ice bead-blasting

An application which makes ingenious use of the properties of solid CO_2 pellets is dry ice bead-blasting. Ordinary sand-blasting, using sand driven by compressed air, is a very useful technique for preparing surfaces for painting or removing old paint from metals. Bead-blasting uses tiny manufactured beads of glass or quartz, allowing a greater control over the process and reducing the contamination of the workpiece by tiny sand particles. The beads can be recovered and reused to some extent.

Unlike sand-blasting, with all its problems of apparatus wear and heavy contamination of the working area with sand, CO_2 pellets are very clean. After they have impinged on the surface being treated, they simply sublime into the air. If humidity is kept low, then condensation of atmospheric moisture can be minimised. The technique is only limited by the necessarily high cost of the process; the beads, which are not recovered, and the additional equipment are expensive. Paint-stripping of aircraft is one application which justifies the extra costs.

3.7.5 Liquid helium and NMR/MRI

The outstanding exception to the rule that liquid helium cryogenics is of no interest to industry is, of course, the use of LHe-cooled magnets for nuclear magnetic resonance (NMR). Although ordinary copper electromagnets can be used for any chosen field strength and size, at high field strengths and volumes (say above 0.1 $\text{tesla}^2\ \text{m}^{-3}$) it becomes economical to employ a superconductor cooled in liquid helium. There are two principal applications for nuclear magnetic resonance: research on molecular structure (NMR) and human body imaging (magnetic resonance imaging, MRI).

NMR scanners are generally required to have a field in excess of approximately 1 tesla and to have a volume sufficient to contain the patient being scanned. Although it is possible to use lower fields, signal-to-noise ratios rise as a power of approximately 3/2 so the higher fields possible without power penalty using superconductors are desirable. The highest fields are used for molecular structure research.

NMR and pharmaceutical research. NMR can be used to understand the chemistry of complex molecules. Many atomic nuclei have a magnetic dipole value. The normal isotope of hydrogen and the rarer stable isotope of carbon ^{13}C are the most common examples. When liquids containing these atoms are placed in a magnetic field and inside an RF oscillating electric field, resonance effects are seen.

The nuclear magnetic resonance frequency of a nucleus depends on the magnetic field at that nucleus. In effect, the nucleus can be considered to be like a little magnetic spinning top, whose precession frequency will depend on the applied magnetic field tending to twist its axis of rotation. The applied field, however, depends not only on the applied field but also on the internal magnetic fields caused by the magnetic dipole moments of the circulating electrons in the molecule in which that nucleus is sat. Different molecular

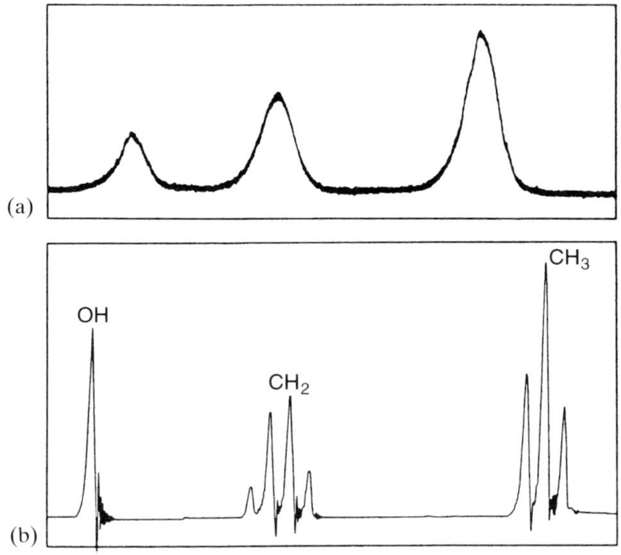

Figure 3.39 NMR scan showing several different hydrogen nucleus resonances from different parts of a molecule of ethanol (CH_3–CH_2–OH) at low (a) and high (b) resolution.

environments lead to slightly different frequencies of resonance and it is these differences which are measured in order to infer chemical structure.

These differences in resonance frequency grow linearly with magnetic field and can be clearly seen only at fields of several tesla. As the desire has arisen to see more subtle molecular effects, fields have had to increase until today the leading-edge equipment is using heroic fields of up to approximately 20 tesla. Higher fields still are needed in order to achieve the maximum resolution in the observation of the environment of each of the NMR active atoms.

Above 20 tesla there is intensive work going on to use HT_c ceramic superconductors as an inner booster solenoid. These booster coils will most likely also be cooled to 4.2 K by liquid He. The HT_c material is used not because it offers a high superconducting transition temperature but because superconductivity in many HT_c materials persists to very high magnetic fields. As such booster coils become available, researchers will be able to carry out NMR biochemical strucutural analysis at higher resolution.

NMR for MRI body scanner magnets. MRI units are designed to image the inside of the human body non-invasively by using nuclear magnetic resonance effect. The part of the person to be scanned must be placed entirely within the scanner so large magnets are required.

As for NMR, the nuclear magnetic resonance frequency of a nucleus such as hydrogen depends on the applied magnetic field. However, by arranging a field which varies slightly across the person being scanned, hydrogen atoms in different spatial locations will differ slightly in resonance frequency and those different parts may be 'addressed' by tuning the detector/transmitter coil, the intensity of resonation and the decay time reflecting the concentration and chemical state of the hydrogen atoms there.

By means of a computer, these many different parts (planes) are addressed sequentially, building up a 'tomograph', i.e. a set of data representing the density of atoms in certain planes or lines. This tomographic data can then be processed by computer into a three-dimensional data set representing the density of atoms versus x, y and z coordinates in the scanned person's body. This can then be displayed in clinically useful form as, for example, cross-sections in different planes.

The NMR body scanner business has grown to be the biggest application for liquid helium and superconductors, with every large hospital in the developed world running them. This would be surprising to a time traveller from, say, 1965, when the predicted main applications for superconductors were power transmission, superconducting levitating trains or superconducting motors. (Examples of all these applications exist but only in prototype form; they have not achieved any commercial success.)

3.7.6 Gaps in the cryogenic spectrum

There are gaps in the cryogenic spectrum, as noted in section 3.7.1 between LHe at 4.2 K and LN at 77 K, and between liquid argon at 87 K and carbon dioxide at 195 K.

In the lower gap there is only neon and hydrogen to choose from. Neon is one of the most exotic cryogens ever used and has been used only in satellites. It is used in satellites not as a liquid but in solid form. Neon is unusual in its small liquid range (25–27 K) and the large amount of gas evolved on vaporisation (1410 volumes gas/volume liquid). Its convenient boiling/melting points and reasonably high latent heat capacity make it a reasonable choice for cooling satellite detectors in the region of 30 K.

Without a suitable boiling liquid, cooling can be carried out by careful control of a lower boiling cryogen, often liquid helium. However, this is very expensive; not only is LHe many times more expensive than LN but its heat capacity per unit gas volume is 64 times lower. In recent years, this practice has been largely supplanted by the use of direct gas refrigerator cooling systems, typically Stirling cycle expansion engines using helium as the working fluid. To provide a steady temperature for even a small heat load requires a surprisingly large cooling engine, however, because of the low heat capacity of gases.

There are in fact few problems today in operating small heat load systems in these gaps using gaseous cooling. However, gaseous cooling is not used unless necessary. A boiling liquid bath provides a large heat capacity and a steady temperature which are difficult and expensive to provide with gaseous systems.

In the higher gap, there are many compounds with reasonable boiling points, although many of them are unsuitable on other grounds, e.g. cost, (krypton, 120 K), flammability or reactivity (methane, 112 K or fluorine monoxide, 128 K). The refrigeration industry swiftly adopted the CFCs once industrial fluorine chemistry had enabled their low-cost production, because they enabled safe operation of cooling plant near and beyond the top of the gap range. The lowest boiling CFCs are useful cryogens; they are mostly very inert, nonflammable and have useful heat capacities. However, the concerns for the ozone layer noted in section 2.5.12 will prevent their use. Compounds such as the HCFCs and others with a low ozone depletion potential (ODP), such as carbon tetrafluoride (halocarbon 14, boiling point 145 K), however, will continue to be widely used.

3.7.7 Cryogenic pipelines

Although now entirely practicable for long-distance liquid natural gas (LNG), cryogenic pipelines are nevertheless almost unknown. A study by Berbesson (1979) showed that the economic factors for transporting natural gas as a liquid are quite favourable. A 1050 mm diameter 100 bar steel gas pipeline could be replaced by a 500 mm diameter low-pressure nickel steel liquid pipeline, resulting in savings of material and pipelaying costs. Other savings occur in running the pipeline because the cost of pumping the gas in liquid form is less than half that of pumping gas itself.

Fortuitously, the strength of the nickel steel increases at a low operational temperature, enabling a thinner material to be used or enabling a longitudinally fixed line to be used. The latter feature avoids the need for compensation of the thermal contraction of the line as it cools down. Interestingly, the insulation needed for the line is a simple 75 or 100 mm layer of foam; the principal heat load in a high flow rate pipeline would be due to friction losses. The relative ease of construction of a methane line contrasts strongly with the complexities of the LHe-cooled pipeline needed for LT_c cryogenic electric power transmission.

3.8 Gases across the breadth of industry

3.8.1 Gases in space

Rocket propellants. Large rockets are propelled mostly by combusting a liquid fuel with a liquid oxidiser. The fundamental parameter of importance in a rocket fuel is its 'specific impulse', I, which has the dimensions of time. The specific impulse is the thrust (in mass units at 1 g) developed by an ideal adiabatic rocket divided by the mass flow rate of the fuel/oxidiser consumed. The reason why this is the standard measure of a rocket fuel's efficiency is that it gives the maximum change in momentum that can be added to the rocket for the expenditure of a given mass of fuel. If a mass Δm of fuel is burned in time t, then thrust F/g will be produced for time t, giving Δp for the rocket of mass M_r as follows

$$\Delta p\ (\simeq M_r \Delta V) = I\mathrm{g}\Delta m$$

where V is the rocket speed and g is the acceleration due to gravity. The specific impulse is dependent on the adiabatic temperature achieved in the fuel/oxidant reaction and the molecular mass of the ejected products of reactions. The exhaust velocity v is given by $v = \surd(kT/m)$, so

$$I = v/\mathrm{g} = (1/\mathrm{g})\surd(kT/m)$$

where m is the mean molecular weight of the products of reaction.

A number of oxidants have been tried and many fuels. The most efficient are rather expensive, considering that each rocket fuelling requires many tonnes of fuel, so cheaper combinations are more popular. The use of LO/diborane, for example, would be at least 100 times more expensive than LO/paraffin. Examples of the specific impulse for some gas mixtures are given in Table 3.13.

The only high-performance propellants which are in use today in large quantities are LO/paraffin and LO/LH. They have the advantage of:

- comparatively good performance
- low cost
- innocuous by-products
- controllability.

Although liquid hydrogen does present some handling problems, it and liquid oxygen are now routinely used on every space launch. In recent years, liquid hydrogen has rarely led to problems. Accidents such as the early Ariane rocket losses were due to motor problems, whilst it was solid fuel motors that caused the NASA Space Shuttle *Challenger* disaster.

Table 3.13 Specific impulse of possible rocket propellants

Oxidant/fuel	Impulse (s)
LO/ethanol	290
LO/paraffin	300
HNO_3/paraffin	270
N_2O_4/methylhydrazine	290
LO/NH_3	320
LO/diborane	360
LO/LH	390

It is interesting to note that, on energy per unit weight grounds, hydrogen is the best source of power for an air-breathing vehicle of any kind. It is about three times better than petroleum fuels in an engine:

$$H_2 \Rightarrow H_2O \quad + 68\ kJ \quad 143\ kJ\ g^{-1}$$
$$CH_4 \Rightarrow CO_2 + 2\,H_2O \quad + 886\ kJ \quad 55\ kJ\ g^{-1}$$
$$C_{10}H_{22} \Rightarrow 10CO_2 + 11H_2O \quad + 6762\ kJ \quad 48\ kJ\ g^{-1}$$

A Boeing 747 Jumbo Jet airliner, for example, fuelled on liquid hydrogen but incorporating no other modifications, could in principle fly around the world non-stop without refuelling.[63]

The example does, however, ignore the problem that would occur with the bulk of hydrogen, which, with LH density around 0.1 g cm^{-3}, is very high. Liquid/solid hydrogen slush achieves a higher energy density than liquid hydrogen and has been used on occasion to slightly uprate the performance of space vehicles.

Spaceborne supplies. Gas technology provides numerous critical elements in the world space programme. Clearly the massive usage of liquid oxygen, and to a lesser extent liquid hydrogen, dominates the gases used in space. However, significant smaller uses for gases have also been found to be essential. Hydrogen and oxygen are used for generating electric power from fuel cells, helium for scientific experiments needing superconductors and hypergolic propellants are used for manoeuvering. Compressed gases, such as nitrogen or CO_2, are used for small orientation manoeuvres and for breathing atmospheres needed for human pilots. The unfortunate and near-fatal explosion of the O_2 tank on one of the Apollo moonshots served

[63] This would give an airline's 'round-the-world' ticket an entirely new meaning; a latter day Phineas Fogg would finally not have to worry about missing his connections! A round-the-world non-refuelled flight was achieved as a record-breaking attempt with conventional fuel, flown by an American couple (the Rutans) in a strange glider-style aircraft.

to emphasise how dependent spacecraft are on these many different gas systems.

3.8.2 Cleaning: thermal blasting of stone

Smooth granite surfaces, carefully polished, are hardwearing and not particularly expensive. Granite and similar stones are now popular for pavements in indoor and outdoor pedestrian areas such as shopping malls. However, when wet, the granite, particularly in heavily used areas where it is particularly smoothly polished, is slippery.

One answer to this it to treat the granite surface to make it slightly rough. Although grinding can achieve this, a neater solution is to use the intense heat of an oxy/acetylene flame to momentarily heat the surface of the granite. The surface layer expands, whilst the cooler body of rock remains unaffected. The resultant stresses cause the surface to spall off. As little as a millimetre can be removed with care. The surface left behind is fairly evenly rough and non-slip but is still attractive to look at.

3.8.3 Firefighting

Several gases have come to prominence in fire extinguishing. Most gaseous firefighting works simply by excluding oxygen from the air by displacement with an inert gas. The brominated fluorocarbon contained in bromochloro-difluoromethane (BCF) extinguishers is a more ingenious choice of gas because the bromine released when the gas contacts a flame actually interferes with the mechanism of combustion; it acts as a negative catalyst. A small quantity of BCF is equivalent to a much greater volume of CO_2 or N_2. However, CFCs and brominated CFCs are serious greenhouse and stratospheric ozone-destroying gases, and they are the subject of severe restrictions on production. Hence, only two gases are likely to be important in firefighting in the future: carbon dioxide and nitrogen. A recent patent from the fire prevention company Wormald International illustrates an application for both gases together. They claim that a 40% Ar, 8% CO_2, 52% N_2 mixture is ideally suited for automatic extinguishing of fires inside indoor equipment such as computers. The mixture is inert and suppresses the fire but any gas which leaks out has the same density as air and will not be displaced upwards or downwards. Also, the CO_2 content will stimulate the breathing of people near the fire, enhancing their ability to survive even with lower concentrations of oxygen in the air locally.

Both CO_2 and N_2 are used as propellants for dry powder extinguishers typically discharging sodium bicarbonate. On its own, CO_2 is a convenient and effective extinguishing agent for small hand-held or wheeled-trolley extinguishers. These can be used on any fire with no danger of electric shock if electrical apparatus is in the fire or near to it, and they will not damage equipment. However, fires extinguished, particularly hot solids on fire, can reignite when the CO_2 stream is switched off as the gas will not cool a conflagration as much as water. The addition of dry inert powder to the gas is useful because the powder will coat the burning solid, helping to prevent reignition.

The most commonly used gas in large fires is nitrogen because of its ease of vaporisation and handling. Ships' holds and coal mines are examples of peculiar circumstances where the use of nitrogen can be a potent weapon in firefighting. For an inert gas to be useful on a big fire, the seat of the fire must be more or less sealed against ingress of more air. Huge quantities of nitrogen can be needed to tackle fires effectively, however, as although they are more or less sealed, many ships' holds and underground workings have huge volumes. Many mines have tunnels several miles long, for example.

The great advantage of nitrogen is that it will not damage equipment or lead to flooding. Coal mine fires are very common; methane, carbon monoxide and finely powdered coal are all flammable hazards. The placing of heaps of chalk on ledges throughout coal mines is still carried out to suppress air/dust explosions with coal. When an explosion starts, dust is shaken from the ledges. Instead of more coal dust being shaken down to exacerbate the flame, chalk dust appears and forms a fire-suppressing cloud. Flooding is always an option with mine fires but flooding of the mine shafts and tunnels can easily turn out to be irreversible as it damages equipment and can block the workings with mud.

Nitrogen-filled soap/water foam has occasionally been used to tackle fires in unusual industrial or marine situations. More commonly, foam inerting is used during demolition of tanks, pipework and chemical plant used for flammable liquids (the procedure is described in section 3.3.7).

3.8.4 Gases in tyres and inflatables

Inflatable emergency lifejackets and inflatable boats are mostly inflated automatically from small cylinders of liquid CO_2. Carbon dioxide cylinders are much lighter than a cylinder with a similar volume of nitrogen. There are potential problems, however, with diffusion gas loss and nozzle blocking. CO_2 is lost by diffusion through plastic and rubber faster than air or nitrogen but this sort of slow loss of inflation is not important in emergency applications. CO_2 can freeze solid in the nozzle of the inflation cylinder. However, careful sizing of inflation system components avoids this.

Nitrogen can be advantageously used for filling tyres. Tyres under arduous conditions of use (at high speed and under heavy load) tend to heat up. The rubber on the inside of the tyre tends to decompose to some extent in these circumstances, being oxidised by the hot compressed air within it.

Degradation of a nitrogen-filled tyre is slowed by the absence of oxygen inside it, giving a longer lifetime. Also, momentary ignition inside tyres, caused by the formation of a flammable mixture of high-pressure air and rubber particles is not unknown. This is prevented by N_2. Aircraft tyres are almost invariably filled with N_2 and the practice is spreading to truck tyres. Automobile tyres have very light loadings, low pressure and low heat build so little is probably gained by the use of N_2.[64]

3.8.5 Gases in glass manufacture

Oxygen is sometimes used in glass tanks for flame enhancement. The productivity of a glass tank can be enhanced if its charge can be heated to melting point more quickly. Most of the cost of glass production is in the melting of the materials at the start of the process and small improvements in this stage yield big advantages in reduced energy consumption. The materials used do not have high thermal conductivity and have high melting points. Furthermore, carbonates of calcium and magnesium used in glasses must be calcined during the melting, driving off carbon dioxide during the melt. High temperatures and a very high heat flux are needed. These are only obtainable from specialised electric furnaces or specialised gas-fired furnaces. Electrical melting is used in small tanks but is uneconomic for large quantities of glass. All this means that melting is ideally carried out with an extremely hot, highly radiating flame.

Unfortunately, the emissivity of a standard gas flame at the most effective absorbtion bands of the glass is not high. Traditionally this has been approached, along with the need to conserve energy from exhaust gases from the melting furnace, by the use of recuperative or regenerative furnaces (section 3.1.9). Most typical is the use of 'checkerwork' or hollow grid brick assemblies that are warmed by exhaust gases and then switched to heat incoming combustion air. The combustion air can be heated to between 1000 and 1300°C, allowing the glass charge to be heated to approximately 1600°C with ordinary natural gas and air. However, huge assemblies of bricks are needed and the switching of hot gas streams is not easy, giving rise to undesirable fluctuations in furnace temperature.

An alternative is O_2-enrichment of the furnace air, giving a hotter flame that has a higher infrared emission for the glass to absorb. O_2 enrichment is controllable, gives steady temperatures, can be used to enhance existing furnace output in retrofit installations and is lower in capital cost. Many

[64] The use of sulphur hexafluoride has been suggested. The potential advantages of SF_6 are not, however, entirely clear. It is an inert gas, so would prevent oxidation and it might enhance convective cooling within the tyre because of its high molecular weight. It is expensive, however.

advanced glass processors are now using oxygen-enriched flames. O_2-based plants can be smaller and lower in capital costs than regenerative plants, whilst existing plants can sometimes be stretched in capacity by the use of oxygen-injected burners or by oxygen enrichment of air supply. The amount of fuel used can be lowered since the hotter oxygen flame radiates very much more brightly (following the T^4 radiation law) and transfers heat to the glass more effectively.

Keeping the flame temperature as before but enhancing the emissivity of the flame by seeding it with other gases is another approach. Acetylene has been tried in this role with some promising results. Flat-flame oxy/fuel burners (burners arranged to produce a horizontal plate-shaped flame) are being tried in glass-tank heating and there are claims that the flat shape further benefits heat transfer to the glass charge.

Glass furnaces have problems with NO_x emissions. The high temperature of the flames needed means that they have much higher NO_x production than most industrial furnaces. The use of pure oxygen, which avoids any NO_x production simply by excluding N_2, rather than air may be the solution ultimately. However, many current furnaces have to solve their NO_x problems by processing downstream. One of the common techniques is that of selective catalytic reduction (SCR). In SCR a carefully metered stream of gaseous ammonia is added to the exhaust stack at a point where it is still very hot, approximately 350°C. The result is a reaction yielding harmless nitrogen

$$4NO + 4NH_3 + O_2 \Rightarrow 4N_2 + 6H_2O$$

Without the catalyst, higher temperatures are needed and N_2O production occurs. The amount of liquid anhydrous ammonia needed is rather modest since NO_x levels are in the tens of ppm range and the injection equipment needed (an ammonia vaporiser and a metering valve, perhaps adjusted by a servo operated by an analyser) is small. Various transition metal oxide catalysts are in use and there is currently active development work on improving these, particularly on reducing the temperature at which the SCR reaction can be undertaken. British Technology Group (BTG) in London together with a Dutch group, have one such improved catalyst. Their patent (Kapteijn and Singoredjo, 1993) describes a process of applying manganese salt in a solution to a porous alumina support medium, at a small percentage loading of manganese, then calcining the catalyst to 400°C for a few hours. They claim useful activity for this catalyst at 100°C with commercial operation proposed to run only a little hotter than this.

Float glass. The production of glass in flat, optically perfect sheets has been pursued for millenia. Only in the last 30 years, however, has a low-cost process been developed and widely adopted for quality glass: the float glass

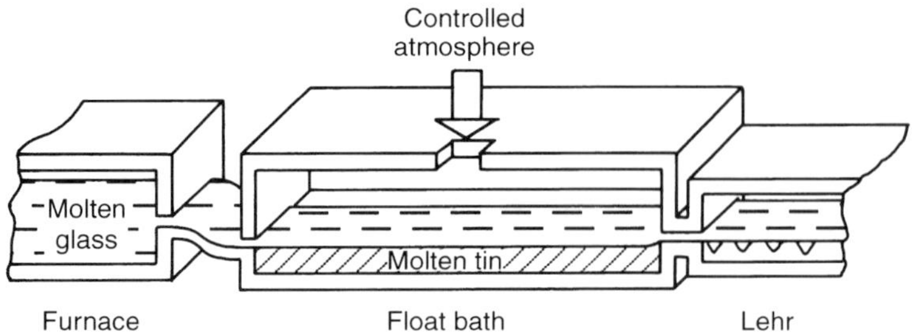

Figure 3.40 Float glass process.

process (Figure 3.40). The modern float glass process involves the addition of heat, raw materials and scrap glass (cullet) to a large tank containing up to hundreds of tonnes of molten glass at up to 1600°C. The glass is drawn off continously over a weir at the cooler end of the glass tank and run out at 1100°C onto the surface of a bath of molten tin, 4 m or more wide. As the 4 m wide strip of glass cools to 600°C while moving across the tin bath, it solidifies and can be drawn off by rollers downstream. There then follows a long (100 m) row of rollers. The continuous sheet of glass cools slowly in this section (the 'annealing lehr') to around 100 or 200°C. Cutting and handling can then be carried out. The production output of a float glass plant is prodigious, 500 tonnes per day or more, and computerised automated cutting stations and warehouses are typically needed to handle it.

The formation of the slightest quantity of dross on the surface of the molten tin bath would give rise to imperfections in the glass produced. The dross can be prevented, as in soldering, by the use of a N_2/H_2 mixture.

Glass bottles are mass-produced by a blow-moulding process not dissimilar to that used for plastic bottle manufacture (section 3.7.3), except that the glass is much hotter. However, the benefits of faster production as a result of internal cooling could also be obtained in blown-glass products by using a cryogenic nitrogen blast on the inside of the parison.

3.8.6 Coated glass

The first coated glass was produced by chemical processes such as the silver mirror reaction used in chemistry as a test for aldehyde compounds. Transparent or semitransparent coatings are difficult, though not impossible, to make in this way. Tin oxide-coated glass is transparent but electrically conductive and can be made in a simple way by heating glass in an atmosphere of vaporised $SnCl_4$. Many glass 'coatings' are in fact carefully glued polymer films. Some safety glass is made in this way. Thin polymer layers can peel off over a long period, however, are not completely UV resistant, and are soft and easily scratched. True coatings of hard metals and minerals are more satisfactory, in general, but more difficult to achieve.

One of the most useful coatings for glass is IR active low-electron multilayer coating. Glass with low-ϵ coating, when used in conjunction with double glazing, allows sunlight in but does not allow heat energy out of a building. It is an important energy conservation measure in cool climates. The coating has alternating layers of high and low refractive index spaced at intervals of a quarter wavelength thick so that light of typical room temperature wavelength (5 to 100 μm) is reflected from the different layers and the reflected waves interfere constructively. The overall result is that the glass has a low effective emissivity in the infrared region of the light spectrum; even when it is hot, it will not radiate much light energy. If it is installed on the inner side of the inner pane of double glazing, it will prevent heat transfer by radiation from the inner pane to the outer pane. (The layer will also work if installed on the inside of the outer pane as it will there act to reduce the absorption of infrared emitted from the inner pane.) Visible light is transmitted almost unaffected because light reflected from the different layers does not interfere constructively. Similar low-ϵ multilayer coatings are also used in tropical climates where the low emissivity also means that heat energy is reflected away. Tropical low-ϵ glass incorporates an overall grey tint to reduce heating from visible solar irradiation.

Although low-ϵ coating can be achieved with a polymer film, a more satisfactory hard coating is achieved by using silane/oxygen-based gas mixtures to deposit silicon dioxide-based CVD layers on glass in a similar way to the CVD glassy layers used in semiconductors (section 3.5.6). These layers can achieve simultaneously high visible transparency and optical quality, low IR emissivity and a reasonably durable, washable surface. The Pilkington 'Kappafloat'™ low-ϵ glass is of this type and glasses such as Kappafloat are now widely used in energy-saving buildings.

A recent patent reveals some of the difficulties of the coatings. The composition of ordinary float glass is dominated by sodium and potassium silicates. As in semiconductors, the alkali metal ions are relatively mobile and will migrate from the glass substrate and bleed through into the circuitry layers below. In the case of fine glass used for liquid crystal display (LCD) screen manufacture this diffusion of alkali ions can kill the circuit. This problem can be prevented by using an oxygen-containing gas which does not react with silane (or carbon dioxide) and an unsaturated hydrocarbon mixed with silane to deposit the first layer at 600°C. The resultant 50 nm layer of quartz-like glass is sufficient to suppress alkali migration. IR-coated glass and other types have for some time been made on glass straight from the float-glass tank. Until recently, fine-coated glass, such as is required for LCDs, has been produced only off-line on cold glass. More patent applications reveal that the aim of some glass manufacturers is to produce LCD quality fine coatings on hot glass in 4 m widths.

Sputter-coating of glass. Thin-film processes using low-pressure argon gas-based sputtering are often nowadays used for coating glass with metals (section 3.5.6). Vacuum-deposited aluminium is used for high-quality mirrors such as the front-silvered mirrors used in lasers. Extremely thin gold-sputtered film made in this way has a brown appearance and provides a useful barrier to unwanted sunlight, and sun heat, in air-conditioned buildings with large glazed areas. 'Solar gain', the heating of a building due to the sun's heat through its windows, is often nowadays controlled by coated glazing. Car windscreens are also commonly coated using a sputtering process. The substrate is either heat-treated (toughened) glass or, for windscreens, more common is a laminate of the 'Triplex'™ type, i.e. glass/polybutyrate/glass. Glass can be sputter-coated with simple opaque metal/metal oxide layers which act as absorbers or reflectors on the glass providing, for example, sun-shade or reduced overall transmission.

An argon (sometimes helium) low pressure (approximately 10^{-3} mbar) glow discharge sputter process is typically used. Although the principles are the same as for semiconductor sputtering machines, glass sheet sputtering is performed in enormous machines at very high speed. The sputter target in machines for flat glass is often not a fixed target but a rotating cylinder. Magnetic confinement of the sputtering ions ('magnetron sputtering') is employed. The magnets are mounted inside the rotating cylindrical target, enabling higher powers and higher coating rates to be used and allowing good utilisation (over 50% or more) of the sputter target material. A large system will use around 800 V on the target, attracting the positive argon ions to sputter off the target. Very high speed coating is possible (several hundred square metres per hour or hundreds of hectares per annum) (see Figure 3.4.1).

A more sophisticated automotive glass product is the conductive windscreen. Simple metal grid patterns are used for rear windscreen demisting applications but for front windscreens the metal lines are unacceptable. For demisting service, front windscreens can be coated with a simple conductive oxide layer such as tin oxide or indium tin oxide (ITO), applied by chemical deposition of vapours such as $SnCl_4$, followed by oxidation in air. However, tin oxide and ITO layers, although both conductive and transparent, have a high resistance and only a small amount of power can be conducted through them; sufficient perhaps for demisting but inadequate for deicing. The approach taken by the AIRCO Solar glass-coating firm, starting 10 years or so ago, has been to use metallic layers of high conductivity silver, gold or aluminium. These are, of course, highly conductive but normally also rather opaque.

Figure 3.41 Large glass-coating production line. Note the vacuum pumps on the left (courtesy of BOC Group).

The opacity of the layer is reduced by two means:

- using a very thin layer (e.g. 10 or 30 nm)
- using matching layers of oxide, intermediate in optical index between the glass and the metal layer.

The latter relies on an effect that is well-known in optical antireflective coating practice. The reflectivity of a transparent surface of refractive index N can be reduced to zero by applying a layer 1/4 of a wavelength thick (around 0.4 μm in common optical practice) of index $\sqrt{N}$. The coatings are applied to the inside of the outermost piece of glass so they are not damaged by scratching in service. They enable a windscreen to have a low electrical sheet resistance (6 Ω per square) while still transmitting 80% of visible light. Coincidentally, the resultant windscreen will also act to restrict solar gain and similar structures can be used on window glass to control solar gain without affecting transparency too much in order to limit air-conditioning load in hot conditions. The completed windscreens must have copper busbars to enable very high currents and voltages (30 A at 60 V) to be applied to the windscreen to melt ice and demist rapidly.

Similar processes are used to coat plastics. It is even possible, surprisingly, to coat plastics with glass-like 'quartz' film as the still relatively new process from AIRCO shows. A quartz layer from 10 nm to a few tens of nanometres thick put down from a plasma-excited chemical vapour (such as the silane derivative tetramethyldisiloxane) or by sputtering will adhere strongly. Such a layer can decrease a hundredfold the O_2 permeability of typical barrier plastics used for food.

Electron-beam coating and printing. The coating of glass substrates with polymers is just one example of the very common industrial requirement to coat sheet materials with polymers. One of the high technology methods of polymer coating is to use electron-beam cured polymer coating. Electron-beam coatings are now quite common, for example a typical application for the process is for applying the multiple coatings used for cardboard fruit juice and milk containers.

Electron-beam printing and coating processes work as follows. The sheet of material to be coated is first stripped of its air by a gas 'knife', then the wet coating is applied. The wet coating has a chemical formulation chosen to cross-link or polymerise, forming a solid under irradiation from high-energy electrons. Electrons are accelerated to 100 keV or so in a vacuum and the powerful beam then exits via a thin membrane into the air approximately 1 mm above the coating surface. High-energy electrons produce showers of Δ rays (lower energy electrons which can knock electrons or hydrogens off polymer molecules and form reactive free radicals), which can then cause the required cross-linking/polymerisation.

Electron-beam printing always employs nitrogen inerting otherwise, the energetic (100 keV or more) electrons would cause the production of copious amounts of ozone. As well as being an environmental hazard, the ozone would damage the coating.

3.8.7 Gases in double glazing

The objective of a double-glazed window is to combine transparency with the minimum of heat loss. Typical construction involves two glass sheets, spaced 5 to 45 mm apart, perhaps with additional thin plastic or glass sheets between, containing a sealed volume of gas and a dessicant material to prevent moisture condensation.

Heat transfer across a window occurs via radiation, conduction and convection. Radiation can be reduced with a low-ϵ layer on the gas side of the panel. This reduces the radiation of long-wavelength (10–50 μm) infrared photons from the warm inner panel to the cold outer panel as described in section 3.8.6. Although gas mixtures with IR absorbing properties have been suggested for this, they are not in use.

Table 3.14 Thermal conductivity and viscosity of gases

Gas (at 300 K)	Thermal conductivity ($W m^{-1} K^{-1}$)	Viscosity (Pa s)
Air	2.41×10^{-2}	18.6×10^{-6}
Argon	1.63×10^{-2}	22.9×10^{-6}
Sulphur hexafluoride	1.3×10^{-2}	15.3×10^{-6}
Krypton	0.87×10^{-2}	25.6×10^{-6}

Conduction and convection can be reduced by using a gas filling other than air. Basically, switching to a higher molecular weight polyatomic gas reduces conduction but increases convection because of the lower viscosity of a polyatomic. The optimum is a heavy monatomic gas, which is superior even in thick windows. For thinner windows, heavy gases offer substantial energy saving advantages.

Argon is now in use as a filling gas superior to air, as is SF_6 for thinner windows. Krypton is superior again, although very expensive (Table 3.14).

3.8.8 Gases in ceramic firing

Much of the making of clay objects is in essence very simple and apparently low technology. Natural clay is moulded while wet and plastic into a shape which is then slowly heated up to convert it into a rigid, durable artefact.

This simplistic description, however, hides a number of subtleties. First, natural clay only occasionally has the correct properties for a range of products and must be blended with other ingredients. Typically other refractory ingredients, often in carefully controlled particle sizes, such as alumina or silica (sand), are added. Carbonaceous additions that will assist with firing and oxidation state and porosity may be needed, for example powdered coal or straw.

Second, heating the clay is a delicate process. First the large amount of moisture in the wet clay mix must be driven off slowly, to avoid cracking, but efficiently, the artefact shrinking as this happens. Then, as temperature increases, the carbonaceous materials must be activated, and perhaps fully oxidised and carried away in gaseous form, i.e. CO_2 and H_2O. At certain higher temperatures various minerals undergo phase changes. These points must be passed slowly to avoid weakening or cracking the clay. The two forms of silicon dioxide, quartz and cristobalite, interconvert, for example. The final result is a strong, though brittle, polycrystalline article. Overfiring must be avoided as fired clay, although very heat resistant, can be distorted or even melted, depending on its composition, at still higher temperatures.

Colour often depends on the state of the oxidation of included iron compounds, with green raw clays containing Fe(II) complexes turning to brick red fired products with Fe(III) coloured minerals. Carbonaceous inclusions are often added to rough clay products such as bricks, as they decrease the energy needed for firing but lead to blue or black spots and core ('black heart'). Spots may be desirable to some extent in 'antique rustic' non-structural facing bricks. However, black heart is undesirable in excess and in structural products as it leads to very weak products. Black heart problems can be avoided by injecting oxygen during firing. The Messer Griesheim company in Germany has shown the advantages that can be gained by judicious injection of oxygen during the firing process.

For many ceramic tiles and bricks (rustic bricks, for example) a dark colour or dark spots are produced by maintaining an atmosphere with little oxygen or even slightly reducing conditions. The traditional way of doing this is to fire bricks in a gigantic 'clamp'. In this method, hundreds of tonnes of bricks are placed above a bed of coal, insulated with turf on the top, and the coal ignited. After a month or so, fully fired, typically rustic style bricks are obtained. The method results in darker colours, and retains black or blue spots due to carbonaceous inclusions, owing to the low oxidation potential of the gas retained in the clamp.

This traditional method is very labour intensive, however, and the brick quality and yield are low. Modern brickworks are now making better quality bricks of similar appearance via low oxidising atmospheres in their conventional continous or batch furnaces provided by inert gas generators (IGGs). IGGs are common in the brick industry for maintaining an inexpensive inert atmosphere. They generate a very crude grade of nitrogen/CO_2 mixture via controlled combustion (see section 2.2.1).

Advanced ceramics production. Some advanced ceramics cannot be fired in air and require special furnace atmospheres for their preparation. Advanced ceramics is, of course, a very wide group. However, some of the most important members of the group are the hardness compounds WC, SiC, SiAlOH, the magnetic oxide compounds, such as ferrites, and thermal shock resistant compounds, such as Si_3N_4.

The removal of binders from the 'green' material is often enhanced by oxygen addition to the initial low-temperature 'burnout' part of the firing cycle.

Ferrite and Si_3N_4 can both be advantageously fired in mixtures of nitrogen with hydrogen. The presence of a reducing atmosphere is essential for the latter as at the extremely high sintering temperature, 1800°C or more, Si_3N_4 is oxidised, even by trace oxygen in an inert gas.

Hydrogen sintering of alumina is used to achieve a high density translucent alumina, which is used in artefacts such as laser discharge tubes and high-pressure sodium lamp tubes where high strength at high temperature is required. The high pressure sodium lamp provides high efficiency in a whitish (actually pink) lamp, so translucence is vital.

3.8.9 Electrical power generation

Hydrogen alternator cooling. The thermal conductivity of hydrogen is very high indeed (seven times greater than that of nitrogen and 20% or so higher even than helium) while its viscosity is less than half that of air or helium. This is important when thermal transfer is considered vital but a liquid coolant cannot be employed. Very large (hundred megawatt or more) electrical generators are just such a situation and many of these are now cooled by circulating hydrogen around the windings and having a water-cooled heat exchanger. The very high performance given by hydrogen is the reason for its continued usage. In less critical applications such as wafer backside cooling in semiconductors (section 3.5.7) the nonflammable nature of helium makes it the gas of choice. In alternators, however, potential safety problems are overcome by stringent precautions.

Sulphur hexafluoride high-voltage circuit breakers. Sulphur hexafluoride is an excellent electrical insulator, giving dielectric strengths. Unlike oil it is also nonflammable. It is therefore used in high-voltage apparatus where a gaseous insulator is desired but where the maximum voltage should be endured before electrical breakdown occurs. This situation occurs in Van de Graaff accelerators of the type used in nuclear physics work. These need a few megavolts d.c. potential differences in a compact laboratory-sized apparatus to accelerate nuclear particles to megaelectronvolt energies. A much more common application, however, is the application to high-voltage electrical circuit breakers.

Electrical arc discharges are initiated by the rapid acceleration of a few free electrons to energies at which they can collide with atoms and ionise them. The additional electrons released are in turn accelerated and an avalanche is built up within microseconds. The halogen atoms of SF_6 tend to mop up any free electrons, preventing the initiating action, and will not easily release electrons in collisions, so preventing avalanche build-up and arc formation.

When a mechanical circuit breaker operates on a high-voltage and high-current electric circuit, current is driven by the inductance of the upstream and downstream circuit to carry on flowing. A high-voltage will build up across the gap and an arc will tend to occur. If allowed to carry on more than momentarily, this arc would destroy the circuit breaker, so a number of techniques have been evolved to extinguish the arc. Breakers can simply use a very large air gap. The air gap, and the overall size of the apparatus, needed is grossly large, however, for 132 kV or higher voltages

Figure 3.42 High-voltage SF_6 rotating arc circuit breaker. Current from inputs (left) is led via the moving contact bars to the outputs (right). The circular coils visible in the lower photograph are the rotating arc devices. The mechanical operating rod and a small sack of moisture absorber are visible at the bottom (courtesy of Hawker Siddeley Switchgear).

and high current. Air-blast circuit breakers reduce the size needed by using a forced blast of air to extinguish the arc. A high voltage circuit breaker using SF_6 is also considerably smaller than a similar air breaker unit because of the higher electric field that can be tolerated. SF_6 breakers can also use a flow of gas but the lower powered gas flow needed has led to the term 'puffer' to describe the gas arc extinguishing mechanism. Puffer breakers are used on 100 kV or more tranmission lines and are typically contained inside long cylindrical ceramic insulator pressure chambers containing the SF_6. The mechanism is elongated to fit inside the insulator chamber and operated by means of a long insulating rod that has a sliding seal in the base of the chamber.

An ingenious use of the dynamo principle is made in 'rotating arc' SF_6 switchgear (Figure 3.42). In this, the arc is transferred to a circular contact connected to a solenoid coil. The arc current itself then creates a magnetic field which causes the arc to move around the circular contact surface: the cool surfaces and cool SF_6 gas, together with the free electron-absorbing action of the gas, extinguish the arc rapidly.

Pressures used vary from 1 bara up to 5 or 6 barg, the gas being typically contained in a metal cylindrical pot pressure chamber. Many designs have only one or two rotating seals penetrating the pot, in order to minimise leakage. The gas is slowly decomposed by the arcing process. Metal fluorides and similar are formed, which are often absorbed, along with moisture, by a cartridge or sack of absorber often molecular sieve. Smaller units using SF_6 can be sealed for life (10 years or so) and slight losses of gas pressure (say 1% per annum) are acceptable.

Other applications in electricity. There are many and varied small applications for gases in the electricity industry. Nitrogen is often used, for example, along with high-grade oils, to insulate underground high-voltage cables. Nuclear power plants have many small applications for gases.

Ultrahigh purity (argon free) oxygen is used in nuclear applications because Ar-40 is activated by neutron absorption to Ar-41, which is radioactive with a half-life of 1.8 h. This would constitute a distinct danger in the event of gas leaking from a gas-cooled reactor. Early low-power reactors used air for cooling and discharged Ar-41 into the atmosphere. This is no longer acceptable and in any case reactors are now much more powerful and require a forced gas circulation system if they use gas cooling at all (most use water). Somewhat higher power early reactors used nitrogen or helium for cooling; the latter, being especially satisfactory as it required lower power to pump for the same cooling effect, has a low cross-section for the absorption of neutrons and is also a neutron moderator. Carbon dioxide is used in most British reactors and to a lesser extent elsewhere, although some of the reactor power is wasted in pumping the coolant gas compared to water-cooled reactors. Carbon dioxide has reasonably good heat-transfer properties and has a low cross-section for the absorption of neutrons.

3.8.10 Gases in lighting

Incandescent filament light bulbs. The filament lamp is a simple concept. Inside the glass envelope is a thin filament of tungsten which glows white hot, emitting light, when it is heated by a relatively high current. The first lamps were very straightforward, simply evacuating the glass bulb around the filament (then of carbon). There is, however, a fundamental compromise in incandescent lamps, i.e. the trade-off between efficiency and lifetime.

The problem of a filament lamp is as follows. The higher the temperature, the higher the efficiency and better (whiter) the light colour. However, as the temperature is raised, the filament evaporation rate rises. Filament evaporation leads to two problems. First, blackening of the envelope as a result of a deposit of tungsten, with consequent loss of light and, second, thinning of the filament, leading to failure at hot spots. A little observation will confirm that failed bulbs are often blackened, indicating how much tungsten has been deposited from the filament, whilst bulbs just before failure often seem to be brighter and whiter than usual because the thinner filament runs hotter.

Hot spot failure occurs as follows. The filament is not perfectly uniform in cross-section. In thinner places, the filament will become slightly hotter than average when current passes. These hot spots soon become worse in a vicious circle if the bulb is run a little above its rated current. The hot spot loses filament by evaporation and the filament becomes slightly thinner still. This leads to a further increase in temperature, which leads to more evaporation and so on, until the filament fails at the hot spot.

For small lamps, this is all there is to it. Bulbs of a few watts or less are evacuated to a good standard with an oxygen/moisture absorber, the 'getter'.

For larger lamps, however, there is another parameter that can be usefully played to improve performance, the gas filling. Argon reduces evaporative loss of filament, though at the cost of increasing convective/conductive heat losses (these rule out gas filling at low wattages). The lower evaporative loss leads to a longer lifetime, or enables a higher temperature to be used, giving a greater efficiency.

For filled bulbs, the higher the pressure of inert gas, the lower the evaporative loss from the filament. This is not a simple effect of gas pressure. At the moderate pressures used (from a few torr to a few bar), Dalton's law of partial pressures would hold approximately, meaning that the presence or absence of argon would not affect the behaviour of the tungsten in evaporating directly. Probably the effect is that a gas filling of any sort simply forces the tungsten evaporating to form an equilibrium gas, with tungsten atoms both being evaporated and condensed, around the filament. At lower vacuum pressures tungsten atoms can simply travel radially and stick to the glass envelope directly. The cooling effect of the gas may also play some part. Hot spots may be suppressed by the conductive/convective heat transfer of the gas.

In higher voltage (e.g. mains 240 V) bulbs, pure argon could lead to arcing, depending on the electrode configuration, so these bulbs usually include 10% nitrogen in the gas to ensure arcing will not occur. Krypton in tungsten filament bulbs provides increased efficiency because of lower thermal losses; the heavier krypton gas has a lower thermal conductivity.

Quartz–halogen bulbs. Quartz–halogen bulbs are, as the name suggests, filament lamps with a quartz envelope and a halogen in the mostly argon gas filling. It is often thought that they can operate at a higher temperature and at a higher efficiency because of a self-healing action on hot spots induced by the halogen but this is incorrect. The halogen, typically iodine, reacts with the evaporating tungsten atoms, preventing them from being deposited on the bulb wall by reacting with them to form iodide. The tungsten iodide molecules diffuse back towards the filament where they decompose, giving up their tungsten back to the hot filament and releasing iodine for further reaction. This regenerative action means that no tungsten is lost from the filament, although hot spots are not prevented because redeposition occurs all along the filament or even preferentially on cool spots. The quartz envelope is necessary to handle the higher temperatures required (250°C) which prevent deposition of tungsten iodide (WI_2) on the bulb.

It is the fact that tungsten is not deposited on the envelope that gives the true advantage of the quartz–halogen lamp. A smaller lamp can be used (as small as 1% of the volume of a conventional lamp), which can then be run at higher gas pressure, further suppressing tungsten loss from the filament and giving longer life and/or greater efficiency. A small lamp also has little convective heat loss.

The quartz–halogen bulbs in use today do not use self-healing but do obtain better performance by the use of halogen. Bulbs with a true self-healing action are possible, however, using fluorine. With fluorine in the bulb, the whole filament reacts at operating temperature, to a small extent, to produce tungsten fluoride. This compound decomposes at very high temperatures to tungsten, decomposing more rapidly in the hotter spots. The hot spots thus become thicker until they reach the same diameter as the rest of the filament, when equilibrium is reached. There are, however, severe materials problems attached to using fluorine inside a lamp. Fluorine attacks both the glass or envelope and the filament supports. Hence there are no mass-produced lamps of this sort available.

Electric discharge, fluorescent and neon lamps. The rare gases, with the exception of argon, were discovered by their optical emission characteristics in an electric discharge and these characteristics are also responsible for one of their main uses. Argon at a few torr pressure, is used in many fluorescent lamps as a gas filling to allow convenient starting. The common fluorescent lighting tube is in fact a mercury vapour tube with much of its light output in the ultraviolet. The ultraviolet emitted is absorbed by the phosphor coating on the inside of the glass and re-emitted at visible wavelengths. A mixture of phosphors emitting different colours ensures that the light finally emitted is approximately white. The argon filling enables a small initial discharge while the heaters warm up the mercury in the lamp ends. Once the mercury is vaporised a little, the light from the low-pressure argon discharge is swamped by the powerful mercury emission.

The argon and a little mercury vapour together form a Penning mixture, in which electrons in the discharge can raise the the majority of Ar atoms to an excited state by collision. The excited Ar atoms can then ionise the Hg atoms in a low-speed collision because the excitation energy of the Ar (11.6 eV) is very similar to, but a little larger than, the ionisation energy of the Hg

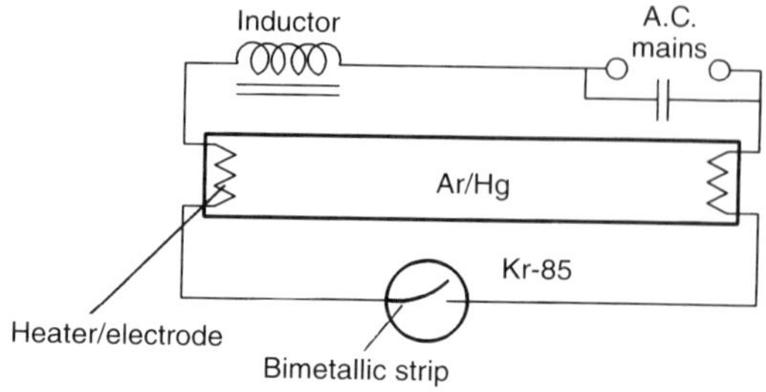

Figure 3.43 Low-cost lamp-starter circuit using Kr-85 tube.

(10.4 eV). This process increases the degree of ionisation induced and reduces the minimum voltage needed to set up a discharge (the striking voltage).

Other common discharge lamps use sodium instead of mercury, giving a bright yellow light, and these are very efficient. High-pressure sodium lamps are also now used, which give a pink light, although severe materials problems had to be overcome in containing high-pressure hot sodium vapour before these became practicable. The discovery that a mixture of metals could be used in a discharge lamp by including a halogen led to the metal–halide lamps. These give a better colour of light than sodium whilst maintaining a high efficiency.

Both sodium- and mercury-based fluorescent lamps require special starting conditions. Heating filaments at each end to warm up the sodium and mercury need to be supplied, along with a striking voltage. In some modern equipment, start-up is controlled by an electronic control system. In older and simpler systems, a gas discharge 'starter' tube may be employed. This includes a minute amount of radioactive gas, either Kr-85 or H-3 (tritium) and a bimetallic switch (Figure 3.43). The starter tube initially has the full mains voltage across it and, with the help of ionisation from the radioactive gas, immediately strikes up a discharge and rapidly warms up the bimetallic switch, which closes, shorting out the starter itself but warming the filaments connected to it. After a few seconds, the bimetallic contacts cool and cut off filament current, allowing the main fluorescent tube to start. Once the latter is running, the voltage available is reduced by the series inductor and is insufficient to close the starter tube contacts.

Other lamps rely wholly on the optical emission of their gas, rather than metal ions. Xenon is an example. It can be used in a high-current 'pinched' discharge to give a bright white light. Another purely gaseous emission lamp is the neon lamp. This actually uses another Penning mixture, i.e. neon with a little argon. Neon discharge tubes are familiar as the glass tubular emitters which can be bent into extraordinary shapes and are used in illuminated signs. The red tubes do actually use neon gas at a few torr pressure but careful selection of different gas/vapour mixtures, often including neon, can give alternative colours. To judge by the night-time lights in neonophile cities such as Tokyo, most colours in the rainbow seem to be available!

Gas lasers. Gas lasers rely on an electrical discharge causing electron/ion/molecule collisions resulting in excited energy states. In this way, with the correct choice of gas mixture and pressure, a 'population inversion' (high energy states more numerous than low energy ones) can be built up. Once the population inversion is strong enough, a single stimulating photon can result in a cascade of photoemission events. All the stimulated emission photons are precisely in phase with the stimulating beam. With reflection at each end of the discharge, the photons can be built up into a strong coherent beam, part of which can be released from the tube by a hole or semitransparent part in the end mirrors.

Carbon dioxide lasers actually use a subatmospheric pressure gas mixture of N_2 (10–40%), CO_2 (5–10%) (and sometimes CO) with balance He, although it is the CO_2 molecule which emits the laser light. These lasers emit at 10.5 μm in the far infrared and can be made exceptionally high in power output; several kilowatts is quite normal. Such high-power beams can easily melt most solids, even in quite thick sections. For this reason, these lasers are the preeminent metal working laser (section 3.2.6).

CO_2 lasers developed fairly early in laser development and rapidly grew in power and efficiency, as knowhow grew, until today hundred kilowatt, 15% efficient machines are available; the most efficient and powerful lasers in regular use. The key discoveries in their development were as follows.

- The discovery that nitrogen vastly increased the power output of a CO_2 laser discharge. This happens because excited nitrogen molecules can transfer their energy efficiently to ground-state CO_2 molecules by collision and produce the required population inversion in the CO_2 energy states.
- The discovery that addition of helium offers additional power increases. These are thought to be due to He's ability to cool the CO_2 molecules more rapidly back to the ground state from which they are excited by N_2 into their lasing levels, and because He aids the electrical discharge process.
- Improved techniques for cooling the whole laser system, both gas and tube. Many new designs for CO_2 lasers focus on improving the cooling, typically by circulating the gas mix more rapidly.

With these developments, the CO_2 laser has become powerful and efficient enough to be a useful tool for industrial cutting of all materials, from dress fabric to steel armour plate, and in the welding of metals.

Other gaseous lasers are possible using similar principles to that of the CO_2 laser. Excited helium can transfer energy to neon atoms, which can then give a laser action. He/Ne lasers rely on an electrode gas discharge in a mixture of about 10% Ne in He. They are the cheapest lasers available, costing only a few tens of dollars for a 2 mW device. They emit red light at 623 nm at up to a few tens of milliwatts. They are often used as a highly convenient source of a narrow light beam, such as in supermarket checkout scanners. These sorts of applications ignore many of the extraordinary qualities of lasers such as monochromaticity and coherence, but are far more important commercially than applications, such as making holograms, which use those qualities.

Other common gaseous lasers are the argon ion systems and the excimer laser. Argon ion lasers rely on transition between two levels in the Ar^+ ion and emit in the green at 440–530 nm. They are commonly used when a powerful, continuous (tens of watts) visible beam is needed. A similar krypton laser can be made with outputs at different frequencies from 460 to 690 nm, and an almost white laser by mixing these gases in the laser. Krypton fluoride and similar 'excimer' lasers rely on transitions in excimers, i.e. pseudo-molecules formed between the first excited state of a noble gas atom (which in this state is as reactive as an alkali metal atom) and a halogen. They emit in the deep UV region at 200 nm and are currently being used mainly in the laboratory, although industrial applications such as photolithography of submicron silicon integrated circuits are now beginning. Excimer lasers use a mixture of 5% fluorine or HCl, which is then blended with neon, krypton or xenon for use in the laser.

3.8.11 Balloons and airships

Although many gases are lighter than air, only three are common in balloons and airships: hot air, hydrogen and helium.

All aerostatic craft work by displacing a greater weight of air than their own weight. This is not an easy goal to achieve, as air only weighs approximately 1.2 kg m^{-3}. Leonardo da Vinci drew a picture of a potential balloon craft based on evacuated copper globes. This is unworkable, however, owing to the impossibility of making the globe simultaneously strong enough to resist collapse due to external air pressure and lighter than the air displaced. The only practical solution is to fill the envelope of the craft with a gas lighter than air. Fortunately, H_2 weighs only 100g m^{-3} and He only 200g m^{-3} so their performance is close to that of da Vinci's vacuum.

The envelope of a balloon, even that of a hot-air balloon, must be selected carefully to minimise gas losses. Hydrogen and helium are both small molecules and diffuse rapidly through most materials. The most obvious flexible materials, rubbers, plastics, and textiles, all have problems. Textiles can be used for hot-air balloons, but only after a plastic proofing compound has eliminated most of their porosity. Elastic rubbers, although convenient, because a rubber envelope can expand and contract easily with pressure changes and packs away more easily, provide little protection against gas

Figure 3.44 Meteorological balloon using hydrogen as a lift gas. The rubber balloon supports a large wire mesh corner reflector from which is suspended a small instrument pack (courtesy of the British Meteorological Office).

diffusion. Low-cost plastics such as polythene are also highly permeable. The least expensive barrier plastics are nylon film and PVDC-coated types. Even better than this is to use a metal film as part of the plastic film composition, metals providing an almost perfect gas barrier.

Toy helium balloons may sound like a frivolous use of gas but they are in fact a business worth hundreds of millions of pounds worldwide. Helium diffuses out of rubber balloons in less than a day so fancy decorative balloons are normally made out of aluminised Mylar^{TM} which will stay filled for a week or so.

Meteorological balloons. For applications in which fire risk is not a consideration, the use of hydrogen offers useful cost savings over helium. Weather balloons are such an application. Balloons can be used to track air currents and relay temperature, pressure and humidity high in the atmosphere back to meteorologists. Despite advances in remote imaging from space and chains of interlinked storm radar units, it is still useful to be able to probe the atmosphere directly with balloons.

Meteorological balloons vary in size from less than 1 m^3 to hundreds of cubic metres (Figure 3.44). The smallest are used to carry simple radar reflectors (corner reflectors similar to those seen on yacht masts), whilst the most sophisticated balloons are used by researchers to study the atmosphere right up to the edge of space. Huge numbers of simple radar reflector units are used. They rise in the air, the rubber envelope expanding as they gain in altitude until the rubber is fully stretched. If the balloon still has enough buoyancy, however, it carries on rising until its internal pressure causes it to burst.[65]

With a plastic envelope that is relatively resistant to stretching, balloons can be made to float aerostatically at a more or less predictable height, i.e. the height at which their density with the envelope fully dilated equals the surrounding air. Very high altitude balloons use a simple thin, lightweight PE or similar envelope, which is filled with only a small volume of gas at ground level. In this way, balloons can easily be made to ascend to over 30 000 m (100 000 ft).

Methane and anhydrous ammonia, despite having a lifting effect only half that of hydrogen, are occasionally used for filling meteorological balloons. Methane, in the form of piped mains gas, is the cheapest filling. Ammonia has the advantage that it can be transported efficiently in lightweight cylinders in liquefied form at only 6 bar pressure. This means that a similar lifting effect to a cylinder of hydrogen can be obtained in a package several times lighter in weight; an advantage for balloons launched, for example, by expeditions. However, ammonia is both flammable and toxic, and contact with moisture must be avoided as the ammonia would dissolve, reducing the balloon volume.

Airships. Ever since the *Hindenberg* airship fire, hydrogen has been regarded as exceptionally dangerous. Yet it is interesting to note that in their 30 years of operation (roughly from 1910 to 1937) airships achieved many millions of passenger-miles, many of them on very long routes such as that of the *Graf Zeppelin* from Germany to South America, with a safety record better than the aeroplanes of the day. The *Hindenberg* incident, which may have been sabotage (Mooney, 1973) killed 13 passengers but 23 escaped, as did two-thirds of the crew. This would be a rare occurrence in an aeroplane crash.

The airships seen occasionally above big cities and above sporting events are helium-filled. The helium is a considerable cost. For each tonne of payload, around 1000 m^3 of He is needed, which, if a price of \$5 per m^3 is assumed, means that their gas filling costs at least \$5000 per tonne. It also needs constant replenishment as losses due to leaks and diffusion must be replaced. These small airships are semi-rigid; their shape is maintained by internal pressure rather than the rigid frame of the earlier Zeppelins.

[65] The lightweight mesh reflectors and instrument packages are occasionally seen descending, apparently from nowhere, after this has happened.

Figure 3.45 High-capacity (13 MW) triple nozzle burner for use in hot-air balloons. The coils of tubing are for rapid vaporisation of the liquid propane fuel (courtesy of Cameron Balloons Ltd).

Although a few passenger-carrying machines are maintained, the main applications for these are for providing a steady aerial platform for television cameras and to present a slow-moving aerial signboard for advertising. In addition to large-scale graphics, some have giant LED display panels covering large areas of the envelope for nocturnal use.

Hot-air balloons. Hot air may seem a curious choice for a balloon gas as it would appear to be very inefficient, because it is only very slightly less dense than cold air. However, if the air inside the envelope can be maintained at a mean 50°C above a 20°C (290 K) ambient, then the ideal gas laws can be used to show that the lifting capacity is $(1 - 290/340) \times 1.2\ \mathrm{kg\ m^{-3}} \simeq 180\ \mathrm{g\ m^{-3}}$, which is around six or seven times less than that of hydrogen or helium. This may sound disastrously less efficient until it is remembered that this means that the balloon diameter must be increased by a factor of the cubic root of six or seven, i.e. less than a factor of two bigger.

The reason for the current popularity of hot-air balloons for sporting purposes is the ease of use and low cost of operation of such an aerostat. After use the balloon may be packed away each day without loss of a valuable gas, whilst a relatively simple but very powerful propane burner provides a readily adjustable means of controlling the balloon's lifting capacity (Figure 3.45).

Modern hot-air balloons rely on a certain amount of up-to-date technology for their function. Although the wicker basket suspended beneath them is old-fashioned, most of the rest is not. They use lightweight aluminium or stainless steel propane cylinders, the gas being burnt in ultrahigh capacity burners with vaporiser coils heated by the flame itself. The envelope construction typically includes Nomex™ fireproof cloth in the lowest part and ultrastrong suspension tapes, cords and envelope fabric.

A combination of hot air and helium has occasionally been used in balloons. A lifting balloon of helium is supplemented by hot air; the former provides most of the basic lift, the latter provides controllable additional lift. Such balloons have been used for altitude and distance record-breaking, and for providing a stable and long-lived high-altitude platform for scientific work.

3.8.12 Aerosol spray cans

An aerosol spray can in its basic form today comprises an aluminium or tinplate can filled with a mixture of propellant, typically 50 to 95%, and a smaller amount of the active product dissolved in that propellant. A dip-tube from the bottom of the can leads to a valve, which when depressed releases a high-pressure stream of product via a spray nozzle, driven out by the gas pressure in the ullage space. Until the late 1980s, the propellants generally favoured were CFCs and these are still used in medical products. With that exception, products today use flammable hydrocarbon mixtures, mainly of isobutane and propane.

A variant of the standard aerosol can is the bag-in-can system in which the product is contained in a flexible bag, pressurised by a propellant below the bag. These cans avoid the need for the propellant to be dissolved in or mixed with the product. This is particularly difficult when oil and water must be mixed, which is what is attempted with hydrocarbon propellant and water-based product. Bag-in-can aerosols also enable corrosive materials to be dispensed because the product never touches the can. They are most commonly used for dispensing gels[66] or other products where the presence of propellant in the product is undesirable. Bag-in-can aerosols have had to overcome a number of technical problems, principally permeation of propellant through the bag and poor spray characteristics because of a lack of propellant to flash-off from the product. A selection of aerosol can types is

[66] Bag-in-can aerosols can also be used for shaving gels, where an additional subtlety is added. These gels are made with a small addition of isopentane, which happens to boil at a little under body temperature. The dispensed gel is clear when emitted, therefore, but rapidly foams up when it contacts warm skin prior to shaving. Although dismissed by some, these shaving gels are now popular.

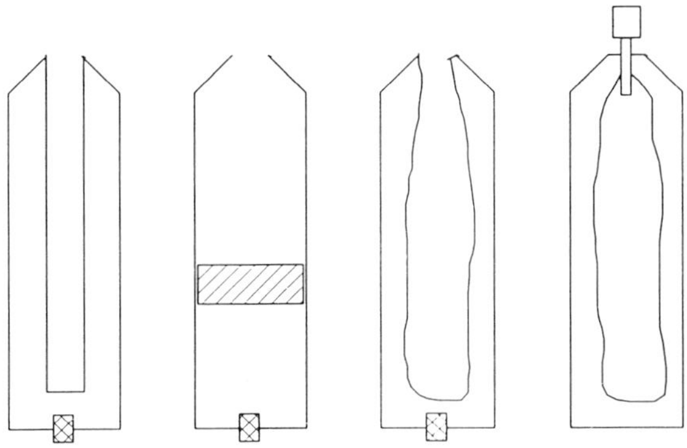

Figure 3.46 A selection of aerosol can types.

shown in Figure 3.46: propellant can be added via the bottom plug or via the valve.

The permeation problem can be solved by using gas-proof coatings such as Al foil on the flexible bag. The spray characteristics can be made acceptable by the use of higher (up to 130–150 psig) pressures and a mechanical break-up nozzle (MBU). The MBU nozzle works by swirling the product in the duct behind the nozzle. As the product is forced through the tiny orifice, angular momentum is conserved, causing the fluid to accelerate its rotation. On exiting the nozzle, a fine spray is formed. MBU nozzles are limited in flow rate so to achieve small particle size (say 60 μm on a water-viscosity product) a low flow rate MBU nozzle is needed. An acceptable compromise, e.g. 80 μm particles and 0.8 $g\,s^{-1}$, can be obtained.

An aerosol spray can is a convenient means of dispensing small amounts of a liquid or fine powder product from a hermetically sealed container. It protects the product being sprayed from degradation by the air, which is normally seen in bottles and similar, and enables the use of smaller quantities of product than might be the case with brushing, for example. With an aerosol can there is no need for solvents to clean brushes. The rapid shut-off of pressure when an aerosol can nozzle is released means that there is little 'spitting' from an aerosol can. Pump spray guns have a problem with spitting as well as having a coarser and less controllable spray particle size.

It is largely as a result of these genuine technical advantages that aerosol packages have grown rapidly from early crude devices for applying insecticide in the 1940s to the present plethora of aerosol cans. The aerosol can has, however, accumulated a bad public image over the years. The first blows against the aerosol undoubtedly came from the aerosol paint can. The many plain concrete surfaces of 1960s and 1970s architecture were ideal substrates for impromptu artwork and slogans, and the aerosol can provided a clean, neat, pocketable package and a high speed of delivery for the unauthorised artist.

More serious objections came in the 1970s and 1980s with the growing realisation that the stratospheric ozone layer was thinning and that CFCs might be primarily responsible for ozone loss (section 2.5.12). There is a range of CFCs and mixtures of CFCs with room temperature vapour pressures of between 15 and 65 PSIG approximately, and virtually all of these have been used. Probably the most popular were CFC-11 and CFC-22. Greener CFCs, such as HCFCs, are now becoming available but are too expensive for a throw-away product, probably have some ozone-destruction potential and certainly would contribute to global warming if adopted on a large scale.

The aerosol industry was persuaded by pressure from environmentalists to move to hydrocarbon propellants. These are basically mixtures of isobutane, butane and propane with small amounts of ethane and other light hydrocarbons dissolved in them. These led to some problems, being less miscible with products than the CFCs, but the addition of suitable co-solvents usually solved these problems. Yet more recently, environmental lobbyists, already looking closely at aerosols, have insisted that spraying a considerable amount of hydrocarbon into the atmosphere, as current spray cans do, is also undesirable as the hydrocarbons contribute to VOC pollution. Ironically, hydrocarbon propellants actually encourage the formation of ozone but in the lower atmosphere, where it is toxic and undesirable.

There are occasional accidents caused by the flammable nature of the hydrocarbon propellants now used.[67] 'Fogger'-type spray cans, for example, have occasionally led to devastating detonations. Foggers are typically designed to dispense their entire contents of paraffin-based insecticide product. In accidents the paraffin is aerosolised and then somehow ignited whilst the air/droplet/vapour concentration ratios are near stoichiometric values. However, arguably the most serious problem with hydrocarbon propellants is their abuse by children as an inhaled drug; an unfortunate practice which is now widespread with frequent fatalities.

Alternative aerosol can systems. There are alternative aerosol packages that do not need a hydrocarbon propellant. A simple replacement proposition is to use an oxygenated solvent with high vapour pressure. Oxygenated solvents still have a flammability problem but are miscible with more products, are not responsible for tropospheric ozone production or stratospheric ozone destruction, and are less toxic when abused as an inhaled drug. However, most oxygenated compounds have too low a vapour

[67] The author assisted a specialist fire test company to conduct tests simulating the effects of a can which is accidentally or deliberately overheated near a source of ignition. On these 'calibrated barbeque' tests some relatively small spray cans produced a violent explosion and a large (3 m diameter), though transient, fireball.

pressure and the only one currently in use is dimethyl ether (DME). DME is today used in many low-volume products such as perfumes but has yet to advance into higher volumes on cost grounds. Perhaps the building of large-scale DME production plants (DME is an unleaded petrol octane improver) currently under way will lower prices and make DME more attractive.

Simple N_2 pressure can also be used. However, N_2 in the headspace has a poor technical performance. There is a large pressure 'droop' when the can is more than two-thirds used up and to counteract this it is necessary to use a large ullage space and higher pressure, thicker-walled cans. The resultant half-filled cans give a 'poor value' marketing image and the thick-walled cans are expensive. Even with these precautions, N_2-powered cans can degenerate to give not a spray but a useless streaming flow; a disaster with a product such as hairspray or paint. Patented systems have been developed, but not so far widely adopted, which employ a pressure regulator inside the aerosol actuator button. This has the advantage of giving a more constant pressure at the nozzle, which can be designed to give optimum atomisation at the lowest pressures of the can.

Carbon dioxide and nitrous oxide both dissolve to a much greater extent in some products and can be used in a simple aerosol can to give a reduced pressure droop relative to N_2 because of the reservoir of gas which remains in the product. CO_2 has always been effective for certain aerosols, for example engine starter and deicer products, which need pressure at very low temperature. Liquid CO_2 under pressure, because it is a good solvent for many organic substances, could be very useful as an aerosol propellant. However, the pressure required at room temperature is too high (over 70 bar must be allowed for) and means that this is not possible with tinplate or thin-wall aluminium cans.[68]

N_2O is favoured for the dispensing of food products. It does not add an acidic flavour, as would be the case for CO_2, but does foam the product to a controllable extent. Filling machines must all be carefully sterilised. Cream is the product most often packaged. The dissolved N_2O expands the liquid cream into a light whipped cream consistency foam on exiting the nozzle, thus conveniently avoiding the need to whip the cream mechanically.

A system called Polygas, currently being developed for using CO_2, uses a flexible bag inside the can pressurised by means of CO_2 absorbed into a solvent/solid absorber mixture in the propellant space. It has the advantage of a pressure droop lower than nitrogen but is operable at higher pressures than hydrocarbons. (The latter cannot be used above about 70 psi because aerosol cans must withstand heating to 55°C, at which temperature

[68] The BOC company offers an 'aerosol' system of this type using conventional high-pressure gas cylinders as the containers for its 'Aromagas'™ system of dispensing smells and for some of its pesticide formulations.

hydrocarbons of above 70 psi room temperature vapour pressure would be in danger of splitting the can.)

Problems with aerosols. There are clearly potential materials problems in aerosols. Typical aerosol products contain from two to twelve ingredients, whilst cans employ deterioration-prone tinplate, plastic and rubber components. Some of the precision valve or button parts can fail if they swell slightly in a solvent additive. Metal corrosion is also a problem; alkaline products are unsafe in aluminium cans and acidic or chlorided products are unsafe in steel cans. Rubber seal failure is not uncommon with new formulations or propellants. Acetone and DME, for example, are powerful solvents, causing strong swelling or shrinking reactions with most cheap rubber seal compounds.

Permeation through plastics is always a potential problem, as mentioned above, for bag-in-can systems. As well as propellant loss, the product can diffuse through cheap, thin single-layer bags into the propellant space. If a bag-in-can system was chosen to protect the metal can from a corrosive product in the first place, this could be serious.

A small percentage of aerosol cans will leak propellant. This is disappointing to the consumer who will be unable to use the product. To avoid this, and to provide a modicum of safety against cans which might explode because of faulty manufacture, cans are legally required to be 'water-bathed'. The water bath is a large tank at 50 or 55°C in which cans are heated for a few minutes. Small leaks cannot be found in a water bath but gross leaks will emit a tell-tale trail of bubbles. Weak or over-gassed cans will split.

3.8.13 UV coatings

Most coatings, paints, inks and varnishes, are still today applied in wet form with the resins that will form the final coating dissolved in a liquid, very often an organic solvent. Wet coatings using water as a solvent are tricky to make waterproof and are slow to dry, hence the attraction of solvent-based formulations. However, there is a drive towards low-solvent emission with regulations on VOC emission becoming steadily more stringent throughout the world. This drive means that other means must be found to form coatings. Hot-melt dipping and dry-film lamination are limited in application. Thus wet coatings must be devised that dry without the aid of solvents. The oxygen-setting resins, such as the time-honoured linseed oil, will no doubt have some niche, along with the catalytically set epoxy coatings. However, wet coatings set by ultraviolet light are growing rapidly and will continue to do so.

UV coatings are formulated from monomers, additives, such as benzophenones, and polymers dissolved in the liquid monomer. UV coatings set by polymerising the liquid monomer with active radicals. The latter are formed when the UV photons strike sensitive molecules, photoinitiators, such as benzophenones or thioxanthones, and split them into two active free radicals, which then start a chain reaction of polymerisation, acting as catalysts. Gas-inerting with nitrogen avoids undesirable side reactions, allowing the use of smaller amounts of photoinitiator and other expensive additives. UV photons have much less energy than the showers of electrons produced in the electron-beam system and ozone production is small compared to electron-beam. However, the oxygen of the air, together with the small amounts of ozone, cause undesirable chain-terminating reactions, leaving the coating sticky.

UV curing is currently used for around 5–10% of the volume of inks and varnishes used in the printing industry. Printers, taken to include all thin-sheet manufacturers (e.g. floor tiles and wallpaper, not just documents), are the largest users of coatings, with wooden products, consumer hardware and car producers not far behind. Total radiation-cured coatings now amount to 2% of the entire mass-production coatings industry and are rapidly expanding by 20% or more per annum. By the end of the century, 8% or so of industrial coatings will be radiation-cured. Of the radiation-cured systems, the most common, over 90%, are UV cured; the remainder use an electron beam for curing. The prospects for the UV curing industry as a whole are exceedingly good. However, it is unclear whether gas-inerted UV coatings will also grow in usage.

Ten years ago, when UV curing was still in its infancy, many laboratories had test rigs using N_2 inerting of the curing unit. However, the high cost of N_2 put them off scaling these units up. Nitrogen improves quality and can reduce costs but is not essential. Early in the UV coatings industry there were problems with gas inerting. Gas prices were high and volumes used were high enough, together with high coating material costs, to put most printers off. High on the wish-lists of UV printers at the time was the elimination of the requirement for nitrogen. This was achieved by the addition of large amounts of photoinitiator and other additives.

Gas inerting could offer the following advantages, however:

- elimination of the need for large amounts of coinitiators like amines, which increase costs and degrade varnish quality and solvent/water resistance;
- faster cures, enabling faster press speeds;
- lower surface tackiness (tackiness is caused by oxygen inhibiting cure at the surface);
- lower UV lamp power needed;
- lower quantities of photoinitiators needed;
- elimination of undesirable ozone production.

If a package of optimised coating/gas supply/UV lamps with guaranteed performance and appreciably lower cost were available the system might be widely adopted. At the cost of increasing photoinitiator or coinitiator levels slightly, it may be possible to employ PSA quality gas in UV cure inerting, particularly if PSA gas purity improves. This would give another supply route option but would only be useful if it enabled reduced overall costs relative to liquid nitrogen. Some new formulations are being tried out which may perform satisfactorily on 99.5% or 99.8% purity nitrogen, matching low-cost PSA nitrogen.

3.9 Medicine and science

3.9.1 Gases for medicine

Some of the earliest applications of gases were to medicine. This may have been because of the tendency of early experimenters, after a few cautious preliminary sniffs, to inhale deep breaths of a new gas to see what happened. Just like today's glue-sniffers and solvent-abusers, the early pioneers sometimes came unstuck doing this. However, the result of this hazardous early work was the rapid development of the medical possibilities of oxygen for improving breathing, and nitrous oxide and others for anaesthesia.

Gases in breathing. Breathing is the process of exchanging CO_2 produced in the body for O_2 that is needed to convert food into energy. The human lungs have about the area of a tennis court, most of this is the enormous area of the alveoli, the tiny terminal branches of the tree structure of the lung tubes. In the tiny delicate alveoli, oxygen from the air can diffuse through a thin membrane into the blood where it is absorbed by haemoglobin complex and carried off to oxygenate the rest of the body. A high-speed mass spectrometer sampling the air at the end of the nose will see the carbon dioxide and oxygen concentrations varying.

O_2/haemoglobin dissociation is the key to blood function; not only does haemoglobin absorb O_2 easily, it also gives it up easily, being somewhere close to physical absorption in its reversibility. The solubility of oxygen in the blood is very high; within the chemistry laboratory its performance can only be approached by exotic perfluorinated compounds.[69]

Although oxygenation is its most difficult technical feat, blood also has a

[69] The latter have been proposed and developed, but not yet accepted, as 'universal' blood substitutes for emergencies. Small creatures, such as mice, can survive for extended periods with fluorocarbon in their lungs instead of air. The first application for these fluorocarbons in humans may also be in lungs rather than in blood, in the treatment of premature babies. A low-viscosity perfluorinated compound is used to treat premature babies with insufficiently formed lungs. Instead of oxygen gas therapy, the fluorocarbon liquid, saturated with oxygen, is used to inflate the infants lungs and provide oxygenation for a few days until the lungs have matured enough to allow gas breathing.

function with respect to carbon dioxide gas. Not only must O_2 be provided to the body but CO_2 must be removed. It turns out that fairly precise CO_2 control is needed. Hyperventilation leads to 'alkalosis', i.e. CO_2 is washed out excessively and the body pH rises. Conversely, hypoventilation leads to 'acidosis', i.e. a fall in pH. Under normal circumstances, animals detect CO_2 variations and as a result the presence of excessive CO_2 is a stimulant for breathing. (However, this works only up to a small percentage in air; beyond this CO_2 is toxic and CO_2 narcosis is possible.)

Oxygen and breathing. One of the first applications for oxygen gas was its use in reviving people with breathing difficulties. Although important, in fact only small volumes of oxygen are used in medicine; O_2 requirements for the human body are about 2 kg per day. The reason why such large gas cylinders seem to be necessary for the maintenance of life is because most gas is administered on a one-pass system, i.e. exhaled air, containing most of the O_2 that was in the inhalation, is simply dumped. For most conditions, pure oxygen is not required, which lowers gas volume requirements to a greater or lesser extent. Actually, despite the fact that oxygen is fundamental to life, pure O_2 at pressures of above around 0.5 bar for extended periods of many hours actually has some toxic effects, in the worst case leading to blindness.

Oxygen is needed for surgery, for breathing by patients in intensive care and for breathing by patients with lung and heart disorders where oxygen uptake is too low. In hospitals, operating theatres and intensive care, patients are generally supplied by gas from a pipeline system. Oxygen, or rather oxygen-enriched air, is often supplied in addition in cylinders for patients needing it in parts of the hospital which are not on the pipeline. For patients needing high concentrations of oxygen there is no alternative to a face mask. For those needing only a small enrichment of the air a less intrusive device is a pair of tubes gently blowing oxygen up the nose.

Continuing oxygen therapy when patients leave hospital is not so easy. Small cylinders for temporary use during travelling and large compressed oxygen cylinders for use at home are the norm. For patients who need only a little oxygen and those who do not need it continuously, this is acceptable. Oxygen concentrators, small oxygen PSA or membrane units, can be used to supplant cylinders of oxygen, at least where the O_2 concentration needed is not too high. They are now in regular use for treating people who need oxygen at home almost continuously, i.e. for more than 16 h per day. For convenience in the home, and for patients who are away from home for considerable periods, a liquid oxygen vaporiser system is possible. Already quite common in the USA, other parts of the world are now looking more seriously at this. A lightweight Dewar flask can supply oxygen for 8 h at 2 or 3 l per minute consumption.

Helium/oxygen mixtures are sometimes used for people with breathing difficulties. The lower dynamic viscosity of helium means that although similar in most respects to air, it can be breathed more easily and deeply than air. The use of helium/oxygen mixtures in diving means that there is a lot of experience with them. By contrast there is as yet little experience with nitric oxide, which is now being tried to assist breathing.

Recently, the role in biology, and possibly also in the treatment of breathing difficulties, of nitric oxide gas has come to the fore. Nitric oxide can only exist fleetingly inside a biological system because it is highly reactive. It might at first be surprising that a small, reactive and toxic molecule should play such a key role. Most chemicals in biology are large, complex and not especially reactive molecules. It now appears that NO is used as a chemical messenger for the control of certain muscles, the 'smooth' muscles which are not under direct voluntary control.[70] NO reacts quickly with oxygen to form nitrogen dioxide and is then further reacted. However, it is used as a messenger in biology probably because its small molecular size allows rapid diffusion and its short lifetime ensures that its action stops promptly.

Arguably, NO is already important in medicine; the use of trinitroglycerine as a remedy for *angina pectoris* relies on the release of NO. As the role of NO becomes better understood, its importance in medicine may increase. Following pioneering work by, among others, Dr Claus Frostel at the AGA company, NO is now in the process of being trialled in the USA in the role of relaxing smooth muscles in the lungs to prevent the occurrence of respiratory distress in newborn babies. The NO is administered via the ventilator machine, such that controlled low levels of NO are achieved in the infant patient's lungs.

N_2O, cyclopropane and anaesthetic gases. Prior to the advent of anaesthetics, surgical intervention amounted to an almost barbaric assault on a patient and only the simplest operations could be carried out. The discovery by Humphrey Davy in England of nitrous oxide and the pioneering work of such clinicians as William Morton in the USA on diethyl ether and James Simpson in Scotland on chloroform led to an entirely new concept for surgery, which allowed operations to be performed satisfactorily.

Many gases are too reactive to be breathed at all. However, for those that are sufficiently inert to be breathed it appears that all act as anaesthetics; it is simply a question of the pressure at which the gas begins to act as an anaesthetic and renders the subject unconscious (Table 3.15). People assisting in a hospital operating theatre are often impressed by the ease with which anaesthesia is induced. Given the anaesthetic nature of so many substances this should not, however, be regarded as impressive. The difficult

[70] The tale of NO elucidation is an interesting one. The action of NO was first noted by physiologists working on the erection of the penis.

Table 3.15 Anaesthetic pressures (for mice)

Gas	Anaesthetic pressure (bara)	Gas	Anaesthetic pressure (bara)
He	190	N_2O	1.5
H_2	85	C_2H_2	1.1
N_2	35	C_2H_2	0.85
Ar	24	cyclopropane	0.11
		$CHCl = CCl_2$	0.05?
SF_6	6.9	chloroform	0.008
CH_4	5.9	halothane	0.017

part, the acid test, of anaesthesia is making sure the patient wakes up again without after-effects.

The first theorists of anaesthetic action noted that the anaesthetic pressure of a gas and its solubility in a lipid (olive oil at blood temperature) were close to proportional. They supposed that the anaesthetics dissolved in the lipid of nerve cell membranes, inhibiting nerve conduction in some way and thus inducing anasethesia (absence of pain) and eventually unconsciousness. It has been shown that the inhibition of nervous signals by anaesthetics comes about via increasing the permeability of nerve membrane to potassium ions. The levels of gas used in clinical practice lead to a measurable (nA) ion flow.

The (1961) Pauling hydrate theory ascribed anaesthetic action to the formation of microcrystals of gas hydrates near nerve membranes but was not proved satisfactorily. More detailed models of anaesthetic action have not yet been proved but the current favourite is the Franks–Lieb model, which ascribes changes in membrane properties to reactions of gases with receptor protein molecules in nerve membranes. The theory seems plausible but is far from being able to predict which compounds will be effective anaesthetics. There is a large commercial anaesthetic vapour research effort but this is based on extensive empirical performance and toxicity testing.

A good anaesthetic should be:

- potent
- non-toxic for the patient (short-term heavy exposure)
- non-toxic for medical staff (long-term trace exposure)
- substantially non-flammable
- easily vaporisable
- swift-acting during induction and swiftly removed during recovery
- a non-irritant to the respiratory tract.

The latest modern drugs, mainly substituted methylethyl ethers, such as isoflurane, fulfil most of these criteria well and are replacing halothane, which in turn replaced chloroform and ether earlier in the century (Table 3.16). However, in hospital practice, these vapours are rarely used alone.

Table 3.16 Modern anaesthetic vapour compounds

Diethylether	$C_2H_5OC_2H_5$
Halothane	$CF_3CHClBr$
Desflurane	$CF_3CF_2OCHF_2$
Isoflurane	$CF_3CHClOCHF_2$
Enflurane	$CF_2ClCF_2OCHF_2$

Although practice varies from region to region, the following is a reasonable guideline:

- first the patient is administered an injected anaesthetic, often an opiate, producing unconsciousness (this is often continued via a drip feed to a vein);
- a further depth of anaesthesia is then induced via a mixture of N_2O, O_2 and a potent vapour;
- if the operation requires it, a muscle relaxant is administered and a mechanical ventilator applied to the lungs;
- unconsciousness is maintained by a low continuous dose of N_2O and vapour, along with oxygen at 20%;
- CO_2 is administered to stimulate breathing during recovery;
- the patient is monitored closely during recovery, which occurs as the injected and inhaled anaesthetics are metabolised or evaporated off in the breath.

For short operations, such as dental anaesthesia, cyclopropane may be used alone, whilst for pain relief without unconsciousness, a 50/50 mixture of nitrous oxide/oxygen at 120 bar (ENTONOX™)[71] can be used. The latter is used especially where, as in childbirth, injected anaesthetics are undesirable.

Equipment for administering anaesthetics. In essence a modern anaesthetic system still implements the functions of the machines devised by Boyle and his colleagues in the 1930s. A manifold takes gases (O_2, N_2O and CO_2) via regulators to the manifold which mixes the gases, humidifies them, and delivers them to the patient. A small rebreathing bag may be used

[71] It is a curious fact that this mixture exists as a gas in a cylinder at room temperature and 200 bar. If the N_2O and O_2 behaved independently according to Dalton's law, then the N_2O should liquefy whilst the oxygen sits on top. This situation can pertain if an ENTONOX cylinder is not mixed in the first place or is allowed to stand upright for a period in severely cold (−20°C) weather. However, under normal circumstances, an N_2O/O_2 cylinder behaves in most respects like a cylinder of permanent gas. This is just one example of the breakdown of Dalton's law for gases at higher pressures. Such mixtures, particularly those involving gases which liquefy easily, are not susceptible to simple theoretical calculations; experimental work is necessary.

before the face mask, whilst check valves ensure that gas flow is unidirectional. Gas is passed through a vaporiser unit to provide a stream of gas saturated with vaporised liquid agent at its saturated vapour pressure. The proportion of each gas supplied is governed by the temperature of the vaporiser and the setting of the gas flow rates. In simple machines flows are governed by needle valves and rotameters.

Modern machines provide pin-index gas cylinders to avoid wrong connections, built-in gas analysis functions (see next section) and provide separate interchangeable vaporiser modules for different liquid anaesthetic agents. Inside each vaporiser module is a temperature-sensitive needle valve[72] which diverts part of the patient gas supply via a tank of vaporisable agent, which has a wick to ensure that gas leaves saturated with vapour. At lower temperatures, more patient gas is bubbled through the tank, and vice versa at higher temperatures. A needle valve on the saturated gas completes the control circuit, giving the clinician the possibility of varying the set proportion of vapour agent in gas. An alternative thermostatically heated tank system, which does not therefore require a temperature-sensitive needle valve, is used in some vaporiser designs.

The modern anaesthetics trolley also provides facilities for directly monitoring patient condition with respect to the anaesthetic using measurements of approximate blood oxygen level via a pulse oximeter, breathing volume monitor and an electrocardiograph (ECG) (Figure 3.47).

Gas analysis in anaesthesia. The closest possible control of anaesthesia is now pursued in order that the smallest possible dose is administered while avoiding the patient being even slightly conscious. For all but the simplest of operations, gas analysers are now in use.

Capnometers monitor exhaled breath, which gives both a breathing waveform and an idea of the amount of CO_2 that has been metabolised. These simple devices are mostly non-dispersive infrared detectors reading 0–20% CO_2. More sophisticated analysers, such as multi-channel NDIR machines, reading CO_2/N_2O and potent anaesthetic agent simultaneously, are also available. Small electrochemical oxygen cells are inexpensive but are rarely used because their slow response time (measured in fractions of a minute) means that they can only measure the average level of O_2 exhaled. The most sophisticated analysers used in the operating theatre are the mass spectrometer-based instruments. These are typically quadrupole analysers capable of response times in the tens of milliseconds, better than fast capnometers, on six or seven gases simultaneously.

Medical mass spectrometers make measurements of peak intensities at a

[72] It should be noted that some ordinary industrial needle valves are unintentionally temperature-sensitive, a feature that can cause problems as 'set' flow rates in apparatus drift off their intended values with temperature changes.

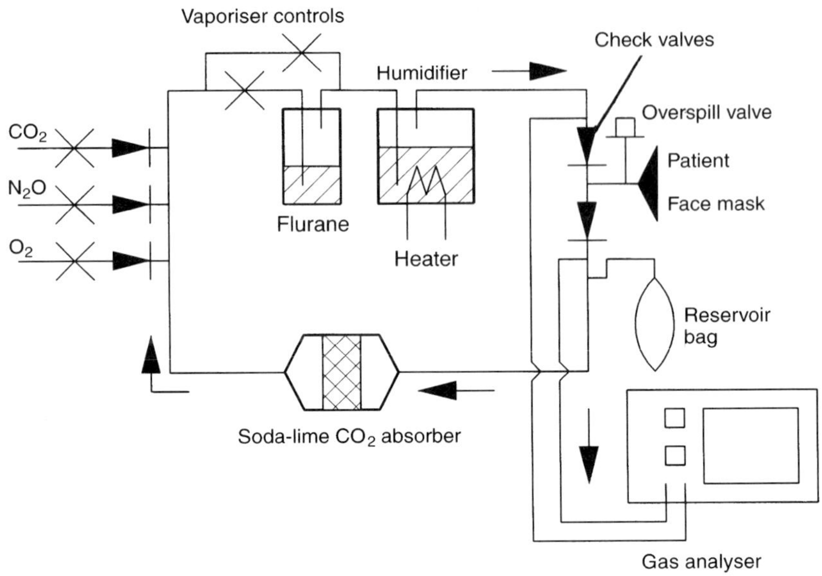

Figure 3.47 Principle of closed-circuit anaesthesia.

spectrum of different mass numbers, with no attempt at resolution better than 1 a.m.u. The data is then used in 'matrix inversion' software to determine gas concentrations. This is necessary because many ion mass/charge ratios overlap; for example, N_2O^+ has the same mass/charge ratio as CO_2^+. Being already in electronic form, the data produced is readily displayed and, if necessary, processed by the built-in computer to indicate metabolism and anaesthetic absorption.

On-line blood-gas analysis is also used occasionally in anaesthesia. Simple two-electrode electrochemical probes inserted into arteries can measure O_2 levels. An alternative means of gas analysis relies on the haemoglobin in the patient's own blood as an indicator. If the patient is receiving too little oxygen, then the amount of red oxygenated haemoglobin is reduced and the purple deoxygenated form is increased. Pulse oximeters measure light transmission (typically at two different red or IR wavelengths) through a thin part of the patient's body and hence oxygenation of the haemoglobin. For adults the part of the body used is usually a finger or earlobe.[73]

Closed circuits. In principle, a closed-circuit anaesthesia system is very desirable because:

[73] For small babies, systems have been devised which look through the top of the baby's head, through the 'fontanelle' or gap in the bone. University College Hospital, London, is one of the institutes which research this remarkable technique.

- smaller amounts of gases escape into the operating theatre;
- less anaesthetic gases needed; only that required for make-up of losses and absorbtion by the patient.

However, such systems have only recently achieved a level of acceptance and most operations are still performed with once-through gases. The problems lie with performance of the soda-lime absorber, which scrubs the CO_2 produced by the patient, and maintaining the balance of gases in the system. The latter dictates a higher level of instrumentation (see Figure 3.47).

3.9.2 Measurement of lung function with gas mixtures

Some measurements of lung function are usefully made with simple apparatus. Huge numbers of people are familiar with the mechanical or electronic peak flow meter which measures the maximum flow from the lungs and is used to indicate when inhaled drugs are necessary in the satisfactory control of asthma. However, with a mass spectrometer or perhaps a simpler gas analyser, together with a cylinder of mixed gas, it is possible to make accurate measurements which are more directly related to lung function.

Oxygen is carried in the blood almost entirely by its reversible reaction with haemoglobin molecules. So, as it happens, is carbon monoxide, which forms the stable bright red complex carboxyhaemoglobin, the haemoglobin being thereby irreversibly poisoned. This latter fact enables a simple test to establish the oxygen-carrying capacity of a person's (or animal's) lung/blood system. A sample of air with a trace amount of CO is breathed in and the amount of inhaled CO is measured. Then the reduced amount of CO breathed out is measured. The reduction measures the capacity of the lung to absorb CO and hence its capacity to absorb oxygen.

The inclusion in a test breathing mixture of a little helium will allow the measurement of the volume of the alveoli (the terminal vessels in the lung from which diffusion into the blood finally takes place) and airways. The He inhaled is diluted by the ratio of the lung volume to the inhaled volume. By measuring flow, the latter can be estimated so that total active lung volume can be derived. This is an important parameter for clinical diagnosis and treatment. (In many diseases, the lungs fill with fluid, reducing lung volume.) More ingeniously, diseases where the lungs have trapped air, air that does not communicate with the airways, can be estimated by comparison of plethysmographic results with He dilution. A medical plethysmograph is a large sealed 'telephone box' in which the patient sits and breathes through a tube. It measures changes in body volume and thus lung volume (including trapped volume) as a function of small applied pressures. Boyle's law is applied to obtain the total lung volume.

One of the standard lung function tests now carries out both helium dilution and CO diffusion measurements simultaneously by using a mass spectrometer and a mixture of 0.3% CO/10% He/21% O_2/balance N_2. It should be added that there are other lung function gases used. Small concentrations of acetylene or SF_6, for example, are occasionally used for similar purposes.

3.9.3 Other medical applications of gases

Cryosurgery. The deliberate use of cryogenic burns for treating some skin lesions, such as warts, is now widespread. It has the advantage of being a fairly pain-free process as the liquid nitrogen inducing the burn to the unwanted lesion also anaesthetises. (Of course this anaesthetic effect wears off as the skin warms up and cryogenic burns can be as painful as any other.) The technique is now more or less mature, with specially made applicator heads and trigger control of the liquid nitrogen flow aiding the medical practitioner in accurate and effective treatment. The necessity of storing LN in a superinsulated vessel and regularly topping this up has to some extent restrained the more widespread use of cryosurgery by general medical practitioners. The use of alternative freezing fluids has not yet achieved much success because many are flammable or ozone-unfriendly and, being higher boiling liquid gases, do not freeze as fast as LN.

An interesting variation of cryosurgery is the possibility of freeze-branding animals. The application of a liquid nitrogen-cooled branding iron of the correct heat capacity and conductivity to an animal results in a permanent alteration in the fur growth, leaving an indelible brand without hurting the animal.

Gases in surgery. Various minor applications of small amounts of gases have developed over the years. In some minor abdominal procedures that involve operating through a small incision, for example, carbon dioxide gas is used to inflate the abdomen and give a clear view to the surgeon despite the smallness of the incision. Perfluoropropane and other stable fluoro gases, such as SF_6 are injected into the human eye as part of a procedure to deal with detached retina problems. The gas is injected into the eye during the operation to form a bubble, which is used to push the retina back into position and prevent fluid from the back of the retina pushing through any break in it. Once in position, the detached retina may be welded back in position with a laser beam. Any mechanical contact with the front face of the retina to push it back into position would result in damage and must be avoided. The use of a gas bubble avoids this damage. Perfluoropropane and SF_6 do not dissolve easily in tissue and are the favoured gases for this procedure.

Cryostorage. There is little doubt that the cryostorage of humans beloved of science fiction writers is still a long way away if, indeed, it is ever possible.

However, this does not diminish the undoubted importance of cryogenic temperature storage of biological samples. At liquid nitrogen temperatures all normal biological processes are suspended. However, all life depends on water as its chemical medium and water freezing causes difficulties with cooling biological samples. Provided the cooling process is carried out quickly, with, if necessary, anti-freeze additives, many small animals and plants, and sections of larger ones, can be preserved without appreciable damage.

Cryogenic dust mite treatment. All around us, but especially in bedding, live small dust-mites: microscopic insects which feed on minute fragments such as the tiny pieces of human skin rubbed off while sleeping. The microscopic faeces of these apparently harmless insects are responsible for much of the asthma and hay-fever and respiratory tract allergic reactions seen in medicine today. Most bedding and clothing can be washed regularly to minimise the amount of allergen to which the asthma subject is exposed. However, carpets and mattresses are not easily cleaned sufficiently thoroughly. The dust mites cling to cloth with claws which are powerful relative to their size and even when dead they are not easily detached.

A treatment which has been successfully applied in Europe for several years now is to spray the bedding and clothing with liquid nitrogen and, while the material is still very cold, suck with a high-efficiency vacuum cleaner. The liquid nitrogen renders the creatures brittle so that their claws and legs break and allow them to be swept up. Very large reductions in mite counts are possible with this treatment.

Sterilisation with ethylene oxide. Ethylene oxide is often mixed with CO_2 for ease of handling and to render it less flammable. It is a powerful room temperature sterilisation agent with few side reactions with common materials and has now supplanted steam in many instances, especially where the equipment to be sterilised may be damaged by heat. Surprisingly perhaps, ethylene oxide can sterilise even plastic prepackaged equipment if the correct packaging is selected, i.e. it has some penetrating power. Pure ethylene oxide is subject to unforeseen polymerisation reactions under some circumstances so a shelf-life of 6 months is often specified. Mixtures with from 10 to 90% CO_2 are used, as well as mixtures with more ozone friendly HCFCs such as HCFC-124.

Ethylene oxide has distinct handling dangers, however, and, even in the small quantities used, is an undesirable addition to our atmosphere. With the ready availability of oxygen on hospital sites, and the convenience of computer-controlled ozonisers, perhaps the use of ozone for sterilisation will become more popular.

3.9.4 *Breathing mixtures: diving and aviation*

The limits of diving gases. The world's oceans have many important natural resources, varying from fish to manganese nodules and petroleum. However, exploitation of these resources is hampered by the need to access the ocean floor. Although midget submarines (with their crews either remote or inside at 1 atm pressure) with manipulator arms can carry out simple tasks, they are too clumsy for many operations that require a diver to operate with a thin rubber suit so that he is at the same hydrostatic pressure as the surrounding water (known as 'aqualung' or 'scuba' operations).

Most scuba-diving takes place in depths of a few metres of water using compressed air via pipes or, more commonly, in gas cylinders. However, hydrostatic pressure increases 1 barg for each 10 m of depth and at the depths usually required for mineral exploitation air is not adequate because of anaesthesia and the 'bends'. As noted above, pure oxygen is toxic and pressurised oxygen is even more so. The only solution is to use mixtures of inert gas with oxygen. The anaesthetic pressures given in Table 3.14 provide ultimate limits to the pressure at which the diver can operate with these inert/oxygen mixtures. With helium/oxygen, the deepest diving gas, anaesthesia can be predicted to occur below 190 bar or about 1900 m. With other gases diluting the oxygen, lower limits will occur; 240 m for argon/oxygen, for example.

In practice, other problems occur which prevent aqualung operation long before these depths. The severe sickness known as 'the bends' occurs when a diver is decompressed too fast. Bubbles of gas, nitrogen with compressed-air breathing mixtures, form in the body, especially at joints, and in blood, causing great pain and tissue damage because the bubbles block capillary blood vessels. Oxygen and carbon dioxide are metabolised and it is the inert gases which cause the bends, although helium causes fewer problems than nitrogen. Viscosity problems occur with higher pressures. At very high pressures, gas viscosities deviate from the ideal gas (which has constant viscosity) and begin to edge upwards, increasing flow resistance in the lungs. Also, the ratio of viscosity/density, the dynamic viscosity shrinks and formerly laminar flows can change into turbulent flow, also giving rise to flow difficulties.[74] In practice, diving at depths beyond air depth (100 m or so) requires helium/oxygen mixtures arranged to give a reasonable partial pressure of oxygen at the operational depth. A higher proportion of oxygen is necessary at lower pressures and depths so that the diver is not rendered anoxic. Rebreathing circuits using soda-lime carbon dioxide absorbers can be used to economise on helium consumption. Use of helium allows free

[74] These flow difficulties are not reflected in many animal experiments. Small animals can in fact breathe easily, even in viscous gas or liquid.

diving from the surface to 150 m and with acclimatisation 500 m or more is possible.

Breathing at high altitudes. Higher performance military aircraft and balloons in the 1930s, and the climbing of very high mountains, quickly exposed the problems of humans breathing at high altitude. Atmospheric pressure falls quickly with altitude (Figure 3.48). Although the fall of temperature with height compensates slightly for the otherwise exponential pressure fall, it is nevertheless impossible to breathe normally above about 5000 m.

Commercial aircraft today operate in the 12 200m (40 000 ft, subsonic aircraft) to 18 300 m (60 000 ft, supersonic aircraft) altitude region. At these altitudes, atmospheric pressure is only 0.19 bara and 0.07 bara, respectively, too low for breathing. To avoid the necessity for oxygen masks for passengers during normal flight, therefore, the cabin is pressurised. Other medical problems which are caused by very low pressure, such as the bends, are also avoided by cabin pressurisation. However, the pressurisation is not taken right up to 1 bara but is reduced 25% relative to sea level pressure to correspond to an altitude of, at most, 2600 m (8500 ft or 0.75 bara pressure). This slightly reduced air pressure causes little discomfort to passengers and reduces the stress on the airframe considerably compared to operating at full atmospheric pressure. Surprisingly, the air pressure reduction actually reduces the weight of a large aircraft appreciably; by about half a tonne in the case of a 747 jet.

Oxygen is, of course, necessary for aircraft operating without a pressurised cabin or in the event of loss of pressure in a commercial aircraft. The latter thus all carry an emergency oxygen system comprising lightweight cylinders of oxygen and a set of oxygen masks which drop automatically out of the cabin ceiling. There must be enough oxygen for passengers and crew to breathe while the aircraft descends normally to 4600 m (15 000 ft, 0.57 bara); the maximum altitude at which it is thought safe to breathe normal air for a long period.

Manned spacecraft cabin atmospheres have varied with different series of spacecraft. Russian craft use a mixture similar in composition and pressure to ordinary air, approximately 1 bar and 20% oxygen, as does the USA Space Shuttle. As with aircraft, however, it is clearly an advantage to use a lower pressure atmosphere to reduce stress and the use of oxygen allows even lower pressures to be used. Early USA spaceflights used pure oxygen at 0.3 bara. After a disastrous fire in an early three-man Apollo mission, however, mixtures with more nitrogen were used. Collins (1988) records that the Apollo 1 was particularly hazardous because of the practice of pressurising the spacecraft to 1 bara pressure whilst on the launchpad. The pressure was reduced to 0.3 bara only after launch. In 1 bara pure oxygen, fires are easily started and will consume even normally non-flammable

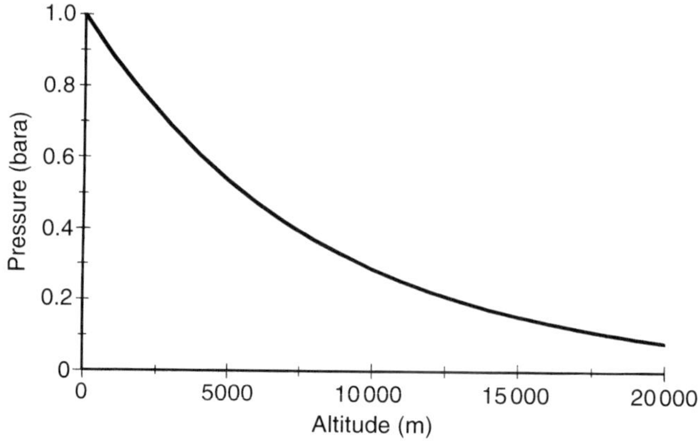

Figure 3.48 Atmospheric pressure versus altitude.

materials such as steel. The procedure on launch is now to use mixtures with 30 or 40% or more nitrogen at 1 bara, the nitrogen concentration being reduced after launch as necessary, while more recent spacecraft use mixtures with more nitrogen anyway (e.g. 28% nitrogen in Skylab and air in the Space Shuttle).

Acclimatisation is possible for people living permanently at high altitude. Effects such as the bends are transient and acclimatisation for even a matter of minutes can eliminate those effects. Other acclimatisation takes much longer, however. Those living high in the Andes, for example, have very high haemoglobin counts; their blood has thus an increased capacity for absorbing oxygen. Athletes in the Mexico and other Olympics held at higher altitudes have always striven to achieve the maximum level of acclimatisation before competing because of this. Even so, record-breaking performance is slightly less likely at high altitude games. Hunt (1914) describes how a closed-circuit oxygen-conserving oxygen enrichment apparatus was developed for use in climbing Mount Everest although the first conquest used straightforward O_2 cylinders. The logistics of getting enough cylinders to a high enough altitude were difficult. However, the wonderful warm glow of well-being that a little oxygen gives at these altitudes is well described. The art and science of acclimatisation have now been carried sufficiently far that climbs without oxygen are now commonplace; something that would perhaps have surprised the early climbers. Groups like the Institute for Aviation Medicine at Farnborough, UK, carry on this research today.

3.9.5 Gases: vital tools of research

Gases of many different types, in the highest practical purity, are a vital component of many research activities in fields as diverse as nuclear physics and zoology. Liquid helium, a technical *tour de force*, is an everyday

Table 3.17 Sensitivity of GC detectors to contaminants

Detector	Contaminant	Contaminant effect
Any	H_2O	Baseline drift
Electron capture (ECD)	Halogen species	Reduced sensitivity, noise
Flame ionisation (FID)	Hydrocarbons	Baseline drift, noise
Photoionisation (PID)	O_2	Reduced sensitivity
Katharometer (TCD)	O_2, H_2O	Filament oxidation, reduced sensitivity

commodity in many laboratories today and has made the performance of almost any experiment that can be performed at room temperature repeatable at a few degrees Kelvin or so[75].

Ultrapure gases. Ultrapure grades of gas are important for research. The problem of using an impure grade is that its impurities must be few, accurately measured, and of known small effect on a particular experiment. If they are not, then results may be obtained which are due to an unknown and uncontrolled impurity effect. It is much the safest course of action in research to start off with the purest materials, including gases, only scaling down on these quality requirements when the subject being investigated is thoroughly mastered.

The use of ultrapure materials in the past has led to discoveries that would never have been made had research been confined to ordinary grades. The whole science of semiconductors depends on silicon, germanium and allied semi-metals being pure at the level of 1 part in 1000 billion.

Fortunately, most gases can be fairly easily prepared in purities that for most chemicals and materials would be regarded as extraordinary. Impurity levels in nitrogen at the ppb level are now achieved. However, even in many other more ordinary purity gases ppm levels of contaminant are the maximum seen. A common nomenclature for the purity of gases is the 'number of nines' system. In this system, the number of 9s in the purity percentage is given, followed by the last figure of the purity. In this way, ultrapure gases can be more concisely and reliably specified. A gas of purity 4.0 in this system is 99.99% pure, whilst a purity 7.0 gas is 99.99999% pure.

Purity of carrier gases for gas chromatography. High gas purity is necessary in GC analysis, not least the GC analysis of gases themselves (section 2.3.4). GC is the standard technique for analysis of vaporisable substances, even

[75] The use of even lower temperatures than those of liquid He-4 may even more legitimately be regarded as a technical *tour de force*. Although lower temperatures are often achieved by adiabatic demagnetisation, a convenient 'refrigerator' from 4K down to a few mK is the 'dilution fridge'. This functions by using the entropy of mixing, as seen in the theoretical air separation energy calculated in section 2.1.1) to cause cooling. The mixing process used is that of the exotic (0.00014% of natural He) and expensive isotope He-3 with normal He-4.

those with vapour pressures in the sub-mbar region, and is in many cases the technique used where some pre-reaction of the substrate can be used to convert it into vaporisable substances. The latter technique, derivatisation, allows for GC analysis of involatile substances such as steroid hormones, for example, in checking athletes for illegal drug-taking.

In using GC for analysing substances of low vapour pressure, however, it is even more important to ensure carrier gas purity. As a general guide, the carrier gas should have total impurities less than the desired level of analysis. To analyse down to the ppm range, a carrier gas in the ppm range or less of total impurity is probably appropriate, whilst for a cruder analysis at the fractional percentage level it may be sufficient to use a carrier gas of approximately 100 ppm total impurities.

Different GC detectors vary widely in their sensitivity to different contaminants (Table 3.17).

Gas-filled nuclear particle detectors. Many nuclear particle detectors rely on the detection in special gas mixtures of the ionisation caused by the passage of a high-energy particle. Some detectors carry out an intrinsic amplification of the original ionisation. They achieve this by applying a high electric field, which causes an avalanche multiplication effect. If the avalanche is limited in extent and duration, and proportional to the original ionisation, the counter is dubbed 'proportional'. If the avalanche carries on to produce a local arc then it is known as a 'spark' tube, while if the discharge engulfs the whole tube, then it is dubbed a 'Geiger' tube. Spark tubes, operated at atmospheric pressure (often with neon), and Geiger tubes, operated at lower pressure, are less popular now that electronic amplification is readily available. A proportional tube with an amplifier recovers and is ready to detect another particle in a matter of microseconds or less, whilst spark and Geiger tubes have a high 'dead time' as they have to wait for charge lost in the large discharge to build up again to operating voltages. The Geiger tube, however, gives a large electrical pulse and can be made very simple and reliable, making it suitable for field and educational use.

Most nuclear detection relies on the following principles:

- argon as the basic component of the mixture, often at 1 atm pressure (mainly argon atoms are ionised by the incoming particle);
- a hydrocarbon which absorbs ultraviolet light by forming free radicals, preventing the spread of the discharge all over the detector tube and giving the detector a fine position resolution (alcohol or isobutane are common choices);
- a halogenated compound such as a CFC, which mops up electrons and allows faster operation by deactivating the detector more quickly, giving a shorter dead time;
- an optional component (methylal) that prevents polymerisation of the

hydrocarbon radicals to a 'gunk', which renders the counter inoperable as it accumulates.

Mixtures of the above basic form, generally containing argon, isobutane, methylal and CFC for proportional detector tubes, are often called 'magic gas'. Developments of proportional tube detectors, many pioneered by George Charpak of CERN, have improved the performance to the point where proportional detector tubes can count particles at rates of tens of millions per second using magic gas.

Not all ionisation detectors employ gas in its gaseous form. Effective detectors for high-energy photons and electrons can be constructed with liquid argon. Ionisation induced in many liquids can be collected on electrode plates, although ion and electron transit speeds are lower. Liquid argon has a reasonably high atomic weight and good electrical and chemical properties. By adding lead collector plates, a liquid argon detector can be made very sensitive to the shower of lower energy photons and electrons produced by an incoming electron or photon in the energy range above about 1 GeV.[76]

Radically different to proportional detectors in principle are Cerenkov detectors. These rely on the fact that the speed of light is minutely slower in gas than it is in vacuum. High-energy charged particles travel so fast that they may exceed the velocity of light in the gas (although still be less than the speed of light *in vacuo*). When this happens, a small amount of light is emitted by a kind of 'shockwave' effect analogous to the supersonic 'bang' of a high-speed aircraft. The light can be detected in photomultiplier tubes and used to define velocity limits on transiting particles. The use of dense gases, such as pressurised CFC, allows the detection of even slow particles, whilst atmospheric argon only allows the detection of much higher energy particles.

Superconductor magnets and nuclear magnetic resonance. Despite the burgeoning numbers of NMR or MRI magnets (see section 3.7.5), the biggest user of superconducting magnets today is probably still the fundamental particle physics research community. A moving charged particle moves in a circular path in a magnetic field. At the immense energies needed to probe fundamental particle interactions, 10 GeV or more, high magnetic fields are needed to keep the particles in their orbits; even with accelerator rings several kilometres across. Because these high magnetic fields must be maintained for hours or days while the particles in the beams are accelerated and kept in their orbits, superconductor magnets are indicated. Although there are only a small number of particle accelerators, such as DESY in Hamburg or Fermilab in Chicago, each contains hundreds of magnets.

[76] Liquid argon detectors have been used in large-scale experiments such the TASSO detector assembly at DESY Hamburg which discovered the first evidence for the subnuclear 'gluon' particle in 1979.

4 The future of industrial gases

4.1 Market pull

In order to make useful predictions about the future of any industry, both sociological and technological factors ('market pull' and 'technology push') need to be accounted for. The recent *Technology Foresight* study (Calder, 1995) in the UK starts with pulling and pushing factors and works towards a SWOT[77] analysis of the industrial and research scene in the chemicals sector and some future predictions. *Technology Foresight* in the UK and similar studies in other countries have highlighted some common social and economic trends such as:

- an increased emphasis on technologies with small environmental impact and big opportunities for technologies which replace those with adverse effects on the environment;
- rapid development of many other state economies towards the highest industrial level (i.e. the USA) – these economies must be expanded with minimal impact on the environment;
- demographic changes towards an older population and smaller families worldwide, especially in the older advanced economies, giving an emphasis to healthcare;
- the rise of individualism will favour technologies for individualised products;
- the gradual movement of the most advanced economies from large-scale commodity products towards higher added-value specialised or higher performance products.

Quite how these general trends will impact on the gases business is, naturally, unclear. However, some predictions can be made safely. In services such as power and water, 'greener' as well as more efficient technologies are indicated. Greatly increased use of electric private automobiles and other vehicles may provide opportunities for gases in the production of their batteries. Future electric traction batteries will certainly be more sophisticated in manufacture and may well be based on completely

[77] SWOT: Strengths Weaknesses Opportunities Threats.

novel technology. Increasingly specific pesticides (used in smaller quantities) may enhance the opportunities for industrial gases in fine chemical production, as will increased demand for medical drugs.

The growth in the use of high technology methods of food preserving and the development of effective but fully biodegradable packaging materials seem certain, given trends towards foods which offer convenience but still minimise environmental impact and meet demands for taste and minimum additives content. As more and more women join men at work, and less time is spent at home, other labour-saving concepts, for example for cleaning, will become more prominent. It is possible, for example, that processes based on chemical vapour deposition could be used to deposit dirt-repellent surfaces (silanisation is already used to some extent) or perhaps an innocuous supercritical fluid such as carbon dioxide could be used to clean clothes without the negative environmental impact of dry-cleaning fluids or soap powders and phosphates.

High-technology materials can enhance efficiency in many industrial sectors, for example in lighter aircraft structures or higher temperature engine parts. Similarly, an increased use of biocompatible materials for prosthetic devices from hip joints to tooth repairs can be expected. Many of these materials, coatings are a typical example, involve gases in their production. Coatings and superstrong fibres, despite their apparently dulled prospects at the moment (lower military spending has had adverse effects), promise growth, leading to further growth in industrial gases needed in the future.

4.2 Prospects for new production processes

Across the breadth of production processes the exceedingly low cost of intelligent computer-based control means that many processes can now be operated much closer to optimum conditions than has hitherto been the case. Further decreases in the cost of computing will probably have less impact. However, the increasing availability of software which combines sensor readings in an intelligent way and models the process being controlled at some level will be important in the future. Similarly, software with a user-friendly graphical interface seems likely to be used on all but the smallest process plant; here the shrinking cost of computer hardware may still make a difference since graphical interface software is notorious for its need for substantial computer power.

There may well be an increased emphasis on sensors in the future; they have become the weakest link in the chain of optimisation in many cases. Now that computational corrections are readily available, problems which have traditionally restricted the use of sensors, such as nonlinearity and cross-sensitivities, will have less prominence. Computer control will allow 'fingerprinting' and further widen the use of sensors.[78] Finally, some sensors which rely intrinsically on computation, such as cross-correlation flow metering, will become more affordable and more widespread.

If the cost of sensors falls steeply enough, this, combined with the already low costs of computing and small-bore pipeline components, will allow the economic construction of smaller gas production systems. Although the strong scale factors in cryogenic plants will mean there is little impact there, the impact may be much greater for techniques such as PSA and membrane separation.

4.2.1 Bulk nitrogen and oxygen

The consequences of the availability of nitrogen and oxygen at significantly lower prices might well be quite far-reaching. Many processes that are currently uneconomic with gases would become more economic.

There seems little room for lowering the costs of the liquefied gases by very much. Liquefaction accounts for the majority of energy costs for LN and LO, and liquefaction can only be improved by improved refrigerator plant. This latter point is being addressed in the latest generation of plant where much higher efficiency liquefiers are now being employed. Beyond these improvements, and admittedly the modernisation or replacement of much of today's cryogenic air separation plant will take many years, only small increments seem possible.

More spectacular improvements may be seen in the lowering of costs for gaseous oxygen from on-site plants. The separation process is, of course, a part of the production costs in cryogenic air separation. Cryogenic separation efficiency has been the subject of a long and sustained drive as separation plant has grown in scale with the requirements of large gaseous oxygen and nitrogen applications in the steel and chemical industries. However, the improvements in efficiency seen in these large plants have not been matched by the efficiency of smaller cryogenic plants; scaling factors favour larger plants heavily with respect to separation efficiency. However, there would seem to be good prospects that the cost of gaseous oxygen supplied from smaller on-site plant could well decrease, perhaps by 50% or more given developments in PSA, and it is interesting to speculate what impact this will have on industrial processes.

[78] Fingerprinting is taken to mean the use of sensors to measure and reproduce operating conditions even where the precise meaning of the measurement is unclear. As an example, consider the use of a simple quadrupole mass spectrometer (MS). A quadrupole MS will normally fragment most ionised incoming molecules, with only a few molecular ions, giving a spectrum with a half a dozen principal peaks. The MS will give a complex mixture of these sets of peaks, many overlapping, if fed with, say, a hydrocarbon gas mixture from a reactor. However, the reactor may still be well-controlled by the MS by defining what set of peaks gives a good result, without knowing which peak corresponds to which molecule.

Humble combustion processes might well be the first beneficiaries, particularly because the use of pure oxygen can eliminate NO_x in exhaust gases; a major consideration with today's environmental constraints. Low-temperature, low-pressure processes for converting alkane feedstocks into more valuable oxygenated products might well become available as complex biological catalysis chains are brought closer to commercial reality. Oxygen will provide a useful intensification of these processes, just as it does in wastewater treatment. As discussed in section 3.3.4, even in high temperature conversion of petroleum to oxygenated products, oxygen may have a role to play in avoiding the costs of preheating nitrogen and allowing simpler recycling conversion schemes.

4.2.2 Pressure swing absorption

PSA/VSA units have now been around for some two decades in more or less their current form. Their progress into the marketplace is still slow; will it accelerate as PSA increases in technical performance? Current efficiencies are still well away from the thermodynamic limit, so there is plenty of room for improvement. There will undoubtedly be improvements in absorber selectivity, greater ingenuity in cycle design and some improvements using vacuum swings. These improvements will all help to lower PSA running costs. However, it seems unlikely that PSA plants will become much simpler or lower in capital costs, although some VSA designs show promise here.

Lower running costs may lead to PSA units of the same scale as cryogenic air-separation units in the future. A pointer to this has been the recent installation of PSA units of tens of tonnes per day at several small steelworks (for example, the Irish Steelworks, Ireland). Efficiencies for early oxygen PSA units has been reported as low as 1.2 kWHr/Nm^{-3} but modern units are reporting results twice as good at 0.7 kWHr/Nm^{-3}. The changeover to vacuum swing absorption (VSA) in oxygen absorption units has led to results around 0.4 kWHr/Nm^{-3}.

The maintenance requirements of a PSA rig are considerable and the reliability of efficient compressors in small sizes is imperfect. By comparison, a liquid cryogen tank, even with vaporiser, pressure regulation and valves, is a very simple and reliable system. It could be speculated that if PSA systems (including compressors) approach the reliability of a vacuum-insulated tank, even if capital and running costs do not go much lower than they are now, then they will make much bigger inroads in the marketplace.

4.2.3 Membrane separation

Great expectations have built up around membrane separation. In principle, membrane units should provide the perfect solution for gaseous N_2 and O_2, at any rate for those applications requiring low to moderate degrees of purity. The more popular polymer membranes are steadily evolving better performance, whilst more revolutionary membranes, such as the ceramic membrane being researched by the Air Products Corporation, may offer step changes in performance. Current membrane units, however, are severely limited in performance and are only capable of success in certain niche applications.

A recent example of the latter is the use of small membrane units for producing nitrogen in beer cellars. In the dispensing of beer it is convenient to use nitrogen top pressure. This can avoid rancidity in the beer by preventing oxygen ingress and can, if used at high pressure, also avoid the necessity for a pump. Formerly, CO_2 was used for this purpose. However, CO_2 dissolves in beer and can make it too fizzy. N_2 can be supplied in cylinders but this, too, has problems. The cellars below bars, alehouses and hotels where beer is stored are relatively inaccessible and it is difficult to put standard $\simeq 5\ m^3$ cylinders there. Instead, smaller cylinders are in use that, although they have a good capacity for CO_2 in liquid form, contain only a very limited amount, typically only $1\ m^3$, of nitrogen. Purity requirements for this application are not high. The high capital and running costs of the membrane unit are justified by comparison to the relatively high cost of N_2 in small cylinders and the elimination of cylinder handling.

There are other potential applications for membrane separation where small amounts of air gases are needed. New applications, for which the use of gas cannot be justified today, may become commercial possibilities with the simplicity of a membrane separation. Gaseous atmosphere control for foodstuffs, lightweight oxygen enrichment devices for medical use and nitrogen for automobile tyres could all use simple, lightweight, membrane separation.

In general, however, membrane units are likely to continue to have problems, particularly in larger sizes, with capital and running costs. They are in principle simpler than PSA but are relatively new and hence of unknown reliability. To this must be added the reliability problems typical of running relatively small compressors.

Transit of food under a controlled atmosphere normally requires large amounts of nitrogen, which cannot be as efficiently stored as CO_2 (which is liquid in the cylinders) or oxygen (which is added by adding air). Crude nitrogen for food atmospheres could easily be supplied by small membrane units, with all-up weight including compressor perhaps not exceeding that of sufficient nitrogen storage, although supplying even the modest amount of power needed to the food storage container will still be inconvenient. Membranes could also provide the means of removing evolved ethylene to restrain ripening.

Membranes could be used to generate nitrogen at the point of use in a factory using the factory's existing air lines. This would have two advantages. First, no dedicated small compressor is needed, instead air from the existing large compressor on the site is used. Second, no nitrogen pipeline will be needed; the already installed air lines can be used.

Replaceable membranes might be a way forward in some applications. Membrane lifetime is always a source of concern in current units. However, in smaller units, where most of the cost is in other components in the equipment, a cost-effective approach might be to design in easily replaceable standard modules. Current membrane costs are very high but this reflects the fact that membranes are an exotic commodity and made in small batches. If there was a steady demand for large numbers of identical modules, as might be the case if a replaceable membrane application caught on, there would seem to be no reason why costs should not fall steeply because raw materials and energy for making membranes are not expensive.

4.2.4 Liquid hydrogen

The use of liquid hydrogen (LH) in space vehicles developed the technology of handling large quantities of LH. Since then, LH has been proposed as a fuel for vehicles. It burns to produce only the most innocuous of by-products, water, and it is the lightest known fuel for a given energy content. As discussed in section 3.8.1, a large airliner running on hydrogen fuel could fly around the world non-stop without refuelling and the BMW company in Munich, Germany, have demonstrated the possibilities for the use of LH as an automobile fuel. Safety problems would have to be attacked successfully, however, before usage became widespread. For example, perhaps aircraft LH fuel tanks could be designed to dump LH very rapidly in an emergency, leaving only a small amount of paraffin fuel for emergency landing. The bulky nature of LH (it is only one tenth as dense as water), however, as well as its flammability may prevent its use in aviation.

Predictions have frequently been made in the past about the rosy future of fuel cells. It is certainly true that hydrogen-fuelled cells are now reasonably reliable and efficient (although it is worth noting that fuel cells were one of the most troublesome pieces of equipment in earlier spaceflight missions such as the Apollo expeditions to the moon). Perhaps advances in the technology of fuel cells for ordinary fuels such as methane, propane or even petrol will come about. Currently, however, cells using hydrocarbon fuels are less efficient and reliable, and even more expensive to make, than hydrogen fuel cells.

With modest improvements in lowering costs and improvements in lifetime and reliability, fuel cells would make adequate power sources for transportation leaving the hydrogen supply as the difficult factor. The *Hindenburg* factor still haunts H_2 even though H_2 handled correctly would probably be no more dangerous than petrol. Also, hydrogen liquefaction will always be expensive, for simple thermodynamic reasons. Hydride storage is an alternative to LH which may have a role to play.

4.2.5 More efficient gas cylinders

The first non-military uses of wrapped gas cylinders are now starting up. Whether these small beginnings presage a major move towards wrapped cylinders is a moot point. The wrappings, although apparently robust enough, may prove to be too easily damaged by sharp point impacts, either on customer premises or during filling and transit (the gas cylinder must be one of the world's most abused objects).[79]

The very slow trend towards the use of higher pressure in cylinders seems likely to continue. Although higher pressure cylinders are not necessarily more efficient in gas/packaging ratio over larger but thinner-walled cylinders at the same pressure, the convenience of a smaller size, and a more robust thick-walled container, means that the higher pressure cylinder will continue to be favoured. As more standard gas equipment becomes available at 300 and 400 bar, the dodge of adding a cylinder valve preregulator may be avoided and the newer higher pressures may become standard. The whole process, however, is restrained by the availability of stronger steels and alloys and by the largest size and weight of object that can be conveniently handled. A cylinder should not exceed 100 kg or so in weight and 1.5 or 2 m in length.

At the other end of the scale from ultrahigh pressure are cylinders storing gas with absorbers present at low pressures. This is a technique which seems ripe for use in many gases. The applications to dangerous semiconductor hydrides and hydrogen may point the way. It would be possible, for example, to use zeolite absorbers on special calibration gases and mixtures used in instruments (such as mass spectrometers) which can easily remove gases from a cylinder at subatmospheric pressure. With subatmospheric pressure storage, in addition to safety advantages, valuable gases such as xenon would not be lost to inadvertent leaks; an economic advantage to the user.

High-pressure absorber systems, other than the storage of acetylene in acetone, may also have their place in the future. For example, vehicles burning methane as fuel have both economic and environmental advantages. Methane, however, must be stored in heavy, low-capacity high-pressure cylinders whereas propane, for example, can be stored in low-cost, lightweight cylinders in liquid form. These disadvantages could be offset by ultrahigh pressure cylinders. A more economical solution, however, might

[79] I have seen gas cylinders variously used, for example, as rollers for moving heavy objects and traffic bollards!

be found in some sort of solid or liquid adsorber for methane, for example an enhanced hydrate compound.

4.2.6 Carbon dioxide

As more and more factories and power stations are forced to clean up their exhaust stack emissions, CO_2 will be easier to extract from stacks. The new cleaner stacks will remove substantially all the sulphur dioxide and nitrogen oxides that currently make extraction of CO_2 from stacks expensive. It is therefore possible that CO_2 may stage a gradual come-back in Europe and challenge N_2 at the lower quality end of the freezing/chilling sector.

4.2.7 Acetylene

The slow decline in acetylene sales has made R&D of more efficient means of production and delivery for the gas an unattractive area for industrial gas suppliers. The use of acetylene by many small users for welding, however, is not contracting as fast as many had predicted. In addition, acetylene will continue to be used for many years by large numbers of users, both large- and small-scale, for oxy/acetylene cutting. The industry could well rationalise further on fewer primary sources of the gas for commercial reasons. Some companies will choose to stay with acetylene production and others will buy it in. If an effective new production process were devised which showed economies of scale stronger than the current processes, this would be especially desirable and profitable for the companies remaining in production. Similarly, improved means for delivering acetylene (smaller, lighter cylinders with increased contents) might succeed in enhancing the attractiveness of acetylene to users and holding up sales of acetylene.

4.3 Developing new applications

Developing new applications for gases is an old-established practice in the industrial gas industry. Ever since the oxygen pioneers such as the Brin brothers ran out of limelights to supply, new applications for gases have been sought. New production processes enable the production of gas in larger quantities at lower cost. Supply then outstrips demand and new applications need to be sought. All the gas majors have gas application laboratories, often duplicating the facilities in several major countries where they operate. The application laboratories often fall somewhere between a true R&D facility and a technical sales and sales support operation. This combination of 'supplier push' along with 'customer pull' has proved a very fruitful way of advancing technology in the past and it is a feature of the industry that will most likely be retained even with major business organisation changes. Fortunately, the structure of the industrial gas supply industry is such that most suppliers are large businesses that should be able to afford to run adequate development facilities for new applications.

It is always dangerous to try to assess what the future will bring but any prediction of new gas applications is yet more fraught because of the vast field of industrial enterprise it must encompass. What follows is a selection of applications that may be important in the future.

4.3.1 Oilfield injection

Although the principles are now well known (section 3.3.3), oilfield injection is currently only employed at a very few sites and then often only on a trial basis. It has the potential of being an enormous user of gas, probably nitrogen, if it becomes a more widespread practice. As enhanced oil recovery itself becomes more widespread, gas injection should become more common.

The availability of very compact, very high-capacity nitrogen generators, however, may well become a factor in oilfield injection. With an increasing number of oilwells operating on platforms offshore, the size of plant is a big issue because of the high cost of space on the platforms. Even with oilwells on land, injection may only be required for a period of months or a few years, not the ten years or more required to achieve payback on a large cryogenic air-distillation unit. Smaller separation plant is thus indicated on portability grounds.

With the cost and size of PSA plants due to decrease, even for very large sizes, and with purity requirements of the application minimal, it seems very likely that they will be used more widely. If membrane separation units prove possible to scale-up drastically, perhaps they will provide a solution. The huge size of large-scale cryogenic separation units would appear to rule them out but perhaps concepts such as the centrifuge distillation tower could so shrink the size of a distillation unit that cryogenic methods, already proven at the 10 000 tonne per day level, could be adapted to offshore oil recovery use.

4.3.2 Transonic wind tunnels

There are now two large transonic wind tunnels, the ETWT (European Transonic Wind) tunnel, Cologne, Germany, and the NASA facility at Langley, Virginia, USA. These facilities avoid the need for still larger facilities. By boiling-off large tanks of liquid nitrogen and operating at low temperatures (90 K), they can operate at lower viscosity and higher density than an air tunnel, allowing higher Reynolds numbers while keeping the tunnel a reasonable size. The linear shrink factor enabled is around $T^{-3/2}$, or about 6, which gives a volumetric reduction of 200 on the size of the facility needed. Where it is necessary to reach Re values of several million, needed for looking at the complex transonic flows around a large airliner, for example, this kind of shrinkage is vital. More wind tunnels, perhaps on a smaller scale, can be expected in the future although the market for such complex research tools is limited.

4.3.3 New applications for liquid nitrogen freezing

Shortly after the invention of the laser, it was described as a solution looking for a problem. Today, lasers are being used in many processes but new applications are still being found almost daily. Similarly, the low cost and almost ubiquitous availability of liquid nitrogen (LN) is a technology looking for a new problem to solve. New applications of LN are almost routine; there is a constant stream of new inventions. There would seem to be no reason why this should not continue in the future, too. New processes will always be found which require the ultrafast freezing, cleanness and convenience that LN freezing conveys.

4.3.4 Supercritical carbon dioxide

The possibility of supplanting organic or organochlorine solvents with supercritical CO_2 was mooted in section (3.4.8). Further to this, there are many industrial reactions carried out in organic solvents which could perhaps be run in the future in CO_2. The cost of CO_2 is lower than organic solvents, so solvent recovery might not be necessary, simplifying processing. Hydrogenation and polymerisation reactions are both candidates for supercritical CO_2 processing.

The way in which food is processed today is increasingly being driven by consumer pressure to achieve higher quality, lower cost and lower environmental impact. Could supercritical fluid extraction (SFE) be employed in the extraction of oils and flavourings from crushed plant material? SFE could offer higher efficiencies and purer product than simple crushing or heating but without the use of organic solvents.

In a similar vein, perhaps, one stage of wood pulp processing could be carried out using CO_2 SFE, extracting undesirable wood components such as polyphenolics. The presence of these compounds gives paper made from such wood pulps a dark colour. Pretreatment with methanol or other solvents can improve brightness but SFE CO_2 might offer advantages in cost and environmental control. Another application of supercritical CO_2 with wood has recently been reported: the preservation of ancient waterlogged wood (Kaye 1995). The steadily increasing importance attached to the preservation of historic and prehistoric artefacts means that further inventions which employ industrial gases in museum conservation work can be anticipated.

4.3.5 New developments in steel production

Even small changes in the way gases are used in steel production may have a large impact on the gas industry because of the large size of the steel industry. Entirely new processes in iron and steel may yet evolve. Direct reduction, for example, was unheard of until the last decade. However, in the main, technical possibilities that already exist will most likely be further developed, driven by the environment in which the manufacturers find themselves. In addition to the usual drives for greater energy efficiency, a number of factors may drive the steel industry to change its methods significantly over the next decade:

1. the slow change of large-scale steel production from (in most countries) a government-run monopoly, or at least a subsidised commercial operation, to a truly commercial business;
2. the modern emphasis on low pollution;
3. the modern emphasis on high levels of recycling;
4. the slow growth in importance of the economies of developing countries.

The last point means that larger and larger volumes of scrap steel will become available, reducing demand for pig iron from blast furnaces and perhaps restraining the growth of direct reduction iron, at least for larger plants. Scrap steel allows for an increased growth of mini-mill operations and mitigates against the leviathan integrated blast furnace/steel refining/rolling mill complexes that have dominated steel in recent decades.

Change away from subsidy may also encourage mini-mill operations by encouraging private capital, often of more modest proportions, into new steel plants. However, breaking away from parochial nationalism may lead to greater rationalisation in large-scale steel production and may mean that a few even larger integrated plants could be constructed in the future to replace the current larger number of smaller integrated units.

The increasing use of scrap will increase the problems of 'tramp' elements (unwanted impurities) and heavy metals. Perhaps novel innovatory approaches using gases will be devised to help deal with these.

An insistence on low pollution in new plants will put strong pressure on new processes to avoid the use of coke. Coke-making plants are now considerably older than their design life in many parts of the world and when these plants are closed they are likely not to be replaced. New iron-smelting methods will instead avoid the need for coke in new plants. The removal of coke from blast furnaces is also being driven by processing cost and shortages of suitable coking coal.

Direct reduction of iron ore to iron is one method of removing the need for coke. An increased number of mini-mills, even if these are largely fed with scrap, may increase demand for direct-reduction feedstock to smooth fluctuations in scrap quantity and quality. Direct reduction, although apparently simple, is still not a mature technology and some further innovations can be expected. A new process is being claimed by the NUCOR/US Steel joint venture in Trinidad based on the reduction of iron ore to iron carbide rather than to spongy iron, the usual product of direct reduction (see section 3.1.3). A mixture of 'fines', finely powdered ore, in this case from Brazil, is heated to 600°C in hydrogen and methane. The result of this is a powdered iron carbide containing 6% C (Fe_4C approximately), which can be heated to drive off some carbon, then melted and refined into steel using the heat generated by the lancing oxygen/carbon reaction in the usual way.

The implications of all these changes for the use of gases in the industry are clear in some cases, less so in others. It is clear, for example, that oxygen use in blast furnaces will increase as coke declines. The increasing numbers of mini-mills will lead to more demand for argon for plasma-melting and for ladle stirring.

4.3.6 Changes in the liquid fuels market

Small changes in the liquid fuels market could also have a large impact on the demand for oxygen and other gases. The South African SASOL project, with its large oxygen plants, shows how big the demand for changes could be.

The use of oxygen in the manufacture of methanol from syngas is currently not a particularly important route to liquid fuel. Methanol is barely accepted as a fuel owing to its current high cost and poor calorific value. (Although, paradoxically, methanol can be used to extract higher powers from internal combustion engines.) However, as oil prices rise, coal, still currently decreasing in production, may once again become more important and its conversion to syngas and thence to methanol offers a relatively elegant route to a convenient, clean-burning liquid fuel. The scale of such a change, if it ever happens, would be enormous. The use of oxygen would be dictated by the reduction in methanol plant size that oxygen can give, otherwise investment needed in the new fuels would be simply too high.

4.3.7 Nitrogen tunnels for semiconductors

There is currently a movement in semiconductor manufacturing plants (wafer fabs) away from the use of large, extremely expensive, Class 1 or better cleanrooms and towards the 'mini-environment' concept. This concept involves using an inexpensive Class 1000 standard of cleanroom but with the silicon wafers being processed inside machines, transfer tunnels or transfer boxes (the mini-environment) where the atmosphere is controlled to better than Class 1 conditions. The mini-environment approach has been taken by several recent wafer fabs, for example GEC Semiconductors in their Roborough, UK, facility and by TSMC in Taiwan. These new facilities are promising very good conditions for high yielding silicon integrated circuit (IC) processing but without excessive cost and with a more pleasant environment for the operators manning the cleanroom. (Class 1 or better cleanrooms require special uncomfortable clothing and special helmets if contamination of silicon wafers by the operators is to be avoided.)

Beyond mini-environments a further development can be envisaged. In a truly ideal semiconductor cleanroom, silicon wafers would not 'see' air, even highly purified air, but only pure N_2 or vacuum. They would travel from machine to machine in a nitrogen tunnel. The mini-environment of each machine would be nitrogen or possibly vacuum. There is still debate on the relative desirability of N_2 and vacuum. The leading ultraclean group of Professor Tadahiro Ohmi at Tohoku University, Japan, prefers nitrogen. Some researchers in the industry think that vacuum is inferior in performance but often superior in practice because it is better matched to the mostly low-pressure processing chambers of a typical wafer fab. The argument for N_2 revolves around it providing protection against the almost instantaneous diffusion of contaminants which occurs *in vacuo*. Perhaps N_2 for a whole wafer fab process may not be necessary but islands or clusters of N_2 handling will be used, together with other clusters where handling is in vacuum.

There are hints that some new research installations will use high purity N_2 backfilled into a low vacuum, e.g. 1 mbar. This is thought to give an optimum low particulate and gaseous contamination. Mean free paths for contaminant vapour molecules are still short but particulates will fall quickly and be extracted by the continuous flow vacuum pumping/gas supply system. It is certainly true that whatever the pressure, flowing N_2 is ideal.

Today it is not always the case that the atmosphere is the major source of wafer contamination. The processing machines themselves, some gas processes, many wet processes and many photolithography processes all yield contaminants, and as the wafers shift to smaller and smaller feature sizes, smaller and smaller contaminant particles become important. However, more attention is being paid to making the processing equipment cleaner, which will therefore focus more attention on achieving a cleaner wafer environment.

Wafer tunnel handling is by no means the only way to achieve the required goal of an ultraclean mini-environment for silicon wafers. Integration of many wafer processes onto one machine or suite of machines linked by a vacuum handling robot arm is an obvious alternative. Creation of a 'cluster tool' is useful for local handling. The use of ultraclean versions of the so-called SMIF boxes or similar is another route. Here the principle is to employ machines that remove wafers from a sealed box, process them and return them to the sealed box. The SMIF box is then moved from machine to machine, not the wafers themselves. Further sophistications involve replacing the passive SMIF box with an active module. This blankets the contained wafers in pure air or nitrogen, which is recirculated via a filter to minimise pick-up of contamination from the SMIF box itself. A project called WAFEC, part of the pan-European JESSI project, is studying this at the LETI microelectronics institute in Grenoble, France. Vacuum is again an alternative, and the MMST group at Texas Instruments, Dallas, Texas, uses vacuum SMIF-type boxes.

4.3.8 HTc superconductor cooling

A large range of devices are projected to be made using the new ceramic oxide-based HT_c superconductors, such as 123 YBCO or 2223 BISSCO. These materials have the remarkable property that they are superconductors at relatively high temperatures: up to 140 K has been claimed. They also are capable, when cooled to liquid helium temperature (4.2 K), of resisting much higher magnetic fields than LT_c superconductors.

The first HT_c superconductor devices just appearing in the market place are mostly liquid helium-cooled.

- High current leads: connect LT_c superconducting coils to room temperature power supplies. The HT_c lead conducts current but not heat and any heat influx to a helium-cooled LT_c superconductor is very expensive.
- High-field magnet inserts: HT_c coils placed inside LT_c coils, operated at liquid helium temperatures. In this way, magnetic fields as high as 30 tesla can be reached. These high fields are especially interesting to NMR chemists (section 3.7.5).

These are very specialised applications and will lead to little additional use of liquid helium or liquid nitrogen. However, a second generation of HT_c devices using liquid nitrogen cooling seems likely in the foreseeable future, perhaps by the turn of the century.

- Large powerful HT_c magnets: for applications from nuclear magnetic resonance imaging (hospital body scanners) to mineral extraction.
- HT_c motors and generators: rotating electrical machines all benefit from superconductors. Many different superconductor machines have been made with liquid H_e cooled LT_c material, some on a very large scale, but they have so far all had problems which have made them uneconomic.
- Fault current limiters: for use as high-speed 'fuses' on high-voltage electric power lines. Varying the gas pressure above a bath of liquid nitrogen containing a piece of HT_c superconductor acting as a fault

current limiter is a possible way of varying the critical current of the superconductor and hence the current limit.[80]

Once thought impossible, technically very difficult LHe-cooled cables have actually been constructed and run for specialist purposes. LN cables will therefore certainly be practicable if HT_c superconductors become practicable large-scale conductors. The economics of LN-cooled cables are now well-known. Basically, all high-powered cables that have to run underground would be replaceable if HTc materials go low enough in cost. Sections of cable, once initially cooled, would be maintained at 77 K by small ASU/refrigerator/cryogen pumping units located at intervals along the underground run of the cable.

4.3.9 Diamond production

The production of diamond thin-film out of a mixture of gases is a perfectly practical, although as yet not fully developed, technology.

The know-how for what is now the standard process for the production of artificial diamonds was developed over 30 years ago by General Electric in the USA. It involves the production of stupendous pressures and high temperatures in a tiny reactor volume. Although now a well-established technology, it can produce a product adequate for specialised uses but at a cost higher than that of natural diamonds. The possibility of coating tough substrates with diamond films was also realised at this time but remained intractable until techniques and equipment borrowed from the semiconductor industry improved the control of diamond formation to the extent where consistent product could be produced. As know-how has grown, deposition rates have gone up, from approximately one micron per hour with early CVD up to a maximum of one hundred microns per hour today.

In the thermal method (e.g. Hirose, 1987) a hot tungsten filament decomposes the gas mixture at low pressure (tens or hundreds of millibars) and a heated substrate (hundreds of °C) is placed below this. All sorts of gases can be used, including many alcohol and ether vapours, methane and acetylene. Plasma-assisted diamond thin-film production relies on the production of methyl radicals and hydrogen atoms in a microwave plasma discharge. The substrate is heated in the low pressure chamber to 800°C or more. The method using an ECR (electron cyclotron resonance) plasma is similar but uses a magnet as well as microwaves, giving a more intense plasma. Many vacuum reactor makers now market systems specifically for diamond film deposition, which can deposit several tens of microns per hour of diamond, whilst a smaller number of suppliers offer plasma-jet systems,

[80] In Downie (1991) HTc material was coated in an enhanced boiling surface to ensure that it and cryogen temperature would track very accurately the temperature corresponding to the applied pressure.

which can deposit several hundred microns per hour. A plasma-jet deposition system might run just below atmospheric pressure with argon/hydrogen/methane.

Purity issues are important in gases used in semiconductor CVD, with levels above a few ppm provably affecting film quality. In an analogous way, purity may well become an issue in achieving higher quality diamond films. Even isotopic purity is important with diamond films. Films made with isotopically pure ^{12}C methane have much higher thermal conductivity than those made with normal methane containing 1% ^{13}C. GE actually sell ^{12}C methane products for this reason.

Practical, though expensive, industrialised technology now has a large turnover, with products enhanced with diamond films worth billions of dollars per year. As the new equipment allows the more widespread use of the technology, this business looks set to grow much further. Other hard coatings are used industrially, such as SiAlON, BN and SiC, and these are being researched and used to some extent in the engineering industry. Research groups are currently looking at CVD or PECVD (plasma enhanced CVD) techniques for laying down very high quality films of these materials and growth can be expected for these coatings, too.

Applications for hard coatings are mostly obvious, e.g. drill bits, wear resistant parts and bearings. One of the principal applications for diamond is less obvious: heat-sinking microelectronic circuits. This relies on the high thermal conductivity of diamond, which is much higher than that of SiO_2, the standard microcircuit thin film insulator. Also, doped diamond is a semiconductor and it is conceivable that thin-film diamond semiconductor devices will one day be produced commercially.

4.3.10 Production of superstrong 'whisker' materials

It ought to be surprising to anyone who has studied chemistry how weak most materials are. After all, if the tensile strength of a single chemical bond is of the order of 10^{-9} N, then materials should have tensile strengths of the order of 100 GPa[81], whereas typical strong materials, such as steel, are only 0.5 GPa in standard form, although 2 GPa or so can be reached in special hard-drawn alloy wires.

When a material is stretched cracks tend to form, concentrating stress in a small part of the material, so that it fails long before the average stress is up in the GPa region. One way of controlling this crack propagation problem and producing much stronger, lighter materials is to fabricate composites. In these strong fibres or whiskers are held together in a matrix, such as a polymer, of modest strength. Cracks cannot propagate from one fibre to

[81] For example, at 100 GPa an ordinary automobile weighing 1000 kg could be suspended by a thread only 0.2 mm in diameter.

another and the matrix is insufficiently stressed to fail. Tensile strengths can, therefore, be high.

Two types of high-strength fibres are now in routine production: those based on aramid plastics, such as Kevlar™ and those based on carbon fibre, such as Graphil™. These are now of sufficiently low cost that they are used for high-value consumer items as well as in industrial and aerospace applications. Tensile strengths of up to 4 GPa for aramid and carbon fibre materials are possible. These are high strengths for such lightweight materials (1–2000 kg m^{-3} as opposed to the 8000 kg m^{-3} of steel). Even higher strength is possible with short fibres or 'whiskers' of other materials such as boron or silicon. In addition, these fibres, being stable at high temperatures, can be used as the reinforcement of a metal matrix.

Carbon fibre is often produced from polyacrylonitrile plastic fibres by stretching, heating in hot air, then pyrolysing in inert gas. Boron and silicon carbide and superstrong whiskers are made by CVD deposition from diborane and silane mixtures, respectively, onto fine tungsten wire. 10 μm wires are built up to 100 μm or so in typical processes. The production of ultrastrong materials is dependent on defence demands. So far only military hardware needing the utmost strength/weight ratio has justified using these expensive materials. They are currently made in very small (and decreasing) amounts and suffer from the high costs of all small production rate materials. However, breakthroughs in ultrastrong fibre production, similar to those seen in carbon fibre, may well be possible and this would lead to a major expansion of gases used in this sector.

4.3.11 Greener processes

With the growing realisation of the importance of environmental issues, it seems highly likely that many new applications for industrial gases will stem from greener processes in industry. A myriad of small industrial processes will switch to greener technology, often involving gases. One example is the use of solvent-free coatings (section 3.8.13). The use of solvent-free coatings relying on radiation curing is expected to grow in the future and will have the benefit of reducing the amount of solvents emitted to the atmosphere causing local and regional pollution. With lower cost nitrogen from membrane and PSA units conveniently available, the use of superior coatings (which require inerting during UV curing) will be possible on a larger scale.

Similarly, changes expected in major industries will lead to increased opportunities where industrial gases can contribute in processes. Some of the processes already described above will achieve major environmental benefits (Table 4.1).

Biochemical methods also offer a route to greener processes where they can supplant methods currently using bulk chemicals. In the case of chemical

Table 4.1

Industry	Process	Gas use
Electric power generation	Desulphurisation	O_2 oxides useless sulphite sludge to useful gypsum or sulphur
Papermaking	Decolourising	$O_2 \rightarrow$ ozone, replacing Cl_2 and avoiding the production of dioxins
Ironmaking	Tuyere injection	Oxy-fuel blast replacing coke ovens

synthesis biochemical methods can offer good yields and specific isomers where classical methods yield mixtures of isomers, some of which may be pollutants. In other industrial areas, biochemical processes can offer a greener solution by avoiding other pollutants.

In many biochemical processes some industrial gases may be needed. The reactor or fermenter vessel will need to have pressures of gases, such as oxygen, carbon dioxide and inert gases, carefully governed by removal or addition as appropriate. In biochemical methods which rely on solid plant or animal tissue cultures or thick beds of single cells, diffusion oxygenation of the cells may be easier with pure oxygen rather than air. In many such processes it will be an advantage to use pure oxygen to accelerate reactions and enable smaller fermenter vessels to be used. One example where intensification with oxygen might be needed for an economic process is oxygenated hydrocarbon production, which could conceivably convert to biocatalytic methods in the future using enzyme-derived catalysts at room temperature. The size of fermenters may become an issue with the more widespread adoption of biochemical methods. For example, the ICI single-cell protein unit at Billingham, UK, is over 100 m high and 10 m in diameter.

4.4 Epilogue

Industrial gas production technology is now so well-established that there would seem to be little room for major innovations. The unspectacular but steady growth of alternative production processes such as PSA in recent years perhaps points the way. Large changes may eventually be achieved but these changes will come over a considerable period of time.

Industrial gas technology is an important and still growing part of the sum total of industrial technology. Whilst the changes in industrial gas consumption figures expressed in tonnes of gas used may still be dominated by relatively small changes in the big consumers in the iron and steel industry, there are many other areas of industry which will see significant changes towards processes using industrial gases. These process changes will affect the gases industry considerably but will have an even more far-reaching impact on the users of industrial gases.

The story of gas technology has had many surprising twists in it and it is hoped that the reader will have gained some appreciation of these by reading this book. Gases are now a maturing industry in many ways. It is certain, however, that the technology of industrial gases still holds some surprises for the future.

Appendix A: Glossary

Gas industry terms

Absolute pressure Pressure referred to zero pressure or vacuum, as opposed to gauge pressure.
AOD Argon–oxygen decarburisation: a process involving blowing argon and oxygen through a ladle of molten iron. It is applied to reduce the carbon content of steel.
APU Argon purifier unit: typically used to describe units which heat argon to a few hundred °C and pass it over a supported copper reducing agent, where oxygen contamination is eliminated.
ASU Air separation unit: normally a cryogenic plant for separating the principal constituents of air, i.e. nitrogen, oxygen and argon.
Atmosphere In the gas industry atmosphere can denote any gas-filled space, such as the inside of a furnace, as well as the ambient air atmosphere.
Bar, bara, barg The principal modern unit of pressure measurement in industry. The earth's atmosphere exerts an average pressure of 1.013 25 bar at sea level. Bar(a) or Bara means pressure referred to zero pressure (vacuum). Bar(g) or Barg means pressure referred to (local) atmospheric pressure. 1 bar = 14.5 psi.
Blanket A gas blanket typically refers to a volume of inert gas that shields a sensitive part of a process from damage caused by air.
Bursting disc A safety relief valve consisting of a thin stainless steel diaphragm, obtainable in various thicknesses that rupture at different specified pressures.
Carbide Calcium carbide is reacted with water to make acetylene.
CFC Chlorofluorocarbon.
Coal gas The gas produced by the pyrolysis of coal.
Compressed gas A generally recognised legal term for any compound which exerts a pressure of more than 0.5 barg on its container at room temperature.
Cryogen Any liquid boiling below *c.* 110 K.
Decant A technique for delivering gases compressed in bulk in which a set of cylinders on the delivery vehicle are allowed to equilibriate in pressure with the user's bank of cylinders.
Dewar A small (less than 200 l) vacuum-insulated vessel for storage of cryogenic liquids at low pressure.

Dip-tube A tube, typically of plastic, copper or mild steel, running from the valve at the top of a gas cylinder to the bottom. A dip-tube can help with the mixing of compressed mixtures during the filling process. It also enables the user to withdraw liquid from the bottom of a gas liquefied by the cylinder pressure without turning the cylinder upside down.
Doping gas A gas, typically a hydride or organic derivative of an atom from Group III or Group V of the Periodic Table, which is used to add controlled amounts of those atoms to a semiconductor to render it suitably electrically conductive.
Filling ratio Used in fixing the maximum fill of gas cylinders, this is the number of kilograms of gas added divided by the cylinder volume in litres.
Flashback flame arrestor A device to prevent propagation of a flame backwards down a gas pipeline from the burner and towards the gas source.
Gauge pressure Pressure referred to ideal atmospheric pressure, 101 325 pascals, or about 15 psia.
Goose-neck eductor A small dip-tube enabling liquid or gas withdrawal, depending on orientation, from a compressed gas cylinder that is held in the horizontal attitude.
GWP Global warming potential: the capacity, relative to CO_2, of a gas for causing global warming via the infrared greenhouse effect.
HCFC Hydrochlorofluorocarbon.
HFC Hydrofluorocarbon.
Kieselguhr The porous silica material (the fossilised remains of microscopic diatoms) used to absorb acetone in acetylene cylinders.
LC50 The concentration by volume of gas in air that will cause the death within 14 days of 50% of a batch of male and female 200–300 g rats after 1 h continuous inhalation.
LN/LIN Liquid nitrogen.
LO/LOX Liquid oxygen.
LPG Liquefied petroleum gas: covers both commercial propane and butane and mixtures of these.
MAP Modified atmosphere packaging: replacement of air in food packaging with a more suitable mixture of harmless gases.
MAPP Methyl acetylene propadiene: a vaporisable liquid used to augment flame temperature and flame emissivity of the lower performing oxy-fuel mixtures.
MCP Manifolded cylinder package: typically 15 standard (*c.* 7 Nm^3) compressed gas cylinders mounted in a frame with a manifold giving one outlet pipe weighing approximately 1 tonne.
Mercaptans A range of analogues of alcohols in which the –OH group is replaced by the –SH group. They have an exceedingly strong smell and are typically added to heating gas in ppm quantities to enable people to smell gas leaks.
MIG welding Metal inert gas welding, electric arc welding in which a

consumable electrode melts and is incorporated in the welded joint. The inert gas prevents the incorporation of too much oxygen in the joint.

MMA Manual metal arc-welding with stick electrodes without the benefit of gas shielding.

Nm^3 The industry standard gas unit (except US), a cubic metre of gas at normal temperature and pressure.

ODP Ozone depletion potential: the amount by which a gas will cause ozone depletion in the stratosphere relative to freon-12.

OFN Oxygen-free nitrogen: a reasonably pure grade of nitrogen that has been filled into a preevacuated cylinder. In principle a cylinder of nitrogen at 200 barg, filled on top of an 'empty' cylinder, actually containing 1 bar air, would contain 1000 vpm oxygen. In fact ordinary nitrogen cylinders are rarely as impure as this.

Permanent gases Gases which cannot be liquefied by application of pressure at room temperature. H_2, He, N_2, O_2 and Ar are the common permanent gases.

Plug flow Occurs where gas A flows into a pipeline to displace gas B with a well-defined front, like a piston displacing gas.

ppm/vpm/ppb ppm = parts per million, usually by weight, vpm = parts per million by volume. The term ppm is often used carelessly, when vpm is in fact meant. vpm is more useful in the sense that for gases the volume ratio is approximately a ratio of the number of molecules and is a partial pressure ratio. ppb = parts per billion, i.e. parts per 1000 million.

PPU Prepurifier unit: typically used in connection with purifiers for air in cryogenic ASUs.

Producer gas A flammable mixture of N_2, H_2, CO and CO_2 produced by the action of steam and air on coke or coal. It is common on large integrated chemical plant. The air ensures that the reaction is exothermic and can be more or less continuous. However, the quality of the gas is very poor with often over 60% consisting of nitrogen.

PSA Pressure swing absorbtion (see section 2.1.6).

psi, psia, psig Pounds per square inch, a unit of pressure still in common use in industry. psi(a) or psia is referred to vacuum. The most common unit is psig, which is referred to local atmospheric pressure.

RHE Reversing heat exchangers: used to clean up air instead of PPUs for elimination of CO_2 and H_2O, which could block up cryogenic ASUs.

SCF Standard cubic foot: a cubic foot of gas at standard temperature and pressure (US standard unit of gas).

SIVL Super-insulated vacuum line: pipework for LN consisting of dual pipe with the annular space filled with evacuated silver-coated Mylar insulation.

Snifting Venting a gas cylinder or gas line momentarily before connection. This ensures that any debris that has accumulated in the valve outlet is ejected. It also checks that there is something in the cylinder. It is not recommended, however, for toxic, reactive or inflammable gases.

Stenching The addition of small amounts of mercaptans or other smelly compounds to help people smell gas leaks.
TIG welding Tungsten inert gas welding, electric arc welding with a non-consumable electrode and an inert gas to prevent oxidation of the tungsten and oxygen in the joint.
Town gas Town gas consists mainly of H_2 and CO. It is still in use in parts of the world where natural gas is expensive and coal is cheap.
Ullage The gas-filled space above the liquid level in a tank.
Vacced cylinder Preevacuated cylinder: a precaution ensuring the minimum contamination from air ingress.
VIE/VIT Vacuum-insulated evaporator/vacuum-insulated tank: a tank for cryogenic liquid storage. VIE refers to a high-pressure unit, as these are used mainly for gaseous supply; VIT refers to a low-pressure unit.
Water gas A flammable mixture of H_2 and CO produced by passing water through hot coke. The reaction is endothermic and heat must be supplied periodically, typically by interrupting the process while air is pumped in.

General terminology

Absolute temperature Temperature referred to the coldest temperature physically possible, normally measured in degrees Kelvin. The Kelvin scale of absolute temperature has degrees the same size as the Centigrade scale. For example, 0°C, the freezing point of water, is 273.16 K, whereas absolute zero is −273.16°C and 0 K.
Adiabatic A process involving no input or output of energy. For example, in the adiabatic expansion of a gas, the gas cools because heat energy has not been supplied from the walls of its container.
Azeotropic mixture A mixture of liquids which when boiled yields a vapour mixture of the same composition, a constant boiling mixture.
CFCs Chlorofluorcarbons: a special selection of the halocarbons. Most of these compounds are now implicated in the destruction of stratospheric ozone and will now not be designed into new equipment or processes. Most processes will use alternative compounds by 1995 or 2005 under the Montreal Protocol international directive and national directives.
Critical pressure The pressure above which a gas, no matter how compressed, will not form a liquid.
Critical temperature The temperatures above which a gas, no matter how compressed, will not form a liquid.
Differential pressure (ΔP) The difference in pressure between two points.
Dry ice Solid CO_2.

Explosive limits (LEL, HEL) The upper and lower percentage byvolume limits of a flammable mixture of a gas or vapour with air. Below the lower explosive limit (LEL) the mixture is too weak (has too much air) to ignite. Above the higher explosive limit (HEL) the mixture is too rich (has too much fuel) to ignite.

Flashback The propagation of a flame back down a pipe containing a mixture of fuel and oxidant. This is a dangerous condition, normally avoided by keeping the fuel and oxidant separate until nearly at the burner and by fitting flashback arrestors. These are typically metal gauze filters which conduct heat away from a flame in the same manner as the gauze around the classic Davy miner's safety lamp.

Fusible plug Fitted to gas cylinders, these melt at a fairly low temperature (approximately 150°C) and release the contents of a gas cylinder before the the cylinder walls rupture.

Gas constant This constant has the value 8.31 J K^{-1} mol^{-1} (symbol R).

Halocarbon A compound of carbon with one or more of the halogens fluorine, chlorine, bromine and iodine.

Joule–Thompson effect The cooling effect seen in all gases below a certain temperature, when they are expanded at a constriction from a high to a low pressure.

Latent heat The heat energy absorbed to make a solid melt or sublime, or make a liquid boil (latent heats of fusion, sublimation and vaporisation respectively).

Mole 6.023×10^{23} molecules.

Montreal protocol The international agreement which controls production and use of ozone depleting substances such as CFCs.

Noble gas The highly unreactive atomic gases of column 0 of the Periodic Table of the Elements, i.e. helium, neon, krypton, xenon, radon.

NTP Normal temperature and pressure, i.e. 20°C and 1.013 25 bar.

OEL Occupational exposure limit: the concentration of an airborne substance, averaged over a reference period (8 h, or 15 min for substances with acute effects), at which, according to current knowledge, there is no evidence that it is likely to be injurious to employees if they are daily exposed by inhalation to that concentration (UK).

Pascal The scientific unit of pressure. Rarely used in industry.

Rare gas *See* noble gas.

Specific heat The heat energy required to raise the temperature of 1 g of substance through 1 K. All gases have a specific heat of between 2.5 and 3.5 times R, the gas constant. The specific heat at constant volume (C_v) of a gas is alway R less than the specific heat at constant pressure (C_p).

Stoichiometric A stoichiometric mixture of A and B is one in which, when reaction occurs forming a compound such as AB or A_2B_3 or similar, no A or B is left over.

STP Standard temperature and pressure, i.e. 0°C and 1.013 25 bar.
TLV Threshold limit value: the maximum level of a gas to which people can be exposed on every working day (USA).
Triple point The triple point is the pressure and temperature at which vapour, liquid and solid can coexist in equilibrium.
Vapour pressure The vapour pressure of a substance is the pressure exerted, at equilibrium, of vapour in contact with that substance in liquid or solid form, where the liquid, vapour and container are all at the same temperature.

Tradenames

Arcton Chlorofluorocarbons (ICI).
Argoshield Argon-based gas mix for MIG welding (BOC).
Astec Helium-based welding gas (Air Products).
Calor An important liquified petroleum gas trademark (Imperial Group).
Coogar Argon-based gas mix for MIG welding (Air Products).
Fomblin Perfluoropolyether oils and greases used with oxygen service and especially with lubricated vacuum pumps and equipment (Montedison).
Freon The key trademark for chlorofluorocarbons (DuPont).
Helishield Helium-based welding gas (BOC).
Polygas Aerosol propellant system based on a microporous polymer, acetone and CO_2 (BOC/Polygas International).

Appendix B: The human nose as a gas detector

Nose anatomy

Inside the nose lies a patch a few square centimetres in area bristling with sensitive sensor cells. Immersed in a thin layer of mucus lie millions of tiny moving hairs, cilia, connected to the olfactory cells. These have an electrical output via their axons that connects them to the nerve cells in the 'olfactory bulb', i.e., the region of the brain that interprets the olfactory cells' signals into smells. Some idea of the complexity of the nose can be gained by noting that there are around 10 million receptor cells.

Nose limitations

The nose is quite limited in its response by its chemical nature and because of its development to serve animals in the wild, where light gases are in general unimportant. There is both an upper and a lower size limit for molecules which can be smelled. The upper limit is set because only those compounds which have an appreciable vapour pressure at room temperature, which naturally includes all gases, will reach the nose. The lower limit seems to be set by the absorption of the compound in the mucus membrane in which the sensitive cilia are set. Very light gases or those with low solubilities, such as methane, hydrogen or argon, are undetectable.

Gases which are highly inert, such as the lighter saturated aliphatic compounds, methane, ethane, propane, butane and pentane, are also undetectable. Another awkward fact about smell is the great variation in its efficiency for different compounds. A mercaptan might be detected in parts per trillion concentration, while ethylene is faint even at a few per cent. A final limitation to the nose performance is that any gas which is continuously present is undetectable, i.e., in electronic engineering terminology, the nose is 'a.c.-coupled'. Gases that are always present in the nose, such as oxygen and water vapour, are undetectable. Any gas present for a long time will become less and less sensitively perceived and finally fade almost completely. The latter is an essential defence mechanism. The brain would otherwise be overloaded with continuous smell signals in many environments.

Nose theory

The details of how the nose works are still to a considerable extent mysterious. A successful theory would have to account for which compounds can be smelled, the enormous variation in sensitivity to compounds, the large number of identifiably different smells, and would ideally predict the smell of new compounds.

Theories which seem to have some viability are as follows.

- Specific receptor theory, i.e. Amoore's theory. Amoore *et al.* (1970) proposed that there are seven different receptors in the nose, each sensitive only to one class of compounds (Figure A.1), according to molecular shape and charge distribution. Over the years since this theory was first presented, however, no such specific receptors have been identified and many problems have arisen in using the theory.
- Membrane puncture theory, i.e. Davies' theory. In this theory (Davies and Taylor, 1954, 1959; Davies, 1965) gas molecules interact with a sensitive membrane in such a way as to puncture it and allow a flow of sodium and potassium through, generating a nerve signal directly. Different membrane components will interact differently with different smelling gases. Again, although attractive as a theory, no sensitive puncturing membranes have been found, although Davies showed a good correlation between the catalytic acceleration of saponin haemolysis of red blood cells and the strength of smell of a range of substances from methanol to musk.
- Matrix theory, i.e. Erickson's theory (Engen and Trygg, 1982). There is some experimental support for this from neurophysiology. If measurements are made at receptor level, there seems to be little predictable difference between receptor response. However, similar measurements at the olfactory bulb, after 'pre-processing' of the raw signals by the brain nerve cells, seem to show more order and selectivity. Basically, this theory maintains that the receptors are similar but do have sensitivities to different molecules that are different. These slightly differing receptor signals are then used as inputs to a matrix, constructed out of neurones, the outputs of which are specific to a smell. (This technique is sometimes used to extract gas analysis results from a mass spectrum of mixtures, section 2.3.6.)
- Gas chromatography theory, i.e. Mozell's theory (Engen and Trygg, 1982). There is some experimental support from this in the form of time delays in the response from different parts of the olfactory sensor area in the nose. Mozell and others have argued fairly successfully that the time delays inherent in gas molecules diffusing through the mucus to the sensor cilia are present and could be used as part of the input information in a matrix type of decoding.

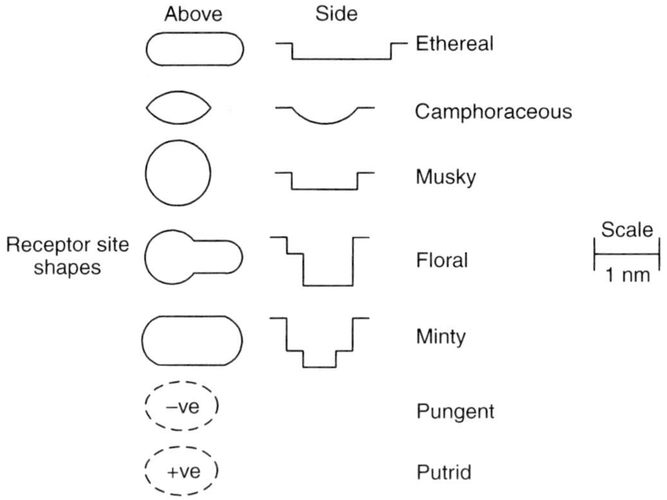

Figure B.1 Amoore theory of smell receptor sites.

The truth of the matter may be a simple but as yet unthought of theory or it may be a complex combination of the ideas above. Whatever the truth, there are some rules of thumb which connect smell to molecular structure. Moncrieff in his seminal book *The Chemical Senses* lists 22 general rules and ennumerates examples connecting chemical make-up with smell over 120 pages. This hardly amounts to a satisfactory theory, however, and even Moncrieff's 22 rules admit to many exceptions!

Using the nose

The limitations of the nose as a gas sensor are many. Unfortunately, many industrial gases fall into undetectable categories, at least in their pure state. It is also unfortunate that oxygen deficiency is not detected. Even amongst those gases which can be smelled, some are only detectable in concentrations at which they are already dangerous. Increasingly, toxic and flammable gas and oxygen deficiency monitoring is installed as a matter of course in places regularly handling gases, even where the gas concerned is easily detectable.

Enhancing the efficacy of the nose is a practice that should perhaps become more common. The prime example (and the only common example) is the addition of minute amounts of powerfully unpleasant smelling thiophane or mercaptans to domestic fuel supplies. (There is another example from Japan where oxygen gas is sold in small amounts in

Table B.1 Usual smells of common industrial gases

Gas	Usual smell
Acetylene	Garlic
Ammonia	Pungent, ammoniacal
Arsine	Garlic
Carbon dioxide	Pungent, acid
Chlorine	Pungent, irritating
Ethylene	Faint, sweet, ether
Ethylene oxide	Sweet, irritating
Hydrogen chloride	Pungent, acid, choking
Hydrogen sulphide	Rotten eggs
Nitrous oxide	Faint, sweet
Ozone	Pungent, 'new mown hay'
Phosphine	Decaying fish
Sulphur dioxide	Pungent, choking, sulphurous

aerosol cans and used for inhaling for its 'refreshing' effect. The gas is normally scented with mint and strawberry essences.)

Is it perhaps too fanciful to suggest that different gases should be coded with different powerful smells, much as gas cylinders are colour-coded with paint now? Flammable gases could use mercaptans, simple asphyxiant gases could use ionone (a powerful violet essence), while toxic gases could be equipped with a pungent ammoniacal addition, for example.

The mercaptans are so powerful that when a small container of this fuel gas 'perfume' was accidentally spilled in Manchester, UK, the fire brigade and gas company offices were flooded with hundreds of calls reporting gas leaks, some from several kilometres away from the spill. (The human nose is reported to have a threshold sensitivity of 30 parts per trillion(3×10^{-11}) for ethyl mercaptan, C_2H_5SH.)

Fortunately, in many gases which would be otherwise be undetectable there are adventitious impurities, which, although present at a low level, give the gas a clear smell (Table A.1). A classic example of this is acetylene, which often smells of garlic if made from calcium carbide (presumably because of phosphorus or arsenic impurities) and of acetone if taken from a cylinder (commercial acetylene is supplied dissolved in acetone). (There is perhaps a hidden danger here that gas users who come to rely on these impurities for giving them warning may come unstuck if the process to make the gas is modernised, giving a purer product, or if the gas comes from a source other than the usual one and the alternative source gas has no smell.)

Appendix C: Use of industrial gases in schools

Experiment is a key principle of scientific research. In chemistry and physics at least, experiment should also be a key part of teaching science. There will always be matters unexplained in textbooks and papers which are only clarified when an actual trial is carried out.

For educational purposes the preparation of gases by chemical reaction may often be valuable as it teaches pupils a preparative reaction as well as the task in hand. This view is a valid one, especially applying to the first time the gas is used. However, there are many applications for which a reasonable degree of purity and controllability is needed and these are better served by an industrial cylinder supply. The preparative reaction can still be carried out but the cylinder supply can be used for the next stage on the 'this is some I prepared earlier' principle often seen, perforce, in the classroom.

Many schools already use industrial gases for a small number of applications but many more schools could use them for more applications. These applications could be aimed both at the didactic element of the school curriculum and at the illustration of current industrial practice.

Industrial gases in compressed cylinders or liquid form are, in many ways, a very good buy for schools. They provide large quantities of very pure gas, at a very low price. For example, carbon dioxide is one of the simplest gases to prepare via a chemical preparation route but some maintenance and materials costs are involved and the gas contains significant impurities of water and acid spray, whilst cylinder gas costs around £1–2 per cubic metre delivered. Even when cylinder rental/maintenance charges (around £20–£30 per annum) are taken into account, industrial cylinder gases still represent a bargain. For gases such as argon and helium chemical preparations are impossible, whilst for oxygen, ethylene or nitrogen they are relatively difficult.

Nitrogen

Inerting chemical reactions

There are many reactions commonly avoided in the school chemistry laboratory that have didactic value but which are rendered possible only with an inert atmosphere. Similarly, in industry, many important reactions are often performed under nitrogen atmosphere, which is generally cheaper to achieve than vacuum conditions.

Dilution

It is frequently necessary to limit reaction rates, for example to avoid excessive temperature rise, and nitrogen is generally used as a carrier gas to dilute an active reactive gas. The classic example of this is combustion, where flame temperature and burning rate are dependent on the nitrogen in an oxygen/nitrogen mixture. Organic oxidation reactions are possible with nitrogen dilution of oxygen, which would proceed to complete oxidation (combustion) otherwise.

Liquid nitrogen

Liquid nitrogen has been traditionally difficult to supply economically in small quantities and this problem will continue to dog attempts to use LN in the classroom, at least for smaller schools. Perhaps it would help if someone (perhaps the gas suppliers) could list universities and other institutes willing to allow schools to have access to and collect their own supplies from their own large storage facilities.

Embrittlement of solids

The embrittlement of rubber and plastics is, of course, an important industrial application. Embrittlement of many materials also makes a spectacular demonstration and a vivid reminder that properties vary with temperature.

Freezing of biological systems

It is a comparatively simple matter to rapidly freeze and then thaw soft fruit, such as raspberries, and compare with the product thawed after a slow freeze in a standard refrigerator. The industrial importance of this needs no emphasis here.

It should also be possible to design 'suspended animation' experiments with yeast or even small insects.

Cooling of YBCO HT_c superconductors

Many schools with keen chemistry departments managed, as early as 1988, to make a crude sintered YBCO pellet, which when cooled to LN temperatures, attains the uniquely interesting state of superconductivity.[82]

Oxygen

The justification for the use of oxygen, rather than air, in most applications revolves around the intensification of the chemical reaction achieved with oxygen.

Oxy-gas flames

There are many operations for which an oxy-gas flame can be used to advantage in the school laboratory. For example, many more metals can be melted in the methane/oxygen flame than with a methane/air flame. The ultimate, of course, is the 3100°C oxy acetylene flame.

Oxygen in chemical reactions

There are a number of reactions in which the regulation of oxygen concentration can be demonstrated to show dependence of reaction rate and thus rate equations.

Argon and helium

Sintering

Many of the achievements of powder metallurgy have been made possible by the advent of pure inert gases. A demonstration of sintering technology should be possible on a small scale using apparatus already available in the laboratory.

Speed of sound

Simple whistles or organ pipes easily demonstrate the dependence of the speed of sound on molecular weight.

[82] The author tested a number of pellets of the higher current HT_c superconductor BISSCO as part of an industrial materials development programme. It was found that with the simplest of apparatus many hundreds of amps could be passed through samples of less than 1 cm^2, using simple welding power supplies.

Argon arcs and plasma

The requirement for a powerful electrical source, typically at RF frequencies in the case of plasma, restricts school experiments with these interesting phenomena.

Welding

An important part of metallurgy studies, frequently undertaken by the chemistry department in schools, should be the study of welding. The preparation of TIG/MIG welds, and their assessment relative to the gas used (CO_2 argon mixtures) and relative to stick electrodes, is rarely undertaken, however, despite the fact that mini-MIG welders are now available for around £100. The involvement of the physics department, observing arc temperature and electrical arc characteristics, could also be useful.

Helium buoyancy

The contrast between the 'transport properties' (i.e. diffusivity, viscosity and thermal conductivity) of a very light gas such as helium and those of air are marked and could be employed, for example, in apparatus for the measurement of mixing, damping of mechanical oscillation and double-glazing performance.

Hydrogen

A simple gas chromatograph with a flame ionisation detector can easily be built for a couple of hundred pounds in school; the parts are readily available and not expensive.[83] Today, with an inexpensive PC-based data system, using a computer that the school will already have, a fairly sophisticated GC is possible. Such a system is easily capable of separating mixtures of organic compounds and is a useful adjunct to the study of organic chemistry, helping with the determination and understanding of side-reactions. Safety aspects are mostly taken care of by attention to leaks and ensuring that the hydrogen is shut off should the FID detector flame fail for any reason.

Fat hardening

Fat hardening is a very important industrial process and could probably be modelled in the school laboratory. The use of finely divided industrial

[83] The author built one whilst at school. It comprised a commercially available packed column with a home-made heating bath and a home-made FID/electrometer, running off a cylinder of hydrogen, with the electrometer fed to a chart recorder. The most expensive item turned out to be the chart recorder.

catalysts and pure grades of oil enables fairly safe low temperatures to be used. Obviously, the combination of heat and hydrogen needs careful attention to safety.

Carbon dioxide

Dry ice can be used in schools for cryogenics and also for gas lubrication.

The unique sublimation of CO_2 at atmospheric pressure has been used for years in the food industry as a uniquely simple freezer/refrigerator. Less well-known are applications in which a thin film of CO_2 derived from dry ice pellets is used to provide a high-pressure gas bearing.

Physics courses have for years used 'ice pucks' for kinematics experiments. These are actually metal pucks using an almost frictionless CO_2 'bearing' to glide frictionlessly across a glass-topped table. The bearing is simply a hollow underneath the puck in which a small piece of dry ice can be placed. As the puck slides along it warms the dry ice, producing CO_2 gas which escapes around the smooth skirt of the puck, raising it a fraction of a millimetre above the glass table.

The cryogenic properties of dry ice can be exploited in many ways. It can be used, for example, in cooling methanol used in making a diffusion (continuous) cloud chamber capable of revealing nuclear particle tracks.

Gaseous CO_2 can be used in respiration demonstrations. The respiration rate (in breaths per minute and also in terms of CO_2 production) of creatures depends on CO_2 concentration. Demonstrating these important physiological effects may require the use of a gas analyser. However, as noted above, a crude CO_2 gas analyser can be quite simple or, alternatively, a straightforward wet chemical technique could be used.

It should be possible to show the effects of a CO_2 gas layer on the heating and cooling of a 'black body' under the sun or a tungsten lamp. This is the basis of the greenhouse effect, which is now a much-discussed environmental issue.

Ethylene and propylene

Ethylene is now important in the school chemistry syllabus, with its emphasis on organic chemistry, and is relatively difficult to make from liquid chemicals. Ethylene is available in relatively high purity at low cost in bulk from industrial gas companies. However, in most countries it is currently a relatively expensive special gas in small quantities because it is used by only a few small users owing to the superiority of acetylene in oxy-fuel and a lack of other small-scale industrial uses. Nevertheless, if a sizeable number of schools began to purchase it, then prices would tumble and it could usefully be provided to schools for classic organic chemistry experiments.

Propylene is a suitable substitute in many instances. It is much more readily available at reasonable cost and can be used to illustrate ethylene-based reactions.

Acetylene

The use of acetylene in flames is obvious. Less obvious, but interesting and still of some industrial importance, are the chemical reactions of acetylene. Acetylene, because of its reactive carbon–carbon triple bond, will react relatively easily in many saturation reactions.

Appendix D: Gas cylinder colour marking

Table D.1 Typical gas cylinder colour marking in the UK

	Colours			
Contents	Body	Top	Horizontal stripe	Additional horizontal stripe
Air	Grey			
Ammonia	Black	Red	Yellow	
Argon	Blue			
Argon/helium	Brown	Blue		
Argon/methane	Blue	Red		
1,3-Butadiene	Silver	Red	Yellow	
Carbon dioxide	Black			
Carbon monoxide	Red	Yellow		
Chlorine	Yellow			
Ethylene	Purple	Red		
Ethylene oxide	Purple	Red	Yellow	
Ethylene oxide/carbon Odioxide	Pink	Red	Yellow	Green
Helium	Brown			
Hydrogen	Red			
Krypton	White	Grey		
Methane	Red		Green	
Methyl bromide	Blue	Black		
Neon	Brown	Black		
Nitrogen	Grey	Black		
Nitrous oxide	Silver	Buff		
Oxygen	Black			
Phosgene	Black	Blue	Yellow	
Propane	Red			
Propane/butane, etc., mix	Silver	Red		
Propylene	Silver	Red		
Sulphur dioxide	Green	Yellow		
Xenon	White	Grey	Green	
Non-toxic/non-flammable	Buff			
Toxic/non-flammable	Buff	Yellow		
Non-toxic/flammable	Buff	Red		
Toxic/flammable	Buff	Red	Yellow	
Dip-tube	White vertical stripe			

Different gas companies use different colour schemes around the world, depending on their own standards and local regulations. The colours also differ for cylinders intended for medical use. All gas users should always ensure that they are fully informed on the colours used in their country by their own suppliers.

There are also different schemes around the world for gas cylinder connections. Connections differ in screw-thread pitch, direction and diameter. They also differ in the diameter and form of the gas sealing surfaces and the use of sealing washers. In total there are hundreds of different connection types and this helps to prevent incorrect gases being connected. As a rule, oxidising gases, such as air or oxygen, use a normal (right-hand) thread, whilst flammable fuel gases use a left-hand thread.

In Europe, DIN (Deutsches Institut für Normung, Germany) or BCGA (British Compressed Gas Association, UK) standard connections are employed, whilst in the USA the CGA (Compressed Gas Association) connections are used. Standard bodies, such as CGA and BCGA, issue up-to-date information on connections, as do the gas suppliers themselves.

References

Agricola (1950, orig. 1556) *De Re Metallica*, Hoover Publications, New York.

Amoore, J.E., Johnston, J.W. Jnr and Rubin, M. (1964) The stereochemical theory of odor. *Scientific American*, **210**(2), 42–49.

Andeen, C. (1971) A capacitive gauge for accurate measurement of high pressure. *Rev. Sci. Instr.*, **42**, 495–496.

Anon (1975) *High Pressure Engineering, Proceedings of the 2nd International Conference*, Mechanical Engineering Publications (Inst. Mech. Eng.), London.

Anon (1980) *Priestley in London*, Royal Society of Chemistry, London.

Anon (1989) *Separation of Gases*, 5th BOC Priestley Conference, Proceedings of the Royal Society of Chemistry, London.

Anon (1993) *The Role of Oxygen in Improving Chemical Processes*, 6th BOC Priestley Conference, Proceedings of the Royal Society of Chemistry, London.

Anon (1995) Happy birthday helium, *New Scientist* **146**(1972), 8th April.

Aykroyd, W.R. (1935) *Three Philosophers*, Heinemann, London.

Berbesson, J. (1979) Long distance transport of LNG via cryogenic pipeline, in *New Technologies for Exploration and Exploitation of Oil and Gas Resources*, Vol 1, Graham & Trotman, London, p. 581.

Bridgman, H.A. (1990) *Global Air Pollution: Problems for the 1990s*, Belhaven Press. London.

Buchanan, R.A. *The Power of the Machine*, Viking, London.

Calder, A. (ed.) (1995) *Technology Foresight – Chemicals*, HMSO, London.

Carr, P. (1994) *In situ* gas analysis. *European Semiconductor*, Sept., 13.

Cherin, A.H. (1983) *Introduction to Optical Fibres*, McGraw-Hill, Tokyo.

Collins, M. (1988) *Liftoff*, Grove Press, New York.

Corbett, M.J. and Smith, R.B. (1993) in *The Role of Oxygen in Improving Chemical Processes*, Proceedings of the 6th BOC Priestley Conference, Royal Society of Chemistry, London.

Coulson, J.M. and Richardson, J.F. (1970) *Chemical Engineering*, Pergamon, Oxford.

Davies, A.C. (1965) *Journal of Theoretical Biology*, **8**, 1.

Davies, A.C. and Taylor, F.H. (1954) *Nature*, **174**, 693.

Davies, A.C. and Taylor, F.H. (1959) *Biology Bulletin*, **117**(2), 222.

Dennis, N.T.M. and Heppell, T.A. (1968) *Vacuum System Design*, Chapman & Hall, London.

Denny, R. (1977), in *Oxygen in the Metal and Gaseous Fuel Industries*, Proceedings of the 1st BOC Priestley Conference, Royal Society of Chemistry, London, pp. 232–255.

Downie, N.A. (1991) Cryogenic system for fault current limiter. *Research Disclosure*, **330**, 829.

Dunbobbin, B.R. (1987) MOLTOX process. *Gas Separation and Purification*, **1**, 23.

Einstein, A. (1924) *Z. Electrochemie* **14**, 235.

Elvin, L. (1971) *Organ Blowing: its history and development*, Laurence Elvin, Lincoln.

Enge, H. (1966) *Introduction to Nuclear Physics*, Addison-Wesley, London.

Engen, T. (1982) *The Perception of Odors*, Academic Press, New York, pp. 17–34.

Fricke, J. (1988) Aerogels. *Scientific American*, May, 92–97.

Gibbs, F.W. (1965) *Joseph Priestley: Adventurer in Science and Champion of Truth*, Thomas Nelson, London.

Gibbs, F.W. (1967) *Joseph Priestley*, Doubleday & Co.

Haldane, J.B.S. (1927) On being the right size, in *Possible Worlds*, Chatto & Windus, London.

Hendrick, D.J. and Sizer, K.E. (1992) 'Breathing' coal mines and surface asphyxiation from stythe (black damp). *British Medical Journal*, **305**, 509.
Hands, B. (1986) *Cryogenic Engineering*, Academic Press/Harcourt Brace, London.
Harper, G.S. and Parrish, G. (1985) *Production Gas Carburising*, Pergamon Press, Oxford.
Hirose, T. (1987) *Thermal Production of Diamond Thin-films*, Proceedings of the Osaka Sanso Conference, Tokyo. Osaka Sanso, Osaka, Japan.
Hrubesh, L.W. (1987) *Optical Characterisation of Silica Aerogel Glass*, Lawrence Livermore Laboratory, July report.
Hunt, J. (1954) *The Conquest of Everest*, Hodder & Stoughton, London.
Irani, C.A. and Funk, E.W. (1977) Separations using supercritical gases, in *Recent Developments in Separation Science*, CRC Press, Florida, pp. 171–193.
Isalski, W.H. (1989) *Separation of Gases*, Oxford University Press, Oxford.
Jolley, S. (1990) European Patent No. 0487183, May.
Jones, T. (1988) *Corporate Killing: Bhopals will happen*, Free Association Books, London.
Kapteijn, F. and Singoredjo, L. (1993) *Catalyst and catalytic reduction*. World Patent No. 9302776.
Kaye, B. (1995) Conservation of waterlogged archaeological wood. *Chem. Soc. Rev.*, **24**, 35.
Kirk-Othmer (1978) *Encyclopaedia of Chemical Technology*, 3rd edn, Wiley 1982.
Lavoisier, A.L. (1952) *Elements of Chemistry*, Encyclopaedia Brittannica/William Benton, Chicago.
Melvin, A. (1988) *Natural Gas*, Adam Hilger, Bristol.
Monnot, G. (1985) *Principles of Turbulent Fired Heat*, Gulf Publishing Co/Editions Technip Paris (original in French).
Mooney, M.M. (1973) *The Hindenburg*, Hart-Davis MacGibbon, London.
Ohmi, T., Okumura, T., Sugiyama, K., Nakahara, F. and Murota, J. (1988) Outgas-free corrosion-resistant surface passivation of stainless steel for advanced ULSI processing equipment. *Extended Abstracts: 174th Electrochemical Society, Fall Meeting*, Paper 396, pp. 579–580. Electrochemical Society.
Perkins, D.H.P. (1972) *Introduction to High Energy Physics*, Addison-Wesley, London.
Reid, R.C. (1977) *The Properties of Gases and Liquids*, McGraw-Hill.
Wightman, W.P.D. (1950) *The Growth of Scientific Ideas*, Oliver and Boyd, Edinburgh.
Williams, T.I. (1981) *A History of the British Gas Industry*, Oxford University Press, Oxford.
Yang, R.T. (1987) *Gas Separation by Adsorption Processes*, Butterworths, London.
Zuwala, M.S., Craig, G.R. and Cass, D.L. (1994) *Cryogenic processing of used tyres and other materials*. World Patent No: 9407670.

Further reading

Ahlberg, K. (1985) *AGA Gases Handbook*, AGA AB, Lidingo, Sweden.
Alexander, W. and Street, A. (1982) *Metals in the Service of Man* 8th edn, Penguin, Harmondsworth, UK.
Anon (1977) *Oxygen in the Metal and Gaseous Fuel Industries*, 1st BOC Priestley Conference, Proceedings of the Royal Society of Chemistry, London.
Anon (1978) *Basic Oxygen Steelmaking: a new technology emerges?*, Proceedings of The Metals Society Conference, May 4–5, The Metals Society, London.
Anon (1987) *Distillation and Absorption 1987*, Institution of Chemical Engineers, Rugby.
Anon (1990) *STEP 1990*, Proceedings of SEMI Conference, Brussels. SEMI, Mountain View, California.
Anon (1991) *RADTECH Europe 1991*, RADTECH International, Northbrook, Illinois.
Anon (1995) *CRC Handbook of Chemistry and Physics*, 76th edn, CRC Press Inc, Boca Raton, Florida.
Atkinson, H.V. and Rickinson, B.A. (1991) *Hot Isostatic Processing*, Adam Hilger, Bristol.
Banks, R.E. (1980) *Fluorocarbons and their Derivatives*, 2nd edn, MacDonald Technical, London.
Barnard, J.A. and Bradley, J.N. (1995) *Flame and Combustion*, 3rd edn, Chapman & Hall, London.
Belsterling, C.A. (1971) *Fluidic Systems Design*, Wiley-Interscience, New York.
Board, K. and Morgan, D.V. (1985) *Semiconductor Microtechnology*, Wiley, Chichester.
Brauer, H. and Varma, Y.B.G. (1981) *Air Pollution Control Equipment*, Springer-Verlag, Berlin.
Braithwaite, A. and Smith, F.J. (1995) *Chromatographic Methods*, 3rd edn, Chapman & Hall, London.
Butrica, A. (1980) *Out of Thin Air*, Praeger Publishers Inc., Westport, Connecticut.
Cayless, M.A. and Marsden A.M. (eds) (1983) *Lamps and Lighting*, Edward Arnold, London.
Cherin, A.H. (1983) *Introduction to Optical Fibres*, McGraw-Hill,
Clarke, K.P., Coulson, D., Gibbs, E.D., Keon, K.B. and Kruijshoop, A.F. *Optical Fibre System* Telstra Corporation patent WO 9422241.
Cowper, C.J. and DeRose, A.J. (1985) *The Analysis of Gases by Chromatography*, Pergamon Press, Oxford.
Davidson, F.H. and Harrison, H. (1963) *Fluidised Particles*, Cambridge University Press, Cambridge.
Davies, A.C. (1992) *The Science and Practice of Welding*, 10th edn, Cambridge University Press, Cambridge.
Desrosier, N.W. and Tressler, D.K. (1977) *Fundamentals of Food Freezing*, AVI Publishing Co. Inc., Westport, Connecticut.
Douglas, J.F., Gasiorek, J.M. and Swaffield, J.A. (1984) *Fluid Mechanics*, Pitman Publishing, London.
Dumas, T. and Bulani, W. (1974) *Oxidation of Petrochemicals*, Applied Science Publishing, London.
Eilbeck, W.J. and Mattock, G. (1987) *Chemical Processes in Waste Water Treatment*, Wiley/Ellis Horwood, Chichester.
Fifield, F.W. and Kealey, D. (1983) *Analytical Chemistry*, Chapman & Hall, London.
Flowers, B.H. and Mendoza, E. (1970) *Properties of Matter*, John Wiley, Chichester.
Fogg, P.G.T. and Gerrard, W. (1991) *Solubility of Gases in Liquids*, John Wiley, Chichester.
Fraas, A.P. and Ozisik, M.N. (1965) *Heat Exchanger Design*, Wiley, Chichester.
Gordon, J.E. (1981) *Structures or Why Things Don't Fall Down*, Penguin, London.

Hausner, H.H. (ed.) (1971) *Modern Developments in Powder Metallurgy*, Plenum, New York.

Heaton, C.A. (ed.) (1984) *Introduction to Industrial Chemistry*, Chapman & Hall, London.

Houghton, J.T., Taylor, F.W. and Rogers, C.D. (1986) *Remote Sounding of Atmospheres*, Cambridge University Press, Cambridge.

Hutzinger, O. (1980) *The Handbook of Environmental Chemistry*, Springer-Verlag, Berlin.

Kaye G.W.C. and Laby T.H. (1995) *Tables of Physical and Chemical Constants*, 16th edn, Longman, London.

Kletz, T. (1992) *HAZOP and HAZAN*, Institution of Chemical Engineers/Hemisphere, Rugby.

Kohl, A.L. and Riesenfeld, F.C. (1960) *Gas Purification*, McGraw-Hill, New York.

Lacey, R.E. and Loeb, S. (1972) *Industrial Processing with Membranes*, Wiley-Interscience, New York.

Ledwith, A. and Moss, S.J. (1987) *The Chemistry of the Semiconductor Industry*, Blackie, Glasgow.

Linan, A. and Williams, F.A. (1993) *Fundamental Aspects of Combustion*, Oxford University Press, Oxford.

Liu, B.Y.H. (ed.) (1976) *Symposium on Fine Particles*, Academic Press, New York.

McMillan, F. *et al. Treatment of Scrap*, Patent WO 8002672, Commonwealth Industrial Gases.

Meyer, E. (1977) *Chemistry of Hazardous Materials*, Prentice-Hall, New Jersey.

Moncrieff, R.W. (1967) *The Chemical Senses*, 3rd edn, Leonard Hill, London.

Moore, W.J. (1978) *Physical Chemistry*, Longman, London.

Morgan, D.V. and Board, K. (1985) *Semiconductor Microtechnology*, John Wiley, Chichester.

Mosely, P.T., Norris, J.O.W. and Williams, D.E. (eds) (1991) *Techniques and Mechanisms in Gas Sensing*, Adam Hilger, Bristol.

Nash, P. (1983) *The Essentials of Foam, Dry Powder, and Gaseous Extinguishing Systems*, Building Research Establishment, Borehamwood.

Newstubb, P.F. (1969) *Mass Spectrometry and Ion–Molecule Reactions*, Cambridge University Press, Cambridge.

Nunes, A.C. (1977) Gas welding origins. *Welding Journal*, June, **56**, 15–23.

Pasek, A.D. (1991) *Pool Boiling on Porous Surfaces*, PhD Thesis, Cryogenics Institute, University of Southampton.

Patterson, H.B.W. (1983) *Hydrogenation of Fats and Oils*, Applied Science Publishers, London.

Perry, R.H. (1984) *Perry's Chemical Engineers' Handbook*, 6th edn, McGraw-Hill, Maidenhead, UK.

Pichot, P. (1986) *Compressor Application Engineering*, Gulf Publishing, Houston.

Roberts, T.A. and Skinner, F.A. (eds) (1983) *Food Microbiology: Advances and Prospects*, Academic Press, London.

Ruelle, D. (1991) *Chance and Chaos*, Princeton University Press, Princeton.

Sale, F. (1979) *Steel Production*, Open University Press, Milton Keynes.

Scott, R.W.W. (ed.) (1982) *Developments in Flow Measurement*, Applied Science Publishers, London.

Scurlock, R. (1992) *History of Cryogenics*, Clarendon Press, Oxford.

Spitzer, D.W. (1990) *Industrial Flow Measurement*, Instrument Society of America, North Carolina.

Strauss, W. (1966) *Industrial Gas Cleaning*, Pergamon Press, Oxford.

Thelning, K.E. (1975) *Steel and its Heat Treatment*, Butterworths, London.

Thorne, S. (1983) *Developments in Food Preservation – 2*, Applied Science Publishers, London.

Walker, M.J. (ed.) (1993) *Concrete in Hot Climates*, Chapman & Hall, London.

Walker, R.D. (1983) *Small-scale steelmaking*, Applied Science Publishers, London.

Yang, R.T. (1987) *Gas Separation by Adsorption Processes*, Butterworths, London.

Yinon, J. and Shmuel, Z. (1981) *Analysis of Explosives*, Pergamon Press, Oxford.

Zemanski, M.W. (1968) *Heat and Thermodynamics*, McGraw-Hill, New York.

Index